ENVIRONMENTAL APPLICATIONS OF INSTRUMENTAL CHEMICAL ANALYSIS

ENVIRONMENTAL APPLICATIONS OF INSTRUMENTAL CHEMICAL ANALYSIS

Edited by
Mahmood M. Barbooti, PhD

Apple Academic Press Inc. | Apple Academic Press Inc.
3333 Mistwell Crescent | 9 Spinnaker Way
Oakville, ON L6L 0A2 | Waretown, NJ 08758
Canada | USA

©2015 by Apple Academic Press, Inc.

First issued in paperback 2021

Exclusive worldwide distribution by CRC Press, a member of Taylor & Francis Group
No claim to original U.S. Government works

ISBN 13: 978-1-77463-369-4 (pbk)
ISBN 13: 978-1-77188-061-9 (hbk)

Library and Archives Canada Cataloguing in Publication

Environmental applications of instrumental chemical analysis / edited by
Mahmood M. Barbooti, PhD.

Includes bibliographical references and index.
ISBN 978-1-77188-061-9 (bound)
1. Chemistry, Analytic. 2. Chemistry, Analytic--Methodology. 3. Environmental chemistry.
I. Barbooti, Mahmood M., author, editor

QD75.22.E58 2015   543   C2015-901599-5

Library of Congress Cataloging-in-Publication Data

Environmental applications of instrumental chemical analysis / [edited by]
Mahmood M. Barbooti, PhD.

pages cm
Includes bibliographical references and index.
ISBN 978-1-77188-061-9 (alk. paper)
1. Environmental chemistry. 2. Chemistry, Analytic. I. Barbooti, Mahmood M.

TD193.E544 2015   577'.14--dc23   2015009009

Apple Academic Press also publishes its books in a variety of electronic formats. Some content that appears in print may not be available in electronic format. For information about Apple Academic Press products, visit our website at **www.appleacademicpress.com** and the CRC Press website at **www.crcpress.com**

ABOUT THE EDITOR

Mahmood M. Barbooti, PhD

Mahmood Barbooti, PhD, is a professional analytical chemist and an Assistant Professor at the School of Applied Sciences at the University of Technology in Baghdad, Iraq. He was a Visiting Professor at Montclair State University in New Jersey, USA, from 2009 till 2012. He has held many roles, including Director of the Environmental Analysis Department of the Ministry of Environment in Iraq from 2004–2006, Director of Petrochemical Industries Research Centre, Director of the Ibnul-Beetar Pharmaceutical Research Centre, Chair of the Quality Control Department of the Ministry of Industry and Minerals, and senior scientific researcher, among others. He is an expert in environmental site assessment (ESA) in Iraq and has led teams in two ESA projects. He managed and lectured a large number of training courses on instrumental analysis, environmental analysis, monitoring, and water treatment. The author of more than 80 published papers and 8 patents, he has also published two books in Arabic and has chaired the Iraqi committee for the rewriting of chemistry books of the secondary schools in 2009.

Dr. Barbooti received a BSc in industrial chemistry in 1974, an MSc in analytical chemistry in 1976, and a PhD analytical chemistry, under the supervision of the late Professor F. Jasim from Iraq and Professors W. Frech and E. Lundberg from Sweden.

CONTENTS

LIST OF CONTRIBUTORS

Suleyman Akman
Istanbul Technical University, Science and Letters Faculty, Chemistry Dept. 34469 Maslak, Istanbul-Turkey

Nidhal Al-Derzi
Department of Applied Sciences, University of Technology, Sinaa Street, Baghdad, Iraq

Ana Isabel Argente-García
Department of Analytical Chemistry, Faculty of Chemistry, University of Valencia, Spain

Mahmood M. Barbooti
Department of Applied Science, University of Technology, Sinaa Street, P.O.Box 35045, Baghdad, Iraq

Asli Baysal
Istanbul Technical University, Science and Letters Faculty, Chemistry Dept. 34469 Maslak, Istanbul-Turkey
Tel: +90 212 285 31 60 Fax: +90 212 285 63 86
E-mail address: baysalas@itu.edu.tr

Nabeel A. Fakhri
Department of Chemistry, College of Education, University of Salahuddin, Erbil, Iraq
Email: havras@yahoo.com

Pilar Campins Falco
Department of Analytical Chemistry, Faculty of Chemistry, University of Valencia, Spain

Farnoush Faridbod
Center of Excellence in Electrochemistry Department of Chemistry, College of Science, University of Tehran, Tehran, Iran

Mohammad R. Ganjali
Dean of College of Science, University of Tehran,
Editor-in-Chief of "Analytical and Bioanalytical Electrochemistry"
Center of Excellence in Electrochemistry, Tehran, Iran

Rosa Herráez-Hernández
Department of Analytical Chemistry, Faculty of Chemistry, University of Valencia, Spain.

Michael Kruge
Department of Earth & Environmental Studies, Montclair State University, 1 Normal Ave, Montclair 07043, NJ, USA
Email: krugem@mail.montclair.edu

Yolanda Moliner-Martinez
Department of Analytical Chemistry, Faculty of Chemistry, University of Valencia, Spain

Carmen Molins-Legua
Department of Analytical Chemistry, Faculty of Chemistry, University of Valencia, Spain

Parviz Norouzi
Center of Excellence in Electrochemistry Department of Chemistry, College of Science, University of Tehran, Tehran, Iran

Oscar Nunez
Department of Analytical Chemistry, University of Barcelona, Barcelona, Spain
oscar.nunez@ub.edu

Mustafa Ozcan
Istanbul Technical University, Science and Letters Faculty, Chemistry Dept. 34469 Maslak, Istanbul-Turkey

Nil Ozbek
Istanbul Technical University, Science and Letters Faculty, Chemistry Dept. 34469 Maslak, Istanbul-Turkey

Morteza Rezapour
College of Science, University of Tehran

Jorge Verdú-Andrés
Department of Analytical Chemistry, Faculty of Chemistry, University of Valencia, Spain

LIST OF ABBREVIATIONS

AAS	atomic absorption spectrophotometry
AdCSV	adsorptive cathodic striping voltammetry
AdSV	adsorptive stripping voltammetry
AE	auxiliary electrode
ALC	alachlor
APAHE	Asian Pacific Americans in Higher Education
Aq	aqueous
ASAP	as soon as possible
ASE	accelerated solvent extraction
ASV	anodic stripping voltammetry
AU	absorbance units
AuNPs	gold nanoparticles
BE	back-extraction
BiFE	bismuth film electrode
BGE	background electrolyte
BOD	biochemical oxygen demand
BPA	bisphenol A
BTEX	benzene, toluene, ethylbenzene and xylenes
CapLC	capillary liquid chromatography
CE	capillary electrophoresis
CEC	capillary electrochromatography
CF	counter-flow
CGE	capillary gel electrophoresis
CIEF	capillary isoelectric focusing
CI	chemical ionization
CITP	capillary isotachophoresis
COC	chain-of-custody
COD	chemical oxygen demand
COL	colorimetric methods
CPE	cloud point extraction
CPEs	carbon paste electrodes
CSV	cathodic stripping voltammetry
CV	cyclic voltammetry
CWEs	coated-wire electrodes
CZE	capillary zone electrophoresis
DAD	diode array detector

DBPs	disinfection by products
DC	direct current
DLLME	dispersive liquid–liquid microextraction
DME	dropping mercury electrode
DO	dissolved oxygen
DOAS	differential optical absorption spectroscopy
DP	differential pulse
DPCSV	differential pulse cathodic stripping voltammetry
DPP	differential pulse polarography
DPV	differential pulse voltammetry
DRIFTS	diffuse reflectance ftir spectroscopy
DOM	dissolved organic matter
DRIFTS	reflectance FTIR spectroscopy
DSPE	dispersive solid phase extraction
E	energy
ECD	electron capture detector
EDL	electrical double layer
EIS	electrochemical impedance spectroscopy
EKS	electrokinetic supercharging
EI	electron ionization
EMF	electromotive force
EOF	electro-osmotic flow
EPA	US Environment Protection Agency
EQS	environmental quality standards
ESI	electrospray
ETAAS	electrothermal atomic absorption spectrometry
EU	European Union
F	flow rate
FAAS	flame atomic absorption spectrometry
FASI	field amplified sample injection
FET	field effect transistor
FFT	fast Fourier transform
FIA	flow injection analysis
FID	flame ionization detector
FIMS	field ionization mass spectrometry
FPD	flame photometric detector
FTIR	Fourier Transform infrared
FTIR-ATR	Fourier transformed infrared spectroscopy—attenuated total reflectance
GC	gas chromatography
GC-MS	gas Chromatography-Mass Spectrometry
GPS	global Positioning System

HETP	height equivalent to a theoretical plate
HF-LPME	hollow fiber liquid phase microextraction
HMDE	hanging mercury drop electrode
HODS	higher-order derivative spectrophotometry
HPCE	high-performance capillary electrophoresis
HPLC	high performance liquid chromatography
HS	headspace
IC	ion chromatography
ICP	induced coupled plasma
ICP-AES	inductively coupled plasma optical emission spectroscopy
ICP-MS	inductively coupled plasma mass spectrometry
ICP-OES	inductively coupled plasma optical emission spectrometry
IR	infrared
I.S.	internal standard
ISAB	ionic strength adjustment buffer
ISE	ion selective electrode
ISFET	ion selective field effect transistors
IT-SPME	in-tube solid phase microextraction
Ksp	solubility product constant
LE	leading electrolyte
LED	light-emitting diode
LIBS	laser Induced breakdown spectroscopy
LIF	laser-induced fluorescence
LLE	liquid–liquid extraction
LOC	lab-on-a-chip
LOD	limit of detection
LOQ	limit of quantification
LPME	liquid phase microextraction
LSV	linear sweep voltammetry
LVSS	large volume sample stacking
M	molarity
MCL	maximum contaminate level
MAE	microwave assisted extraction
MECC	micellar electrokinetic capillary chromatography
MEKC	micellar electrokinetic chromatography
MFE	mercury film electrode
MIPs	molecular imprinted polymers
MP-AES	microwave plasma atomic emission spectroscopy
MPM	matched potential method
MS	mass spectrometry
MSM	mixed solution method

MSPD	matrix solid phase dispersion
MSPE	magnetic solid-phase extraction
MWCNTs	multiwall carbon nanotubes
MSWV	multiple square-wave voltammetry
N	normality
NACE	non-aqueous capillary electrophoresis
nm	nanometer
NPD	nitrogen phosphorous detector
NPEC	nonylphenolethoxycarboxylates
NPs	nanoparticles
NTs	nanotubes
NTU	nephelometric turbidity units
NVO	nonvolatile organic
OBD	optical bacteria detector
OFAT	one-factor-at-a-time
OMWCNT	oxidized multi-walled carbon nanotubes
OPs	organic pollutants
OPPs	organophosphorous compounds
PAD	pulsed amperometric detection
PAHs	polycyclic aromatic hydrocarbons
PAM	polyacrylamide
P&T	purge and trap
PBDEs	polybrominated diphenyl ethers
PCA	principal components analysis
PCBs	polychlorinated biphenyls
PCDD	polychlorinated dibenzo-p-dioxins
PCDFs	polychlorinated dibenzo-furans
PDA	photodiode array
pI	isoelectric point
PDMS	polydimethylsiloxane
PID	photoionization detection
PLE	pressurized liquid extraction
PM2.5	particulate matter, with particle diameter of 2.5 μm and smaller
PM10	particulate matter, with particle diameter of 10 μm and smaller
ppb	parts per billion
PPCPs	pharmaceuticals and personal-care products
PPM	parts per million
PTV	programmed temperature vaporization
Py-MAS-TOF-MS	metastable ion time-of-flight mass spectrometer
Py-FIMS	pyrolysis field ionization mass spectrometry

Py-GC	pyrolysis gas chromatography
Py-GC/MS	pyrolysis gas chromatography Mass Spectrometry
Py-MS	pyrolysis – mass spectrometry
PVC	poly(vinyl chloride)
QA	quality assurance
QA/QC	quality assurance/quality control
QDs	quantum dots
QuEChERS	quick, easy, cheap, effective, rugged and safe method
R	resolution
RE	reference electrode
RSD	relative standard deviation
RSP	respirable suspended particulate matter;
SBSE	stir-bar sorptive extraction
SCE	saturated calumel electrode
SCOT	support coated open tube
SDME	single drop microextraction
SEC	size exclusion chromatography
SFE	supercritical fluid extraction
SHE	standard hydrogen electrode
SIM	selected ion monitoring
SMDE	static mercury drop electrode
SOM	soil organic matter
SPE	solid-phase extraction
SPME	solid phase microextraction
SPMD	semi permeable membrane device
SSM	separate solution method
SVO	semi volatile organic
SWCNTs	single walled carbon nanotubes
SWV	square wave voltammetry
SWASV	square-wave anodic stripping voltammetry
TCD	thermal conductivity detector
TD	thermo desorption
TE	terminating electrolyte
THF	tetrahydrofuran
THMs	trihalomethanes
tITP	transient isotacophoresis
TMA	trimethylaniline
TMAH	tetramethylammonium hydroxide
TOC	total organic carbon
TPHs	total petroleum hydrocarbons
tR	retention time
TRIS	trishydroxymethylamino methane

TSP	total suspended particulate matter
TXRF	total reflection x-ray fluorescence spectrometry
UAE	ultrasonic assisted extraction
UHPLC	ultrahigh performance liquid chromatography
UNEP	United Nations Environmental Program
USAEME	ultrasound-assisted emulsification microextraction
UV	ultrviolet
VGII	vinylguaiacol Indole Index
Vis	visible
VO	volatile organic
VOCs	volatile organic compounds
Vol%	volume%
Wt%	weight%
WB	peak width
WCOT	wall coated open tube
WE	working electrode
WHO	world health organization
WQI	water quality index
WR	retention volume
Wt/vol%	weight/Volume %
WWTPs	wastewater treatments plants
XRD	X-ray diffraction
XRF	X-ray fluorescence

LIST OF SYMBOLS

A	eddy-diffusion parameter
a_x	unknown activity
C	analyte concentration
C_0	original concentration of the constituent
C_A	conjugate base
C_{HA}	concentrations of the acid
C_{max}	concentration of the constituent at peak maximum
C_{trans}	cell current
d	capillary inner diameter
D_O	diffusion coefficient
E	applied electric field
E_{cell}	cell potential
F	flow rate of the gas
g	gravitational constant
h	Plank's constant
i_d	current in μA
i_p	peak current in ampere
k	force constant of the bond in the Hook's law
l	corresponding results from the second method
L	length of column
n	number of atoms
q	ion charge
r	capillary radius
R	correlation coefficient
r	ion radius
t	injection time
V	applied voltage
v_s	volume of the added standard
v_x	sample volume
x	results of method

GREEK SYMBOLS

μ_{ap}	apparent electrophoretic mobility of the analyte
ρ	BGE density

ε	dielectric constant of the BGE solution
Δh	height differential of the reservoirs
μ_{EOF}	mobility of the electro-osmotic flow
ΔP	pressure difference across the capillary
υ	vibration quantum number
ζ	zeta potential
μ_e	electrophoretic mobility
η	viscosity of the solution
μ	reduced mass

PREFACE

The global concern about man's interference in the environment is increasing. The acceptance of modern technology is now accompanied by raising the hands to stop the degradation of our environment. Along with the studies of improving the performance of services and the introduction of products with better quality, much concern is now directed toward solving environmental problems and cleaning of polluted areas. Top universities are involved now in introducing elementary and advanced course studies for the environment covering legalization, monitoring and remediation of contaminated sites. Environmental studies benefit from all branches of knowledge, ranging from history, law, science, and engineering though the common feature of all environmental studies is to precisely estimate the concentration of pollutants in air, water, and soil as the main receptors of the contamination.

The last four decades witnessed the introduction of new analytical methodologies and the improvement of the existing methods to cover a wider range of applications and to reach extremely low concentrations of the contaminants. A variety of chemical analytical techniques are now invented and developed to determine not only the overall amounts of the analytes but to differentiate between the various species of each material that vary in their environmental impact. The best example in this respect is chromium, which is now included in diet recipes as a supplement and at the same time is considered a toxic material depending on the chemical state at which it is present. The manufacturers of chemical instrumentation are in a real race to develop methodologies for the determination of low concentrations of pollutants and to adapt automation to handle the increasing numbers of the environmental samples. Automation is also necessary to reduce the efforts of sample pretreatment steps and the size of the sample requirement necessary to fulfill the environmental concern.

Few books are available on the topic of environmental chemical analysis and its adoption of various approaches. However, a good analytical practice is not enough to reach acceptable and accurate results without a reliable method of sampling. This reflects the importance of the sampling process and the degree to which a sample may be representative of the whole lot. Also, of equal importance are the safe procedures of preservation, transportation, and pretreatment of samples to prepare them for the analysis. Thus, a successful publication on environmental application of instrumental analysis must give a reliable description of sampling and sample handling together with coverage of most recent modernization of chemical analysis methods.

As an editor of this book I must admit that I was very lucky to have responses from a selection of distinguished scientists to participate in tackling of the various aspects of environmental analysis. Actually the process was fruitful, and this will encourage us to go deeper in each analytical technology.

A short introduction on environmental chemistry was included in the book as a link chapter between analytical chemistry and the environmental applications. A comprehensive chapter is included on the sampling process to fulfill this important task. Some modern instrumental techniques were either introduced or modified during the last decade, such as capillary zone electrophoresis, ion chromatography, analytical pyrolysis, flow injection, pyrolysis-gas chromatography, and portable X-ray fluorescence. These techniques and their environmental applications were given in this book as chapters or specific section in the relevant chapters. We hope that our book will be a valuable addition to environmental analysis. We, herewith, present the extract of the experience of hard worker-researchers in a growing field. Researchers as well as students will find this book an indispensible reference. We never assume perfection, and thus, we are open to all comments from lecturers and scientists and their criticisms on the material as well as the approach of the book.

INTRODUCTION

This book **Environmental Applications of Instrumental Chemical Analysis** is presented as a new addition to the field of environmental analysis, which is growing progressively. The first part of the book is a review of the most used instrumental analysis techniques in the environmental work, and thus the reader will find no mention of techniques like Raman spectroscopy and conventional C, H, and N analysis. As a link between analytical chemistry and environmental work, a chapter was included on "Environmental Chemistry" to introduce the most important features of the environment and pollution sources. At the end of the applications chapters, the contributors presented selected methods of analysis for environmental samples including water, soil, and air to enrich the applied part of the book. Sampling is an essential part of environmental analysis to obtain reliable data about the pollution level. A chapter was included in the book specifically dealing with the sampling requirements, sampling processes, preservation, transportation, documentation, and finally the sample preparation for analysis.

PART I

INSTRUMENTAL CHEMICAL ANALYSIS

CHAPTER 1

FUNDAMENTALS OF CHEMICAL ANALYSIS

MAHMOOD M. BARBOOTI

Department of Applied Chemistry, School of Applied Sciences, University of Technology, P.O. Box 35045, Baghdad, Iraq; Email: brbt2m@gmail.com

CONTENTS

1.1   ANALYTICAL CHEMISTRY

Analytical chemistry is the study of the chemical composition of natural and synthetic materials and determination of various elements and ions in media of economic health and industrial importance. Thus, analytical chemistry is not specialized with certain compounds or reactions like other chemistry branches like inorganic or organic chemistry. Analytical chemistry studies the physical properties like the structure and atom configuration and the chemical properties like the chemical composition and determination of elemental composition of compounds. Analytical chemistry participated n the development of other scientific fields specifically chemistry, biology, geology, soil and expanded the development of:

- Medicine and biochemistry by establishment of precise methods for the determination of important components of tissues and body fluids and urine to help physicians in the diagnosis and assessment of deficiency as well as excess of some elements and compounds related to health, Genetic analyzes and food metabolism in life sciences research benefited much from analytical chemistry techniques.
- Environmental studies through monitoring concentrations of elements and compounds that are effective on human health and establishment of reliable methodologies to analyze water, soil and air.
- Quality control for industrial products especially drugs.
- Evaluation of raw materials and purity of an industrial product and its suitability for use and to monitor the various stages of industrial processes.
- Forensic investigations and
- Geological surveys studies for mineral resources to evaluate their chemical composition.

The analytical chemistry methods are divided into two main categories: qualitative and quantitative. Quantitative analysis can be done by traditional methods (gravimetric and volumetric) and instrumental methods. A variety of methods were established for the identification of chemical species in water, air, and soil samples. Visual indications are mostly employed in this respect. Spot tests represent well-defined category of qualitative test methods [1]. These methods use specific chemical reactions that yield colored products, precipitate formation and sometimes gas evolution to indicate the presence of analytes at certain concentration levels. Qualitative analyzes are often employed for screening a group of samples prior to their transportation to the laboratory for detailed quantitative determinations.

1.2   TYPES OF CHEMICAL ANALYSIS

1.2.1   *QUALITATIVE CHEMICAL ANALYSIS*

The operations by which the elements or compounds present in in a sample are identified and detected. The sample may be a solid mixture of materials or dissolved in
a solvent regardless of their concentrations.

1.2.2   *QUANTITATIVE CHEMICAL ANALYSIS*

The determination of quantities of components or elements contained in a sample or
a compound employing a variety of methods. The methods are preceded by qualitative analysis to ensure the presence of the specific components or elements in the
sample. Quantitative analysis is carried out by one of the following two methods.

1.2.2.1   *GRAVIMETRIC ANALYSIS*

The analysis is performed by using weighing of a material which is first precipitated
from the unknown sample solution and its determination as a single element or certain known derivative and separation from the bulk of the solution by precipitation
of centrifugation, washing drying and weighing. The weight of the required analyte
can then be calculated from the weight of the derivative and its composition precisely. As an example, chloride can be determined precisely in a sample of sodium
chloride by dissolving certain weight of the salt in water and addition of silver nitrate solution to precipitate it as silver chloride. The precipitate is washed, dried and
then weighed to calculate the quantity of the chloride in it and then its percentage in
the original sample.

1.2.2.2   *VOLUMETRIC ANALYSIS*

The analysis is performed by using direct or indirect methods to determine the concentration of certain material or some its components. It includes volumetric calibration methods, where solutions of precisely known concentrations are used and
specified volumes of these solutions are measured. This will react quantitatively
with the required component in the solution of the unknown sample until certain
point is reached (equivalence or end point). This can be determined by certain indicators giving a sharp change in the properties of the solution such as color or tur-
bidance which is clear enough to be noticed by naked eye or using simple physical
measurements like potential difference or electrical conductivity. The solution of
the known concentration is termed the calibration standard, the solution in which
a specified weight of the solute is contained in certain volume. The addition of the
calibration standard from a burette into a certain volume of the solution of the un-

known material contained in a conical flask until the end of the reaction is termed: the titration. The weight of the unknown component or its percentage in the sample can then be determined by using laws of chemical equivalence and the evaluation of the volume of the calibration standard solution used in the titration either directly or indirectly.

However, quantitative determinations are not exclusively done by chemical reactions. Natural separation methods played an important role in the performance and establishment of chromatographic analysis of the components of a mixture followed by identification and quantification by chemical methods. Although volumetric methods require the availability certain conditions and experience to overcome defects and faults, they are practical and more favorable over the precise gravimetric methods which take longer times to finish the analysis that may exceed few hours or even days to attain a reliable values. Such time requirement is not always available to carry out practical needs especially in quality control of the industrial processes to take over the necessary action of correcting a chemical process to get high quality results.

1.2.3   INSTRUMENTAL CHEMICAL ANALYSIS

As mentioned earlier, the chemical analysis is the science and the art of material composition determination indicated by the elements and/or compounds contained in it. The introduction of instrumentation into analysis methods resulted in significant developments and widens the capabilities of analytical techniques quantitatively and qualitatively. The invention of the precise balance was the first step to develop the chemical analysis. The invention of the spectroscope at the end of nineteenth century led to the introduction of very important concepts. The introduction of colorimetric and turbidimetric methods could gradually participate in boosting qualitative methods because the spectroscope is useful in qualitative aspects of the analysis. Later, the use electrochemical measurements proved themselves as efficient candidate methods to detect the end points of titration procedures and considered, then, a revolutionary achievement in chemical analysis. Today the analyst is surrounded by dozens of instrumental methods that are available to help him in his search.

Any physical characteristic of an element or a compound may be used as a basis of a successful analytic tool for its determination. Materials can be determined by measuring some of their physical or chemical characteristics like density, refractive index, color, electrical conductivity and thermal and electrical changes, etc.

The basic features of instrumental analysis methods are identical and thus, we may discuss them here to prevent any repetition. They are similar in the calibration procedure, the need for standard solutions, and calculation of sensitivity and detection limits in addition to concepts like the standard addition.

Instrumental methods of chemical analysis include:

1.2.3.1 SPECTROPHOTOMETRIC ANALYSIS

A group of analytical methods that are based on the interaction of matter with light. The useful interaction of matter may be represented by excitation of atoms or molecules through their electrons or the induced vibrations or rotation of functional groups within the molecules. Practically all ranges of light found useful applications in analytical chemistry. The various spectral ranges of light require specific instrumentation for the generation or detection. The analytical signal comprises the measurement of the extent to which the incident radiant energy is absorbed by the chemical species.

1.2.3.2 CHROMATOGRAPHIC SEPARATION ANALYSIS

A group of analytical methods that are based on the separation of sample components according to chemical and/or physical characteristics. The separation is aided by the interaction of the various components with a stationary phase by electrostatic or hydrogen bonding which vary with the molecular weight, boiling point, polarity, and the existence of certain functional groups. A mobile phase usually a solvent or an inert gas is allowed to induce the elution of the components and hence their separation and move the them till the end of the column or the surface where they were adsorbed. A detection tool is placed at the end to generate a measurable signal. The methods are grouped in accordance with the range of boiling point as in gas chromatography and liquid chromatography or the ionic character as in ion chromatography. Size exclusion is another tool for the separation of sample components. Electrophoresis uses elecric field to aid the transport of the various species with no stationary phase.

1.2.3.3 ELECTROCHEMICAL ANALYSIS

A group of analytical methods that are based on the electrochemical properties of solutions. The analytical work is performed in electrolytic cell with electrodes using various approaches. Potentiometric methods involve the use of working and reference electrodes. They are extensively used as a tool for the determination of ionic species and based on the potential measurement. The most known analytical measurement, the pH measurement is an example of the potentiometric methods employed in almost all laboratory and field measurements. Voltametric methods use the current as a measure for the concentration of metallic ions and some organic species suffering reduction at specific potentials.

1.2.3.4 OTHER METHODS

There are other instrumental methods that furnish some quantitative and qualitative data on the samples that allows the study of chemical composition, structure, and decomposition that aid the characterization. The x-ray diffraction is a highly specific method for the identification of crystalline compounds. Thermal methods of analysis also found analytical application to aid the characterization of natural and synthetic materials to give quantitative information on the sample and often correlated with composition. They are employed for proximate analysis of fossil like coal and characterization of crude oils [2]. The combination of pyrolytic procedures with chromatographic separation detecting opened a wide range of application of this combined method in the characterization of sediments pollution with organic matter [3].

Frequently, a combination of two methods like gas chromatography and mass spectrometry are used. Some of these instrumental methods are capable of detecting and determination compounds or elements with relatively high sensitivity that approaches parts of a billion. Environmental work usually needs the determination of variety of organic and inorganic compounds, anions, and cations. The evaluation of drinking water quality necessitates the utilization of all kinds of quantitative methods to estimate the levels of many analytes [4]. Classic titrimetric and gravimetric methods are used along with sophisticated instrumental methods to analyze unknown samples. Portable equipment and kits are usually indispensible items accompanying environmental chemists during field measurements and test. Barbooti et al., trapped nitrogen dioxide in a hydrogen peroxide to measure the removal efficiency of the gas to form nitric acid which could be estimated by titration against 0.01 N sodium hydroxide solution using phenolphthalein indicator [5]. Tap water is occasionally tested for residual free chlorine content with the aid certain tablets that give distinctive color, where the intensity of the color can be indicative of the free chlorine level in the water.

1.3 KINETIC ASPECTS OF CHEMICAL ANALYSIS

Most chemical analyzes are carried out on systems that are at equilibrium. This is true whether the analysis method is spectrometric, electrochemical, chromatographic, calorimetric, or even radiochemical in nature. Incomplete equilibrium means that the analysis will be performed on systems under quasi steady state conditions. Chemical wise the preparation of samples for analysis necessitates a sufficient period of time for the system to be at or very near equilibrium.

However, in kinetic methods of analysis, one measures the rate of an ongoing chemical reaction to determine the quantity or concentration of the analyte of interest, are becoming increasingly popular. The first, and probably most important, advantage of kinetic methods in general is speed. You don't have to wait for the reaction to go to completion to make your measurement. Kinetic methods based on reaction-rate measurements are usually done only during the first 3 to 10% of the reaction, before possible back-reactions start to become significant. In this situation, one measures the so-called initial reaction-rate with little error.

Of particular importance in the kinetic methods is the applicability in the presence of other species that may interfere in the determination, at various concentrations. A kinetic method may be developed where only the analyte reacts and not the interfered. It will be applicable and measurements can be performed on modern instruments with microprocessors. We may end up with a decent analysis by following up the rate of change in absorbance of a sample while the reaction is in progress. Thus, any steps for the separating the analyte from the matrix or removal of interfering species will be omitted saving time and effort.

The areas of analytical interest in which kinetics play direct or indirect important roles.

1. Formation of precipitates from solution: successful filtration of gravimetric precipitates depends on theories of nucleation and crystal growth.
2. Solvent extraction: Kinetic aspects are important in slow solvent extraction processes, including the rate of mass transfer across the interface between the solvent and the original solution [6].
3. Ion Exchange: experimental conditions are essential factor in the determination of rate of exchange including particle size, film diffusion.
4. Electrode kinetics: Current-potential dependence of electrode processes provides characteristic information on electrode kinetics which depends on the rate of electron transfer at the interface as well as pure chemical kinetics in and in the vicinity of the double layer.
5. Continuous flow systems: kinetics is the heart of continuous flow analyzers. Modern flow injection systems include much of kinetics [6].

1.4   SPECIATION IN CHEMICAL ANALYSIS

Modern environmental studies indicated that the properties and the health effects of an element are basically dependent on the form in which the element occurs in the sample. The basic example in this context is the chromium, which is top toxic element when it is in the hexavalent form. In the meantime, the trivalent form is not toxic. Thus, the species of the element and not the element itself is the toxicity-determining factor. This has spurred a remarkable development in the analysis termed the speciation analysis and it is of growing interest. IUPAC defines a chemical species as a specific and unique molecular, electronic, or nuclear structure of an element [7]. Speciation analysis studies the forms at which an element is present in the sample or its distribution among various species. Analytical techniques are combined to aid the speciation where the first separates the individual species and the other detects the composition of it.

Specific analytes are targeted by the speciation analysis like organoarsenic, alkyl lead or some selenium species. Gas chromatography with its high separation efficiency and low detection limits may serve as the separation technique especially with organometallic species that have some volatility. The atomic spectroscopic techniques are used for the detection of the species eluting from the column [8].

The ability of some plants and grasses to live and reproduce in heavily contaminated environment leads to the utilization of such plants in the remediation of soil. They are termed the hyperaccumulators [9]. These plants evolve efficient metal homeostasis. Speciation analysis is particularly essential in such remediation studies by furnishing information about the binding ability as well as the nature of bonding and whether it is with proteins or other biological molecules [10]. The absence of the standards for the studied species necessitates the use of molecular detection or specification like mass spectrometry.

Bouyssiere, et al. [11] reviewed the recent advances in the application of the hyphenated (coupled) techniques for species-selective determination volatile organometallic anthropogenic contaminants and nonvolatile organometallic compounds and heavy metal complexes in environmental matrices.

1.5   NUMBERS IN ANALYTICAL CHEMISTRY

Analytical chemistry is a quantitative science. When the analyst tries to determine the concentrations or evaluate the equilibrium constants or draw the relation between the composition of a compound and its activity, he performs measurements and calculations. There are some important aspects in the use of numbers in analytical chemistry.

1.5.1   BASIC MEASUREMENT UNITS

Units are essential part of an analytical procedure or report. They precisely express the meaning of the measurement, which consists of a number and a unit. They describe the quantity of the measured material. A procedure "transfer 7.5 of material A into a 50-mL calibrated flask and dilute to the volume" is not comprehensive because it misses the unit. Currently, the basic units of the various physical quantities must follow the international System of Units, *Système International d'Unités, termed the "SI System."*

The mole contains 602213670000000000000000 particles and some analytical techniques are capable to detect 0.000000000000001 g of a compound. However, some other units are practically available for use to measure the quantities of Table 1 or other quantities. Power may be use to prevent the use of zeros or the decimal points.

1.5.2 SIGNIFICANT NUMBERS

The recording of the measurement gives information about the quantity and the degree of its uncertainty. A reading of weight with a 4-digit balance indicates that the reading is true with the exception of the last digit. We propose that the uncertainty in the last digit is ∓ 1, the weight is accurate with uncertainty degree that never exceeds ∓ 0.0001. Significant numbers are used to express the degree of uncertainty in the readings. The uncertainty in the readings and values is transferred to summation operation where the product of summing a value with three digits with a value of two digits will be a value with only two digits. The handling of the data depends on the measuring tool and the number of digits given by the instrument. Writing the data with the wright number of digits is the way to keep the wright degree of uncertainty. As an example, in the division of 101 on 99 a product of 1.02 is acceptable because both numbers are correct within $\pm 1\%$ (101 ± 1 and 99 ± 1). To prevent any mistakes due to approximation, it is wise to keep one extra digit, where the approximation will give the right significant numbers.

1.5.3 UNITS OF CONCENTRATION

Concentration represents the amount of the solute in the medium or in the solution. It is usually expressed by a number of methods. Table 1.1 shows the units of concentration. The molar concentration is identical to formal concentration for nonionizable materials while it is different for the ionizable materials.

TABLE 1.1   Units of Concentration

Unit	Symbol
Molarity	M
Formality	F
Normality	N
Molality	M
Weight%	Wt%
Volume %	Vol%
Weight/Volume %	Wt/vol%
Parts per million	ppm
Parts per billion	ppb

The dissolution of one mole of $MgCl_2$ means the formation of one mole of Mg ions and two moles of chloride ions leaving the $MgCl_2$ concentration at zero M. In the meantime, the formal concentration of $MgCl_2$ will be 1 because the formal represents the amount of the dissolved salt. The square bracket $[Mg^{2+}]$ is used to express the molar concentration. However, the Molar concentration is no longer predominant in modern literature.

1.5.3.1   NORMALITY, N

It is an old expression of concentration that uses the equivalent weight. Normality is an expression for the reaction and not for the concentration. The normality of sulfuric acid of certain molar concentration (specified in the label on the bottle) is dependent on its reaction. In the precipitation reaction of lead iodide, the number of the equivalents represents the amount of the material, which enters the reaction, and also the charge of the cation or the anion, which enters the reaction. For lead the number will be 2 and for iodine the number will be 1.

$$Pb^{2+}(aq) + 2I^-(aq) \leftrightarrows PbI_2(s) \tag{1}$$

For the acid-base reaction the reacting unit will be the number of hydrogen ions donated from the acid or accepted by the base. For the reaction of ammonia with sulfuric acid:

$$H_2SO_4(aq) + 2NH_3(aq) \leftrightarrows 2NH_4^+(aq) + SO_4^{2-}(aq) \tag{2}$$

We find that for the acid it is 2 and for ammonia it is 1.

For complex formation reactions, the reacting unit is identical to the number of the electron pairs accepted by the metal ions or donated by the ligands. For the reaction of silver ions with ammonia.

$$Ag^+(aq) + 2NH_3(aq) \leftrightarrows Ag(NH_3)_2^+(aq) \tag{3}$$

N will be 2 for silver ion and 1 for the ammonia.

For redox reactions the reacting unit will be identical to the number of electrons given by the reducing agent or those accepted by the oxidizing agent.

$$2Fe^{3+}(aq) + Sn^{2+}(aq) \leftrightarrows Sn^{4+}(aq) + 2Fe^{2+}(aq) \tag{4}$$

Normality is defined as the number of equivalent weights in unit volume, which is similar to the formal concentration being independent on dissolution or dissociation. Molar and normal concentrations are related as follows:

$$N = n \times M \tag{5}$$

Example 1

Calculate the normality of a 6 M phosphoric acid solution using the following reactions:

$$H_3PO_4(aq) + 3OH^-(aq) \leftrightarrows PO_4^{3-}(aq) + 3H2O(l) \tag{6}$$

$$H_3PO_4(aq) + 2NH_3(aq) \leftrightarrows HPO_4^{2-}(aq) + 2NH_4(aq) \tag{7}$$

$$H_3PO_4(aq) + F^-(aq) \leftrightarrows H_2PO_4^-(aq) + HF(aq) \tag{8}$$

Solution
The number of equivalents of phosphoric acid will be represented by the number of hydrogen ions donated to the base. For the given three reactions the number of equivalents will be respectively: 3, 2 and 1. Thus, the calculated equivalent weights and the normalities will be,

$$W = \frac{FW}{n} = \frac{97.994}{3} = 32.665\,N = n \times M = 3 \times 6.0 = 18N \tag{9}$$

$$EW = \frac{FW}{n} = \frac{97.994}{1} = 48.997\,N = n \times M = 2 \times 6.0 = 12N \tag{10}$$

$$EW = \frac{FW}{n} = \frac{97.994}{1} = 97.994\,N = n \times M = 1 \times 6.0 = 6N \tag{11}$$

For gasses, the part per million concentrations will be represented by volume ratios. A helium concentration of 6 ppm in air means that one liter of air contains 6 uL of helium.

Example 2
The maximum permissible chloride concentration is 2.5×10^2 ppm. Express this in molar concentration.

$$\frac{2.5 \times 10^2\,mg\,Cl^-}{L} \times \frac{1g}{1000\,mg} \times \frac{1\,mol\,Cl^-}{35.453\,g\,Cl^-} = 7.05 \times 10^{-3}\,M \tag{12}$$

Example 3
A concentrated solution of ammonia 28.0% by weight and density of 0.899g.mL calculate the molar concentration:

$$\frac{28.0\,g\,NH_3}{100\,g\,solution} \times \frac{0.899\,g}{mL} \times \frac{1\,mol\,NH_3}{17.04\,g\,NH_3} \times \frac{1000\,mL}{L} = 14.8\,M \tag{13}$$

1.5.4 THE USE OF THE P FUNCTION

The expression of concentration with some units will frequently become not inconvenient when large changes (many orders of magnitudes) occur after a chemical reaction. The plotting of progress of the reaction as a function of time or the added volume of the reactant will be difficult as the concentration of hydrogen ions will change from 0.01 M to 5×10^{-13} after the addition of 75 mL of the titrant and hence the presentation of reactant change will be in narrow ranges. Here the utilization of the logarithmic expressions which is known as the p = function.

$$pX = -Log (X) \tag{14}$$

the pH value for the starting concentration will be =1.

$$pH = log[H^+]\, 0 - log\,(0.1) = 1.0 \tag{15}$$

At the end the pH of the solution will be 12.3

$$pH = log[H^+]\, 0 - log\,(5.0\times10^{-13}) = 12.3 \tag{16}$$

Therefore, the use of this function will give more accurate and clearer expression for the change in concentration during the reaction progress.

Example

Calculate the pNa for sodium phosphate solution Na_3PO_4, 1.76×10^{-3}.

Solution

Since the dissolution of this salt will give three sodium ions, the Na concentration will be

$$\left[Na^+\right] = \frac{3\,mol\,Na^+}{mol\,Na_3PO_4} x1.76x10^{-3} = 5.28\ x10^{-3}\ M \tag{17}$$

From which pNa can be calculated

$$pNa = -log\,[Na^+] = -log\,(5.28 \times 10^{-3}) = 2.277 \tag{18}$$

Similarly pH may be calculated for the solution by knowing the molar concentration.

1.5.5 STOICHIOMETRY

A balanced chemical reaction is indicative of the quantitative relations between the moles of the reactants and the products. Such relations furnish the basis of many analytical calculations. Let's consider the case of determination of oxalic acid in certain herb material. One of the methods depends on the following reaction where oxalic acid is oxidized into carbon dioxide:

$$2Fe^{3+}_{(aq)} + H_2C_2O_{4(aq)} + 2H2O_{(aq)} \rightleftarrows 2Fe^{2+}_{(aq)} + 2CO_{2(g)} + 2H_3O^+_{(aq)} \tag{19}$$

The balanced reaction equation provides chemical relation between the moles of the consumed ferric ions and the moles of oxalic acid provided that the data of the ferric ions are known.

In this specific example, oxalic acid is present in the form required by the method of analysis. For various applications the analyte must be transformed into an analyzable form prior to the analysis. As an example, for the determination of some pharmaceuticals like $C_{10}H_2ON_2S_4$ we need to oxidize its sulfur content into SO_2 and driving the gas produced through hydrogen peroxide to produce sulfuric acid, followed by titration of this acid with a suitable case like NaOH. It is possible to write balanced chemical equations that describe these steps. Principally, the rules of reservation of reaction units are obeyed here. Such transformations are essential to convert the analyte into a form, which responds to the analysis. The original weight calculations must be kept in mind by establishing the mathematical relations between the measured and the original forms of the analyte.

1.5.6   ANALYSIS VERSUS ESTIMATION AND MEASUREMENT

The analysis provides the physical and chemical information about the sample. The important components are termed the *Analytes*, and the rest of the sample will represent the medium or matrix. During the analysis the identity, the concentration, and the characteristics of the analyte are determined. To carry out the estimation we measure one or more physical or chemical properties of the analyte.

1.6   BASIC TOOLS AND EQUIPMENTS

1.6.1   TECHNIQUES, METHODS AND PROCEDURES

To develop analytical method for a specific material we must think with four levels and establish the technique, method and the procedure.

1.6.1.1   TECHNIQUE

Any physical or chemical principle that can be used to study the analyte. Many techniques are used for the determination of a material like atomic absorption by changing the required metal into atoms that will absorb radiation at certain wavelength and measure the absorbance. Thus, the basis is chemical (atomization) and physical (light absorption).

1.6.1.2   METHOD

The application of the technique to determine the analyte in certain medium. The method of lead determination in a blood sample differs from that used for soil. For certain samples atomic absorption could be the best technique and other techniques like potentiometer using selective electrodes may be preferred for other samples.

1.6.1.3   PROCEDURE

A set of written steps that describe the details of sampling of certain material and handling of the interfering materials and approval of results. The method may end up with more than one procedure because analysts and directories will adapt the analytical methods in accordance with their requirements.

1.6.2   THE CHOICE OF ANALYTICAL METHOD

A method is the application of a technique on a certain analyte in a certain matrix. The analysis requirements determine the best method. Among the basic considerations:

1.6.2.1   ACCURACY

Measures of how close an experimental result to the expected value (Fig. 1.1). The difference between the result and the expected value is divided by the expected value and the product is written as the percentage error:

Error % = (result–expected value) x 100

Analytical methods are classified in accordance with the percentage error into three categories:

When the error is within 1%, the method is considered of high accuracy;

When the error is between (1–5%), the method is considered of moderate accuracy

When the error exceeds 5%, the method is considered of low accuracy.

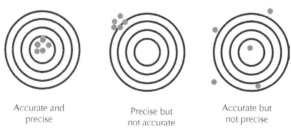

Accurate and precise   Precise but not accurate   Accurate but not precise

FIGURE 1.1   Schematic representation of accuracy and precision.

1.6.2.2   PRECISION

When a sample is analyzed for several times, the results is rarely identical. The results, instead, are distributed randomly. Precision will be a measure of this variation of the results. As the agreement of the results increases, the precision of the method increases.

1.6.2.3   SENSITIVITY

Chemical analysis is characterized by the discrimination between two samples with respect their content of certain analyte. The sensitivity of a method is defined as the measurement of the ability of a method to indicate a clear difference between two samples. Sensitivity must not be mixed with the concept of detection limit, which refers to the lowest amount of the analyte that can be determined with sufficient confidence. Detection limit, therefore, is a statistical term.

1.6.2.4   SELECTIVITY

An analytical method is considered selective if the measured signal is a function of the analyte present in the sample. Selectivity of the method is determined by the selectivity coefficient of the interfering material in comparison with the analyte, which may be positive or negative depending on the effect of the interference on the signal. The coefficient is $> +1$ or < -1 whenthe method is more selective for the interferent material than the analyte itself. The estimation of selectivity coefficient helps in the determination of effect of the interfering materials on specific analyzes. However, the analytical procedures based on chemical activities are less selective and hence will be more susceptible for interferences. The selectivity problem will be greater if the analyte is present at very low concentrations.

1.6.3   COST AND TIME ASPECTS OF INSTRUMENTAL ANALYSIS

The work in chemical analysis laboratories includes both classical and instrumental methods depending on the analyte under investigation and the available resources. For comparative purposes, the methods are evaluated on basis time requirement of an analysis degrees of training requirement to run new instrumentation. These are all translated into the cost of a single analysis and its time requirements. Classical volumetric methods will require only simple equipment and reagents that can be easily handled. The time requirement for analyzing a single sample is not significantly different for the variety of methods by considering the sample preparation and preparing the instrument for measurement as the main time consuming step. After the samples are prepared and the instrument is set for the analysis, time may be used to determine the number of analyzes per hour as a possible comparison criterion between the various methods.

Many factors determine the cost of the analysis including cost of supplying the instrument, reagents and the cost of the working staff and the number of samples per hour. Highly sophisticated instruments necessitate keeping them running continuously with a steady flow of some inert gases, which contributes in the cost of analysis. In general, the cost of instrumental methods is usually more than manual methods.

1.6.4 VALIDATION

For an analytical method to be recommended for a certain determination there must be some data on its reliability to give acceptable results. Validation is an approval for the method reflecting the precision, accuracy and convenience for a specific problem. Validation is the way the detailed procedures are written so that analysts from different laboratories will apply the steps of method and obtain comparable results. A standard sample is used with a concentration close to the unknown sample for which the method was designed. The repeatability of the results of the various laboratories is compared to evaluate the method with regard to the precision, accuracy. The new method may also be compared with other certified methods if other standard samples are available.

1.6.5 PROTOCOLS

The protocol is a set of restricted guidelines and steps for a precise procedure to be applied and its results to be accepted. The protocol must also contain a methodology for quality control/ quality assurance, QA/QC, internally and externally. The internal QA/QC is essential to evaluate the method with respect to accuracy and precision and carried out in the laboratory. While the external QA/QC means that the laboratory is applying for the accreditation certificate from an external agency or authority.

1.7 HANDLING OF THE ANALYTICAL DATA AND STATISTICS

The random error is distributed symmetrically around the mean values. The standard deviation(s) is a measure of how the readings surround the mean value. As s gets smaller the data are close to each other in the distribution.

For a number of results, n, a value of $(n-1)$ is used to calculate the standard deviation. The square of the standard deviation, s^2, represents the variation. The expression of the standard deviation as a ratio from the mean value is the relative standard deviation or the coefficient of variation.

$$X = \frac{-\sum_i X_i}{n} \tag{20}$$

$$S = \sqrt{\frac{\sum(x_i - X)}{(n-1)}} \tag{21}$$

1.7.1   REJECTING DATA BASED ON IMPRECISION

Sometimes, a value within a set might appear aberrant. Although it might be tempting to reject this data point, it must be remembered that it is only abnormal in respect of a given law of probability. There exists a simple statistical criterion for conservation or rejection of this outlier value. This is Dixon's test, which consists of calculating the following ratio (on condition that there are at least seven measurements):

$$Qcalc = \frac{|suspect\ value - nearest\ value|}{Larg\ est\ value - Smallest\ value} \tag{22}$$

Q calculated in this manner is compared with a table of critical values of Q as a function of the number of data (Table 1.2). If $Q_{calculated}$ is greater than $Q_{critical}$ the value in question can be rejected.

Q_{tab} is looked up in a table and compared with Q_{calc}. If $Q_{calc} > Q_{tab}$, the outlier data point can be rejected at the specified confidence level. It should be noted that in order to continue to statistically treat your data after a point has been rejected, you should have at least four data points to begin with (n=4). The Q-test can only be used to reject one data point from a dataset.

TABLE 1.2   Tabular summary of critical values for Q (Dixon's test).

Number of measurements, n	Confidence level	
	95%	99%
3	0.94	0.99
4	0.779	0.89
5	0.64	0.78
6	0.56	0.70
7	0.51	0.64
8	0.47	0.59
9	0.44	0.58
10	0.41	0.53

1.7.2 CALIBRATION GRAPH

Quantitative chemical analysis is based on comparison methods. As an example, the sample which contains a quantity of the analyte equal to that contained in another sample or a standard solution will give identical signals, when the same conditions and the same instrument is employed for the measurements. However, the calibration must be done with lab equipment and instruments. Balances and pipets are examples of the essential equipment employed in the quantitative analysis and hence must be calibrated for reliable measurements to be obtained. Volumetric pipets are calibrated by carefully filling with water, dispensing the water in a tarred beaker and measure the mass of he water. The density of water is used to calculate the volume of the dispensed liquid. Similarly, volumetric flasks are calibrated.

Lab instrumentation also need to be calibrated to have a reliable signal which can be correlated to a given amount or concentration of the analyte. In most cases reference is made to a single or multiple point standardization. Multiple point standardization is more reliable method of standardization. A set of standard solutions with known concentration is prepared using the calibrated glassware and their signals are measured. A calibration graph can be drawn for the analytical signals as a function of concentration. The graph can then be employed to determine the concentration of the unknown sample or to calibrate the response of certain instrument. Traditionally the concentration is used as the x-axis and the signal as the abscissa. The most desirable case is when the calibration graph is linear where the slope can then be used to calculate the concentration of the unknown sample solution.

We assume that any difference between the matrix of the standards and that of the samples has no effect on the response and hence on the slope of the calibration graph. When the difference in the matrix composition between the standard solutions and the sample solution cannot be neglected, the relationship of the concentration and the measured signal will lead to erroneous result. This case is graphically shown in Fig. 1.2. It can be seen that the result from normal calibration graph is less than the actual value. Thus, matrix matching is an important step for reliable analysis.

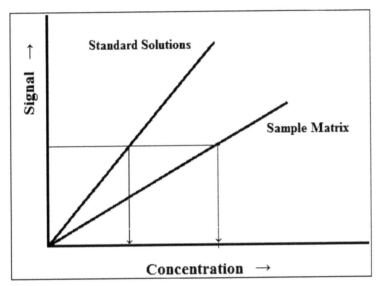

FIGURE 1.2   Calibration errors when the sample matrix affects the measured signal.

1.7.3   REGRESSION ANALYSIS

The purpose of the regression analysis an analytical chemistry is to find a relation between signal and concentration, so that the obtaining of a signal is sufficient to produce the concentration. Thus, regression analysis steps farther than simple relation and allows the establishment of a mathematical model between two or more variables. The modeling performed for the signal produced on a certain instrument (y-axis) as a function of concentration (x-axis). The model is applied such that Y value can be evaluated as x value and the function is known. Any uncertainty in the results will be the accumulation of uncertainty degrees related to the model and the measurement selected for the analysis.

A number of mathematical models are employed in the software of chemical analysis. Here some preliminary results of simple linear regression which is the most suitable method for quantitative analysis.

As the response of many detectors is linear as a function of the measured variable, taking account of the differences due to experimental conditions as well as of the instrument, the goal is to find the parameters of the straight line that best fits.

The *least squares regression line* is the line, which minimizes the sum of the square or the error of the data points. It is represented by the linear equation

$$Y = ax + b \tag{23}$$

The variable x is assigned the independent variable, and variable y is assigned the dependent variable. The term b is the y-intercept or regression constant (the value of when x=0), and the term a is the slope or regression coefficient. Coefficients a and b, as well as the standard deviation on a, may be given by several formulae and specialized software.

Moreover, the dimensionless Pearson *correlation coefficient* R gives a measure of the reliability of the linear relationship between the x and y values. If R=1 it exists an exact linear relationship between x and y. Values of R close to 1 indicate excellent linear reliability. If the correlation coefficient is relatively far away from 1, the predictions based on the first order relationship, y = ax + b will be less reliable.

$$a = \frac{n\sum x_i y_i - \sum x_i \sum y_i}{n\sum x_i^2 - (\sum x_i)^2} \qquad (24)$$

$$b = \frac{\sum x_i - a\sum x_i}{n} \qquad (25)$$

$$R = \frac{n\sum x_i y_i - \sum x_i \sum y_i}{\sqrt{[n\sum x_i^2 \sum(x_i)^2].[n\sum y_i^2 - (\sum y)^2}} \qquad (26)$$

A straight line supposes that the errors in "y" follow the law of Normal distribution. R^2 is the *determination coefficient* that conveys information about how the variations of x overlap variations in y.

In classic calculations the experimental error is considered to affect the y value exclusively and not the concentration recorded in x. If this is not the case, the data points will not have the same quality for the regression line hence comes the idea of according less value to data more distant from the line. Through iterative calculations the equation of a straight line is reached which takes account of the weighting of each point.

In order to compare whether two methods of analysis are highly correlated, a series of *n* standards of different concentrations are analyzed by two pathways. Each standard is next represented by a point on a graph where *x* bears the results of method *I* and *y* the corresponding results from the second method. The correlation coefficient R is determined and Eq. (27) is applied. If the *t* value obtained is greater than the value found in the table for n−2 degrees of freedom, there is a strong correlation between the two analytical methods.

$$t = R\sqrt{\frac{n-2}{1-R^2}} \qquad (27)$$

1.7.4   OPTIMIZATION THROUGH THE ONE-FACTOR-AT-A-TIME (OFAT) EXPERIMENTATION [3]

When a measurement depends upon a signal (absorbance or intensity of fluorescence), which is itself influenced by several factors, then in general, it is customary to seek overall conditions, which will lead to the maximum signal. If the factors involved in an analysis are independent (which is rarely the situation), a common practice is to experiment with *one factor at a time* while holding all others fixed. Then the influence of each one can be studied on the result by using a simple repetitive method.

However, the results are often misleading and fail to reproduce conclusions drawn from such an exercise. This method needs some understanding of basic *design of experiments* (DOE) principles.

This repetitive method is attractive, as it is simple and supported by common sense, but it does not always constitute the best approach to the problem. If the iso-response curves form a complex envelope, showing notably a ridge–the mathematical illustration of the interaction between the two factors–the preceding method can lead to a false optimum depending upon the parameters selected at the beginning. A more effective method for these situations is to study their effect simultaneously by setting the DOE statistical technique. This different approach has the aim of discovering the optimal conditions using the minimum number of attempts. This is illustrated by the sequential simplex method of optimization and by different designs described in specialized textbooks.

Anderson and Whitcomb [13] made a comparison of the performance of OFAT with factorial design"—a tool that allows work with many variables simultaneously. The simplest factorial design involves two factors, each at two levels. The equivalent OFAT experiment is shown at the right. The points for the factorial designs are labeled in a "standard order," starting with all low levels and ending with all high levels. For example, runs 2 and 4 represent factor A at the high level. The average response from these runs can be contrasted with those from runs 1 and 3 (where factor A is at the low level) to determine the effect of A. Similarly, the top runs [3 and 4] can be contrasted with the bottom runs [1 and 2] for an estimate of the effect of B.

The advantage of factorial design becomes more pronounced as you add more factors. For example, with three factors, the factorial design requires only 8 runs (in the form of a cube) versus 16 for an OFAT experiment with equivalent power. In both designs, the effect estimates are based on averages of 4 runs each: right-to-left, top-to-bottom, and back-to-front for factors A, B and C, respectively. The relative efficiency of the factorial design is now twice that of OFAT for equivalent power. The relative efficiency of factorials continues to increase with every added factor.

Factorial design offers two additional advantages over OFAT:
- Wider inductive basis, i.e., it covers a broader area or volume of X-space from which to draw inferences about your process.

- It reveals "interactions" of factors. This often proves to be the key to understanding a process, as you will see in the following case study.

1.7.5 CORRECTION WITH BLANK SOLUTION

The presence of a blank solution is essential for calibration methods to correct the measurements from any signal due to nonanalyte components. A successful correction of sample measurements must be based on the right blank solution and calibration. The blank solution must contain all reagents and treated in a similar manner as the sample solution. However, for a more precise analysis, the correction must pay some attention to the shifts that may result from any interaction between the analyte and the reagents.

KEYWORDS

- **Basic tools**
- **Choice of Methods**
- **Data Handling**
- **Qualitative Analysis**
- **Quantitative Analysis**
- **Units**

REFERENCES

1. Feigel, F., & Anger, V. (1988). Spot Test in Inorganic Analysis, 6th Ed., Elsevier Science, Amsterdam.
2. Barbooti, M. M., & El-Sharifi, T. H. (1989). Chemical Evaluation of Crude Oils by Programmed Thermogravimetric Analysis, Thermochim Acta, *153*, 1–10.
3. Kruge, M., & Permanyer, A. (2004). Application of Pyrolysis-GC/MS for Rapid Assessment of Organic Contamination in Sediments from Barcelona Harbor, Organic Geochem, *35(11–12)*, 1395–1408.
4. Barbooti, M. M., Bolzoni, G., Mirza, I. A., Pelosi, M., Barilli, L., Kadhum, R., & Peterlongo, G. (2010). Evaluation of Quality of Drinking Water from Baghdad, Iraq, *Sci. World J.*, *5(2)*, 35–46.
5. Barbooti, M. M., Ibraheem, N. K., & Ankosh, A. (2011). Removal Of Nitrogen Dioxide and Sulfur Dioxide from Air Streams by Absorption in Urea Solution, *J. Environmental Protection*, *2*, 175–185.
6. Mottola, H. A. (1988). Kinetic aspects of Analytical Chemistry, *97*, in "Chemical Analysis", Winefordner, J. D., Kolthoff, I. M., Eds. Wiley Interscience, New York.
7. Templeton, D. M., Ariese, F., Cornelis, R., Gent, U., Danielsson, L., Muntau, H., Leeuwen H. V., & Lobinski, R. (2000). Guidelines for Terms Related to Chemical Speciation and Fractio-

nation of Elements, Definitions, Structural Aspects, and Methodological Approaches (IUPAC Recommendations). *Pure and Applied Chemistry, 72(8)*, 1453–1470.

8. Bouyssiere, B., Szpunar, J., & Lobinski, R. (2002). Gas Chromatography with Inductively Coupled Plasma mass Spectrometric detection in Speciation Analysis, *Spectrochim Acta, B, 57*, 805–828.

9. Prasad, M. N. V., & Hagemeyer, J. (Eds.) (1999). Heavy Metal Stress in plants from Molecules to Ecosystem, Springer, Heidelberg.

10. Bouyssiere, B., Lobinski, R., & Szpunar, J. (2004). Hyphenated Techniques in Environmental Speciation Analysis, Chap 19 in "Organic Metal and Metalloid Species in the Environment," Hirner, A. V. Ed., Springer-Verlag, Berlin.

11. Harvey, D. (2000). Modern Analytical Chemistry, McGraw-Hill Higher Education, A Division of the McGraw-Hill Companies, Boston.

12. Rouessac, F., & Rouessac, A. (2007). Chemical Analysis Modern Instrumentation, 2nd Edition, Chapter. 22, West Sussex, E-mail: cs-books@wiley.co.uk.

13. Anderson, M., & Whitcomb, P., Chapter 3: Two-Level Factorial Design, in DOE Simplified: Practical Tools for Effective Experimentation: 2nd Ed, http://www.statease.com/pubs/doesimp2excerpt-chap3.pdf.

CHAPTER 2

SPECTROCHEMICAL METHODS OF ANALYSIS

MAHMOOD M. BARBOOTI[1*], ASLI BAYSAL[2], and NIDHAL ALDERZI[1]

[1]Department of Applied Chemistry, School of Applied Sciences, University of Technology, P.O. Box 35045, Baghdad, Iraq; *E-mail: brbt2m@gmail.com

[2]T.C. Istanbul Aydin University, Health Services Vocational School of Higher Education, 34295 Sefakoy Kucukcekmece - Istanbul, Turkey; E-mail: aslibaysal@aydin.edu.tr, baysalas@itu.edu.tr

CONTENTS

2.1   INTRODUCTION

Spectrochemical Methods are a group of analytical methods based on the interaction between light and matter. Light is an electromagnetic radiation with two components, electrical and magnetic. Electromagnetic radiation spectrum ranges from the high-energy γ-rays through the visible light (colored radiations) to the radio waves of the low energy. The energy of the radiation increases as the wavelength decreases. The main characteristics of waves are the wavelength, frequency and amplitude as can be seen in Fig. 2.1.

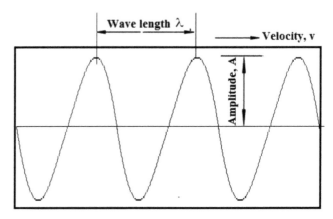

FIGURE 2.1   Light wave.

Wavelength, λ, is the linear distance between two wave maxima. Frequency, ν, is the number of waves per second (wave.s^{-1} = Hertz). The wave number is another term for the frequency and represents the number of waves per one centimeter and expressed as cm^{-1}). The wave number is specifically used in the infrared region.

Wave number # of λ per cm $\left(v=\dfrac{1}{\lambda}\right)$ Velocity of light = 2.99782×10^8 m.s^{-1}

Example

To calculate the frequency and wave number for nickel emission line 231.604-nm.

$$v=\frac{c}{\lambda}=\frac{2.00\,x10^8\,m\,/\,s}{231.604\,c10^{-9}\,ms}=12.95\,x10^{14}\,s^{-1} \tag{1}$$

$$v=\frac{1}{\lambda}=\frac{1}{231.604\,c10^{-9}\,ms}\,x\frac{1\,m}{100\,cm}=4.317\,x10^{14}\,s^{-1} \tag{2}$$

The energy of the photon may be expressed as

$$E = hv = \frac{hc}{\lambda} \tag{3}$$

where, h is the plank's constant with a value of 6.626×10^{-34} J.s. The energy of the photon of the nickel emission line 231.604 nm may be calculated as follows:

$$E = \frac{hc}{\lambda} = \frac{6.626 \times 10^{-34} \times 3.00 \times 10^{8}}{231.604 \times 10^{-9}} = 8.58 \times 10^{-19} J \tag{4}$$

Spectrophotometric methods are useful whenever there is an interaction between light and matter, which leads to characteristic change in some of its properties. Spectral absorption occurs as a result of the transition between energy levels for atoms and molecules. There are only limited energy levels (states) for atoms, ions and molecules. E° represents the ground state, while E^1, E^2 and E^3 represent the excited states (Fig. 2.2). Atoms and molecules can be excited electronically, vibrationally, and rotationally by absorption of light energy. Energy levels of molecules are different from atomic energy levels.

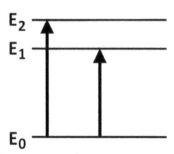

FIGURE 2.2 Energy levels.

1. When the energy state of atoms and molecules changes they emit or absorb energy, which is equal to the difference between the first and second energy levels E^1 to E^2 as in the figure.

$$\Delta E = E^1 - E^{\circ} \tag{5}$$

2. The wavelength or the frequency of the emitted or absorbed radiation is proportional to the energy difference between the two levels ΔE.

$$\Delta E = h.v = \frac{h.c}{\lambda} \tag{6}$$

2.1.1   EMISSION SPECTRA

When a salt crystal is dropped in a flame an orange light is noticed. What is this light? It is the emission. When emission intensity is plotted against the wavelength the emission spectrum is obtained. Emission spectra of atoms comprise linear spectrum as a result of the transition of internal electrons to the ground state. In the meantime, molecular emission spectra occur as a result of vibrational and rotational transitions within the molecule and in the form of wide bands and not linear.

2.1.1.1   CONTINUUM SPECTRA

When a bar of iron is placed in a flame, it becomes red first and then with sever heating it turns white, why? This phenomenon is called the black body radiation (Fig. 2.3). It is a group of very wide bands resulting from the thermal excitation and relaxation occurring in the molecules at various vibrational and rotational levels.

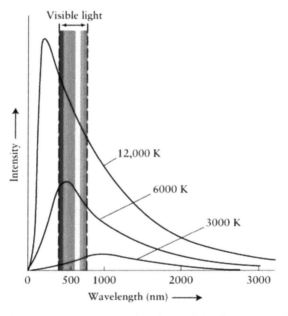

FIGURE 2.3   Black body radiation (Reproduce from Wikipedia Free Encyclopedia) [1].

2.1.2   ABSORPTION SPECTRA

Molecules and atoms absorb selective and quantized light energy. The absorbed energy must be exactly identical to the difference between the two energy levels.

When the absorbed radiation energy is plotted against wavelength the absorption spectrum is obtained (Fig. 2.4). Thus, the maximum of the absorption band occurs at the wavelength representing the energy difference between the two levels within the absorption spectrum.

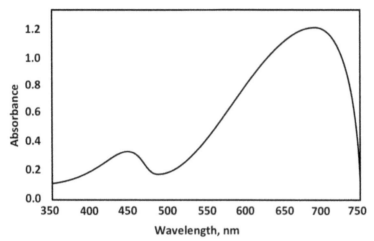

FIGURE 2.4   Absorption spectrum.

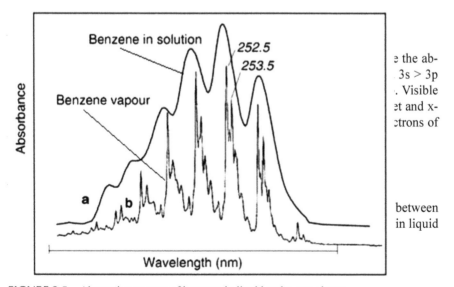

FIGURE 2.5   Absorption spectra of benzene in liquid and vapor phases.

$$\Delta E = \Delta E_{electronic} + \Delta E_{vibrational} + \Delta E_{rotational} \qquad (7)$$

There exist many vibrational states for each electronic state and for each vibrational state there are several rotational states. Absorption spectra are influenced by:

1. Presence of many atoms (spectrum becomes more complex as the number of atoms increases;
2. Solvent molecules (decreases the resolution);
3. The chemical state (with what the molecule is bonded).

2.1.3   FLUORESCENCE

The emission radiation results when the molecule leaves the excited state, which was reached by thermal or spectral excitation down to the ground state. Resonance emission occurs when the wavelength of the emission is identical to the wavelength of the excitation. In the meantime, the molecule may lose some of the absorbed excitation energy by an internal nonspectral conversion. The remaining energy will be emitted at a wavelength longer than that of the excitation energy and the process is termed fluorescence. The process is illustrated in Fig.2.6. Many organic and inorganic molecules exhibit fluorescence in the visible region when they are irradiated with UV light. Atoms also show fluorescence. Fluorescence spectra are mostly used for structural studies. The analytical applications of molecular and atomic fluorescence are limited in comparison with spectral absorption.

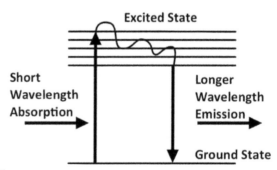

FIGURE 2.6   Fluorescence.

2.2   SPECTROPHOTOMETRY

2.2.1   FUNDAMENTALS

When a beam of light with an intensity of I_0 passes through a medium with a path length **b** containing a material of **c** concentration, it undergoes absorption and the passing portion I will be of lower intensity than the incident radiation (Fig. 2.7). The expression of transmittance T% is used to represent the ratio between the intensities of the passing and incident radiation through the material I/I_0. Absorbance is now used to express the change in the radiation intensity as it passes through certain material.

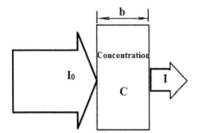

FIGURE 2.7 Light interaction with matter.

$$T\% = I/I_0 \times 100\% \qquad (8)$$

Absorbance, A:

$$A = \log_{10} I_0 / I$$

$$A = \log_{10} 1 / T$$

$$A = \log_{10} 100 / \%T$$

$$A = 2 - \log_{10} \%T = \qquad (9)$$

Examples for absorbance and transmittance values:

$$T=1.00\ (100\%T),\ A=0.00$$

$$T=0.10\ (10\%T),\ A=1.00$$

$$T=0.001\ (0.1\%T),\ A=3.00 \qquad (10)$$

2.2.2 DETERMINATION OF CONCENTRATION FROM ABSORBANCE MEASUREMENTS:

The importance of the interaction of radiation with matter is expressed from the proportionality of absorbance A with material concentration c placed in the absorption cell with path length b by the so called Beer-Lambert's law:

$$A = \varepsilon bc$$

$$A = a.b.c \tag{11}$$

The unit of the constant a (absorption coefficient) is L/g.cm and if the concentration is measured by mol/L the unit of the molar absorption coefficient ε will be L/mol.cm.

$$A = \varepsilon bc = -\log T = -\log(I/I_o) = \log I_o/I \tag{12}$$

2.2.3 MULTI-COMPONENT ANALYSIS: (ABSORBANCE AS AN ADDITIVE PROPERTY)

When the solution contains two absorbing materials, the overall absorbance at any wavelength is the sum of individual absorbance values of the two components (Fig. 2.8). The continuous lines in the figure represent the absorbance of each material while the dotted line represents the absorbance of the mixture in the solution.

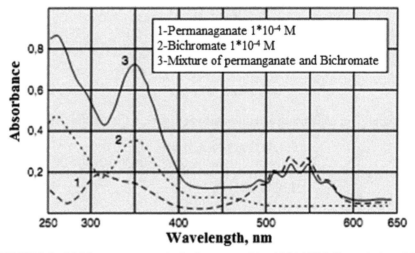

FIGURE 2.8 Multi-component analysis. Spectrum of 1×10^{-4} M KMnO$_4$,a solution of 0.8 $\times 10^{-4}$ M K$_2$Cr$_2$O$_7$ and a solution containing 1×10^{-4} M KMnO$_4$, a solution of 0.8×10^{-4} M K$_2$Cr$_2$O$_7$. (Reproduced under permission from the publisher).

By applying the equation:

$$A_{total} = A_1 + A_2$$

$$= \varepsilon_1 bc_1 + \varepsilon_2 bc_2 \tag{13}$$

and substituting the concentrations and absorbance values for the two wavelengths

$$A_{\lambda 1} = \varepsilon_1 \lambda 1.b.c1 + \varepsilon_{2\lambda 1}.b.c_2 \tag{14}$$

$$A_{\lambda 2} = \varepsilon_{1\lambda 2}.b.c_1 + \varepsilon_{2\lambda 2}.b.c_2 \tag{15}$$

from the concentrations and absorbance of the standard solutions we calculate the ε value for both substances at each wavelength and substitution in the absorbance equation of the mixture at each wavelength to get two simultaneous equations with two unknowns (the concentration of the two substances in the mixture), and solving them to calculate the concentrations.

2.2.4 LIMITATIONS OF BEER-LAMBERT LAW

In accordance with Beer-Lambert's law the calibration curve constructed between absorbance values A and analyte concentration values must give a straight line without an intercept. It is found frequently that this plot is not linear. The deviation from linearity is attributed to three main reasons:

2.2.4.1 BASIC

The law is valid for the performance of absorption of dilute solutions. At high concentration, ions and molecules do not behave as individual species and interact and associate in a manner that affect the absorptivity, ε, values. In addition the absorbance and ε values depend on the refractive index of the analyte. Thus, the dependence of refractive index on the concentration will be reflected on the values of A and ε. While at low concentration the effect of refractive index will be constant as indicated by the linearity of the plot.

2.2.4.2 CHEMICAL LIMITATIONS

The deviation from Beer-Lambert' law occurs when the absorption species are in equilibrium reactions. As an example let's consider a weak acid HA. A number of solutions must be prepared that contain various concentrations and the absorbance measured at the same wavelength. Since the acid is weak, it is,

$$6HA + H2O = H_3O^+ + A^- \tag{16}$$

If HA and A⁻ absorb at the same selected wavelength, Beer's law will be written as follows

$$A = \varepsilon_{HA} b C_{HA} + \varepsilon_A b C_A \qquad (17)$$

where C_{HA} and C_A are the concentrations of the acid and the conjugate base. Since the overall concentration equals the sum of HA and A- concentrations.

$$C_{tot} = C_{HA} + C_A \qquad (18)$$

Thus, the change of HA and A- concentrations will lead to deviation from the law.

2.2.4.3 INSTRUMENTAL LIMITATIONS

In the spectrophotometric analysis, the selection of wavelength is the most important limiting factor. It is worthwhile that the best designed monochromators suffer from stray of radiation at wavelengths very close to the analytical wavelength depending on the width of the selected band in the spectrophotometer.

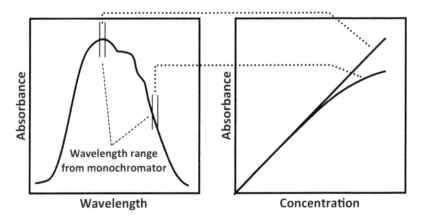

FIGURE 2.9 Effect of wave length selection on linearity of calibration graph.

The use of a multicolor radiation will lead to negative deviation from Beer's law but the deviation decrease if the ε value remains constant (Fig. 2.9). The figure indicates that it is preferable to measure the absorbance at the wavelength of the peak absorption of a wideband, where the absorbance values are identical. The deviation from Beer's law will be low if the width of the effective bandwidth is smaller than one tenth of the natural width of the absorption band. Further, the use of narrower bandwidth will improve the linearity of calibration curve.

The stray of the beam is the second contributor in the instrumental causes of deviation from Beer-Lambert's law. The stray is caused by defects in the wavelength selection kit. Radiations other than that which passed through the sample will reach the detector. Thus, at low concentrations the transmitted and the incident light intensities are large in comparison with any stray radiation that might reach the detector. At high concentrations, the intensity of the transmitted light is relatively low and the contribution of the stray light in the recorded signal will be relatively high causing more negative deviation.

2.2.4.4 PHOTOMETRIC ERROR

At very low concentrations the measured values of the absorbance are low and the intensity of the transmitted light is very close to the incident light intensity

$$I_{incident} \approx I_{transmitted} \qquad\qquad (19)$$

The relative photometric error will be high. In the meantime, at high concentrations the absorbance is high and the transmitted light intensity will be low and absorbance may not be measured precisely. The relative error in absorbance reading will be at its minimum values when the absorbance reading lies in the range 0.2–0.8. The minimum standard deviation in the readings will be at absorbance values of 0.4–1.0.

2.3 SPECTROPHOTOMETERS

Spectrophotometers are the measuring tools of the spectral data of the chemical species. The spectrophotometers used for the various spectral regions are similar in the basic components and differ in the nature of these components depending on the spectral range used. They all consist of radiation source, sample cell, wavelength selector and the detector. Data handling systems are now included in modern instruments to perform saving, concentration calculation and writing reports.

2.3.1 RADIATION SOURCE [1]

Two radiation sources are commonly used for UV/Vis spectrophotometers. Tungsten lamp is successfully employed in the visible range (350–700 nm), while deuterium arc lamp is the source of UV radiation 200–400 nm). A deuterium lamp is a low-pressure gas-discharge light source. A tungsten filament is used and anode placed on opposite sides of a nickel box structure designed to produce the best output spectrum. An arc is created from the filament to the anode. Because the filament must be very hot before it can operate, it is heated for approximately 20 seconds before use. The emission spectrum of a UV deuterium lamp shows continuum emission in the ~160–400 nm region and band emission between around 560 to 640 nm.

The arc created excites the molecular deuterium contained within the bulb to a higher energy state. The deuterium then emits light as it transitions back to its initial state. This continuous cycle is the origin of the continuous UV radiation. It is a molecular emission process, where radiative decay of excited states, in this case of molecular deuterium (D_2), causes the effect.

Normal glass would block UV radiation. Therefore, a fused quartz is used. The typical lifetime of a deuterium lamp is approximately 2000–5000 h.

2.3.2   SAMPLE CELL

Visible and near infrared radiation (400–3000 nm) is transmitted through borosilicate glass which can be used to manufacture the lenses and prisms and sample cells. Normal transparent glass can also be used but some care must be taken to handle some organic solvents like benzene and toluene and chlorinated compounds.

In the UV and near IR ranges (200–3000 nm) quartz must be used to handle the sample solution and the manufacture of the optical components. There are specific plastics that may be used in this region with some care for the handling of certain solvents. A list of solvents and their allowed spectral ranges is given in Table 2.1.

TABLE 2.1   Common Solvents with Their Spectral Cut-Off Limits

Solvent	Cut-off wavelength (nm)
Acetonitrile	190
Water	191
Cyclohexane	195
Hexane	201
Methanol	203
95% ethanol	304
1,4-dioxane	215
Ether	215
Dichloromethane	220

2.3.3   WAVELENGTH SELECTION

It is preferable whenever possible to work with narrow bands of the spectrum and hence specific wavelength. Here it is necessary to use the monochromator. Radiation may be already in linear form from the source and even though we need to isolate specified wavelength radiation. The old method involves the use of filters. Modern instrumentation uses the dispersion/selection by prisms or gratings.

Filters are devices transmitting light of certain wavelengths and absorb the remaining either partially or totally. Colored glass is the traditionally used filter especially in the visible regions (colorimeters). Interference filters is a modified version where interference is used to obtain certain wavelengths. Very thin metallic layer (semitransparent) is deposited on the faces of a transparent glass film. A thin layer of transparent magnesium fluoride is then added and another layer of semitransparent-semireflecting silver layer. Every silver layer will reflect half of the incident radiation and reflect the other half and so on. In accordance with rule that the thickness of the layer is equivalent to half of the wavelength we may produce many wavelengths by changing the angle of the incident radiation.

2.3.3.1   MONOCHROMATORS

Monochromators are efficient tool for the isolation of specific wavelengths from a wide range radiation (Fig. 2.10). The operation may be made manually or mechanically to isolate the desired wavelength. Monochromators consist of a tool to disperse radiation (Prism or grating) with two slits for the entrance and exit of the light. Entrance slit may be designed to select a narrow band that fall on the surface of the dispersion tool. The dispersion tool will reflect certain band of light at certain angle depending on the required wavelength or color. Practically a set of mirrors and lenses are used to focus the radiation and hence increase the wavelength selection.

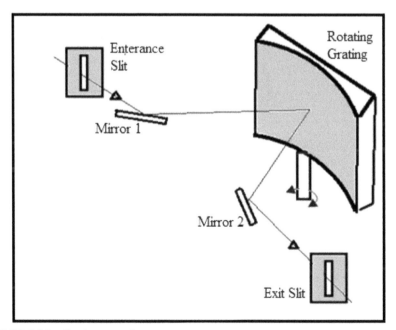

FIGURE 2.10   Grating monochromator.

2.3.4 RADIATION DETECTORS

As we noticed in the previous paragraphs, the spectrophotometric work involves a change in the light intensity. Detectors convert light energy into a measurable electrical signal, which can be further processed. Detectors vary in accordance with the type of the radiation used. Radiation sources produce huge quantities of photons. The energy of the photons of the visible light is sufficient to trigger the production of measurable quantities of electrons in the phototube or the photo transistors. In the UV range the number of the photons produced from the source compared with the visible range. Thus, a more efficient device must be used, the photomultiplier to obtain measurable electric current. Phototubes are evacuated devices containing a big anode coated with a layer of light reflecting material, cadmium sulfide. A positive voltage of 90 V will attract the electrons extracted by the photons. The resultant current is proportional to the number of photons that enter the phototube (Fig. 2.11).

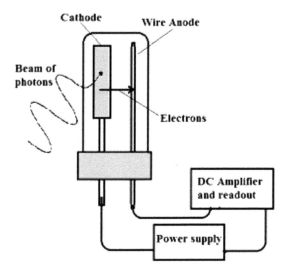

FIGURE 2.11 Schematic diagram of the phototube.

Photomultipliers contains the anode coated with the material that emits electrons. Electrons fall on a second surface to produce a number of electrons from each incident electron and direct them to a third surface and so on. The device contains 9–12 of such surfaces with applied voltage value of 90 V more than the preceding surface to facilitate electron attraction. Figure 2.12 shows a diagram of the photomultiplier. The signal is successively amplified as the electron beam moves from one surface to another. Finally the amplified electron beam falls on the anode and a signal is produced.

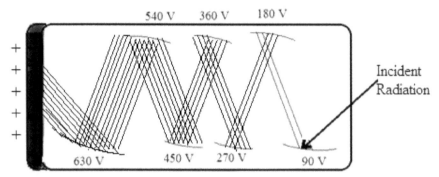

FIGURE 2.12   Schematic diagram of the photomultiplier tube.

2.3.5   SIGNAL PROCESSOR AND READOUT

The electrical signal generated by the transducer is sent to a signal processor where it is displayed in a more convenient form for the analyst. Examples of signal processors include analog or digital meters, recorders, and computers equipped with digital acquisition boards. The signal processor also may be used to calibrate the detector's response, to amplify the signal from the detector, to remove noise by filtering, or to mathematically transform the signal.

2.3.6   DOUBLE BEAM SYSTEMS

Double beam spectrophotometers (Fig. 2.13) are employed to prevent any negative effects of the unstable electricity supplied to the spectrophotometers on producing variable light energy. The basis of operation of this system may be summarized as follows:

1.   Radiation is first segmented into two parts using a rotating mirror to pass light through the sample and reflects the other part to pass through a reference cell.
2.   The two beams are then allowed to reunite with the aid of another rotating mirror synchronized with the first mirror.
3.   The reunited radiation beam to fall on the detector where the signal of the reference cell is subtracted from the sample signal before it goes to the readout system.

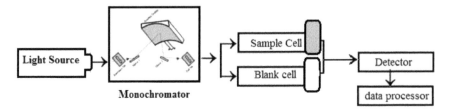

FIGURE 2.13   Schematic diagram of a double beam spectrophotometer.

2.4   UV/VISIBLE SPECTROPHOTOMETRY

2.4.1   FUNDAMENTALS

Visible light ranges between 350 and 700 nm. The mole energy of the photon ranges between 170 and 340 kJ.mol. This compares well with the C-C bond energy, which is 350 KJ.mol and the C-H bond energy of 421 KJ.mol. Such energy amount is sufficient for the transition of the electrons within the molecule and in some cases to break the bonds and cause ionization. The UV radiation covers the wavelength range of 200–350 nm. At the 200 nm oxygen strongly absorbs the energy during the process of ozone formation in the upper atmosphere. This is a real limitation for the utilization of such low wavelength range and vacuum is necessary to overcome such limitation and thus, the 100–200 nm is called the vacuum UV. The energy in this range, 340–595 KJ.mol, is sufficient for the ionizations of materials.

UV/Vis spectra consist of a small number of wide bands, because the electronic transitions are predominant in this region. Organic molecules as well as inorganic ions can be efficiently studied by UV/Vis spectrophotometry.

2.4.2   ORGANIC COMPOUNDS

Table 2.2 shows some of the transitions and their approximate locations (wavelengths) and their molar absorptivity values for some common organic compounds. Qualitatively the UV absorption spectrum shows one or more of unsaturated bonds in a chemical compound. Further, some functional groups possess specific absorption bands in the UV/Vis. It is worthwhile that compounds with saturated bonds absorb only within the vacuum UV. Unsaturated bonds absorb radiation due to the π–π^* transition. The energy difference between these two levels is relatively small and the molar absorptivity is very high.

TABLE 2.2   Some Electronic Transition in the UV Range

Transition	Maximum wavelength
$n > \sigma *$	173–250
$\pi > \pi*$	165–220
$n > \pi*$	204–290
Aromatic $\pi > \pi*$	180–270

UV/Vis spectrophotometry can be used for quantitative and qualitative analyzes. For identification purposes the technique can be used to study the structure of organic compounds. The presence of electrons in molecular bonding orbitals (π-bonding) as in double bonds:

\ / \
C =C C = O–N = N— C = C -
/ \ /

or nonbonding atomic orbitals as in:

| | |
- C–Br–C–OH–C =NH
| |

which allows the molecule to absorb visible and UV light which will excite the bonding and nonbonding electrons to antibonding molecular orbital $\pi*$. This helps the utilization of absorption in this region to prove the existence of electrons of this type. In addition, the conjugation between bonds for alternating double bonds will increase the intensity of light absorption and may result coloration of organic compounds. A good example in this respect is phenolphthalein [2] where the alkaline form is red colored and the acid form is colorless. How is this color change related to changes in the molecule?

The structures of the two differently colored forms are:

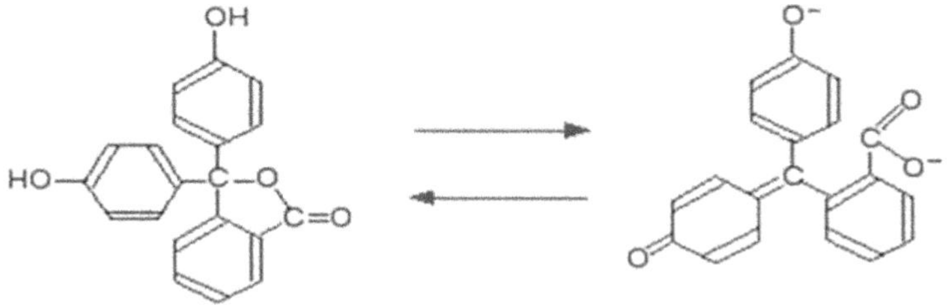

Acid Medium   Alkaline Medium

$$\text{Colorless form} \rightarrow \text{Magenta form} \tag{20}$$

Both of these absorb light in the ultra-violet, but the one on the right also absorbs in the visible with a peak at 553 nm (the green region of the spectrum). Green is the complementary color of magenta. The effect of pH on the spectrum of phenolphthalein is shown in Fig. 2.14.

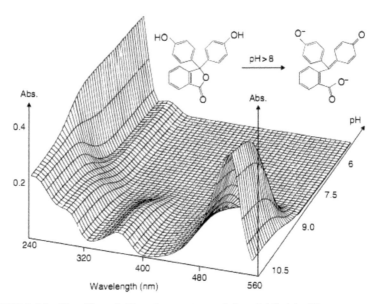

FIGURE 2.14   The effect of pH on the spectrum of phenolphthalein [3].

A shift to higher wavelength is associated with a greater degree of delocalization of electrons. In the structure of the form in acidic solution–the colorless form, the delocalization is broken around the central carbon, and it is not over the whole molecule. In the red form, there is delocalization over each of the three rings–extending out over the carbon-oxygen double bond, and to the various oxygen atoms because of their lone pairs. This delocalization lowers the energy gap between the highest occupied molecular orbital and the lowest unoccupied pi antibonding orbital. It needs less energy to make the jump and so a longer wavelength of light is absorbed [3].

2.4.3   INORGANIC IONS

Most of the transition metal ions are colored (absorb within UV/Vis) as a result of the transition of d electrons to the d* energy level. Transition metal ions can be efficiently determined by spectrophotometric methods when they are allowed to react with certain ligands to produce complexes of distinctive colors. This made the basis of establishing efficient methods for the determination of transition metals in a wide range of media including environmental samples as we will see in the following section.

2.4.4   DERIVATIVE SPECTROMETRY [4]

The principle of derivative spectrometry consists of calculating, by a mathematical procedure, derivative graphs of the spectra to improve the precision of certain measurements. This procedure is applied when the analyte spectrum does not appear clearly within the spectrum representing the whole mixture in which it is present. This can result when compounds with very similar spectra are mixed together. The traces of the successive derived spectral curves are much more uneven than the one of the original spectrum (called *zeroth order spectrum*). These derivative plots amplify the weak slope variations of the absorbance curve (Fig. 2.15) [4]. Modern UV/Vis instruments are equipped with derivation facility to obtain the 1st, 2nd, and 3rd derivative spectra.

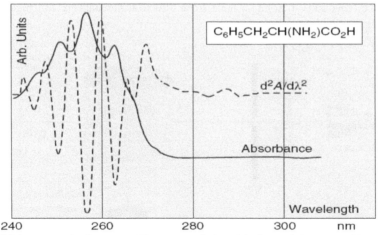

FIGURE 2.15   Derivative curve (UV spectra of phenylalanine and the curve of its second derivative).

The curve of the *second derivative* matches the points of inflection in the zeroth order spectrum by regions of zero slope, which correspond to maxima or minima. The calibration graph is established from a few standard solutions of different concentrations to which the same mathematical treatment has been applied, as to the sample solution.

2.5   TURBIDIMETRY

Turbidimetry measures the effect of light scattering caused by the solids suspended in solution. When the intensity of scattered light increases the turbidity increases. Clays and fine organic and inorganic matter suspended in water and the planktons are considered among the major contributors in water turbidity. The measurement is quantitative mainly for inorganic suspended matter.

Figure 2.16 shows a schematic diagram of a turbidimetric system. The technique is independent of the electrons, molecular structure or the functional groups and is mainly affected by the shape and size of the particles, which change the direction of incident light. It is expected that the wavelength and particle size ate the most effective parameters on the measurements. Raleigh theory, for particles smaller than the wavelength used and Tyndall theory for larger particles is applied for the turbidity measurement. The measurement of the intensity of the scattered light(at right angle to the direction of the incident light as a function of concentration of the dispersed phase is the basis of nephelometric analysis (Nephele in German is Cloud). Nephelometric analysis is most sensitive for very dilute suspension ($\sim$100 mg.L^{-1}). For reproducible cloudiness or turbidity values care must be taken in its preparation. Care must be given to the order and method of mixing as well as the concentrations of the ions, which combine to form the precipitate. However, temperature and the presence other salts in solution have some effect on the stability and fineness of precipitate particles.

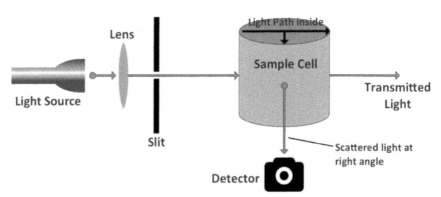

FIGURE 2.16   Schematic diagram of a turbidimetric system.

Environmental Protection Agency, EPA, has published Guidance Manual for Turbidity Provisions, 1999. Specifically the method is important for water quality evaluation. Turbidity in water comes from suspended matter or impurities that interfere with the clarity of the water. The main sources of suspended matter are the waste disposal, algae or aquatic weeds and products of their breakdown in water, humic acid and high iron concentrations.

2.6 INFRARED SPECTROPHOTOMETRY

2.6.1 GENERAL

Infrared, IR, spectrophotometry is a spectral method for the determination of structure of compounds depending on the absorption of radiation in the IR region (2–50 micron); 1.0 micron = 10^{-6} m. It is preferable to use the frequency unit, the wave number, number of waves per one centimeter, by dividing the wavelength by 1000 and represented as cm^{-1}. The spectral ranges and the related wavelengths and frequencies are listed in Table 2.3.

TABLE 2.3   The Spectral Ranges of Infrared

Range	Energy, KJ.mol^{-1}	Wave number, cm^{-1}	Wave length, μ
Near IR	50–150	4000–12800	0.78–2.5
Mid IR	2.5–50	200–4000	2.5 -50
Far IR	0.1–2.5	10–200	50–1000

The absorption of IR radiation causes the excitation of vibration and rotation in molecules and multiatom ions. The vibrational energy levels quantized, that is only a radiation with specific frequency will excite or initiate absorption for a specific vibration. For example the single bond C–C absorbs radiation with energy amount which is less than the bond of C = C because single bond is weaker than the double bond. The number of the allowed vibrations for a linear molecule is (3n–5) where n is the number of atoms. In the meantime, for a nonlinear molecule, the number of the allowed vibrations is (3n–6). Infrared spectra show a large number of bands that are related to various vibrations. Even for a simple molecule like benzene, the expected number of vibrations is nearly 30 but only some of them show up in the IR spectra. For a vibration to cause IR absorption there must be a change in the dipole moment due to the vibration or rotation. Thus, molecules with permanent dipoles (μ) are active in the IR region.

$$\overset{\delta+\ \ \delta-}{H - Cl} \qquad\qquad O - O$$

$$\longmapsto\!\!\blacktriangleright \qquad\qquad \text{No dipole moment}$$

HCl, H$_2$O, NO   Atoms, O$_2$, H$_2$, Cl$_2$

IR active   IR inactive

2.6.2   TYPES OF MOLECULAR VIBRATIONS:

Molecular vibration may be stretching or bending. In stretching, there will be a change of the bond length either symmetrically or asymmetrically. Bending involves a change in bond angle (scissor, rocking, twisting and wagging).

Example: the v (C-O) occurs at 1034 cm^{-1} in methanol, 1053 cm^{-1} in methanol and at 1105 cm^{-1} in butanol. For the nonlinear molecule H$_2$S three absorption bands are expected to show up in accordance with the 3n-6 rule. The possible vibrations are symmetric and asymmetric stretching and scissor.

2.6.3   ORIGIN OF IR SPECTRA

The dependence of IR absorption on the vibrations or rotation within the molecule suggests that the importance of bond characteristics like the force constant of the bond and the reduced mass for the bonded atoms in determining the energy of the absorption.

$$E_v = \left(v + \frac{1}{2} \right)\left(\frac{h}{2\pi} \right)\left(\frac{k}{\mu 2\pi} \right)^{1/2} \tag{21}$$

where   is the vibration quantum number, h is the Plank's constant, k is the force constant of the bond in the Hook's Law, and μ is reduced mass which is calculated as follows:

$$\mu = \frac{\left(m_1 + m_2 \right)}{\left(m_1 + m_2 \right)} \tag{22}$$

The force constant of some common bonds is listed in Table 2.4.

TABLE 2.4   Force Constant of Some Common Bonds

Bond Type	Force Constant, millidyne.A^{-1}
C-H	5
C-F	6
N-H	6.3
O-H	7.7
C-Cl	3.5
C-C	10
C = C	12
C ≡ C	15.6
C ≡ N	17.7

The exchange of some bonds with their isotopes also affects their position in the IR spectrum.

2.6.4   INSTRUMENTATION

2.6.4.1   RADIATION SOURCES

The commonly used source for IR radiation is the Nernst Glower. It is a hollow rod of about 1 inch long and 0.1 inch in diameter composed of rare earth oxides. It is operated at 1000 to 1800°C. The intensity of the radiation from the Nernst glower is steady for long periods of time.

2.6.4.2   OPTICS AND SAMPLE HOLDERS

The IR region is uniquely characterized with a huge absorption band for the hydroxyl group, O-H, and thus, glass and water cannot be used. In this respect sodium chloride is the preferred material for the manufacture of sample holders and it is transparent for the IR through the far UV range. Sample holders with thicknesses between 0.1–1.0 mm can be manufactured from sodium chloride and other alkali halides

2.6.4.3   DETECTORS

The IR radiation is a low energy and is not sufficient to release electrons from the photosensitive surfaces. For the detection of IR radiation, the photoconductor, Lead sulfides PbS, or Cadmium Mercuric telluride, which is sensitive to light or the resistive thermometer at 77 K are employed.

The IR spectrophotometry is widely used in qualitative analysis for gasses, liquids and solids. Thus, it is for the rapid identification and not for the quantitative work.

2.6.5   FOURIER TRANSFORM INFRARED

Normal IR, spectrophotometer operates by scanning the absorption of the various wavelengths to detect the functional groups of a compound. This implies that the exposure time of the functional groups will be limited and consequently the absorption intensities will also be relatively small in magnitude even with slow scanning process. All of the infrared frequencies need to be measured simultaneously to improve absorption intensities. The introduction of the interferometer improved this situation. A schematic diagram of the FTIR system is shown in Fig. 2.17.

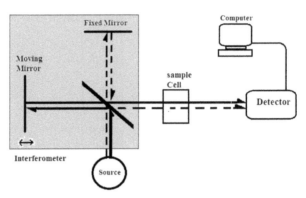

FIGURE 2.17   Schematic diagram of FTIR spectrophotometer.

The light passes through a beam splitter, which sends the light in two directions at right angles. One beam goes to a stationary mirror then back to the beam splitter. The other goes to a moving mirror. The motion of the mirror makes the total path length variable versus that taken by the stationary-mirror beam. The radiation bands recombine and pass through the sample, which absorbs all the different wavelengths characteristic of its spectrum, and this subtracts specific wavelengths from the interferogram. The interferometer produces a unique type of signal, which has all of the infrared frequencies "encoded" into it. The measurement of the signal is fast and may take 1–3 s. Thus, larger number of samples can be analyzed in a given time. The Michelson interferometer is the most common configuration for optical interferometry and was invented by Albert A. Michelson. The interferometer is a fundamentally different piece of equipment than a monochromator. The different paths may be of different lengths or be composed of different materials to create interference fringes on a back detector [6, 7].

A mathematical function called a Fourier transform allows us to convert an intensity-vs.-time spectrum into an intensity-vs.-frequency spectrum. A computer system is essential for the transformation of the complex signal into a spectrum. In addition to the simultaneous measurement of absorption at all wavelengths, FTIR is characterized by simpler optical system in comparison with normal IR spectrophotometry. A laser beam is superimposed to provide a reference for the instrument operation.

2.6.5.1   DIFFUSE REFLECTANCE FTIR SPECTROSCOPY (DRIFTS)

When the IR beam enters the sample, it can either be reflected off the surface or transmitted through the particles. The IR energy reflecting off the surface is typically lost. The next particle may reflect or transmit the IR beam, which passed through the first particle. This transmission-reflectance event can occur many times in the sample, which increases the path length. Finally, such scattered IR energy is collected by a spherical mirror that is focused onto the detector. The detected IR light is partially absorbed by particles of the sample, bringing the sample information (Fig. 2.18). The main advantages of DRIFTS spectroscopy are: fast measurement of powdered samples, minimal or no sample preparation, ability to detect minor components, ability to analyze solid, liquid or gaseous samples, is one of the most suitable methods for the examination of rough and opaque samples, high sensitivity, high versatility, capability of performing of the measurements under real life conditions.

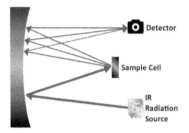

FIGURE 2.18   Schematic diagram of the DRIFTS spectrophotometer.

Three methods are used for sample preparation. For solid powders, the mixture of sample powder and KBr is poured to fill the microcup. The focusing mirror of the diffuse reflectance accessory aids the focus of the IR beam on the sample surface. Thus, the microcup must be full to keep the focus. Alternatively, the sample surface is scratched with a piece of abrasive silicon carbide (SiC) paper and then measuring the particles adhering to the paper. For liquids or solutions, drops of solution are placed on a substrate. Suspensions and colloids or solutions in volatile solvents may be evaporated to allow the measurement of the remaining residues on the substrate.

The particle size is an important factor in a transmission measurement with the pellet method. Large particle will results in the scattering of the energy, leading to the shift of the spectrum baseline and the broadening of IR bands. The scenario becomes worse in a diffuse reflectance measurement, because the infrared light travels in the sample for a long period and the optics collects a large portion of the distorted energy. It is important to grind the sample particles to 5 microns or less.

2.6.6   SAMPLE PREPARATION

Liquids are simply placed in a short radiation path length cuvette (0.015–1 mm). Solids must be dissolved in appropriate solvent. The solid sample (2–4 mg) is ground with potassium bromide and the mixture is pressed to make a thin semitransparent disc or mixed with liquid nujol (a hydrocarbon oil) to form a paste. A drop of this paste is placed between two sodium chloride discs and squeezed to form the sample cell. Water is aggressive for the IR cuvettes.

For the qualitative analysis, IR spectra are entirely different from those of the UV/Vis. IR spectra consist of a number of peaks that are highly different from one compound to another. Thus, the IR spectra are typical for the identification and qualitative analysis of organic compounds. The IR spectrum comprises two parts: the 4000–2500 cm⁻¹ region represents the group frequency region and the 2500–200 region cm⁻¹ which represents the finger print region. In the group frequency region, some zone can be precisely identified to characterize the absorption of specific functional groups. The nitrile group, CN, for example, has a sharp absorption line at 2260–2240 cm⁻¹ and OH group has a wide and big absorption peak at 3000 cm⁻¹.

Many peaks show up in the finger print region that allows the differentiation of compounds with similar functional groups. All alcohols have wide peak at 3000 cm⁻¹, but each alcohol is characterized by some specific peaks within the finger print region. The first step after recording the IR spectrum is to identify the functional groups (group frequency region). The second step is to compare the spectrum with some reference spectra (finger print region). The absorption of the organic groups occurs within certain regions that can be estimated by calculations of the bond strength and the weights of the bonded atoms. Some standard tables are available to assist such calculations to predict the composition of the unknown compound. The C-H group absorb as in Table 2.5A. The O-H group absorb at 3500–3200 cm⁻¹ depending on its bonding (alcohol, phenol or carboxylic acid) as in Table 2.5B. The N-H group absorbs at 3650–3300 cm⁻¹ for amines and amides. The absorption of the C-C group depends on the type of the bond, whether single, double or triple (Table 2.5C). The C-N and C-O also depend on the type of compound (Table 2.5D).

TABLE 2.5 Group Absorption Frequencies

Bond	Type of Compound	Frequency Range, cm⁻¹
A: C –H Bonds		
C-H	Alkanes	2850–2970
		1340–1470
C-H	Alkenes	3010–3095
		675–995
C-H	Alkynes	3300
C-H	Aromatic Rings	3010–3100
		690–900
B: O-H Bonds		
O-H	Monomeric alcohols, phenols	3590–3650
O-H	Hydrogen bonded alcohols, phenols	3200–3600
O-H	Monomeric carboxylic Acids	3500–3650
O-H	Hydrogen bonded carboxylic acids	2500–2700
C: C-C Bonds		

C = C	Alkenes	1610–1680
C ≡ C	Alkynes	2100–2260
C = C	Aromatic Rings	1500–1600
D: C-N and C-O bonds		
C-N	Amines, Amides	1180–1360
C≡ N	Nitriles	2210–2280
C -O	Alcohols, ethers, carboxylic acids, esters	1050–1300
C = O	Aldehydes, ketones, carboxylic acids, esters	1690–1760

Table 2.6 shows typical IR absorption frequencies for a variety of organic compounds [8].

TABLE 2.6 Typical IR Absorption Frequencies

Functional Class	Range (cm^{-1})	Intensity	Assignment	Range (cm^{-1})	Intensity	Assignment
		Stretching Vibrations			Bending Vibrations	
Alkanes	2850-3000	str	CH_3, CH_2 & CH 2 or 3 bands	1350-1470 1370-1390 720-725	med med wk	CH_2 & CH_3 deformation CH_3 deformation CH_2 rocking
Alkenes	3020-3100 1630-1680	med var	=C-H & =CH_2 (usually sharp) C=C (symmetry reduces intensity)	880-995 780-850 675-730	str med med	=C-H & =CH_2 (out-of-plane bending) cis-RCH=CHR
	1900-2000	str	C=C asymmetric stretch			
Alkynes	3300 2100-2250	str var	C-H (usually sharp) C≡C (symmetry reduces intensity)	600-700	str	C-H deformation
Arenes	3030 1600 & 1500	var med-wk	C-H (may be several bands) C=C (in ring) (2 bands) (3 if conjugated)	690-900	str-med	C-H bending & ring puckering
Alcohols & Phenols	3580-3650 3200-3550 970-1250	var str str	O-H (free), usually sharp O-H (H-bonded), usually broad C-O	1330-1430 650-770	med var-wk	O-H bending (in-plane) O-H bend (out-of-plane)
Amines	3400-3500 (dil. soln.) 3300-3400 (dil. soln.)	wk wk	N-H (1°-amines), 2 bands N-H (2°-amines)	1550-1650 660-900	med-str var	NH_2 scissoring (1°-amines)
	1000-1250	med	C-N			NH_2 & N-H wagging (shifts on H-bonding)
Aldehydes & Ketones	2690-2840(2 bands) 1720-1740 1710-1720	med str str	C-H (aldehyde C-H) C=O (saturated aldehyde) C=O (saturated ketone)	1350-1360 1400-1450 1100	str str med	α-CH_3 bending α-CH_2 bending C-C-C bending
	1690 1675 1745 1780	str str str str	aryl ketone α, β-unsaturation cyclopentanone cyclobutanone			
Carboxylic Acids & Derivatives	2500-3300 (acids) overlap C-H 1705-1720 (acids) 1210-1320 (acids)	str str med-str	O-H (very broad) C=O (H-bonded) O-C (sometimes 2-peaks)	1395-1440	med	C-O-H bending
	1785-1815 (acyl halides) 1750 & 1820 (anhydrides) 1040-1100 1735-1750 (esters) 1000-1300 1630-1695(amides)	str str str str str str	C=O C=O (2-bands) O-C C=O O-C (2-bands) C=O (amide I band)	1590-1650 1500-1560	med med	N-H (1¡-amide) II band N-H (2¡-amide) II band
Nitriles	2240-2260	med	C≡N (sharp)			
Isocyanates,Isothiocyanates, Diimides, Azides & Ketenes	2100-2270	med	-N=C=O, -N=C=S -N=C=N-, -N_3, C=C=O			

The compound composition especially organic compounds can be determined from the study of its IR spectrum. As an example let's consider the carbonyl group, C=O, which is characterized by a well known stretch process and affected by the neighboring groups in the molecule. In carboxylic acids the absorption occurs at 1650 cm^{-1}, 1700 in ketones and 1800 cm^{-1} for carbonyl chloride.

The introduction of computer software into the chemical instrumentation was a great move in the identification of organic compounds in their mixtures with other compounds. The spectral data of the pure compounds are stored in the memory of the computer and the work is carried out systematically to extract the spectrum of a specific compound from the spectrum of the mixture or the impure sample to estimate the purity or the concentration of the compound in the mixture. This is termed the spectral searching. Figure 2.19 shows the results of spectral searching of a mixture suspected to contain cocaine and mannitol by subtracting the spectrum of pure mannitol and pure cocaine (already stored in the memory) to achieve the composition of the mixture.

Inorganic compounds also have their characteristic group absorption like sulfate, nitrate and silicate.

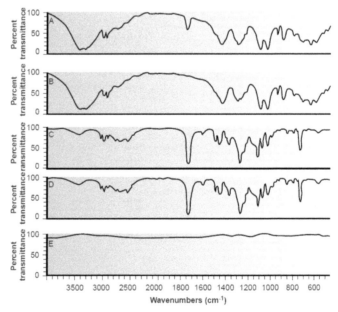

FIGURE 2.19   Spectral search for a mixture of mannitol and cocaine. A, mixture spectrum; B, mannitol spectrum; C, result of subtracting mannitol spectrum from mixture spectrum; D; cocaine spectrum; and E, result of subtracting cocaine spectrum from spectrum C (9).

2.7   ATOMIC ABSORPTION SPECTROMETRY

2.7.1   GENERAL

Atomic absorption spectrometry, AAS, is an analytical technique for the measurement of metal ion concentrations. It is very sensitive and may be useful for trace concentrations in the order of part per billion, ppb, which is equivalent of $\mu g.L^{-1}$. The method uses the selective absorption of metal atoms of radiation at certain wavelengths. The finds a wide range of application in medical examinations for specific metals in blood or urine, in environmental investigations for toxic metals in water, air, and land, in pharmaceutical industry to search for residual metals that might be contained in the drugs compositions like platinum in cancer preparations and industrial samples for major and trace metal contents.

The AA spectrophotometer consists of a radiation source (hollow cathode lamp), sample introduction device (Flame or graphite furnace), wavelength selection system (monochromator), detector (photomultiplier) and read out system (digital or plotter). Figure 2.20 shows a schematic diagram of a flame atomic absorption system. The two unique components of the AA spectrophotometer are the sample introduction system and the radiation source. The wavelength selection, detection and read out systems are similar to those of UV/Vis spectrophotometer. The following discussion will focus on these two components.

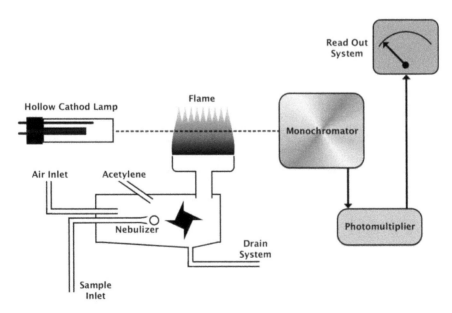

FIGURE 2.20   Schematic diagram of a flame atomic absorption system.

2.7.2 ATOMIZATION

The production of neutral atoms is the main task in the AAS analysis because free atoms are the absorbing species. Atoms do not exist free in nature and thus, they must be produced by applying large amount of heat energy to break down the compounds into atoms.

2.7.2.1 FLAME ATOMIZERS

Flames are the traditional technique, which accompanied the invention of the method in the 1950's. The inventors noticed the exceptionally high energy supplied by the oxy-acetylene flame used in the welding and metal cutting as a first candidate source of heat energy to produce free atoms from compounds. The sample solution needed to be introduced to the flame to be atomized. Now a range of flames are available for this purpose. The: atomization steps are shown in Fig. 2.21.

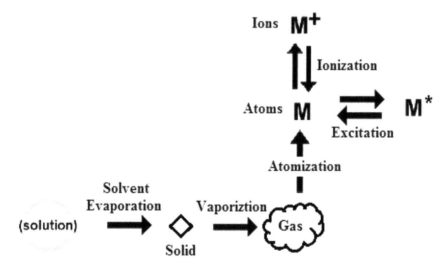

FIGURE 2.21 Atomization steps.

2.7.2.1.1 PREMIX BURNER

It was noticed that the most efficient method of atomization of samples is by mixing them, 1–2 mL, with the oxidant and fuel gases and feeding the mixture into a burner from which the gases and the sample aerosol through a narrow and long slot on the top. Air-acetylene was found adequate to produce temperature ranges of 2400–2700

K, while nitrous oxide acetylene flame produces 2900–3100 K. The latter was found suitable to analyze for refractory metals like aluminum and vanadium.

Burning will supply the required energy for the evaporation of the solvent and the dissociation of molecules into free atoms that will absorb radiation. The heat energy may exceed the dissociation energy requirement and cause some ionization of easily ionized atoms. In some case the excess energy will be used to excite the atoms. As the excited atoms return to the ground state they will emit radiation in a specific wavelengths that may interfere with the analytical signal.

2.7.2.1.2   NEBULIZATION

The sample solution is transported from the flask to the flame via a capillary tube by the action of stream of flowing oxidant (air) gas to be mixed with the flame components for the fine droplets to find their way to the top of the mixing chamber to be burned. The excess sample is removed from the premix chamber through a drain. The drain uses a liquid trap to prevent combustion gasses from escaping through the drain line. Impact devices are used to reduce droplet size further and to cause remaining larger droplets to be deflected from the gas stream and removed from the burner through the drain. Two types of impact device are used typically, impact beads and flow spoilers.

Impact bead systems are normally used to improve nebulization efficiency, the percentage of sample solution converted to smaller droplets. The design and positioning of the impact bead are critical in determining how well it will work. Properly designed impact bead systems will improve nebulization efficiency and remove many of the remaining large droplets from the spray.

Thus the control of the atomization efficiency through the flame temperature control is essential to get reliable AAS analysis Temperature control may be achieved by the control of mixing ratio of oxidant and fuel gasses entering the burner. Flame is proved as a reliable method for the atomization of all types of samples. The burner is designed to give a stable laminar flame in the path of the monochromatic radiation and proved more efficient that normal circular slit burners in which most of the atomized matter is lost being not in the radiation path. Flame atomization is reliable and characterized by high precision (~1%), but the residence time is limited and the incomplete volatilization of sample will reduce the sensitivity for some element.

Flame consists of different temperature regions. The sensitivity is expected to vary if the resonance radiation passes through the different regions. The best flame region for certain metal may the worst for another metal. The proper selection of the height above the burner base is essential to obtain reliable and sensitive analysis. Silver atomic absorption is improved as we move higher from the base, while the chromium analysis necessitates the passage of resonance radiation with a region that is close to the burner base.

In this "premix" design, sample solution is aspirated through a nebulizer and sprayed as a fine aerosol into the mixing chamber. Here the sample aerosol is mixed with fuel and oxidant gases and carried to the burner head, where combustion and sample atomization occur. Fuel gas is introduced into the mixing chamber through the fuel inlet, and oxidant enters through the nebulizer sidearm. Mixing of the fuel and oxidant in the burner chamber eliminates the need to have combustible fuel/oxidant in the gas lines, a potential safety hazard. In addition to the separate fuel and oxidant lines, it is advantageous to have an auxiliary oxidant inlet directly into the mixing chamber. This allows the oxidant flow adjustments to be made through the auxiliary line while the flow through the nebulizer remains constant. Thus, for a burner system with an auxiliary oxidant line, the sample uptake rate is independent of flame condition, and the need to readjust the nebulizer after every oxidant flow adjustment is eliminated [10].

2.7.2.2 GRAPHITE FURNACE ATOMIZER

To improve the sensitivity, some electrothermal atomization devices were developed using a graphite furnaces acting as an electrical resistance heated when a large current passes through it. Figure 2.22 shows a schematic diagram of the graphite tube furnace in the optical system of AA spectrophotometer. A drop (1–100 µL) of sample solution is placed inside the furnace and a predesigned heating program is applied with three main steps:

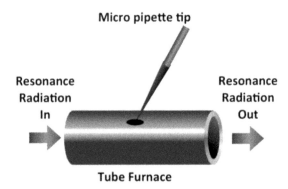

FIGURE 2.22   Schematic diagram of the graphite tube furnace in the optical system of AA spectrophotometer

A flow of an inert gas (nitrogen or argon) is maintained throughout the heating cycle to prevent the burning of the graphite furnace. With such a system nearly complete vaporization/atomization of the sample can be achieved and thus the sensitivity is much improved in comparison with flame atomization.

The mechanism of atom formation in the graphite furnace differs from that of the flame atomization. In the flame heat is the main effective factor in the dissociation of molecules, while graphite plays an important role in the process of atom formation. The successive steps in graphite furnace allow the evaporation of solvent first and then the transformation into inorganic ash. The inorganic form of the ash may be in the form of metal oxides or some thermally stable salts. In the atomization step the oxides are reduced by the action of incandescent graphite into the free metal as a main mechanism of atomization. The glowing graphite surface may also aid the evaporation of metals from the inorganic form that might be present at the end of the ashing step.

$$MO + C \rightarrow M + CO \qquad (23)$$

Metals are then transported into the gaseous state by the action of the hot and reducing atmosphere inside the graphite furnace to absorb the resonance radiation.

2.7.2.3 SPECIAL ATOMIZATION SYSTEMS

To improve the sensitivity of AAS work for some metals it was found beneficial to use some of their physical properties.

2.7.2.3.1 COLD VAPOR SYSTEM

For mercury being liquid at room temperature, a specific system was developed, the Cold Vapor atomization. The idea is based on nebulizing stannous chloride, $SnCl_2$ solution with the mercuric sample solution into the air acetylene flame where a great enhancement in the sensitivity was obtained. A system was later developed to replace the flame with a glass cell with quartz windows. A reaction cell was added to first reduce mercuric ions in the sample solution to the metallic state with $SnCl_2$ and then to transport the formed liquid metal to the glass cell with quartz windows within the radiation path with a stream of inert gas. The signal was could be amplified greatly and hence much lower concentrations could be analyzed. Reducing agents like sodium borohydride was also proved effective in this system. The method has be automated now and the steps timing and dose rates can be adjusted for various sample composition to enhance the sensitivity.

2.7.2.3.2 HYDRIDE GENERATION

Some elements have the ability of forming volatile compounds that can be thermally dissociated into the elemental form, like As, Se, and Te which form volatile hydrides by reaction with sodium borohydride in an acid medium.

$$As^{3+} + NaBH_4 \rightarrow AsH_3 \text{ (Arsenic trihydride, Arsine)} \qquad (24)$$

The produced hydride is transported with a stream of air or nitrogen to a heated absorption cell to decompose into elemental As and hydrogen. The cell is placed in the resonance radiation path of arsenic electrodeless discharge lamp where stomix absorption can be recorded with much improved sensitivity. In this manner the analyte is separated from its original solution and thus, any sources of interferences are removed. The hydride generation and atomization system is shown in Fig. 2.23.

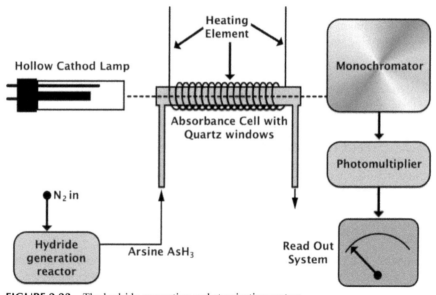

FIGURE 2.23   The hydride generation and atomization system.

Automation was introduced to this method and the flow rate of carrier gas, doses of reagents addition as well as the dissociation temperature can be fully controlled to achieve the best sensitivity.

2.7.3   RADIATION SOURCE

Continuum source of radiation are not useful in AAS work because the absorbing species are the atoms, which are characterized by having fine absorption lines from the continuum spectrum and not as wide bands like in UV/Vis spectrophotometry. Thus, the fraction absorbed radiation (absorption line) by the atoms from the overall incident radiation will be very small. In the meantime, the absorption lines will be a considerable part of the incident radiation when line spectral source is used. The sensitivity will be greatly improved by the application of line sources.

2.7.3.1 HOLLOW CATHODE LAMPS

The design of hollow cathode lamp is unique from other spectral sources and comprises a wire anode made of platinum to aid the closure of the electrical circuit and a cathode manufactured from the analyte metal or one of its alloys placed in a glass casing and the whole tube is filled with an inert gas at a low pressure. Closing the electric circuit will lead to extraction of the charge from the atoms of inert gas. The charged atoms of the gas will attack the metal cathode to extract excited atoms of the metal to form a cloud inside the lamp. The life time of these excited atoms is limited and, thus, they tend to return to the ground state. By doing so, the atoms emit radiation of a characteristic wavelength for the metal analyte. For each metal there is a specific lamp and this will improve the sensitivity of the method. A schematic diagram of the hollow cathode lamp is shown in Fig. 2.24.

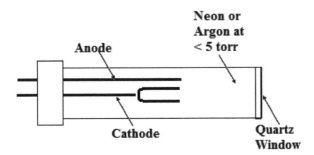

FIGURE 2.24   Schematic diagram of the hollow cathode lamp.

2.7.3.2 ELECTRODELESS DISCHARGE LAMPS

For certain elements, As, Se, P, etc., the radiant intensities from the source can be improved by one to two orders of magnitude by replacing the hollow cathode lamps with electrodeless discharge lamps. The lamp is constructed from a sealed quartz tube filled with inert gas at a low pressure (few torrs) and a small quantity of the metal (or its salt). The lamp is energized by an intense field of radio-frequency or microwave radiation. Ionization of the argon occurs to give ions that are accelerated by the high-frequency component of the field until they gain sufficient energy to excite the atoms of the metal whose spectrum is sought. Electrodeless discharge lamps are available commercially for 15 or more elements.

2.7.3.3 CONTINUUM SOURCES

At the beginning of year 2000, new generation high resolution continuum source atomic absorption spectrometer (HR-CS-AAS) is developed. HR-CS-AAS is equipped with continuum source which is high intensity xenon short-arc lamp, high resolution double monochromator, CCD detector. The continuous source lamp emits radiation of intensity over the entire wavelength range from 190 nm to 900 nm. With these instruments, aside from the analysis line, the spectral environment is also recorded simultaneously, which shows noises and interferences effecting analysis. Improved simultaneous background correction and capabilities to correct spectral interferences, increase the accuracy of analytical results. With high resolution detector, interferences are minimized through optimum line separation. With these instruments, not only metals and nonmetals, for example, F, Cl, Br, I, S, P can be determined by their hyperfine structured diatomic molecular absorption [11].

2.7.4   INTERFERENCES

Interference is a phenomenon that leads to changes in intensity of the analyte signal in spectroscopy. Interferences in atomic absorption spectroscopy fall into two basic categories, namely, spectral and nonspectral. Non-spectral interferences affect the formation of analyte atoms and spectral interferences result in higher light absorption due to presence of absorbing species other than the analyte element.

2.7.4.1   BACKGROUND CORRECTION

For a reliable analysis by atomic absorption methods there must be removal of matrix absorption or emission signals that interfere with the analyte signal. Metal atoms in the flame or the graphite furnace atomizer are present together with other nonatomic species like free radicals and molecules and products of the fuel burning and heat. Such an environment encourages the superimposition of the absorption signal with nonselective absorption of the resonance radiation. This will lead to erroneous results from the amplified absorption signal, especially for wavelengths lower than 250 nm.

The use of a second source of continuous spectrum, a deuterium lamp, installed at a right angle to the radiation path of the hollow cathode lamp for the removal of such interference. A group of mirrors and lenses are employed to focus deuterium radiation parallel to the resonance radiation. The system is designed to allow passing the continuum radiation through the sample atomic cloud with the resonance radiation at a time difference. The detector therefore will receive an alternating signal from the two sources. An electric circuit allows the subtraction of the continuum signal (mostly the background) from that of the resonance radiation (analyte and the background) to end up with a clean atomic absorption signal of the analyte. A schematic diagram of the background correction system is shown in Fig. 2.25.

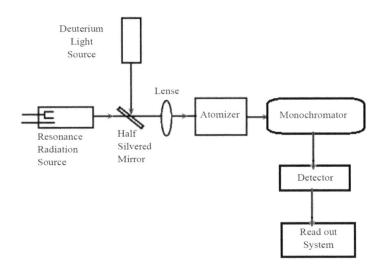

FIGURE 2.25   Schematic diagram of the background correction system.

2.7.4.2   IONIZATION

The use of high temperatures in the atomization process will cause the ionization of some metallic atoms with relatively small ionization potentials, like sodium.

$$Na \rightarrow Na^+ + e^- \qquad (25)$$

This will directly reduce the atomic absorption signal. The addition of easily ionizable material like potassium chloride or cesium chloride will furnish electrons that will shift the equilibrium of analyte ionization back.

2.7.4.3   CHEMICAL INTERFERENCE

The determination of some metals by atomic absorption becomes difficult in the presence of other materials due to the formation of refractory products in the atomizer, which is difficult to be thermally dissociated into free atoms. The determination of calcium for example is difficult in the presence of phosphate due to the formation of the thermally stable calcium phosphate. The free calcium atom formation will be sharply reduced. To correct for this chemical interference another material, like lanthanum chloride, is added to react with phosphate to form lanthanum phosphate that exceeds the thermal stability of calcium phosphate. This exchange of roles will help in the release of calcium free atoms and hence to recover the atomic absorption signal.

$$Ca_3(PO_4)_2 + 2LaCl_3 \rightarrow 3CaCl_2 + 2LaPO_4 \tag{26}$$

2.7.4.4   MATRIX MODIFICATION

The ashing step may sometimes result in the loss of volatile elements like lead and cadmium. Further, the matrix material may consist of salts that volatilize during the ashing step. Their vapor may severely interfere with the analyte absorption signal as in the case of sodium chloride interference in the determination of some elements by emitting a strong radiation, which dominates over the lead signal. A compound is added in such a case that modifies the nature of the interfering species into a safe material. The product of the reaction may be easily decomposed during the ashing step to release the lead atoms and yield a free atomic absorption signal. One of the best examples on matrix modification is the addition of ammonium nitrate:

$$NaCl + NH_4NO_3 \rightarrow NaNO_3 + NH_4Cl \tag{27}$$

The ammonium chloride and sodium nitrate produced by the reaction can be dissociated much easily than the original sodium chloride and at a lower temperature that prevents any possible losses of lead from the analysis volume.

Nickel is added during the determination of selenium by graphite furnace AAS. The purpose of nickel is to form an alloy with selenium and inhibits any possible losses of selenium, due to volatility, during the ashing step. The ashing temperature can, therefore, be increased to allow the efficient removal of matrix materials and hence an improved AA signal can be obtained for selenium.

2.7.4.5   STANDARD ADDITIONS

The composition of the sample solution is frequently very complicated and contains interfering species which lead to erroneous results and with only limited degree of confidence. In such a situation, known portions of the standard solution are added to the solution and the absorption is measured for the treated and untreated solutions and simple calculation is carried out to determine the concentration of the unknown solution. Alternatively, the absorbance values for a series of sample solution to which different amounts of the standard has been added are plotted against the values of the added standard. The concentration of the analyte is calculated by extrapolating the plot to the negative side of concentration axis where it will dissect the concentration axis (Fig. 2.26).

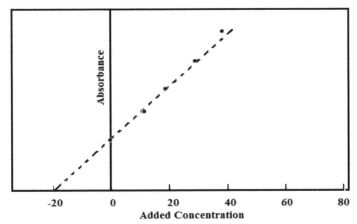

FIGURE 2.26   Standard additions.

2.7.5   ELEMENTS MEASURABLE WITH AAS

The AAS method can be said to be specific for the determination of metals. The simplicity of performance and the ease of sample preparation in addition to the freedom from interferences led to the rapid progress in the instrumentation and extremely large number of publications covering the improvement of the instrumentation and methodologies. Flame atomization, which accompanied the invention and progress of the technique, is still the most attractive and simplest means of atom formation. Table 2.7 shows the common elements that are measurable by AAS. It is clear that the range of air-acetylene flame is very wide. In the meantime, the higher temperature range provided by the nitrous oxide- acetylene flame made it more specifically then air acetylene flame for the metals forming refractory oxides like aluminum, titanium, molybdenum and rare earth elements. The sensitivity varies from parts per billion for elements like copper and cadmium up to few parts per million like lanthanides.

TABLE 2.7   Detection Limits for Some Elements by Atomic Absorption ($\mu g.L^{-1}$) [12]

Element	atomization		Remarks
	Flame	**Graphite furnace**	
Ag	1.5	0.005	
Al	50	0.1	N_2O-C_2H_2 flame
As	150	0.03	Hydride generation
B	1000	1	N_2O flame
Ba	15	0.4	N_2O flame

Bi	30	0.05	
Ca	2	0.01	
Cd	0.8	0.002	
Co	7	0.15	
Cr	5	0.004	$N_2O\text{-}C_2H_2$ flame
Cu	1.5	0.01	
Fe	5	0.06	
Hg	300	0.009	Cold vapor
Mg	0.5	0.004	
Mn	1.5	0.005	
Mo	50	0.03	$N_2O\text{-}C_2H_2$ flame
Na	0.5	0.005	
Ni	10	0.07	
Pb	10	0.05	
Sb	45	0.05	
Se	100	0.03	Hydride generation
Si	90	1	$N_2O\text{-}C_2H_2$ flame
Sn	150	0.1	
V	100	0.1	$N_2O\text{-}C_2H_2$ flame
Zn	2	0.02	

2.8   INDUCTIVELY COUPLED PLASMA

2.8.1   INDUCTIVELY COUPLED PLASMA -OPTICAL EMISSION SPECTROMETRY (ICP-OES OR ICP-AES)

Inductively coupled plasma-optical (or atomic) emission spectrometry (ICP-OES or ICP-AES) is an analytical technique used for the determination of trace metals. This technique uses a plasma source to excite atoms in the sample. These excited atoms produce light of a characteristic wavelength, and a detector measures the intensity of the emitted light, which is related to the concentration. There are main three advantages of ICP.

1.   Samples heated by plasma, which reaches a very high temperature (7000–8000 K), atomize successfully

2. With ICP, multi element analysis can be accomplished. 60 elements can be analyzed in a single sample run of less than a minute simultaneously, or in a few minutes sequentially.

3. The instrument is only optimized for one time for a set of metal analysis. The high operating temperature lowers the interference. Determinations can be accomplished in a wide linear range and refractory elements can be de-termined at low concentrations (B, P, W, Zr, and U).

On the other hand, consumption of inert gas is much higher than it is for AAS techniques, which causes high operating costs.

ICP instruments can be 'axial' and 'radial' according to their plasma con-figuration. In radial configuration, the plasma source is viewed from the side. Emissions from the axial plasma are viewed horizontally along its length, which reduces background signals resulting in lower detection limits. Some instru-ments have both viewing modes [13]. A schematic representation of inductively coupled plasma torch and a block diagram of ICP-OES are depicted in Figs. 2.27 and 2.28. Generally, a radio frequency (RF) powered torch as a source, polychromatous as a wavelength selector, a photomultiplier (PMT) or charge capacitive discharged arrays (CCD) as detectors are used.

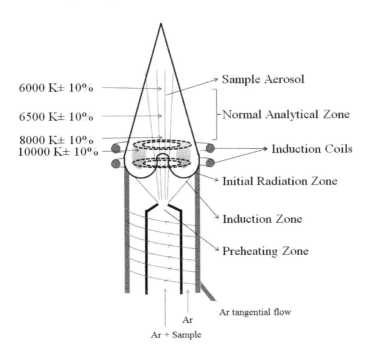

FIGURE 2.27 Schematic representation of inductively couple plasma torch.

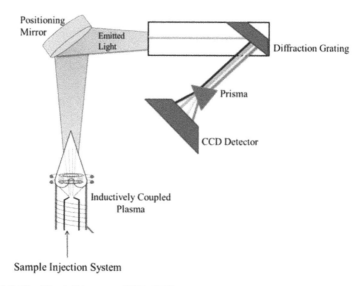

FIGURE 2.28   Block Diagram of ICP-OES.

2.8.2 INDUCTIVELY COUPLED PLASMA MASS SPECTROMETRY (ICP-MS)

Inductively coupled plasma mass spectrometry (ICP-MS) is similar to ICP-OES; the only difference is the usage of a mass spectrometer as the detector. The mass spectrometer separates ions according to their mass to charge ratio. ICP-MS is a very sensitive technique and has the lowest detection limits in ppt (part per thousands) range. Samples can be aerosol, liquid or solid. Solid samples can be directly detected using a laser to form aerosols or dissolved prior to the analysis. It is a fast technique, which can analyze in less than a minute for all elements but the method development stage requires a highly skilled operator. The schematic diagram of ICP-MS instrument is shown in Fig. 2.29. The main difference from optical emission ICP spectrographs is the usage of quadruple mass spectrometers instead of wavelength selectors for detection.

Some improvements have been made in ICP or ICP-MS coupling with other instrumentation to eliminate the matrix effect or obtain a low detection limit. Laser ablation ICP-MS (LA-ICP-MS) is one of the improvements and is applicable to trace element analyzes of magnetite but has not been widely employed to examine compositional variations. Another one, which was developed, is a method using liquid chromatography hyphenated to an inductively coupled plasma mass spectrometer (HPLC-ICP-MS).

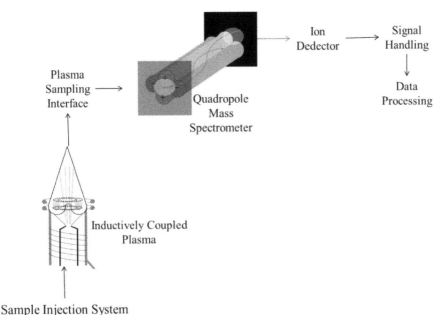

Sample Injection System

FIGURE 2.29   Scheme of ICP-MS instrument.

2.8.3   MICROWAVE PLASMA ATOMIC EMISSION SPECTROSCOPY (MP-AES)

A recent advance in atomic spectroscopy has been the development of different plasma sources, like microwave plasma sources (MP). MP source is used instead of ICP in atomic emission spectrometry. Usage of this fundamental microwave plasma-emission spectrometry (MP-ES) has developed recently for multielemental analysis. This technique is based on microwave magnetic field excitation to generate a highly stable atmospheric pressure N_2-plasma. Using this method the MP source excites the analytes and acts as a radiation source like ICP. This technique has some extra advantages, such as it eliminates flammable and expensive gases, and it runs entirely on air. Although there are a lot of fundamental and operational differences in MP when compared to other plasmas, for an analyst in a routine analytical laboratory, MP-AES would appear to be similar to the widely used ICP-OES. It provides low cost, improved laboratory safety, ease of use, high performance, robustness and reliability [14, 15]. The scheme of MW-AES is depicted in Fig. 2.30.

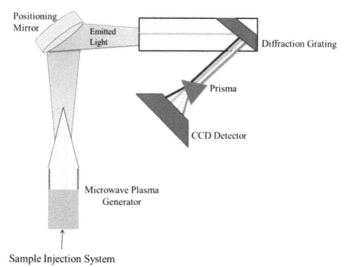

FIGURE 2.30   Scheme of an MW-AES.

2.8.4   LASER INDUCED BREAKDOWN SPECTROSCOPY (LIBS)

Laser Induced Breakdown Spectroscopy (LIBS) is a type of atomic emission spectroscopy, which uses a highly energetic laser pulse as the excitation source. It analyzes atomic absorption lines of a sample generated by laser pulse where the very high field intensity initiates an avalanche ionization of the sample elements, giving rise to the breakdown effect. Spectral and time-resolved analysis of this emission is suitable for identifying atomic species originally present on the sample surface [16]. The limitations of this method are the power of the laser, sensitivity and wavelength range of the spectrometer. Commonly this technique is used for solid samples because there is no preparatory step for them but there are various applications of liquid samples in the literature. Due to the complex laser-plasma generation mechanisms in liquids, there are some drawbacks to using lasers on liquid samples. In addition to the main problem of the laser-plasma generation mechanism, splashing, waves, bubbles and aerosols caused by the shockwave accompanying the plasma formation affect precision and analytical performance in liquid samples [17]. Therefore, many procedures like analyzing the surface of a static liquid body, the surface of a vertical flow of a liquid and the surface of a vertical flow of a liquid or falling droplets, the bulk of a liquid or a dried sample of the liquid deposited on a solid substrate have been used to defeat these problems [18]. A block diagram of LIBS is depicted in Fig. 2.31. The laser generates spark and plasma light is collected by a fiber optic and directed into a spectrograph.

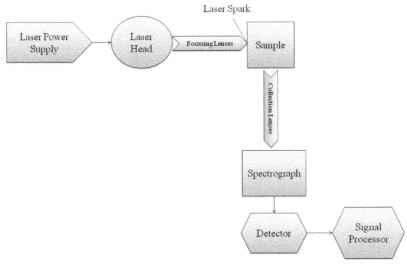

FIGURE 2.31   Scheme of an LIBS instrument.

2.9   X-RAY FLUORESCENCE ANALYSIS

2.9.1   INTRODUCTION

X-ray spectrometry in its various forms is now a powerful well-established technique for environmental analysis. The technique comprises two distinct methods of analysis: the x-ray diffraction and the x-ray fluorescence, XRF. However, the environmental applications of X-ray diffraction (XRD) are limited to the identification of duct components, which is a narrow field in the environmental work. Thus, the following discussion will focus only on the x-ray fluorescence. XRF offers multielement analysis capability, high speed and easy operation. The XRF is used for fast qualitative and quantitative analysis. It provides useful information for scientists working in environmental field, whether it be for analysis of contaminants in soils, identification of radioactive elements (e.g., uranium), composition analysis of air particulate, soils, and to follow recycling operations, etc. [19, 20].

The determination of heavy and toxic elements permits the study of their distribution, the pollution level as well as the risk assessment in the investigated ecosystem. Undoubtedly, the most popular method for this purpose are atomic absorption spectrometry, (AAS) and inductively coupled plasma-atomic emission spectrometry (ICP-AES). The need for exhaustive the sample preparation for (AAS) and ICP-AES) has lead to increasing interest towards X-ray spectrometry (XRF) in environmental investigations [21].

During the past several years, there has been a marked improvement in the technology of the field-portable X-ray instruments, which has made possible their use for measurement of pollutants in air [22], and in dust and soils [23, 24]. It offers a number of significant advantages including minimal sample preparation, rapid analysis times, multielement detection and true field use using hand-held analyzers. These capabilities make the portable X-ray instruments powerful tools for screening toxic elements and rapidly responding to emergency situations that require identification and quantitation of toxic elements.

The process is similar to other techniques in the sense that the specimen is bombarded with a high energy radiation causing individual atoms ionization. As the atom resolves towards relaxation, electrons will relocate from a high to a low orbital emitting photons of a characteristic energy or wavelength. By counting the number of photons of each energy emitted from a sample, the elements present may be identified and quantified.

Following the discovery of X-rays in 1895 by Röntgen, many experiments were made to build an X-ray tube, which can be used to bombard samples with high energy electrons. In 1912, Moseley discovered a mathematical relationship between the element's emitted X-ray frequency and its atomic number. XRF instruments are now capable of analyzing solid, liquid and thin-film samples for both major and trace component in ppm-level. The analysis is rapid and usually sample preparation is minimal or not required at all.

When an element is bombarded with high energy electrons, one of the results of interaction is the emission of photons with a broad continuum of energies. This radiation called bremsstrahlung or (continuous radiation), is the result of the deceleration of the electrons inside the material. The spectrum obtained in the X-ray region is similar to that shown in Fig. 2.32. This illustrates the main features of the spectrum and will be immediately obvious that the spectrum consists of a broad band of continuous (white) radiation superimposed on top of which are discrete wavelengths of varying intensity.

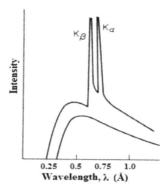

FIGURE 2.32   Intensity distribution from a Mo and X-ray tube of different voltages.

The identification of elements by X-ray methods is possible due to the characteristic radiation emitted from the inner electronic shells of the atoms under excitation
conditions. Characteristic radiation arises from the energy transferences involved
in the rearrangement of orbital electrons of the target element following ejection
of one or more electrons in the excitation process [25]. Figure 2.33 illustrates this
concept and indicate the more important of the transitions, which are involved. For
example, if any of the electrons in the inner shells of an atom can be ejected; there
are various electrons from the outer shells that can drop to fill the vacancy. Thus
there are multiple types of allowed transitions that occur which are governed by the
laws of quantum mechanics, each transition having its own specific energy or line.
The three main types of transitions or *spectral series* are labeled K, L, or M, corresponding to the shell from which the electron $\lambda(\text{Å})$was initially removed. K series
lines are of the highest energy, followed by L and then M [26].

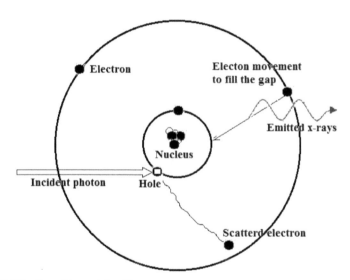

FIGURE 2.33   Transitions giving X-rays.

Each time an electron is transferred, the atom moves to a less energetic state and
fluorescence radiation is emitted whose energy is equal to the difference in energies
of the initial and final states. Detecting this photon and measuring its energy allows
the determination of the element and the specific electronic transition from which it
originated [27]. The ratio of the useful X-ray photons arising from a certain shell to
the total number of primary photons absorbed in the same shell, is called the fluorescence yield which increases with increasing atomic number [25].

Fluorescence, however, is not the only process by which the excited atom may
relax. It competes with the emission of a second photoelectron to regain stability.

This phenomenon is called Auger effect and is rather akin to the auto-ionization effect found in optical spectra [28].

High energy electrons are not the only particles, which can cause ejection of photoelectrons and subsequent fluorescent emission of characteristic radiation. High energy X-ray photons, with the output of an X-ray tube or any source of photons of the proper energy, can create the same effect. In consequence all conventional X-ray spectrometry is now based on the fluorescence technique that primary X-rays be excited from the target of an X-ray tube, and these primary X-rays then used to excite secondary (fluorescent) radiation from the sample.

When X-rays impinge upon a material, will either be absorbed by the atoms of the matrix or scattered. The absorbed photons will give rise to photoelectrons from the matrix atoms, which will appear either as x-ray photons or Auger electrons. When an X-ray is scattered with no change in energy this is called *Rayleigh* or coherent scattering, and when a random amount of energy is lost, the phenomenon is *Compton* or incoherent scattering [29].

2.9.2 *INSTRUMENTATION*

Most of the XRF instruments in use today fall into two categories: energy-dispersive (ED) and wavelength-dispersive (WD) spectrometers. In energy dispersive spectrometers (EDX or EDS), Fig. 2.34, the detector allows the determination of the energy of the photon when it is detected. Here the entire polychromatic spectrum from the sample is incident upon a detector that is capable of registering the energy of each photon that strikes it. The detector electronics and data system then build the X-ray spectrum as a histogram, with number of counts versus energy.

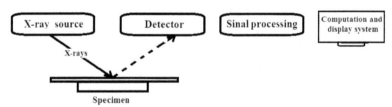

FIGURE 2.34 Schematic arrangement of ED spectrometer.

A schematic arrangement of the wavelength dispersive spectrometers (WDX or WDS) is shown in Fig. 2.35. The instrument operates based on the principle of Bragg diffraction of a collimated X-ray beam,($n\lambda = 2d\sin\theta$). The photons are separated by diffraction on a single crystal before being detected [30].

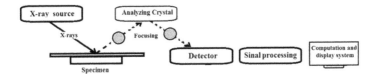

FIGURE 2.35 Schematic arrangement of WD spectrometer.

2.9.2.1 X-RAY SOURCE

The source consists of an evacuated chamber with a heated cathode, which is usually a tungsten filament, and an anode. A potential difference of several tens of kilovolts is applied between the filaments relative to the cathode. Electrons striking the anode produce X-radiation as well as X-ray lines characteristic of the anode material are emitted. A significant portion of these photon passes through a beryllium window build on the side of the tube. X-rays is very inefficient process, and only about 1% of the total applied power emerges as useful radiation. The majority of the remaining energy appears as heat, which has to be dissipated by cooling the anode.

2.9.2.2 DETECTOR SYSTEMS

A detector is used to convert X-ray energy into voltage signals; this information is sent to a pulse processor, which measures the signals and passes them onto an analyzer for data display and analysis. The two main types of XRF spectrometers (WD and ED) differ completely in their detection systems. EDXRF systems depend on semiconductor-type detectors, which receive the entire emitted spectrum from the sample and convert it into a number of counts versus photon energy. WDXRF spectrometers, however, use an analyzing crystal to disperse the emitted photons based on their wavelength and place the detector in the correct physical location to receive X-rays of a given energy [26, 31].

2.9.3 TOTAL REFLECTION X-RAY FLUORESCENCE SPECTROMETRY (TXRF)

Total reflection X-ray fluorescence (TXRF) is an analytical technique for elemental chemical analysis, developed for silicon wafer quality control. The technique is based on the principle that, when an exciting X-ray beam impinges on a thin sample positioned on a very flat support below the critical angle of total reflection on the support, this radiation is totally reflected. In this way, matrix effects can be neglected and TXRF allows the detection of extremely small amounts of elements [32].

The applicability of TXRF in most of the environmental fields is still under exploration. Studies about many different kinds of samples, such as natural waters, deionized water, heavily polluted waters and landfill leachates, waste waters, wastes and wastes leachates sediments, soils, soils related materials such as fertilizers and plants, have been reported [32].

2.9.4 PORTABLE X-RAY SYSTEMS

X-ray fluorescence spectrometry is perhaps the first spectroscopic technique, which can be successfully applied in the field and in industrial environments for the analysis of various materials. The most attractive features of XRF are its speed and simplicity, with minimal sample preparation required, analysis times as short as minutes or less and multielement detection. It is also one of the atomic spectrometric techniques that can be adapted for true field portable use. A relatively recent development has been the availability of portable instrumentation, *which* can be used for both the direct nondestructive analysis of samples, and also is readily transportable to field sites for use in a mobile laboratory style of operation. More specifically portable EDXRF has increasingly been applied, in the last 20 years, to the analysis of aerosols, waters, sediments, soils, solid waste and other environmental samples.

The portable XRF analysis can be briefly explained as follows:

1. an incident X-ray photon produced from a radioisotope source excitation device of the portable XRF creates an inner shell vacancy in which an electron leaves the inner shell;
2. when the atom relaxes to the ground state, an outer shell electron falls to make up for the inner shell vacancy;
3. then photons are given off when an energy in the X-ray region of the electromagnetic spectrum that is equivalent to the energy difference between the two shells;
4. the energy level and intensity of these emitted X-rays identify the elements and their concentrations, respectively [33]

Analysis using portable XRF technique can make an essential contribution to a wide range of applications such as:

- Analysis of soils, particularly in the assessment of agricultural land and contaminated land.
- Geochemical mapping and exploration to locate mineral deposits.
- Environmental monitoring related to air pollution studies and contamination of the work place.
- The on-line control of industrial processes for the production of raw materials.
- Sorting scrap metal alloys and plastics to increase the value of recyclable materials.

Major advantages of portable XRF over conventional laboratory-based analysis include: (i) immediate analytical results, which is important for interactive measure-

ment programs, e.g., assessing sites contaminated with heavy metals, (ii) nondestructive analysis of objects that can neither be sampled nor removed to the laboratory for analysis (e.g., museum samples, works of art and archaeological samples.

2.10   DIFFERENTIAL OPTICAL ABSORPTION SPECTROSCOPY (DOAS)

The DOAS is a method to determine concentrations of trace gases in the atmosphere
by measuring their specific narrow band absorption structures in the UV and visible
spectral region. A typical DOAS instrument consists of a continuous light source,
that is, a Xe-arc lamp, and an optical setup to send and receive the light through the
atmosphere. It is also possible to use the sun or scattered sun light as light source.
The typical length of the light path in the atmosphere ranges from several hundred
meters to many kilometers.

Figure 2.39 shows a typical long-path DOAS instrument. The light of a Xe-arc
lamp is collimated into a parallel beam, which is sent through the atmosphere onto
an array of quartz cube corner retro-reflectors. The retro-reflectors send the light
back into the telescope where it is received by the central part of the double New-
tonian telescope and focused onto a quartz fiber. The fiber is part of a quartz fiber
mode mixer whose other end serves as entrance slit for the grating spectrometer.
The spectrum is recorded by a photodiode array and stored in a personal computer.
The telescope, together with the lamp and the fiber entrance, is mounted on a
frame which can be rotated $\pm 45°$ horizontally and $\pm 20°$ vertically by two step-
per motors. By rotating, the instruments can be aimed at different retro reflector
arrays mounted at different heights. Therefore the instrument can probe different
air masses. Many atmospheric gases can be detected with DOAS system like O_3,
NO_2, BrO, IO, OClO, SO_2, H_2O, HCHO, O_4, and O_2.

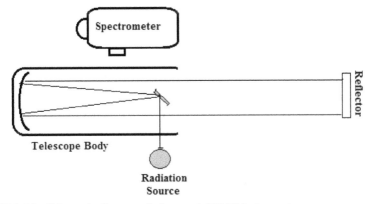

FIGURE 2.36   Schematic diagram of a long path DOAS instrument.

KEYWORDS

- **Atomic Absorption**
- **Induced Coupled Plasma**
- **Infrared**
- **Turbidimetry**
- **UV/Vis**
- **X-ray**

REFERENCES

1. *Kaufmann, W. J. (2011). Universe, 9th Edition, Freeman, W. H. Ed., NewYork.*
2. http://en.wikipedia.org/wiki/Deuterium_arc_lamp
3. http://www.chemguide.co.uk/analysis/uvvisible/theory.html
4. Rouessac, F., & Rouessac, A. (2007). Chemical Analysis, Modern Instrumentation Methods and Techniques, Chapter 9, translated by Rouessac, F., Rouessac, A., & Brooks, S., 2nd Ed, John Wiley & Sons Ltd, West Sussex.
5. EPA Guidance Manual, Chap 7, Importance of turbidity, April 1999 7–1.
6. http://en.wikipedia.org/wiki/Interferometry
7. Gable, K. http://chemistry.oregonstate.edu/courses/ch361–464/ch362/irinstrs.htm.
8. Reusch, W. Infrared Spectroscopy, http://www2.chemistry.msu.edu/faculty/reusch/VirtTxtJml/Spectrpy/InfraRed/infrared.htm.
9. Harvey, D. (2000). "Modern Analytical Chemistry", McGraw Hill, Chapter 10, Boston.
10. Beaty, R. D., & Kerber, J. D. (1993). "Concepts, Instrumentation and Techniques in Atomic Absorption Spectrophotometry" 2nd Edition, the Perkin Elmer Corporation.
11. Welz, B., Becker-Ross, H., Florek, S., & Heitmann, U. (2005). High Resolution Continuum source AAS, Wiley-VCH, Weinheim, ISBN, 3–527–30736–2.
12. Risby, T. H. (2007). In Modern Instrumentation, G. McMahon, Ed., Wiley, Chichester.
13. www.thermo.com.
14. Jankowski, K., & Reszkeb, E. (2013). Recent Developments in Instrumentation of Microwave Plasma Sources for Optical Emission and Mass Spectrometry: Tutorial Review. *J. Anal At Spectrom, 28*, 1196–1212.
15. Hettipathirana, T. D. (2013). Determination of Boron in High Temperature Alloy Steel using Non-Linear Inter-Element Correction and Microwave Plasma Atomic Emission Spectrometry, *J. Anal At Spectrom, 28*, 1242–1246.
16. Barbini, R., Colao, F., Fantoni, R., Lazic, V., Palucci, A., Capitelli, F., & Van der Steen, H. J. L. (2000). Laser Induced Breakdown Spectroscopy for Semi-Quantitative Elemental Analysis in Soils and Marine Sediments, Dresden/Frg. Proceedings of Earsel-Sig-Workshop Lidar.
17. Stiger, J. B., De Haan, H. P. M., Guichert, R., Deckers, C. P. A., & Daane, M. L. (2000). Determination of Cadmium, Zinc, Copper, Chromium and Arsenic in Crude oil Cargoes, Environment Pollutions, *107*, 451.
18. Stronge, L., Kwong, E., Sabsabi, M., & Vadas, E. B. (2004). Rapid Analysis of Liquid Formulations Containing Sodium Chloride Using Laser-Induced Breakdown Spectroscopy, *J. Pharmaceut Biomed., 36*, 277–284.

19. Van Grieken, R. E., Markowicz, A. A. (Eds.) (1993). Hand book of X-Ray Spectrometry, Mercel Dekker, New York.
20. Potts, P. J., Ellis, A. T., Kregsamer, P., Marshall, J., Streli, C., West, M., & Wobrauchek, P. (2001). Atomic Spectrometry Update, X-Ray Fluorescence Spectrometry, J.Anal at Spectrom, *16*, 1217.
21. Jenkins, Ron, Gould, R. W., & Gedcke, Dale **(2000)**. *Appl Spectrosc. Rev., 35(1, 2)*, 129–150.
22. Morley, J. C. (1997). Evaluation of a Portable X-Ray Instrument for the Determination of Lead in Workplace Air Samples, Master's Thesis, University of Cincinnati, Department of Environmental Health; Masters Abstracts International, *36(1)*.
23. Clark, S., Menrath, W., Chen, M., Roda, S., & Succop, P. (1999). Use of a Field Portable X-Ray Analyzer to Determine the Concentration of Lead and Other Metals in Soil Samples, Annals Agriculture Environ. Medicine, *6*, 27–32.
24. Sterling, D. A., Evans, R. G., Shadel, B. N., Serrano, F., Arndt, B., Chen, J. J., & Harris, L. (2004). Effectiveness of Cleaning and Health Education in Reducing Childhood Lead Poisoning Among Children Residing Near Superfund Sites in Missouri; Archives of Environmental Health, *59(3)*, 121–131.
25. Basic Theory of X-Ray Fluorescence (2011). http://www.learnxrf.com/Basic XRF Theory. htm. LearnXRF.com.n.d.web.
26. Jenkins, R. (1999). X-Ray Fluorescence Spectrometry, John Wiley & Sons, Inc., Chichester, QD96.X2J47 ISBN 0–471–83675–3.
27. Anzelmo, John A., & Lindsay, James R. **(1987)**. *J. Chem. Educ., 64(8)*, A181–A185.
28. Skoog, D. A., Holler, F., James, N., & Timothy, A. (1998). Principles of Instrumental Analysis, 5th Ed., Thomson Learning, Inc. CA.
29. Compton and Alison, (1935). X-Rays in Theory and Experiment; Van Nostrand, New York.
30. Knoll, G. F. (2000). Radiation Detectors for X-Ray and Gamma-Ray Spectroscopy, *J. Radioanal, Nucl, Chem., 243(1)*, 125–131.
31. Jenkins, R., Gould, R. W., & Gedcke, D. (1995). *Quantitative X-Ray Spectrometry, 2nd Ed.*, Marcel Dekker, Inc., New York, QD96.X2 J46 ISBN 0–8247–9554–7.
32. Danel, A., Kohno, H., Veillerot, M., Cabuil, N., Lardin, T., Despois, D., & Geoffrey, C. (2008). Comparison of Direct-Total Reflection X-Ray Fluorescence, Sweeping Total Reflection X-Ray Fluorescence and Vapor Phase Decomposition Total Reflection X-Ray Fluorescence Applied to the Characterization of Metallic Contamination on Semiconductor, Spectrochim. Acta, Part B, *63*, 1375–1381.
33. Melquiades, F. L., & Appoloni, C. R. (2004). Application of XRF and Field Portable XRF for Environmental Analysis, *J. Radioanal Nucl Chem., 262(2)*, 533–541.
34. Melamed, M. L., Basaldud, R., Steinbrecher, R., Emeis, S., Ru´iz-Suarez, L. G., & Grutter, M. (2009). Detection of Pollution Transport Events Southeast of Mexico City using Ground based Visible Spectroscopy Measurements of Nitrogen Dioxide. Atmos. Chem. Phys., *9*, 4827–4840.

CHAPTER 3

CHROMATOGRAPHIC METHODS OF ANALYSIS

MAHMOOD M. BARBOOTI

Department of Applied Chemistry, School of Applied Sciences, University of Technology, P.O. Box 35045, Baghdad, Iraq; Email: brbt2m@gmail.com

CONTENTS

3.1 INTRODUCTION

There are a number of analytical techniques for the detection and determination of concentration of the components and ions in certain samples based on the separation of the components using the principle of the distribution of solutes between two phases stationary and mobile in accordance with physical and chemical concepts.

3.1.1 THEORETICAL BASIS

Many theoretical models are introduced to explain the separation processes that mix thermodynamics and chemical kinetics. The aim of such models is at the understanding this phenomenon and to improve the methods of separation to obtain pure materials from the possible impurities or the separation of the components of a chemical mixture for quantitative determinations. Moreover, the establishment of qualitative data about the samples is another important outcome of the separation methods. For more comprehensive theoretical treatment the readers may consult other textbooks of instrumental analysis and chromatographic separation books [1–4].

Separation methods are divided into two categories: equilibrium and nonequilibrium. The equilibrium in this respect is chemical and not physical for most cases. The equilibrium referred to here is between ions in solutions and between two phases, (e.g., gas and liquid) which is usually studied in early chemistry learning. Chemical equilibrium exceeds the consideration of masses to the molecular properties and their interactions. Consider a bottle of aftershave (water and ethanol). The equilibrium in this example indicates that the composition of the gaseous phase is different from the composition of the liquid, which is in equilibrium with it. The gaseous phase (vapor) is rich in ethanol in comparison with the liquid phase. At constant temperature, the mole fraction of ethanol in the gaseous phase will be greater than that in the liquid phase. Alcohol is more volatile than water. Also, the ethanol will reach certain concentration in gaseous phase. If ethanol is replaced with methanol we will find that the alcohol content in the gas will increase. Note that the equilibrium here is not influenced by temperature and the case is a chemical equilibrium.

3.1.2 PHYSICAL METHODS OF SEPARATION

The physical separation methods can be attributed to equilibrium although most of them do not follow equilibrium rules. Equilibrium rules of separation are not obeyed for example in filtration nor in centrifugal separation where a precipitate is separated from the liquid in which it is suspended via the creation of an artificial attraction field. However, some separation methods may not behave in the same manner like gel filtration. In the meantime, the separation by the ultracentrifugal force of a salt gradient is among the physical methods of separation based on equilibrium.

One of the earliest desires of human being is the separation of some components present in water like clays and silt and bacteria for health and sensual purposes. The use of sand as a filtration medium is an old process. Also the use of the dish in the separation of gold particles from sand and other materials by the miners uses the large difference in density between sand and gold as a basis for the separation where sands are washed away from gold, which appears dark just like the sand. And again this is not a filtration process but a nonequilibrium physical separation method.

An important group of compounds are those with optical activity. Chemists need to get these compounds at high purity because the difference in optical activity results in a difference in their biological activity as in pharmaceutical compounds. Purity has to be determined precisely with the modern analytical techniques. It is beneficial that some diastereomers differ in some of their physical properties like solubility. In the meantime, enantiomers are identical in their physical and chemical properties. Thus, reliable methods must be designed for the separation of optically active compounds. The early attempts to separate these compounds were performed under microscope with aid of a polarized light. This is an example of the utilization of a physical property with a chemical basis.

3.1.3 CHEMICAL METHODS OF SEPARATION

Chemical methods of separation are based on equilibrium. Solubility is a major chemical parameter. Precipitation and filtration are physical separation methods and occur whenever a solubility product of a compound attains a value that permits its removal from solution. However, distillation and extraction are chemical separation methods based on differences in equilibrium constants. Chemical separation processes take place as a result of equilibrium between two heterogeneous phases over the boundary between them. In the true distillation the equilibrium occurs during the process of reflux where the condensed material returning to the boiling pot is in equilibrium with the evolving vapor. Thus, it is a conjugation of two phases, liquid and gas.

During the extraction, equilibrium is established across the liquid junction. It is a liquid-liquid junction. If some particles of a solid material are introduced to the liquid phase, there will be a distribution of molecules between solid surface and the liquid. Simple extraction is a forward operation. In liquid extraction (liquid-liquid), the solute is extracted from the phase containing it to the phase into which extraction is to be carried out. To extract a material **A** in the presence of an interfering material **I** and their distribution ratio is 5 and 0.5, respectively. For one step of extraction we may transport more than 75% of material **A** and about 30% of material **I** into the extracting phase. If the original concentrations of the two materials are identical, the ratio between their concentrations will be 5:2. A second extraction process of the extract with a new quantity of the extracting phase will result in improvement of the extraction. This is because the material A have a greater distribution ratio and will be more extracted than in the first extraction step and to a less quantity in the second step. We may end up with a ratio of about 7:1 of A:I. A technique of solute extraction and its reextraction to a new quantity of the original phase was developed (reverse stream extraction). This concept is one of the bases of modern chromatography.

Chromatographic separation is established by passing a material free of the analyte (mobile phase) over another one also free from the analyte (stationary phase). The sample is injected or placed in the mobile phase. By moving with the samples components of the sample may be separated between the two phases. The components for which the distribution ratio is greater in the stationary phase will take longer time to pass the system. With sufficient amount of the mobile phase and the time, the components with identical distribution ratio will be separated.

Chromatographic separation was used at the end of 19th century by Russian botanist, Mikhail Tsvet. The technique developed substantially as a result of the work of Martin and Synge during the 1940s and 1950s. They established the principles and basic techniques of partition chromatography, to win Noble prize in chemistry in 1952. Their work encouraged the rapid development of several chromatographic methods: paper chromatography, gas chromatography, and what would become known as high performance liquid chromatography. Since then, the technology has advanced rapidly to reach the present advancements when electrophoresis was introduced to perform the separation without a stationary phase.

3.2   CLASSIFICATION OF SEPARATION METHODS

Separation methods are classified on three bases, namely, the physical state of the mobile phase, the contact method between the mobile and stationary phases and the mechanism of the component separation. Gas chromatographic separation is established by a mobile gas phase over a liquid phase and termed the gas–liquid-chromatography or over a solid phase and then termed the gas–solid-chromatography.

3.2.1   COLUMN CHROMATOGRAPHY

This is a technique in which the stationary phase is contained in a column and mobile phase (liquid) is passed downward to transport the components in accordance with their binding with the stationary phase and their solubility in the mobile phase. After certain time bands will be separated along the column representing the components of the sample. If the solvent addition continues the components will be completely separated and collected at the end of the column for analysis.

As the sample moves in the column, the solutes start to separate and the sample band will expand attaining a Gaussian profile (a bell shape). The parameter **a** is the height of the curve's peak, **b** is the position of the center of the peak, and c (the standard deviation) controls the width of the "bell"). The solutes will start to separate, if their binding forces with the stationary phase are different, in the form of single bands. The progress of the chromatographic operation can be traced by using certain indicator located at the end of the column. The plot of the indicator signal against time termed the chromatogram, which consists of a number of peaks representing the separated solutes bands.

The chromatographic peak is characterized by the retention time, t_R, the time elapsed from the sample introduction till the peak maximum is attained. This can also be measured as the required volume of the mobiles phase from sample introduction until the appearance of the peak maximum of the solute and termed the retention volume, V_R.

The peak base width is another criterion of the chromatographic signal and may be expressed in time or volume units. A peak appears in the beginning of the chromatogram refers to the signal of the mobile phase or any component not retained by the stationary phase and termed the column voids time (Fig. 3.1).

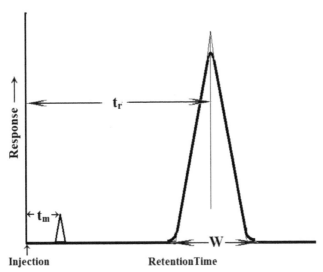

FIGURE 3.1   Fundamental criteria of chromatographic signal.

3.2.2. SURFACE CHROMATOGRAPHY

In this technique, the stationary phase (A solid or a dense liquid) is fixed on a flat surface. A drop of the mixture is placed near one of the ends and mobile phase (a suitable solvent), placed in a glass container in which the plate holding the stationary phase is dipped, is allowed to move upward by the capillary action carrying away the components according to their bonding to the stationary phase and their solubility in the solvent (Fig. 3.2).

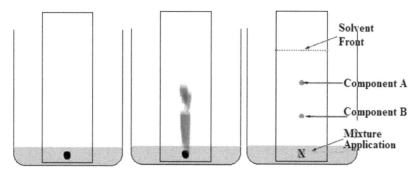

FIGURE 3.2   Chromatographic separation.

When the mechanism of the separation is used as a basis for the classification, ion exchange and size exclusion technique may be considered. The presence of charged centers in the stationary phase will attract certain group of the components in accordance with their charges to be separated. The stationary phase consists of a solid support covalently bonded to ionic functional groups, negative like $[-SO_3]$ or positive like $[-N(CH_3)_3]$. The ionic components are electrostatically attracted to these centers leaving the original matrix.

In size exclusion separation, some kinds of porous gels are used as stationary phases to allow the separation of the components of the sample selectively depending on the differences in sizes. Large volume solutes are inhibited from entering the pores of the stationary phase and pass the column quickly, while small solutes will be retained and their movement inhibited (their residence time in the column will be longer).

Stationary phase is not essential for all separation methods and the solute ions may, for example, travel under the effect of an electric field. The separation in this case is a function of the speed of ion migration.

3.2.3   RESOLUTION

The aim of the chromatography is at the separation of mixture components into a series of chromatographic peaks each representing a single component of the sample. Resolution is a quantitative measure of the degree of separation between chromatographic peaks. For two-component system, A and B, the resolution is defined as:

$$R = \frac{tt_{r,B} - t_{r,A}x^2}{0.5(w_B - W_A)} = \frac{2\Delta t_r}{W_B + W_A} \tag{1}$$

In accordance with Fig. 3.3, the degree of separation is improved with the increase of resolution. If the two signals are equal, a resolution value of 1.5 will be due to the overlap of the two signals by 0.13%. Thus, the resolution can be indicated as a criterion of the effect of controlling the analytical conditions on the efficiency of separation.

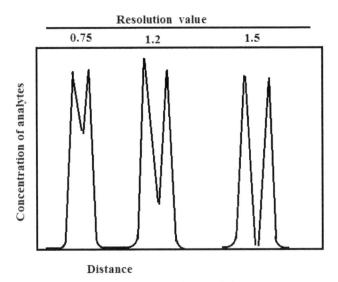

FIGURE 3.3   Three examples on chromatographic resolution.

Example
During the analysis of lemon oil, the limonene signal appears at 8.63 min with a width at the base of 0.96 min. γ-turpentine signal appears at 9.54 min with a width at the base of 0.64 min. What is the resolution between the two peaks?

Solution
Using Eq. (2) above the resolution can be calculated:

$$R = \frac{\Delta t_r}{W_B + W_A} = \frac{2(9.54 - 8.36)}{0.64 + 0.96} = 1.48 \qquad (2)$$

Accordingly, the resolution is improved either by the increase in the difference between the two retention times or by reducing the width of the peak base. The increase in retention time difference can be achieved by improvement of the interaction between the solute and the column or increase of the column selectivity towards one of the solutes. The width of the peak base is related to the movement of the solute between the stationary and the mobile phases. These factors are affected by the so called the column efficiency.

3.3   GAS CHROMATOGRAPHY

3.3.1   INTRODUCTION

Gas chromatography is a method for the separation of components of a homogeneous mixture (mainly organic) to determine the composition. The method is based on the injection of the sample (mixture) into a column packed with certain materials (stationary phase) and passing a gas (mobile phase) in the column. The stationary phase will retain the components with different forces depending on the chemical structure and the physical properties of the components (molecular weight, boiling point and polarity). At the same time the mobile phase tries to travel carrying the components of weaker association with the stationary phase to be eluted at the end of the column as the first arriving components. The components may be eluted according to their binding force with the stationary phase and capability of the mobile phase to carry them. The ability to form hydrogen bonding with the stationary phase may also be among the acting forces that determine the difficulty of elution. The components that were in a homogeneous mixture will appear at different time intervals at the end of the column.

There are two kinds of gas chromatography: gas solid (solid stationary phase) and gas liquid (the stationary phase is a dense liquid strongly fixed inside the column or bound to a solid support. Retention volume, V_R: is the volume of the carrier gas, which succeeds in the elution of certain component. The retention time, t_R, is the time required for the elution of certain component,

$$VR = t_R.F \tag{3}$$

where, F is the flow rate of the gas. The value of F may be specified at the beginning of the experiment by measuring the volume in unit time. To improve or control the separation efficiency, many methods can be used:

- The length of the column;
- Column material;
- Flow rate of the carrier gas (mobile phase);
- Temperature.

3.3.2   THEORETICAL CONSIDERATIONS

Martin and Synge [5] presented a theoretical model for the GC separation process to allow for the improvement of the separation efficiency. In their model they suggested that the GC column consists of a large number of thin sections called, the theoretical plates. The number of theoretical plates, n, is related to the ratio between t_R and the half peak width of analyte, $W_{1/2,}$ assuming that the peak shape is symmetric.

$$n = 5.45(t_R/W_{1/2})^2 \tag{4}$$

Thus, the n value is dependent on the analyte, the packing of the column, the length, and the flow rate of the mobile phase.

3.3.2.1 HEIGHT EQUIVALENT TO A THEORETICAL PLATE

Since n depends on the length of the column, another parameter is used to express column efficiency. It is the height (length of column) equivalent to a theoretical plate, HETP, or just H

$$H=L/n \qquad (5)$$

where L is the length of column in cm or mm. Thus H is the length of column, which represents one theoretical plate in units of cm/plate or mm/plate. The effect of flow on column efficiency is usually shown by plotting H versus flow rate or linear velocity. Such a plot is shown in the figure. Note that the H line goes through a minimum. The minimum occurs at the optimum flow velocity. The van Deemter equation is the simplest relation between the resolving power (HETP) of a chromatographic column with the various flow and kinetic parameters the equation which cause peak broadening [6].

$$H=A+B/v + Cv \qquad (6)$$

where A is the Eddy-diffusion parameter, related to channeling through a nonideal packing, m; B, is the diffusion coefficient of the eluting particles in the longitudinal direction, resulting in dispersion [$m^2 s^{-1}$], C is the resistance to mass transfer coefficient of the analyte between mobile and stationary phase [s]; and the v I the linear Velocity [$m s^{-1}$]. In open tubular capillaries, A will be zero as the lack of packing means channeling does not occur. In packed columns, however, multiple distinct routes ("channels") exist through the column packing, which results in band spreading. In the latter case, A will not be zero.

The kinetic resistance is the time lag involved in moving from the gas phase to the packing stationary phase and back again. The greater the flow of gas, the more a molecule on the packing tends to lag behind molecules in the mobile phase. Thus this term is proportional to v.

3.3.3 INSTRUMENTATION

Figure 3.4 summarizes the instrumentation employed in the gas chromatographic separation. The instrument in gas chromatography consists of:
1. A cylinder of carrier gas (helium, nitrogen, or hydrogen) with valves to control the flow rate within 25–150 mL.min^{-1} for packed columns and 1–25 mL.min^{-1} for the open ended columns.
2. Injector to introduce the sample.

3. Column: 2–50 m glass, steel, or Teflon tube coiled to save the space. Capil-
    lary columns are also used to improve the separation especially in the hy-
    drocarbon analysis.
4. Oven: to be operated up to 400°C precisely and with accuracy of <1°C.
5. Detector: flame ionization, thermal conductivity or electro capture. The out-
    put of the gas chromatograph (Fig. 3.4) may be connected to a mass spec-
    trometer to give a highly advanced technique for analysis and detection, the
    GC-MS.

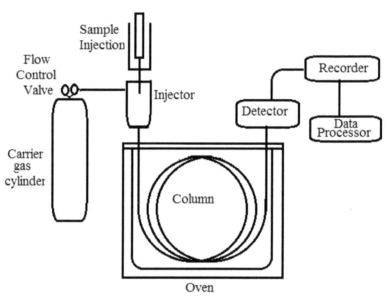

FIGURE 3.4   Schematic diagram of gas chromatograph.

3.3.3.1   GAS FLOW

With a gas as the mobile phase to be moved in the column under pressure, the
determination of the flow rate may be difficult, the type of the instrument and the
pneumatic parts may have important role. From the general law of gases we know
that volume, pressure and temperature are interrelated and all affect the flow rate.
Thus, the measure the flow rate for packed columns by simple bubble flow meters
connected to the instrument. In capillary columns, it is advisable to press a gas that
cannot be retained, air or helium, to determine its retention time and calculate the
linear flow rate of the gas using the length of the column, cm, and the retention time.

3.3.3.2 SAMPLE INJECTION

The injector must be designed to ensure rapid sample introduction and vaporization. The design of the injection chamber assists the transportation of the sample vapor with the carrier gas to enter the column and prevent any losses outside the system. The detailed design of the injector for a GC system is shown in Fig. 3.5. A micro syringe is employed for the injection of sample solution into the injector, which is a metallic component, surrounded by a heater and a rubber disc to insert the syringe. The preheating of the injector is essential to ensure the transformation of the sample into the vapor phase once it enters the injector to facilitate its handling during the course of analysis. The size of the injected samples differ in accordance with the type of the column in use, being 1–20 µL for packed columns and 10^{-3} µL for capillary column. The carrier gas is allowed to enter the system from a hole next to the injection hole where the components of the sample will travel across the column. A detector is fixed at the end of the column to indicate the arrival of the separated columns.

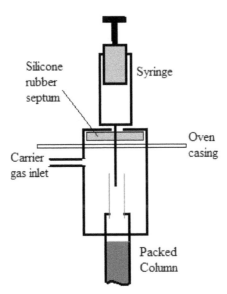

FIGURE 3.5 The schematic diagram of the injector.

3.3.3.3 SPLIT OF VAPOR

A fraction of the sample vapor, which fails to enter the column, will be vented through a hole. A needle valve is used to control the amount of the inlet carrier gas and determines the split ratio, which monitors the gas flow rate inside the column.

A split ratio of 1: 100 means that 10 µL is driven into the column from each 1.0ml injected. The split valve is often closed to increase most of the injected sample into the column. The split valve is essential in capillary gas chromatographic work.

3.3.4 HEAD SPACE INJECTION GC

The headspace sampling is a method for the extraction of volatile components from a nonvolatile matrix. Whenever volatile components can be partitioned from the sample to the gas space above the sample in a vial, they may be sampled and directly injected into the GC column. The method is suitable for low boiling point volatiles and no interference may be expected from volatile components of higher boiling points and semivolatiles. Automation was introduced to this method to expand the application for quality control and sample screening. Thus, there will be no need for the time and solvents consuming procedure of the extraction of the volatiles from a sample. The method prevents the contamination of the GC system and column with the relatively heavier components. Currently, the method is widely accepted and applied for forensic analysis (alcohol in blood), water quality measurements and determination of residual solvents in pharmaceutical products.

The sample sealed in a closed vial, where constant temperature condition is maintained. The volatile components leave sample phase and enter gas phase (Fig. 3.6). There will be equilibrium between liquid and gas phase. The headspace gas phase is sampled by inserting the needle of a GC syringe into the rubber seal of the vial to reach the head space to draw some of the gaseous phase above the sample.

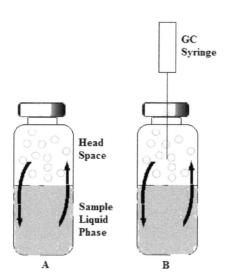

FIGURE 3.6 Preparation of the head space sample (A) and the sampling process (B).

3.3.5   CHROMATOGRAPHIC COLUMNS

Packed and open ended columns are used in gas chromatographic methods with various diameters including the capillary columns. The chromatographic column is the container of the stationary phase. The following are some important details affecting the analysis.

3.3.5.1   STATIONARY PHASE

The separation of the components of a sample involves the sorption on the surface of a solid. The application of gas solid chromatography is limited due to the tail effect (the signal does not come to an end because of the slow removal of the adsorbed material from the solid). The best application, therefore, will be the analyzes of volatile hydrocarbons and gases. The introduction of liquids as the stationary phases may be considered as a main developments in this respect. A support is used and the dense material is loaded onto its particles as a thin layer to be the mechanical medium for the liquid phase. Diatomaceous clays, however, are among the best supporting materials and it is available in various forms. Two methods are used for the preparation of the porous support:

- Treatment with alkaline solution and roasting to change it into a white material from the alkaline flux;
- It may also be prepared in an acid medium (pink colored like the fire brick powder).

The diatomaceous material contains hydroxyl groups in their aqueous form. These groups may act as locations on which various components molecules are adsorbed. This will present a real problem for the GC analysis and result in the occurrence of tailed signal with the use of polar liquid phase to associate with the O-H groups. A better solution for this is the treatment with organo-silicon derivatives to change the VIOH centers into Si-OH.

\ \ \
O O O
/ $Si(CH_3)_2Cl_2$ / ROH /

$$-Si\text{-}OH \rightarrow -Si\text{-}O\text{-}Si(CH_3)_2Cl + HCl \rightarrow -Si\text{-}O\text{-}Si(CH_3)_2OR + HCl \qquad (7)$$

\ \ \
O O O
/ / /

3.3.5.1.1 LIQUID STATIONARY PHASES

Many substances have been studied to function as liquid stationary phases. They are types of waxes, rubbers or glasses at room temperature and change into liquids at the operating conditions of the chromatographic columns. They differ in their polarizability and the working temperature ranges. Some of the common stationary phases are listed in Table 3.1, where there are many choices. The minimum working temperatures of these liquid phases are determined by the ranges at which they are transformed into glasses or highly viscous liquids.

The polarity of the materials is determined by their ability for the separation of polar materials. Light hydrocarbons for example can be separated efficiently on squalane. In the meantime, the compounds accumulate and appear as an overlapping signal, when polar stationary phase is used for the separation. The stationary phase is loaded on the support by soaking the support particles in a solution of the station-ary phase in an open container to allow the evaporation of the solvent. The loaded particles are poured in metallic tubes with some tapping to allow the homogeneous packing. Fittings are then fixed on the tube ends to allow its instillation within the gas chromatographic system. The separation is improved by the reduction of par-ticle size to be loaded by a thin layer of the stationary phase.

3.3.5.2 CAPILLARY COLUMNS

The packing of chromatographic columns with granular materials can be avoided using the long capillary tubes made of glass or fused silica and their internal walls act as the support for the liquid stationary phases. An internal diameter of 0.2 mm and a length between 50 and 100 m are typically used for an efficient separation pro-cesses. Capillary columns were successfully employed for the separation of com-plex mixtures. However, some precautions must be considered regarding the sample handling and connection with the detectors.

Capillary columns are available in two types: the wall coated open tube, WCOT, and the support coated open tubes, SCOT. The WCOT contains a thin layer of the stationary phase (0.25 micron on the inner wall. The SCOT contains a thin layer of diatomaceous clay as a solid support and coated with a layer of the stationary phase on the inner wall. The coating of the inner wall of the capillary columns is a com-plex process. Most liquids tend to form droplets and not thin layers. Thus, the inner surface needs to be rough, scratched, (with hydrochloric acid or other acids). The tube is filled with a solution of the liquid phase in a volatile solvent and the solvent is then allowed to evaporate under low pressure and high temperature.

The efficiency of the capillary column chromatography is relatively high al-though the right polarity is not available. Also, there is no need to install another column with different polarity to cover all types of analytes. Another type of pack-ing is introduced which is considered midway between the solid particles of the gas solid chromatography and the liquid stationary phases. The packing consists of porous beeds of an cross linked organic polymer like the copolymer of styrene with divinyl benzene and used in the range of 250°C. The sample components are distributed between the gas and the amorphous beeds, which will act as a solvent rather than an adsorption medium. The efficiency of this material has proved superior and is available currently in many commercial names like the Poropak supplied from Waters and the Chromosorb 100 series from John Manvel. Chromatographic c columns need conditioning before use by passing the carrier gas at the maximum allowed temperature to remove any volatiles that may interfere with the analysis. Packed columns are typically used for the analysis large volume samples although the analysis is relatively slow. In the meantime, capillary columns is the best choice for low volume samples and is considered faster than the coated tube.

3.3.6   DETECTORS

When the separated materials arrive the end of the column there must be a method for its detection within the components of the gas chromatographic instrumentation. Several detectors are available now and they are all characterized by:
- Sensitivity of 10^{-8}–10^{-15} g of material.s^{-1};
- Working range between 0 and 400°C;
- Stable and reliable;
- Linear response (independent on concentration).

The detector must be of wide range and fast response and simple and have identical response to almost all materials. Some of the detectors are discussed here.

3.3.6.1   FLAME IONIZATION DETECTOR

Figure 3.7 shows a schematic diagram of the flame ionization detector. The separated components traveling across the column to enter a hydrogen/air flame to burn and dissociate into free radicals that can be ionized by flame. The detection of these ions can be done by electrical conductivity to indicate the arrival of the material. Flame ionization detector is characterized by:
- Sensitivity of 10^{-13} g.s^{-1};
- Wide working range;
- Low sensitivity to carbonyl compounds, alcohols and amine; →
- Not sensitive for noncombustible materials like H2O, CO_2, SO_2, NO_x;
- Typically useful for hydrocarbon applications;
- Destructive (the components burn and destructed during detection).

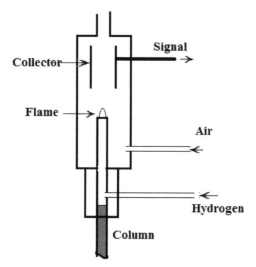

FIGURE 3.7   Schematic representation of the flame ionization detector.

3.3.6.2   THERMAL CONDUCTIVITY DETECTOR (TCD)

The thermal conductivity of the heat of a filament by the carrier gas is affected (reduced) when another material arrived the detector with it, because the thermal conductivity of all materials are significantly less than that of the carrier gas. Thus, any reduction in the thermal conductivity indicates the arrival of a component to the detector. Thermal conductivity detector is characterized by lower sensitivity then the flame ionization detector and in the range of 10^{-8} g.cm^{-1} and varying response for various materials but with wide range of applications.

3.3.6.3   ELECTRON CAPTURE DETECTOR (ECD)

The improvement of separation processes and the use of smaller sample volumes especially for capillary gas chromatography necessitate the introduction of more precision detector. The ionization chamber was developed to allow the detection by radiation. The materials leaving the column are directed to a chamber, which is permanently bombarded by a flow of beta ray. A nano ampere current flows across the chamber transported by the gaseous ions. The current interacts with the organic compounds leaving the column and ionize them. The speed of these ionized molecules is much less than that of nitrogen and hence there will be a decay of this small current. The electron capture detector is specifically advantageous for the detection and determination of halogen-compounds and hence, mainly used in the analysis of pesticide residues in water and soil.

3.3.6.4 PHOTOIONIZATION DETECTOR (PID) [7]

The eluted solutes from a GC column are bombarded by high-energy photons in the UV range to be broken into positively charged ions. UV light excites the molecules, resulting in temporary loss of electrons in the molecules and the formation of positively charged ions. The gas becomes electrically charged and the ions produce an electric current, which is the signal output of the detector. The intensity of the current depends on the concentration of the component, the more ions are produced, and the greater the current. The current is amplified and displayed on an ammeter. The photo-ionization of other components of gas stream is negligible that their impact on the aromatics signal is also negligible.

The PID responds to all molecules whose ionization potential is below 1.6 eV including aromatics and olefinic compounds. The PID is nondestructive and ions recombine to the original molecules. It is often used in series with FID. The detectability of aromatics can be in the ppb range. PID consists of a UV lamp of 1.6 eV, with a small volume thermostated flow cell ~100 µL.

3.3.6.5 GAS CHROMATOGRAPHY MASS SPECTROMETRY

Qualitative GC analysis must be supported by the injection of standard components for reference and evaluation of analytical peaks, because all the detectors are seeing the eluted components as electronic signal. Coupling of mass spectrometry, MS, to gas chromatographic technique allows direct identification of the eluted components from the analytical columns. The combined technique offers a sensitive and highly specific method in comparison with other chromatographic detectors. The GC-MS instrumentation is a regular feature of most environmental laboratories, where complex soil and water extracts require identification e.g. pesticide residues. In principle, the mass spectrometer converts analyte molecules into a charged species by ionization. Molecules may be fragmented and ionized and ions are detected according to their mass to charge ratio (m/z). Several systems are available for the ionization and ion analysis. Modern instrumentation are equipped with libraries for a large number of compounds to help the identification of sample components.

3.3.7 SEPARATION TECHNIQUES

The gas chromatographic separation involves the separation of components in a column by the carrier gas. The mixture is fractionated between the carrier gas and the nonvolatile solvent (the stationary phase) supported on the inert solid particles (the solid support). The heavy solvent selectively hinders the components of the sample in accordance with the diffusion coefficient and changes them into separate bands. The bands travel in the stream of the carrier gas and recorded as a function of time by the detector. The elution has some practical advantages:

- it continuously regenerates the column by the mobile gaseous phase;
- the sample components are completely separated and mix only with the carrier gas, and consequently facilitate the accumulation of each component and hence can easily be quantitatively determined;
- the loading time is short.

3.3.8   TEMPERATURE PROGRAMMING

It frequently happens that retained components move very slowly or even do not move. By careful programming of the heating process, the elution time can be reduced significantly. During the analysis of mixtures containing components with various boiling points, some analytical problems arise. With low temperature column the signals of the high boiling components will appear after relatively long time because the vapor pressure of liquids increases with the temperature. In the meantime, the use of high temperature column will result in the elution of the low boiling component rapidly and with relatively very low resolution and hence the separation will be bad. To solve this problem, the column temperature must be increased gradually during the progress of the analysis. This is termed the temperature programming, which will lead to successful separation of components of various polarities or various boiling points or various vapor pressures.

3.3.9   ANALYTICAL CONSIDERATIONS

3.3.9.1   ACCURACY

The degree of accuracy of gas chromatography is significantly changing for the various samples. In routine analysis, the accuracy ranges between 1–5%. When the analyte exists at very low concentrations especially for complex samples or for samples that need some treatment before the analysis, lower accuracy must be acceptable. For the determination of halomethanes in water, results with error levels as high as $\mp25\%$ are acceptable.

3.3.9.2   PRECISION

Many factors contribute to the degree of precision in gas chromatographic analysis like: sampling method, sample preparation and the instrument used. The chromatographic process contributes with not less than 1–5%. The limitations are due to the detector noise, reproducibility of the measured volumes. The use of internal standard is an effective measure to correct for errors in volume measurements.

3.3.9.3   SENSITIVITY

The detector characteristic is the main factor in the evaluation of the sensitivity of gas chromatographic methods, which can be calculated from the slope of the calibration graph. More specifically, the linear range of the detector (the concentration range where the plot is straight line) is effective in efficient quantitative measurements. Thus, flame ionization and thermal conductivity detectors can be used for the analysis of samples with wide concentration ranges under almost the same conditions of separation and analysis. However, the linear range for other detectors like the electron capture is relatively narrow.

3.3.9.4   SELECTIVITY

The chromatographic methods involve separation processes that make them ideal with regard to selectivity. The separation method may be typically designed to allow the elution of various analytes. The selectivity can be improved by using specific detectors like the electron capture detector, which fails to respond for other compounds.

3.3.9.5   QUANTITATIVE CALCULATIONS

For quantitative evaluation of the chromatographic signals, the peak area or the peak height are employed to calculate the concentration. However, the use of peak height is very limited, because the height is inversely proportional to the width of the peak base. Extreme care must be taken to keep stable analysis conditions to reduce the variations of peak heights. Otherwise, the precision and accuracy of the quantitative determinations will be low. Thus, the utilization of peak area is a safer option because the peak area is directly proportional to the quantity of the injected material and hence any changes in column efficiency will not influence the accuracy and precision.

Calibration graphs are constructed by analyzing a series of external samples and plotting the signal as a function of the known concentration, keeping the same sample size for the injected analyzed samples and the standard compounds. The calibration graph ensures getting precise results that are very close to the real values. Unfortunately, best analysts suffer from inaccuracy in repeated measurements of $\mp 5\%$ or worse. Thus, the quantitative work by gas chromatographic analysis requires high precision and the use of standard additions technique.

3.3.9.6   EVALUATION OF METHODS

The careful selection of gas chromatographic conditions including the right column and the detector enable the analyst to estimate the components concentrations of the major components as well as those with trace levels and avoids the need for dilution. With flame ionization and thermal conductivity detectors large sample volumes may be analyzed. In the meantime, other detectors like the electron capture are able to work with very small samples.

Although the sample sizes are in the range of one microliter, there are sufficient amount of the components to be detected by the available detectors. In trace analyzes, the amounts of the materials are in the picogram level. During the determination of halomethanes in water, a sample volume of three microliters of water with about 1.0 parts per billion of chloroform is injected. This is equivalent of 15.0 picogram) assuming a complete extraction of chloroform.

3.3.9.7   DATA PROCESSING

Modern instrumentation allows the handling of the analytical data and performs electronic integration of peaks to estimate the peak areas and printing the retention times and peak heights for quantitative calculations. Computer system introduction added the possibility of saving and recall of the data. Computer systems can control more than one chromatograph as a network, which helps in the management of laboratories. More peak parameters can also be displayed like marking the beginning and end of peaks and plotting calibration graphs.

3.4   LIQUID CHROMATOGRAPHY

3.4.1   INTRODUCTION

Liquid chromatography is a method for the separation of solid or high boiling point liquid components and of relatively high molecular weights from a mixture for the determination of its composition. The method is based on injecting a solution of the sample into a column packed with certain materials (the stationary phase) and passing a liquid (mobile phase) through the column. The components are held onto the stationary phase differently depending on their properties and the analytical conditions. The main factors for the separation are:
- the boiling point or the molecular weight;
- the polarity; and
- the ability of forming hydrogen bonding with the stationary phase.

Consequently, the components will appear at the end of the column after certain time intervals, the retention time, t_R.

3.4.2   SOLVENT SELECTION

The type of the solvent and its characteristics are very important for efficient separation and analysis. Polarity in particular is a main property in this respect. The exchange of the nonpolar solvent with another one of relatively high polarity will result in the separation of polar materials at high efficiency. However, the presence of polar and nonpolar components in the same sample presents a problem in the analysis because each component will need a solvent of different polarity. Modern high performance liquid chromatographic instruments are equipped with a system for the gradient elution of the mobile phase. Two containers are used of two solvents with entirely different polarities like the polar methanol and the nonpolar hexane. The program is performed to draw the two solvents gradually, so that it starts with hexane followed by withdrawal of methanol in increasing amounts to be mixed with the hexane to end up purely with methanol. This procedure helps to drive solvent with increasing polarity to aid the separation of components of various polarities successively.

3.4.3   INSTRUMENTATION

Figure 3.8 shows a schematic diagram of the high performance liquid chromatograph. The high performance liquid chromatography, HPLC, instrument consists of at least one pump for the delivery of the mobile phase into the tightly packed column. The pressure will increase at the injector up to a value of 200 bar depending on the flow rate and the viscosity as well as the particle size of the stationary phase. The pump is designed to ensure constant flow rate and avoid pulsating of the mobile phase whatever the composition of the solvent. The solvent must be freed of any gaseous components like nitrogen, oxygen or carbon dioxide to avoid any disturbance in the separation process as they will affect the compressibility of the solvent and lead to the formation of bubbles. However, oxygen influences the working life of the column and retards the electrochemical of spectrochemical detection process. Gases are removed from the solvents by ultrasonic method or bypassing helium at high flow rate or by diffusion across low diameter polymeric tube permeable for the gases to act as a membrane.

The HPLC instrument is equipped with two compressors driven by a single motor, so that when the first in a state of pushing the second will be in a state of refill. The pump is connected with the mixing chamber. The pump can deliver single solvent (isocratic) or the changing solvent (Gradient elution). The second case necessitates the compensation of the continuous change of the solvent compressibility.

The application of the gradient elution methods may be avoided on working with large number of samples and the usual practice is to employ certain elution solvent with appropriate stable composition. This will reduce the preanalysis time and the equilibration between the two phases after gradient elution and passing considerable amounts of the solvents.

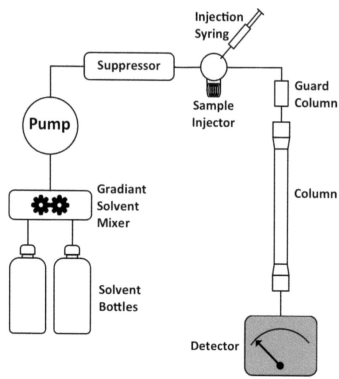

FIGURE 3.8   Schematic diagram of the high performance liquid chromatograph.

3.4.3.1   SAMPLE INTRODUCTION

High pressure is used in HPLC and hence samples cannot be injected in the same manner as in the gas chromatography. The sample instead is introduced using the loop injector as in Fig. 3.9. Loop injectors are currently available with various volumes (0.5–2 µL). When the injector is in the loading state, the loop is isolated from the mobile phase and opened to the ambient atmosphere. A syringe with capacity exceeding the loop volume is employed to introduce the sample solution in the loop. Any excess amount of the sample will be vented out of the loop towards the waste line. After the injection of the sample the injector is turned to the injection state. The sample will move to the stream of the mobile phase.

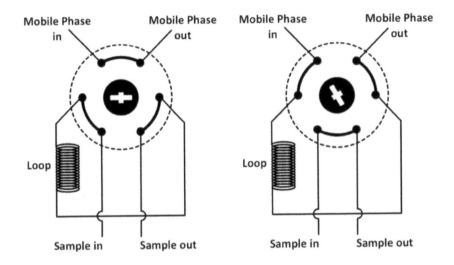

FIGURE 3.9   Loop injector in the loading (a) and the injection positions (b) [3].

3.4.3.2   HPLC COLUMNS

The liquid chromatograph consists of two columns, analytical and guard columns. The guard column is place before the analytical column.

3.4.3.2.1   ANALYTICAL COLUMNS

The most popular analytical column used in HPLC is made of steel with internal diameter of 2.1 to 4.6 mm and 30–300 mm in length. They are packed with porous silica particles (diameter 3–10 μ) and characterized with high efficiency. The packing must be very well packed to avoid any possible channeling, which results in wide chromatographic peaks and hence lowers the separation efficiency.

3.4.3.2.2   GUARD COLUMNS

Liquid chromatographic columns suffer from particles two problems that shorten their working life. The first is the irreversible binding of the mobile phase with the stationary phase, which deteriorates the column by gradual removal of the stationary phase. The second is connected to the particles that may enter the column with the sample and result in blocking of the column. The use of guard column before the analytical column will reduce the effect of these two problems. The guard column contains the same packing material of the analytical column but shorter in length

(~ 7.5 mm) and lower in cost (about one tenth). Obviously they must be replaced periodically as they are the sacrifice for the analytical column.

3.4.4 DETECTORS

The detection in liquid chromatographic analysis is based on some tools that measure a bulk property of the mobile phase and its change when a component reaches the end of the column like the refractive index or density. Other methods are based on measuring a property, which is not related to the mobile phase like the UV absorption, fluorescence and infrared absorption.

3.4.4.1 SPECTROPHOTOMETRIC DETECTION

The absorption of radiation in the UV range can be used to detect the separated components with the aid of a modified spectrophotometer including a flow cell. A deuterium lap is used as source and a monochromator to select the desired wavelength and focus it on the flow cell to detect and measure the quantity of the separated components. A schematic diagram of the HPLC UV detector is shown in Fig. 3.10.

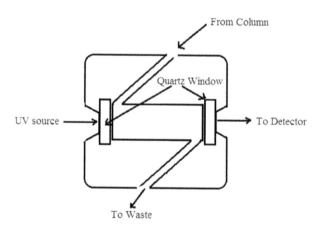

FIGURE 3.10   HPLC ultraviolet detector.

3.4.4.2 REFRACTIVE INDEX DETECTOR

This method is based on the difference in the refractive index between the sample cell and a reference cell in the mobile phase (solvent). It gives a global detection of very large number of materials. This method has relatively low sensitivity (0.01–0.1 mg) when compared with spectrophotometric method and the response is prone to change with any changes in the temperature and flow rates. The method is considered as the best for materials with low spectrophotometric properties like sugars, lipids, organic acids, drug materials and polymers. The detector is equipped with a constant temperature chamber and automatic injection of a reference solvent continuously.

3.4.5 TYPES OF LIQUID CHROMATOGRAPHY

Liquid chromatographic work is carried out in two systems.

3.4.5.1 NORMAL PHASE

The normal phase HPLC involves passing nonpolar solvent into a polar separation column in accordance with adsorption principles. The use of silica as a stationary phase after modification by binding them with functional groups like ammine, cyanide, hydroxide and aided by hexane or dichloromethane as nonpolar solvents (mobile phase). The sample components are soluble in oil. For a homogeneous mixture of three components with increasing polarity: A > B > C. Typical separation can be obtained when a mobile phase of low polarity is employed. The increase in the polarity of the solvent results in the appearance of the components at retention times close to each other.

3.4.4.2 REVERSED PHASE

The reversed phase HPLC employs the principle of repulsion of water from the silica modified by octadecane, octane, or polymer as stationary phase and elution by mixtures of water/methanol or water/acetonitrile or methanol mixture with certain buffer solution as mobile phases. Similarly the polarity of the solvent is important to determine the order of appearance of the three components. The use of high polarity mobile phase will result in typical separation and the order will be: C, B and then A. In the meantime, the use of medium polarity solvent will result in closer peaks. The sample here contains hydrocarbons with various chain lengths. Using the two systems the analyzes may be compared as in Table 3.1.

TABLE 3.1 Comparison of Analyzes with Normal and Reverse Phases

Parameter	Normal phase	Reverse phase
Stationary phase	High polarity	Low polarity
Mobile phase	Low polarity	High polarity
Interaction	Adsorption	Water repulsion
Order of elution	Low to high polarity	Short chain to long chain hydrocarbons

The polarity of solutes increase in the order:
Hydrocarbons < Ethers < Esters < Ketones < Aldehydes < Amines < Alcohols.
The stationary phase is chosen on basis of the resemblance of the polarity to ensure greatest interaction.

3.4.6 CHOICE OF THE MOBILE PHASE

The mobile phase must be carefully selected to ensure the strongest interaction in the HPLC system. High polarity mobile phase will have stronger interaction with high polarity components. The separation will be faster but with low resolution. In addition to the polarity, the elution power of the solvent is also affected by its viscosity. Figure 3.11 shows the elution power of various solvents for normal and reversed phase HPLC conditions. The properties of the solvent can be improved by employing solvent mixtures to control the viscosity and polarity. Water viscosity can be changed significantly by mixing with methanol, ethanol and acetonitrile at various mixing ratios. Although the viscosity of the three solvents are less than that of water, their mixtures with water results in higher viscosity solvent mixtures than water viscosity.

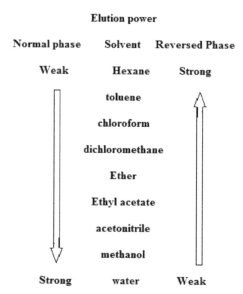

FIGURE 3.11   The elution power of various solvents for different HPLC systems.

3.4.7 EFFECT OF TEMPERATURE

Modern HPLC instruments are equipped with special columns with stabilized temperature to give reliable and reproducible results. High temperatures result in shorter analysis time and some partial overlap of the peaks. The high temperature liquid chromatography (HTLC) reveals interesting chromatographic properties but even now, it misses some theoretical aspects concerning the influence of high temperature on thermodynamic and kinetic aspects of chromatography to improve the method. Guillarme, et al. [8] studied the effect of temperature on solute behavior using various stationary phases and mobile phases. It was shown that the efficiency is improved significantly when the temperature is increased. The temperature control for thermostating columns is a significant source of peak broadening. Three main parameters such as heat transfer, pressure drop and the decrease in solvent viscosity must be considered.

3.4.8 EVALUATION OF THE METHOD

The principles of the evaluation of HPLC are not different from those discussed under GC methods. HPLC involves the injection of relatively larger sample volumes in comparison with GC because of the greater capacity of HPLC columns. A pump is required to drive the mobile phase inside the system. When compared with gas chromatography, HPLC has only a few differences in the scale of operation; accuracy; precision; sensitivity; selectivity; and time, cost, and equipment necessary. Further, the use of sampling loops allows better sample volume injection and hence contributes to the improvement of the precision of HPLC methods. HPLC has wide range of applications as it can deal with solid and liquid samples.

3.5 ION CHROMATOGRAPHY

This chromatographic technique [1] is concerned with the separation of ions and polar compounds. Stationary phases contain ionic sites that create dipolar interactions with the analytes present in the sample. If a compound has a high charge density, it will be retained a longer time by the stationary phase. This exchange process is much slower when compared with those found in other types of chromatography.

This mechanism may be associated, for molecular compounds, with those already dealt with by HPLC when equipped with RP-columns. For HPLC, some columns contain ion exchange packing but they are used in significantly different ways. They are not considered to be ion chromatographic columns. They require concentrated buffers that cannot be suppressed and so are not compatible with conductivity detection. Applications with these columns use more traditional HPLC detection methods (such as UV or fluorescence). The detailed principles and applications of ion chromatography will be discussed in Chapter 8.

3.6 ELECTROPHORESIS

Electrophoresis [3] is another class of separation techniques in which analytes are separated based on their ability to move through a conductive medium, usually an aqueous buffer, in response to an applied electric field. In the absence of other effects, cations migrate toward the electric field's negatively charged cathode, and anions migrate toward the positively charged anode. More highly charged ions and ions of smaller size, which means they have a higher charge-to-size ratio, migrate at a faster rate than larger ions, or ions of lower charge. Neutral species do not experience the electric field and remain stationary.

There are several forms of electrophoresis. In slab gel electrophoresis the conducting buffer is retained within a porous gel of agarose or polyacrylamide. Slabs are formed by pouring the gel between two glass plates separated by spacers. Typical thicknesses are 0.25–1 mm. Gel electrophoresis is an important technique in biochemistry, in which it is frequently used for DNA sequencing. Although it is a powerful tool for the qualitative analysis of complex mixtures, it is less useful for quantitative work.

In capillary electrophoresis the conducting buffer is retained within a capillary tube whose inner diameter is typically 25–75 µm. Samples are injected into one end of the capillary tube. As the sample migrates through the capillary, its components separate and elute from the column at different times. The resulting electropherogram looks similar to the chromatograms obtained in GC or HPLC and provides both qualitative and quantitative information. Only capillary electrophoretic methods receive further consideration in this text.

The principles of capillary zone electrophoresis will be discussed in details in Chapter 4.

3.7 SIZE EXCLUSION CHROMATOGRAPHY

A slightly different process setup is used in Size Exclusion Chromatography (SEC) [9]. In SEC, the matrix consists of porous particles and separation is instead achieved according to size and shape of the molecules. The technique is sometimes also referred to as gel filtration, molecular sieve chromatography or gel-permeation chromatography. The matrices used in SEC are often composed of natural polymers such as agarose or dextran but may also be composed of synthetic polymers such as polyacrylamide. Gels may be formed from these polymers by cross-linking to form a three-dimensional network. Different pore sizes can be obtained by slightly differing amounts of cross-linking. The degree of cross linking will define the pore size. Many gels are now commercially available in a broad range of porosities. In contrast to other types of media the selectivity of a SEC matrix is not adjustable by changing the composition of the mobile phase. Optimally there is no adsorption involved, and the mobile phase should be considered as a carrier phase and not one, which has a large effect on the chromatography. However, the sample may require a buffer solution with a well defined pH and ionic composition chosen to preserve the structure and biological activity of the substances of interest. However, the applications of SEC are mostly for the separation of large molecules like proteins. The environmental applications are limited and thus, we shall not discuss it in details.

KEYWORDS

- **Detectors**
- **Gas Chromatography**
- **Gradient Elution**
- **Ion Chromatography**
- **Liquid Chromatography**
- **Separation Methods**
- **Size Exclusion Methods,**
- **Temperature Programming**

REFERENCES

1. Rouessac, F., & Rouessac, A. (2007). Chemical Analysis-Modern Instrumentation, 2nd Edn, West Sussex, cs-books@wiley.co.uk, John Wiley & sons
2. Braithwaite, A., & Smith, F. J. (1999). Chromatographic Methods, 5th Edn, Cluwer Academic Publishing, Dordrecht.
3. Harvey, D. (2000). Modern Analytical Chemistry, McGrawhill, Boston.
4. Skoog, D. A., James Holler, F., & Crouch, S. R. (2006). Principles of Instrumental Analysis, 6th Edition, Cengage Learning, New York.
5. Martin, A. J. P., & Synge, R. L. M. **(1941).** A New Form of Chromatography employing Two Liquid Phases, *Biochem J., 35*, 1358–1368.
6. http://en.wikipedia.org/wiki/Van Deemter Equation, VanDeemter, J. J., Zuiderweg, F. J., & Klinkenberg, A. (1956). Longitudinal Diffusion and Resistance to Mass Transfer as Causes of Non Ideality in Chromatography, Chem. Eng. Sci., 5, 271–289.doi:10.1016/0009–2509(56)80003–1.
7. Smith, P. A., Jackson Lepage, C., Harrer, K. L., &. Brochu, P. J. (2007). Handheld Photo Ionization Instruments for Quantitative Detection of Sarin Vapor and For Rapid Qualitative Screening of Contaminated Objects, *J. Occ. Environ. Hyg, 4*, 729–738.
8. Guillarme, D., Heinisch, S., & Rocca, J. L. (2004). Effect of Temperature in Reversed Phase Liquid Chromatography, *J. Chromatog. A, 1052(1–2)*, 39–51.
9. Hedhammar, M., Eriksson Karlström, A., & Hober, S. (2013). Chromatographic Methods for Protein Purification, Royal Institute of Technology, AlbaNova University Center, Deptartment of Biotechnology, SE-106 *91* Stockholm, Sweden.

CHAPTER 4

FUNDAMENTALS OF ELECTROPHORETIC ANALYSIS

OSCAR NÚÑEZ

Department of Analytical Chemistry, University of Barcelona.
Martí i Franquès, 1-11, 08028, Barcelona, Spain; E-mail: oscar.nunez@ub.edu

CONTENTS

4.1   INTRODUCTION

Capillary electrophoresis (CE) is a family of related separation techniques in which an electric field is used to achieve the separation of components in a mixture. Electrophoresis in a capillary is differentiated from other forms of electrophoresis in that it is carried out within the confines of narrow-bore capillaries, from 20 to 200 μm inner diameter (i.d.), which are usually filled only with a solution containing electrolytes (typically, although not always necessary, a buffer solution). The use of capillaries has numerous advantages, particularly with respect to the detrimental effects of Joule heating. Capillaries were introduced into electrophoresis as an anti-convective and heat controlling innovation. The high electrical resistance of the capillary enables the application of very high electrical fields (100 to 500 V/cm) with only minimal heat generation. In wide tubes thermal gradients caused band mixing and loss of resolution, however, with narrow-bore capillaries the large surface area-to-volume ratio efficiently dissipates the heat that is generated. The introduction in 1981 of 75 μm i.d. capillary tubes by Jorgensen and Lukacs [1, 2] was the beginning of what is today known as modern high-performance CE (HPCE). The use of high electrical fields results also in short analysis times and high efficiency and resolution. In addition the numerous separation modes in CE which offer various separation mechanisms and selectivities, the minimal sample volume requirements (in general few nanoliters, nL), the on-capillary detection, the potential for both qualitative and quantitative analysis, the automation, and the possibility of hyphenation with other techniques such as mass spectrometry (MS), is allowing CE to become one of the premier separation techniques in multiple fields. CE is then becoming very popular to perform high efficiency separations of both large and small molecules. These separations are facilitated by the use of high voltages, which may generate electroosmotic and electrophoretic mobilities of the buffer and the ionic species, respectively, within the capillary.

One of the key features of CE is the simplicity of the instrumentation configuration used. A scheme of the components of CE instrumentation is shown in Fig. 4.1. All that is required is a narrow-bore fused-silica capillary with an optical viewing window, a controllable high voltage power supply, two electrode assemblies, two buffer reservoirs, and a detection system (typically an ultraviolet (UV) detector). The ends of the capillary are placed in the buffer reservoirs that contain, in general, the same buffer solution that is filling the capillary. The reservoirs also contain the electrodes used to make electrical contact between the high voltage power supply and the capillary. After filling the capillary with the buffer solution, the sample can be introduced onto the capillary by replacing the inlet buffer reservoir with a sample reservoir and applying either an electric field or an external pressure. Then capillary inlet end is placed again in a buffer reservoir, the electric field is applied and the separation is carried out. The optical window viewing included in the capillary is aligned with the detector and detection is carried out directly through the capillary close to the outlet capillary end. Several theoretical and practical aspects regarding sample injection, separation, detection and quantitative analysis in CE will be discussed in this chapter.

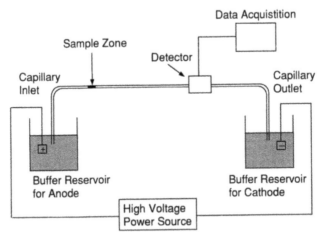

FIGURE 4.1   Schematic diagram of the main components of a typical CE instrument. Image licensed under the Creative Commons Attribution-Share Alike 3.0 Unported license.

CE is offering quite a novel format with characteristics that resembles a cross between traditional polyacrylamide gel electrophoresis and modern high performance liquid chromatography (HPLC). Among the most important characteristics we can find:

- electrophoretic separations are performed in narrow-bore fused silica capillaries;
- utilization of very high electric voltages (10 to 30 kV) generating high electric field strengths, often higher than 500 V/cm;
- the high resistance of the capillary limits the current generated and the internal heating;
- high efficiency (theoretical plates $N > 10^5$ to 10^6) on the order of capillary gas chromatography or even greater, with short analysis times;
- relatively small sample requirement (1 to 50 nL injected);
- easy automation for precise quantitative analysis and very easy to use;
- limited consumption of reagents and, in general, operates in aqueous media;
- presents numerous modes of operation to vary selectivity and is applicable to a wider selection of analytes compared to other analytical separation techniques;
- simple method development and automated instrumentation.

Because of all these characteristics CE is today one of the most promising separation techniques and it is being used in multiple application fields, such as in bio-analysis [3–6], food control and safety [7, 8], and even environmental applications [6, 9] where low analyte concentrations are expected. However, one of CE handicaps is sensitivity due to the short path length (capillary inner diameter) when UV-detection is used. For these reasons, many CE applications will require of off-line and/or on-line preconcentration methods in order to increase sensitivity. In this chapter, the fundamentals of some on-column electrophoretic-based preconcentration methods will be discussed.

4.2   FUNDAMENTALS OF CAPILLARY ZONE ELECTROPHORESIS (CZE)

4.2.1   ELECTROPHORETIC MOBILITY

It is well known that charged species in solution will move under the effect of an electric field seeking the electrode with opposite charge. This is the foundation of CZE separations, which is based on the different velocity experimented by charged analytes through the capillary under the application of an electric field. The velocity of a given ion (v) can be represented by

$$v = \mu_e E \tag{1}$$

where μ_e is the electrophoretic mobility and E the applied electric field.

The electric field generated will depend on the applied voltage (V, in volts) through the capillary and the total length (L, in cm) of the capillary according to the next equation:

$$E = \frac{V}{L} \tag{2}$$

The electrophoretic mobility (μ_e) is a constant value characteristic of the ion in a given medium, and can be represented by

$$\mu_e = \frac{q}{6\pi\eta r} \tag{3}$$

where q is the ion charge, r is the ion radius, and η is the viscosity of the solution (background electrolyte, BGE). Small and highly charged molecules will have higher electrophoretic mobilities than large and minimally charged species.

In addition to the capillary voltage and the capillary length that directly influence the ion velocity, other physical parameters can also play an important role in the electrophoretic separation by indirectly modifying the electrophoretic mobility of a given ion, and the analysts can play an important role in creating a medium that exploits these differences between the molecules in order to achieve separation. The BGE pH is an example and will play an important role on the separation of molecules with acid-base properties. Other parameters, such as the temperature at which the separation will be carried out can also affect the electrophoretic mobility of a given ion as it will directly modify the viscosity of the separation medium.

4.2.2 ELECTROOSMOTIC FLOW (EOF)

A fundamental aspect of CZE is the phenomenon known as electroosmosis, electroosmotic flow, electroendoosmotic flow, or simply by EOF, and it is present in any electrophoretic system. EOF is the bulk flow of liquid in the capillary as a consequence of the surface charge of the capillary internal wall. EOF results from the effect of the applied electric field on the solution double-layer generated at the capillary wall (Fig. 4.2).

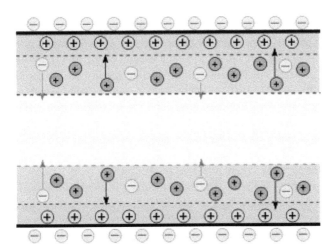

FIGURE 4.2 Diagram of the electric double layer generated within the capillary in CZE. Image modified from the one created by guillaumepaumier.com and multilicense with GFDL and Creative Commons CC-BY-SA-3.0.

In general, fused silica capillaries are frequently used in CZE as well as other CE techniques. These capillaries present ionizable silanol groups in contact with the background electrolyte (BGE) solution within the capillary. The isoelectric point of fused silica capillaries is difficult to determine although it is considered to be close to 1.5, so its degree of ionization will be mainly controlled by the pH of the BGE solution. EOF will become significant at pH values above 4.0. It should be pointed out that because the surface to volume ratio is very high inside a capillary, EOF will become a significant factor in CE.

The velocity of the EOF through a capillary is given by the Smoluchowski [10] equation:

$$v_{EOF} = \frac{\varepsilon \zeta}{4 \pi \eta} E = \mu_{EOF} E \qquad (4)$$

where ε is the dielectric constant of the BGE solution, ζ is the zeta potential (in Volts) of the capillary wall, η is the viscosity of the BGE solution (in Poises), and E is the applied electric field (Volts/cm). μ_{EOF} is then the mobility of the electro-osmotic flow.

The negatively charged wall will attract positively charged ions that are hydrated from the BGE solution up near the surface to maintain charge balance, creating an electrical double layer (Fig. 4.2), and consequently a potential difference very close to the wall (zeta potential). When a voltage is applied across the capillary, these hydrated cations forming the diffuse double-layer migrate toward the cathode, pulling water along and creating a pumping action. The result is a bulk flow of BGE solution through the capillary towards the cathode (EOF).

The zeta potential increases with the charge density in the capillary wall. Consequently, the higher the zeta potential the higher will be the EOF [11]. As previously commented, in fused silica capillaries the charge density on the capillary surface will change with the BGE pH. At high pH values, where silanol groups are predominantly deprotonated, the EOF is significantly greater than at low pH values where they become protonated. Depending on the conditions, it is possible to achieve EOF variations by more than one order of magnitude between pH value of 2 and 12. Although it can be difficult to completely suppress EOF in fused silica capillaries, it can be considered close to zero at pH values bellow 2.

The zeta potential is also dependent on the ionic strength of the BGE solution, as it is described by the double-layer theory. EOF decreases with the square root of the concentration of the electrolyte, so increasing ionic strength results in a double-layer compression, a decrease in zeta potential and a reduction in EOF velocity.

EOF direction is always toward the electrode that has the same charge as the capillary wall. For this reason, when using fused-silica capillary a cathodic EOF is generated. But other type of positively charged or even noncharged capillaries can be used. Additionally, the surface of fused-silica capillaries can be chemically modified in order to suppress or reverse EOF direction, for instance by using surfactants [12] (some properties of surfactants will be explained in Section 4.3.2). Depending on the surfactant charge, EOF can be increased, reduced, or reversed [13]. For instance, the addition of cationic surfactants such as cetyltrimethylammonium bromide (CTAB) to the BGE solution can reverse EOF in fused-silica capillaries [14]. CTAB monomers adhere to the capillary wall through ionic interactions. A capillary wall positively charged results then from hydrophobic interaction of free CTAB molecules with those bound to the wall. This procedure was described for instance for the analysis of quaternary ammonium herbicides in water samples by sample stacking with matrix removal-capillary zone electrophoresis [15].

One of the most important characteristics of EOF in CE techniques is the fact that in the narrow confines of the capillary the velocity of the BGE solution is nearly uniform across the internal diameter of the capillary resulting in a flat profile of the flow [1, 2], as can be seen in Fig. 4.3b. Since the driving force of the flow is uniformly distributed along the capillary there is no pressure drop within the capillary, and the flow is nearly uniform throughout. This flat profile is very beneficial since it does not directly contribute to the dispersion of the solute zones. This is in contrast to the laminar flow generated by pumped systems such as in HPLC techniques (Fig. 4.3a), which creates a velocity profile across the diameter of the tube with an important flow rate drop at the wall, that will be traduced in a higher peak broadening due to the dispersion of the solute zones.

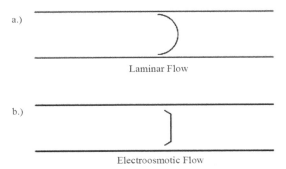

a.)

Laminar Flow

b.)

Electroosmotic Flow

FIGURE 4.3 (a) Laminar flow generated in pumped systems such as HPLC. (b) Flat flow generated by the EOF in capillary zone electrophoresis. Figure released into the public domain by its author, Apblum at the Wikipedia project.

4.2.3 MIGRATION TIME AND APPARENT ELECTROPHORETIC MOBILITY

Regarding terminology, there are significant differences between nomenclature in chromatography and CE, and one characteristic example is the retention time in chromatography. In CE, the analogous parameter will be related to the time required for an analyte to migrate from the injection point to the detection point through the capillary, and this time is known as migration time (t_m), which can be calculated by the quotient between the migration distance (effective length of the capillary, from injection point to detection point, l) and the velocity of the ion:

$$t_m = \frac{l}{v}$$

(5)

As previously indicated in Section 2.1 the velocity of a given ion under the influence of an electric field (E) can be represented by:

$$v = \mu_{ap} E \qquad (6)$$

where μ_{ap} is the apparent electrophoretic mobility (that is the real electrophoretic mobility of the ion), which will take into consideration both the electrophoretic mobility μ_e and the electroosmotic flow:

$$\mu_{ap} = \mu_e + \mu_{EOF} \qquad (7)$$

From these equations, the apparent electrophoretic mobility can also be expressed as:

$$\mu_{ap} = \frac{l}{t_m E} = \frac{lL}{t_m V} \qquad (8)$$

where V is the applied voltage and L is the total capillary length.

It is important to see that there are two different capillary lengths, the effective length (l) and the total capillary length (L), and both must be controlled since the migration time and mobility are defined by the effective length, whereas the electric field is a function of the total capillary length. In general, when on-capillary detection is carried out in CE, the effective length is typically 5 to 10 cm (depending on the instrument configuration) shorter than the total length. In contrast, when off-column detection if performed, such as in the case of using mass spectrometry detection, the two lengths are equivalent.

So, the combination of both electrophoretic mobility and EOF will determine the apparent mobility for a given ion and, consequently, the migration time and the migration order, as represented in Fig. 4.4.

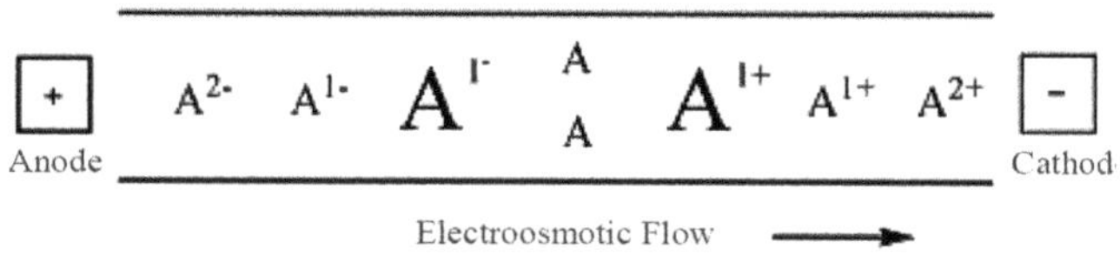

FIGURE 4.4   Schematic representation of migration order in capillary zone electrophoresis when EOF is present. Image licensed under the Creative Commons Attribution-Share Alike 3.0 Unported license.

As can be seen, when working under a cathodic voltage (cathode in the outlet position) as in Fig. 4.4, the cations will migrate first from the capillary with an apparent mobility $\mu_{ap} = \mu_e + \mu_{EOF}$ with both factors contributing in the cathodic direction. The cation migration order will then depend on their specific electrophoretic mobility, and as previously described, this will be higher for lower ion size and higher ion charge. Then, neutral species will migrate from the capillary but they will not be separated because all of them will be moving at the EOF velocity ($\mu_{ap} = \mu_{EOF}$, for any noncharged molecule). Finally, the anions will migrate from the capillary with an apparent mobility $\mu_{ap} = \mu_{EOF} - \mu_e$ (their electrophoretic mobility will be in the opposite direction than EOF). But the ions will only be detected under a cathodic separation such as the one described in Fig. 4.4 if $\mu_{EOF} > \mu_e$ (mode known as counter electroosmotic flow separation). Again, the migration order of these anions will depend on the magnitude of their electrophoretic mobility. In this case anions with lower ion charge and higher ion size will migrate first from the capillary because their electrophoretic mobility will be opposing in a lower magnitude to the EOF than anions with higher ion charge and lower ion size.

A good selection of the voltage configuration (cathodic or anodic separation) and the magnitude of the EOF velocity (by changing BGE composition–buffer type, concentration, pH– and separation temperature) will enable the analyst to achieve good electrophoretic separations under capillary zone electrophoresis.

4.2.4   PARAMETERS AFFECTING EOF IN CZE

EOF is usually beneficial to achieve good electrophoretic separations, but it needs to be controlled for good CZE performance. The magnitude of EOF can most easily be modified by changing the electric field, as previously described. However, fundamentally, the control of EOF requires the alteration of the capillary surface charge or the BGE viscosity.

EOF results as a consequence of the surface charge of the capillary internal wall and the double-layer generated at the capillary wall when an electric field is applied. As previously commented, EOF magnitude will depend on the number of silanol groups ionized in contact with the background electrolyte. So, the most convenient and useful method of controlling EOF in CZE is by working with buffer solutions (controlled pH) as BGEs in order to change the capillary surface charge. At high pH values, however, the EOF is very fast which can result in the elution of the analytes from the capillary before the separation is achieved. In contrast, when working at low or even moderate pH, the negatively charged wall can cause adsorption of cationic analytes through columbic interactions, which becomes very problematic, for instance, for the separation of basic proteins. Additionally, when modifying pH it must be taken into account that this parameter will also affect analytes with acid-base properties and, consequently, the analyte total charge and its electrophoretic mobility. Knowledge of the analyte pI, and pKa values, is often useful in selecting the appropriate pH range of the BGE.

A wide range of buffers has been employed in CZE. The most commonly used are phosphate, borate, citrate, acetate, and Tris (trishydroxymethylamino methane) (TRIS). However, the selection of a buffer for CZE should be based on several factors [16]: the pH value desired (the effective buffering range for any buffer is defined as within one pH unit of the buffer pKa); the operating temperature (buffer molecules exhibit a temperature coefficient, so the pH of a buffered system will change with temperature); the charge of the buffer relative to the analytes and the capillary wall (interaction between the buffer and the analytes can change the effective charge on the analyte molecules, changing their migration velocity. Ion pairing can also occur between the buffer molecules and the analytes affecting the separation); and the effects on detection (buffers differ widely in their absorbance spectra).

The second most convenient way of modifying EOF is by changing BGE viscosity, which can be easily achieved by adding organic modifiers. Usually, the addition of organic modifiers changes zeta potential and it results in a decrease of EOF. However, the effect of the organic modifier in the electrophoretic separation of a given family of compounds can be very complex, as it may affect also the analytes and alter the selectivity, so it is frequently studied experimentally. For instance, Janini et al. [17] studied the effect of organic modifiers on mobility and selectivity using test analytes having different functionalities. The results showed that the EOF was mainly influenced by buffer viscosity, and the importance of the dielectric constant, zeta potential and modifier-capillary wall interactions, which was significant at trace levels of organic modifier, was very much diminished as the organic content increased.

EOF can also be influenced by changing the ionic strength of the BGE. This can be achieved by modifying the concentration of the buffer used in the BGE or by adding additional electrolytes such as, for instance, sodium chloride. In general, the increase of BGE ionic strength produces a decrease on zeta potential and, consequently, on EOF. Although modifying BGE ionic strength is also very common to control EOF, it must be taken into account that capillary current will increase with the ionic strength and, consequently, the Joule heating effect, so capillary temperature control becomes mandatory. In contrast, a low ionic strength could also be problematic because of sample adsorption. Another effect that must be considered when modifying BGE ionic strength is the possibility of peak shape distortions if important differences between BGE and sample zone conductivities are present. BGE ionic strength will also play an important role when on-line sample preconcentration methods based on sample stacking are used, but this will be commented later.

Finally, EOF can also be controlled by the modification of the capillary wall by means of increasing, decreasing or reversing the capillary surface charge. This can be achieved by using dynamic or covalent coatings. Dynamic coatings imply the use of surfactants that can adsorb to the capillary wall via hydrophobic and/or ionic interactions. In general, anionic surfactants can increase EOF while cationic surfactants will decrease or reverse it. However, it must be considered that the use of surfactants may also alter selectivity. Covalent coatings are usually obtained by a chemical bonding to the capillary wall, for instance by chemical polymerization.

There are multiple possibilities of wall modification depending on the hydrophilicity and the charge of the coating used, but the coating stability is often a handicap.

4.2.5 SAMPLE INTRODUCTION INTO THE CAPILLARY

An important operational aspect in CZE is the introduction of the sample into the capillary. In general, in any CE mode of operation only small volumes of the sample are loaded in order to achieve high efficient separations. But due to the small volumes of the capillaries used in CZE, the injection plug length is a more critical parameter than the sample volume in order to prevent sample overloading. The sample overloading can have two significant effects, both of them detrimental to resolution [18]. Injection lengths longer than the diffusion controlled zone width will proportionally broaden peak widths. Additionally, it can result in distorted peak shapes caused by mismatched conductivity between the BGE and the sample zone. As a rule, sample plug lengths lower than 1 to 2% of the total length of the capillary are used, which means an injection length of a few millimeters (sample volumes of 1 to 50 nL), depending on the capillary length and inner diameter. This is one of the advantages of CE modes because small sample volumes are required since 5 µL of sample will be enough to perform several injections. However, these small sample volumes introduced into the capillary will also become a handicap when developing CE methodologies because of the decrease in sensitivity (the effect of a small sample volume together with the short inner capillary diameters used as UV-path length), especially for low concentration samples. So, for numerous CE applications off- and on-column preconcentration methods will be required, especially in the environmental field. Some examples will be discussed later.

Sample introduction into the capillary tubes can be accomplished by two main methods, known as hydrodynamic and electrokinetic injection.

4.2.5.1 HYDRODYNAMIC INJECTION

Hydrodynamic injection is the most widely used method to accomplish the introduction of a small sample volume into the capillary. Three basic strategies are available for that purpose: (i) the application of a positive pressure at the injection end of the capillary (inlet vial); (ii) the application of a vacuum at the exit end of the capillary (outlet vial); and (iii) by gravity (or siphoning action) obtained by inserting the inlet end of the capillary into the sample vial and raising the vial and capillary relative to the outlet end. This last strategy is also known as hydrostatic injection. With hydrodynamic injection, the quantity of sample loaded into the capillary is nearly independent of the sample matrix.

(i) The application of a positive pressure is the most frequently used method in hydrodynamic injection. Positive pressures up to 100 psi (7 bar) can be used for filling and rinsing the capillaries, while injection pressures ranging from 25 to 100 mbar are employed for sample loading. This pressure is delivered either from a source of compressed gas, such as nitrogen, or from an on-board air pump that applies pressure to the headspace of a vial. The sample volume loaded is generally not a known quantity in CE, but it can be calculated. It will be a function of the capillary dimensions, the viscosity of the BGE, the applied pressure, and the injection time (usually, from 0.5 to 5 s). This volume can be calculated using the Hagen- Poiseuille equation:

$$Sample\ volume = \frac{\Delta P d^4 \pi t}{128 \eta L} \qquad (9)$$

where ΔP is the pressure difference across the capillary, d is the capillary inner diameter, t is the injection time (time that the positive pressure is applied), η is the BGE viscosity, and L is the total capillary length. Entering the Hagen-Poiseuille equation into a spreadsheet program simplifies fluid delivery calculations such as these. A free-available computer program called "CE Expert" from Beckman Coulter Inc., can easily help to perform these calculations [19].

(ii) Vacuum delivery is limited to 10 psi or less, and is not used as frequently as positive pressure, but can be useful for drawing fluid from containers that cannot be made pressure tight.

(iii) For gravity or siphoning injection, the pressure differential ΔP used in the previous equation is given by:

$$\Delta P = \rho g \Delta h \qquad (10)$$

where ρ is the BGE density, g is the gravitational constant, and Δh is the height differential of the reservoirs. A typical siphoning injection is obtained by raising the sample reservoir 5 to 10 cm relative to the exit reservoir for 10 to 30 sec. But reproducible injections by gravity require that the vial be raised to the same height for the same duration of time at each injection. Generally, it is only used in CE systems without pressure injection operational capabilities.

Generally, if sensitivity is not a problem, the smallest injection lengths possible should be used. However, injection reproducibility is usually diminished with short injection times due to instrumental limitations. Reproducibility can be improved significantly by using integrated pressure/time profile with active feedback control to compensate for system rise-time effects and variations in the applied pressure. In well-engineered CE systems a 3 s injection at 1 psi and a 10 s injection at 0.3 psi should give identical results, because both are 3 psi-second injections. In practice, the longer time/lower pressure injection usually gives better performance because it allows a longer time for the system to respond to variances.

4.2.5.2   ELECTROKINETIC INJECTION

Electrokinetic injection, also known as electrophoretic or electromigration injection, does not conform to the Hagen-Poiseuille equation. In this method of sample introduction the inlet of the capillary is inserted into the sample vial and the outlet into a BGE vial, and a capillary voltage is applied. The sample is drawn into the capillary through a combination of both electrophoretic migration and the pumping action of the EOF. Usually field strengths 3 to 5 times lower than the one used for separation are applied. A unique property of electrokinetic injection is that the quantity of analyte loaded is dependent on the electrophoretic mobility of the individual compounds, so discrimination occurs for ionic species since compounds that migrate more rapidly in the electrical field will be overrepresented in the sample introduced into the capillary compared to slower moving components [20].

The quantity of a given analyte injected, Q, can be calculated by:

$$Q = \frac{\mu_{ap} V \pi r^2 C t}{L} \tag{11}$$

where map is the apparent electrophoretic mobility of the analyte (see Section 4.2.3), V is the capillary voltage applied, r is the capillary radius, C is the analyte concentration in the sample, t is the injection time, and L is the capillary total length. As described by this equation, sample loading is dependent on the EOF, the sample concentration, and sample mobility. Variations in conductivity, which can be due to matrix effects, such as the presence of ions such as sodium or chloride, could result in differences in the voltage drop and the quantity loaded [21]. Because of these phenomena, electrokinetic injection is generally not as reproducible as hydrodynamic injection. To maximize the quantity injected, the sample should have a considerably lower ionic strength than the BGE. Additionally, subsequent injections will show reduced peak areas because each injection delivers salts from the capillary BGE into the sample vial, raising the ionic strength of the sample. This effect can be minimized (and the injected quantity increased) by preinjecting a water plug immediately prior to the sample injection, but this aspect will be discussed later.

Despite quantitative limitations, electrokinetic injection is very simple, requires no additional instrumentation, and is advantageous when viscous media, or gels, are employed in the capillary and when hydrodynamic injection is ineffective, for instance in order to increase sensitivity although it will always depend on the analyte.

4.2.6 DETECTION IN CE

In order to obtain useful information from CE as separation technique it is necessary to detect and measure the analytes. Most CE detection is carried-out on-capillary, which means that a section of the capillary is linked to the detection device and the capillary itself is the detection cell. It is also possible to couple CE systems to detectors that are outside of the separation capillary or other systems such as mass spectrometry (CE-MS) but this will require specialized interfaces and will not be addressed in this chapter. On-capillary detection in CE is a significant challenge as a result of the small dimensions of the capillary. For this reason, although CE requires small amounts of sample (few μL), it is not considered a "trace" analysis technique since relatively concentrated analyte solutions or the application of preconcentration methods are often necessary. From this point of view it could seem that CE is not a suitable separation technique in environmental analysis where complex matrices with compounds at very low concentration levels need to be analyzed. However, nowadays a great variety of on-column electrophoretic-based preconcentration methods applicable to any CE instrument without any modification are available, allowing CE to be considered as a good option and alternative to other separation techniques such as HPLC even in the environmental field.

Absorbance detectors are the most commonly used in CE instrument systems. They rely on the absorbance of light energy by the analytes. This absorbance creates a shadow as the analytes pass between the light source and the light detector. The intensity of the shadow is proportional to the amount of analyte present. For absorptive detectors the absorbance of an analyte is dependent on the path length, b, the concentration of the analyte, C, and the analyte molar absorptivity, ε, as definite by Beer's law:

$$A = bC\varepsilon \tag{12}$$

However, detection through the capillary is complicated by the curvature of the capillary itself [16]. The capillary and the fluid it contains make up a complex cylindrical lens. The curvature of this lens must be accounted for in order to gather the maximum amount of light and thereby maximize the signal-to-noise ratio. Generally, the effective length of the light path through the capillary is actually about 63.5% the stated capillary i.d. Thus, a 50 μm i.d. capillary has an effective path length of only 32 μm. Taking into account that typical HPLC detectors have light paths in the 5–10 mm range, the absorbance signal obtained in CE systems is very small, and a peak with an absorbance of 0.002 AU is a significant peak.

Detector design is then critical due to the short optical path length. The optical beam should be tightly focused directly into the capillary to obtain maximum throughput at the slit and to minimize stray light reaching the detector. These aspects are very important to both sensitivity and linear detection range.

Photodiode array (PDA) detection is an alternative to single or multiple wavelength detection. This detector consists on an achromatic lens system to focus the entire spectrum of light available from the source lamp into the capillary window. The light passing the capillary is diffracted into a spectrum that is projected on a linear array of photodiodes. An array consists of numerous diodes each of which is dedicated to measuring a narrow-band spectrum. In this manner it is possible to record the entire absorbance spectrum of analytes as they pass by the detection window. One of the advantages of using this kind of detector is that allows confirming the identity of analytes by using the spectral signature. Moreover, by comparing the change in spectral signature across a peak it is also possible to estimate peak purity.

Several novel approaches have been proposed in order to increase the path length (and the sensitivity) in CE detection. Special capillary designs can be used to extend the optical pathway without increasing the overall capillary area. One such design is the "bubble cell" configuration [22], which offers a unique method to extend the pathway with nearly no degradation of separation efficiency and resolution. It is made by forming an expanded region, a bubble, directly within the capillary. Since the bubble is located only in the detection region no increase in current occurs. In the region of the bubble the electrical resistance is reduced and thus the field is decreased. At the same time, a proportional decrease in flow velocity due to the expanded volume of the bubble occurs. When the analyte zone front enters the bubble its velocity decreases and the zone concentrates. As the sample zone expands radially (across the capillary) to fill the increased volume, it contracts longitudinally (along the capillary). Thus the sample concentration remains constant but the path length increase. The use of "bubble cell" capillaries have been described in environmental CZE applications [23, 24]. For instance, Turnes et al. [24] described a method for the monitoring of pentachlorophenol (PCP) in water samples by capillary zone electrophoresis. By means of sample stacking techniques (that will be addressed later) and the use of extended light path capillaries, the determination of PCP at the ng/L range required by environmental and toxicological international regulations was possible.

4.3   MODES OF OPERATION IN CE

A CE system can be operated in several different modes and the versatility of this technique derives from its numerous modes of operation. The separation mechanisms of each mode are different and thus can offer orthogonal and complementary information when approaching an analytical problem. The choice of mode will be based on the analytical problem under consideration. This section will describe briefly the major modes of capillary electrophoretic separations that are currently in use. The most common methods of operation in CE are capillary zone electrophoresis (CZE), micellar electrokinetic capillary chromatography (MECC), capillary gel electrophoresis (CGE), capillary isoelectric focusing (CIEF), capillary isotachophoresis (CITP), and capillary electrochromatography (CEC) [16].

4.3.1 CAPILLARY ZONE ELECTROPHORESIS (CZE)

Capillary zone electrophoresis, also known as free solution capillary electrophoresis, is the most widely used mode in CE due to its simplicity of operation and its versatility. In CZE the capillary is only filled with a buffer solution, and the separation mechanism is based on differences in the charge-to-mass ratio of the analytes. CZE is characterized for the homogeneity of the buffer solution and the constant field strength generated throughout the length of the capillary. After sample injection and application of capillary voltage, the analytes of a sample mixture separate into discrete zones and at different velocities. Separation of both anionic and cationic solutes is possible by CZE due to the EOF and the differences in the charge-to-mass ratio of the analytes, as explained in Section 4.2.3 (Fig. 4.4). In contrast, neutral solutes do not migrate and all of them will comigrate with the EOF.

4.3.2 MICELLAR ELECTROKINETIC CAPILLARY CHROMATOGRAPHY (MECC)

Micellar electrokinetic capillary chromatography (MECC or MEKC) is maybe the most intriguing mode of CE for the determination of small molecules and it is considered a hybrid of electrophoresis and chromatography. The use of micelle-forming surfactant solutions can give rise to separations that resemble reversed-phase LC with the benefits of CE techniques. Introduced by Professor Shigeru Terabe in 1984 [25], MECC is today, together with CZE, one of the most widely used CE modes, and its main strength is that it is the only electrophoretic technique that can be used for the separation of neutral analytes as well as charged ones.

A suitable charged or neutral surfactant, such as sodium dodecyl sulfate (SDS), is added to the separation buffer in a concentration sufficiently high to allow the formation of micelles. The surfactants are long chain molecules (10–50 carbon units) that possess a long hydrophobic tail and a hydrophilic head group. When the concentration of surfactants in the buffer solution reaches a certain level, known as critical micelle concentration (8 to 9 mM for SDS, for example), they aggregate into micelles, which are, in the case of normal micelles, arrangements that will have a hydrophobic inner core and a hydrophilic outer surface. These micelle aggregates are formed as a consequence of the hydrophobic effect, that is, they rearrange to reduce the free energy of the system. For this reason micelles are essentially spherical with the hydrophobic tails of the surfactant oriented towards the center to avoid interaction with the hydrophilic buffer, and the charged heads oriented toward the buffer. A representation of a normal micelle is shown in Fig. 4.5a. Micelles are dynamic and constantly form and break apart, constituting a pseudostationary phase in solution within the capillary. It is the interaction between the micelles and the solutes (neutral or charged ones) that causes their separation. For any given analyte, there is a probability that the molecules of that analyte will associate within the micelle at any given time. This probability is the same as the partition coefficient in classical chromatography. For neutral compounds, it will only be partitioning in and out of the micelle that effects the separation. When associated with the micelle, the analyte will migrate at the velocity of the micelle. When not in the micelle, the analyte will migrate with the EOF (if present). Differences in the time that the analytes spend in the micellar phase will determine the separation. For charged compounds, variations in micelle electrophoretic mobility when the analyte is associated with the micelle and the analyte electrophoretic mobility when not associated with the micelle will play an important role, together with their partitioning in and out of the micelle, in the separation. The overall MECC separation process is depicted schematically in Fig. 4.5b.

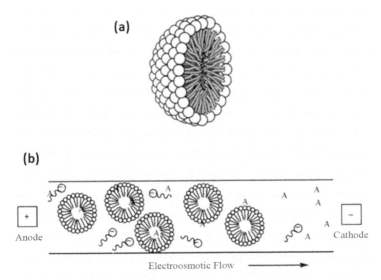

FIGURE 4.5 (a) Representation of a normal micelle containing a hydrophobic core and a hydrophilic outer surface. (b) Schematic of a MECC separation process with partitioning between a solute A and a micelle. These figures were released into the public domain by their authors.

4.3.3 CAPILLARY GEL ELECTROPHORESIS (CGE)

Gel electrophoresis is a classical separation method principally employed in the biological sciences for the size-based separation of macromolecules such as proteins and nucleic acids. The size separation is obtained by electrophoresis of the solutes through a suitable polymer medium such as polyacrylamide or agarose. In CGE the capillary is filled with a gel or viscous solution. EOF is often suppressed so that the migration of the analytes is solely due to electrophoresis. After sample injection, as charged solutes migrate through the polymer network they become hindered, with larger solutes being more hindered than smaller ones, so the separation is effectively based on the molecular size. This CE operation mode is ideal for the analysis and separation of DNA and SDS-saturated proteins that cannot be separated without using a gel since they contain mass-to-charge ratios that do not vary with size. That is, with DNA for example, each additional nucleotide added to a DNA chain adds an equivalent unit of mass and charge and does not affect the mobility in a free solution.

4.3.4 CAPILLARY ISOELECTRIC FOCUSING (CIEF)

Capillary isoelectric focusing (CIEF) is an electrophoretic mode basically used for the separation of peptides and proteins according to their isoelectric point (pI). The isoelectric point is defined as the pH value at which a particular molecule or surface carries no net electrical charge. In the case of amphoteric molecules known as zwitterions (such as peptides and proteins) they contain both positive and negative charges depending on the functional groups. The net charge on the molecule will be affected by the pH of its surrounding environment and can become more positively or negatively charged due to the gain or loss, respectively, of protons. The pI will be then the pH value at which the molecule carries no electrical charge or the negative and positive charges present are equal. So the fundamentals of CIEF are based on the fact that a molecule will migrate as long as it is charged. Should it become neutral, it will stop migrating in the electric field. CIEF is carried out under a pH gradient through the capillary where the pH is low at the anode and high at the cathode, as can be seen in the scheme presented in Fig. 4.6.

The pH gradient is generated with a series of zwitterionic chemicals known as carrier ampholytes. When a voltage is applied through the capillary, the ampholyte mixture will separate according to their charge, those negatively charged migrate towards the anode and those positively charged will migrate towards the cathode. The pH then will be changing through the capillary, decreasing at the anodic section and increasing at the cathodic section. The ampholyte migration will cease when each ampholyte reaches its isoelectric point and is no longer charged (or the net charge is zero).

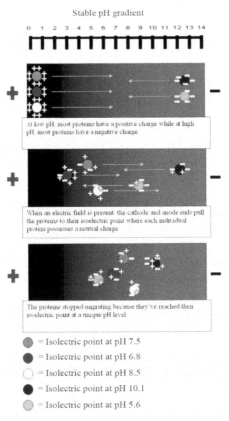

Stable pH gradient

= Isolectric point at pH 7.5

= Isolectric point at pH 6.8

= Isolectric point at pH 8.5

= Isolectric point at pH 10.1

= Isolectric point at pH 5.6

FIGURE 4.6   Representation of a capillary isoelectric focusing separation of five proteins with different isoelectric points. Image licensed under the Creative Commons Attribution-Share Alike 3.0 Unported license.

4.3.5   CAPILLARY ISOTACHOPHORESIS (CITP)

Capillary isotachophoresis is based in the separation within a heterogeneous carrier electrolyte. In this CE mode the sample is introduced between two different buffer solutions. The first one, known as leading electrolyte, has a higher mobility than the sample components. The second one, known as terminating electrolyte, is introduced in the capillary after the sample and has mobility lower than any of the sample components. Separation will occur in the gap between the leading and terminating electrolytes based on the individual mobilities of the analytes. When the voltage is applied the ions in the sample form discrete zones that are not separated into peaks. The concentration of the analyte within the zone will be constant and the length of the zone will be proportional to the concentration within that zone.

4.3.6   CAPILLARY ELECTROCHROMATOGRAPHY (CEC)

Capillary electrochromatography (CEC) is a CE mode considered hybrid between liquid chromatography and electrophoresis. In CEC a chromatographic stationary bed has been introduced into the capillary, and the mobile phase (in fact, the carrier electrolyte) is driven through the capillary by electroosmosis with the application of a capillary voltage. The separation is achieved by electrophoretic migration of solutes and differential partitioning between the stationary phase and the mobile phase. It has the advantage that CEC uses electroosmotic flow to drive the mobile phase, and the resulting plug flow improves the separation efficiency compared to that of a laminar flow obtained with pressure driven systems such as the ones used in HPLC.

4.4   ON-LINE PRECONCENTRATION METHODS

As previously commented, CZE is the most widely used mode in CE due to its simplicity of operation and its versatility, and today is appearing as an alternative to other separation techniques, such as liquid chromatography, in multiple application fields. Despite the low sensitivity characteristic of this technique (see Section 4.2.6), CZE has been proposed for multiple applications, including environmental ones. However, target analyte concentrations in environmental samples are usually very low, so several CZE applications will require of off-line and/or on-line preconcentration methods.

Today the use of on-line preconcentration methods in CZE is increasing, and also in the environmental field, because no special requirement but a CE instrument is necessary for their application. Most of these methods are based in electrophoretic principles and play with the sample injection in order to increase the amount of analyte introduced into the capillary without losing separation efficiency. The number of on-line electrophoretic-based preconcentration methods is huge, so this chapter will focus only on some of them based in the stacking phenomena such as large volume sample stacking (LVSS), field amplified sample injection (FASI), and electrokinetic supercharging (EKS).

4.4.1   LARGE VOLUME SAMPLE STACKING (LVSS)

The mechanism for analyte preconcentration in the electrophoretic-based methods, which can also be called on-column preconcentration methods, is based on the principle of stacking analytes in a narrow band between two separate zones in the capillary where the compounds have different velocities [26]. One of the simplest ways is called sample stacking, which is based on the differences of conductivity between the sample and the BGE. When the predetermined separation voltage is applied, analytes will stack-up in the boundary between the sample zone and the BGE due to their huge decrease in electrophoretic mobility when going from the sample region to the BGE region. Specifically, in large volume sample stacking (LSVV), the analytes with negative charge stack-up at the boundary between the sample zone and the BGE while the large volume of sample previously introduced into the capillary by hydrodynamic injection, is electroosmotically pushed out of the capillary after the application of a reversed voltage, as can be seen in the schematic representation on Fig. 4.7. When almost the entire sample has been removed from the capillary, voltage is switched and separation takes place in counter EOF conditions. A critical point of this method is to know when to switch capillary voltage, moment that is determined by monitoring the capillary current. In general, capillary voltage is changed when capillary current reaches 80–95% of the current observed under normal conditions with the working BGE. It should be pointed out that LVSS is a selective method and only analytes with electrophoretic mobilities lower than EOF and in the opposite direction (anions when working with fused-silica capillaries) can be preconcentrated. However, the application of LVSS for the analysis of cations has also been described by using capillary wall surfactant modifiers such as cetyltrimethylammonium bromide (CTAB) in order to reverse EOF direction in a fused-silica capillary. For instance, for the analysis of quaternary ammonium herbicides in water samples by CZE [15].

LVSS is an in-line preconcentration method able to achieve sensitivity enhancements in the analysis of environmental water samples up to 40-fold when compared with conventional hydrodynamic injection in CZE. Additionally, It should be pointed out that in most cases extended path length capillaries were used [27–30].

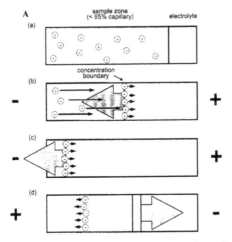

FIGURE 4.7 Schematic representation of LVSS with polarity switching. (a) Sample has a lower conductivity than BGE and fills up to 95% of the capillary. (b) Reverse voltage is applied so the analytes stack on the rear boundary between the sample and the BGE, which moves back towards the inlet due to the EOF. (c) When the optimal current is reach (80–95% of normal), the voltage is reversed. (d) The analytes separate normally in counter EOF mode. Reproduced from Ref. [26] with permission of Wiley-VCH.

4.4.2   FIELD AMPLIFIED SAMPLE INJECTION (FASI)

Among in-line enrichment procedures, field amplified sample injection (FASI) is very popular since it is quite simple only requiring the electrokinetic injection of the sample after the introduction of a short plug of a high-resistivity solvent such as methanol or water. A schematic representation of the application of FASI for the in-line preconcentration of positively charged analytes is shown in Fig. 4.8. Compared to LVSS, this method takes advantage of the higher amount of analytes introduced into the capillary when electrokinetic injections are used.

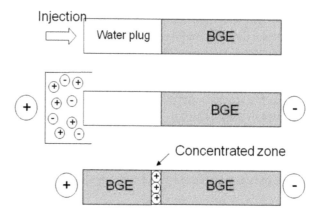

FIGURE 4.8 Schematic representation of a field-amplified sample injection (FASI) procedure for the preconcentration of positively charged analytes.

The preinjection of a short plug of a high-resistivity solvent such as water allows the enhancement of the sample electrokinetic injection because of the conductivity differences between sample and the water plug. Once the analytes enter into the capillary they will stack-up in the boundary region between the high-resistivity solvent and the BGE, and separation will take place.

4.4.3   ELECTROKINETIC SUPERCHARGING (EKS)

As previously commented, the development of new in-line preconcentration strategies to achieve CE sensitive methods able to analyze environmental samples at low concentration levels is needed. A recent in-line preconcentration method for CE that has great potential is electrokinetic supercharging (EKS). This method is the combination of electrokinetic injection under field-amplified conditions (FASI) and transient isotachophoresis (tITP) and was first described for the analysis of rare-earth ions by the group of Professor Hirokawa [31, 32]. EKS was developed to extend the range of FASI and is performed by hydrodynamic injection of a leading electrolyte (L), followed by an electrokinetic injection of the analytes, and finally hydrodynamic injection of a terminating electrolyte (T). Figure 4.9 shows a schematic representation of the steps used in EKS. Upon applying the separation voltage, the diffuse band of analytes introduced during electrokinetic injection is stacked between the leading and the terminating electrolytes by tITP until the ITP stage destacks and the analytes are allowed to separate by conventional CZE. EKS is an exceptionally simple but powerful approach to on-line sample preconcentration and has been shown to improve the sensitivity of analytical response by several orders of magnitude.

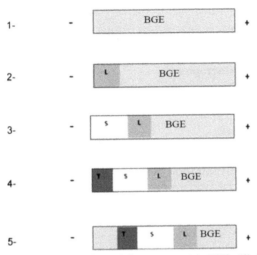

FIGURE 4.9 Schematic representation of the steps used in EKS: (1) filling the capillary with BGE, (2) hydrodynamic injection of leading electrolyte (L), (3) electrokinetic injection of sample (S), (4) hydrodynamic injection of terminating electrolyte (T), and (5) starting tITP-CZE. Reproduced from Ref. [35] with permission of Elsevier.

4.5 SUMMARY AND CONCLUSIONS

Fundamental aspects of capillary zone electrophoresis regarding theoretical principles (electrophoretic mobility and electroosmotic flow), sample introduction (hydrodynamic *vs* electrokinetic injection), separation (including other CE operational modes), and detection have been addressed. CZE is becoming a popular separation technique because of the simplicity of the instrumentation required and its versatility of applications. A good selection of the voltage configuration (cathodic or anodic separation) and the magnitude of the EOF velocity (by changing BGE composition – buffer type, concentration, pH – and separation temperature) will allow analysts to achieve good electrophoretic separations of complex mixtures under CZE.

Multiple CZE applications in a variety of fields (bio-analysis, food safety, environmental analysis) can be found in the literature. In the case of environmental applications where low concentration samples are expected to be found, CZE requires the application of off-line and/or on-line preconcentration methods to improve sensitivity. However, nowadays the low sensitivity characteristic of conventional CZE techniques, where UV-detection is carried-out by using the inner-diameter as optical path length, is not considered a real handicap for many analysts. Many electrophoretic-based in-line preconcentration methods are available, and the fundamentals of some of them based on stacking phenomena, such as LVSS, FASI and EKS, have been presented and discussed.

KEYWORDS

- **Capillary Zone Electrophoresis**
- **Electrochromatography**
- **Gel Electrophoresis**
- **Isoelectric Focusing**
- **Isotachophoresis**
- **Micellar Electrokinetic Capillary Chromatography**
- **On-line Preconcentration Methods**

REFERENCES

1. Jorgenson, J. W., & Lukacs, K. D. (1981). Zone Electrophoresis in Open-Tubular Glass Capillaries: Preliminary Data on Performance, HRC CC, *J. High Resolut. Chromatogr, Chromatogr, Commun, 4*, 230–231.
2. Jorgenson, J. W., & Lukacs, K. D. (1981). Zone Electrophoresis in Open-Tubular Glass Capillaries, Anal Chem., *53*, 1298–1302.
3. Masetto De Gaitani, C., Moraes De Oliveira, A. R., & Bonato, P. S. (2013). Capillary Electro migration Techniques for the Analysis of Drugs and Metabolites in Biological Matrices: A Critical Appraisal. Capillary Electrophor, Microchip Capillary Electrophor, 229–245.
4. Locatelli, M., & Carlucci, G. (2010). Advanced Capillary Electrophoresis Techniques in the Analytical Quantification of Drugs, Metabolites and Biomarkers in Biological Samples, *Global J Anal* Chem. *1*, 244–261.
5. Tseng, H. M., Li, Y., & Barrett, D. A. (2010). Bioanalytical Applications of Capillary Electrophoresis with Laser-Induced Native Fluorescence Detection, Bio-analysis, *2*, 1641–1653.
6. Garcia-Campana, A. M., Gamiz-Gracia, L., Lara, F. J., Iruela, M. D. O., & Cruces-Blanco, C. (2009). Applications of Capillary Electrophoresis to the Determination of Antibiotics in Food and Environmental Samples, Anal Bioanal Chem, *395*, 967–986.
7. Pinero, M. Y., Bauza, R., & Arce, L. (2011). Thirty Years of Capillary Electrophoresis in Food Analysis Laboratories: Potential Applications, Electrophoresis, *32*, 1379–1393.

8. Herrero, M., Garcia-Canas, V., Simo, C., & Cifuentes, A. (2010). Recent Advances in the Application of Capillary Electro migration Methods for Food Analysis and Foodomics, Electrophoresis, *31*, 205–228.

9. Bald, E., Kubalczyk, P., Studzinska, S., Dziubakiewicz, E., & Buszewski, B. (2013). Application of Electromigration Techniques in Environmental Analysis Springer Ser. Chem. Phys., *105*, 335–353.

10. Smoluchowski, M. V. (1905). Elektrosche Kataphorese, Physik Z, *6, 529.*

11. Zhou, M. X., & Foley, J. P. (2006). Quantitative Theory of Electroosmotic Flow in Fused-Silica Capillaries Using an Extended Site-Dissociation Site Binding Model, Anal Chem, *78*, 1849–1858.

12. Melanson, J. E., Baryla, N. E., & Lucy, C. A. (2000). Double-Chained Surfactants for Semipermanent Wall Coatings in Capillary Electrophoresis, Anal Chem., *72*, 4110–4114.

13. Wang, C., & Lucy, C. A. (2004). Mixed Cationic/Anionic Surfactants for Semi-permanent Wall Coatings in Capillary Electrophoresis, Electrophoresis, *25*, 825–832.

14. Beckers, J. L., & Bocek, P. (2002). Multiple Effects of Surfactants Used as Additives in Background Electrolytes in Capillary Zone Electrophoresis: Cetyltrimethylammonium Bromide as Example of Model Surfactant Electrophoresis, *23*, 1947–1952.

15. Núñez, O., Moyano, E., Puignou, L., & Galceran, M. T. (2001). Sample Stacking with Matrix Removal for the Determination of Paraquat, Diquat and Difenzoquat in Water by Capillary Electrophoresis, *J. Chromatogr A, 912*, 353–361.

16. Whatley, H. (2001). Basic Principles and Modes of Capillary Electrophoresis, In Clinical and Forensic Applications of Capillary Electrophoresis; Petersen, J. R., & Mohammad, A. A. Eds, Humana Press Inc.: Totowa, NJ. *21.*

17. Janini, G. M., Chan, K. C., Barnes, J. A., Muschik, G. M., & Issaq, H. J. (1993). Effect of Organic Solvents on Solute Migration and Separation in Capillary Zone Electrophoresis, Chromatographia, *35*, 497–502.

18. Zhu, M., Hansen, D. L., Burd, S., & Gannon, F. (1989). Factors affecting Free Zone Electrophoresis and Isoelectric Focusing in Capillary Electrophoresis, *J. Chromatogr, 480*, 311–319.

19. CE Expert Computer Program, Beckman Coulter Inc. Available at: https://www.beckman-coulter.com/wsrportal/wsr/research-and-discovery/products-and-services/capillary-electro-phoresis/ce-expert-lite/index.htm.

20. Jorgenson, J. W., & Lukacs, K. D. (1983). Capillary Zone Electrophoresis, Science, *222*, 266–272.

21. Huang, X., Gordon, M. J., & Zare, R. N. (1988). Bias in Quantitative Capillary zone Electrophoresis caused by Electrokinetic Sample Injection, Anal Chem, *60*, 375–377.

22. Gordon, G. B. (1991). Capillary Zone Electrophoresis Cell System, United States patent *5,061, 361*, October 29, 1991.

23. Luo, J., Hausler, R., & Waldron, K. C. (2000). Determination of Trace Aluminum in Water and Wastewater by Capillary Electrophoresis, Enviro Anal 2000, Proc. Bienn Int. Conf. Monit Meas. Environ., 3rd, 101–106.

24. Turnes, M. I., Mejuto, M. C., & Cela, R. (1996). Determination of Pentachlorophenol in Water Samples by Capillary Zone Electrophoresis, *J. Chromatogr A., 733*, 395–404.

25. Terabe, S., Otsuka, K., Ichikawa, K., Tsuchiya, A., & Ando, T. (1984). Electrokinetic Separations with Micellar Solutions and Open-Tubular Capillaries, Anal Chem., *56*, 111–113.

26. Breadmore, M. C., & Haddad, P. R. (2001). Approaches to Enhancing the sensitivity of Capillary Electrophoresis Methods for the Determination of Inorganic and Small Organic Anions, Electrophoresis, *22*, 2464–2489.

27. Cugat, M. J., Borrull, F., & Calull, M. (2001). Large Volume Sample Stacking for on-Capillary Sample Enrichment in the Determination of Naphthalene- and Benzenesulfonates in Real Water Samples by Capillary Zone Electrophoresis, Analyst (Cambridge, U.K.) *126*, 1312–1317.

28. Soto-Chinchilla, J. J., Garcia-Campana, A. M., Gamiz-Gracia, L., & Cruces-Blanco, C. (2006). Application of Capillary Zone Electrophoresis with Large-Volume Sample Stacking to the Sensitive Determination of Sulfonamides in Meat and Ground Water, Electrophoresis, *27*, 4060–4068.

29. Quesada-Molina, C., Garcia-Campana, A. M., Olmo-Iruela, L., & Del Olmo, M. (2007). Large Volume Sample Stacking in Capillary Zone Electrophoresis for the Monitoring of the Degradation Products of Metribuzin in Environmental Samples, *J Chromatogr A., 1164*, 320–328.

30. Quesada-Molina, C., Olmo-Iruela, M., & Garcia-Campana, A. M. (2010). Trace Determination of Sulfonylurea Herbicides in Water and Grape Samples by Capillary Zone Electrophoresis Using Large Volume Sample Stacking, Anal. Bioanal Chem., *397*, 2593–2601.

31. Herrera-Herrera, A. V., Ravelo-Perez, L. M., Hernandez-Borges, J., Afonso, M. M., Palenzuela, J. A., & Rodriguez-Delgado, M. A. (2011). Oxidized Multi Walled Carbon Nanotubes for the Dispersive Solid phase Extraction of Quinolone Antibiotics from Water Samples using Capillary Electrophoresis and Large Volume Sample Stacking with Polarity Switching, *J Chromatogr A, 1218*, 5352–5361.

32. Bernad, J. O., Damascelli, A., Nunez, O., & Galceran, M. T. (2011). In-Line Pre-concentration Capillary Zone Electrophoresis for the Analysis of Haloacetic Acids in Water, Electrophoresis, *32*, 2123–2130.

33. Okamoto, H., & & Hirokawa, T. (2003). Application of Electrokinetic Supercharging Capillary Zone Electrophoresis to Rare-Earth Ore Samples, *J. Chromatogr A, 990*, 335–341.

34. Hirokawa, T., Okamoto, H., & Gas, B. (2003). High-Sensitive Capillary Zone Electrophoresis Analysis by Electro kinetic Injection with Transient Isotachophoretic Pre-concentration: Electro kinetic Supercharging, Electrophoresis, *24*, 498–504.

35. Dawod, M., Breadmore, M. C., Guijt, R. M., & Haddad, P. R. (2008). Electrokinetic Supercharging for Online Preconcentration of Seven Non-Steroidal Anti-Inflammatory Drugs in Water samples, *J Chromatogr, A, 1189*, 278–284.

CHAPTER 5

PRINCIPLES OF ELECTROANALYTICAL METHODS

FARNOUSH FARIDBOD, PARVIZ NOROUZI, and
MOHAMMAD REZA GANJALI*

Center of Excellence in Electrochemistry, Faculty of Chemistry, University of Tehran,
Tehran, Iran
*E-mail: Ganjali@khayam.ut.ac.ir

CONTENTS

5.1   ELECTROANALYTICAL METHODS: AN OVERVIEW

Electrochemistry is the science of studying chemical reactions at the interface of an electron conductor called electrode and an ionic conductor called electrolyte. The reactions involve electron transfer between the electrode surface and the electrolyte or a substance in the solution. Charge transfer between species is called oxidation-reduction or redox reaction.

An electrochemical cell is a device in which electrochemical reactions occurs in it. A common electrochemical cell is made of:

1. *Electrode*; is electron conductor device where the electrochemical reactions occur on its surfaces. It is a place for charge transfers. At least, two conductive electrodes (anode and cathode) should be used in an electrochemical cell. Anode is a place for oxidation and cathode for reduction. Any sufficiently conductive materials, such as metals, graphite, semiconductors, and even conductive polymers can be used as an electrode.
2. *Electrolyte*; is an ionic conductor which contains ions that can freely move between electrodes.
3. *A compartment*; is a place where the electrodes and the electrolytes are placed.
4. *External electronic circuit*; which connects the electrodes and detect or control the current, voltage or conductance of the circuit.

There are two kinds of electrochemical cell. If an electrochemical reaction is driven by an externally applied voltage (like the process occurs in electrolysis), it is an electrolytic cell. If a voltage spontaneously creates by a chemical reaction (like the thing occurs in a battery), it is a galvanic cell. In a galvanic cell cathode has positive charge and anode is negative. In contrast, in an electrolytic cell, anode is positive and cathode is negative to attract the species, which are going to oxidize and reduce, respectively.

Electrochemical cell potential is defined as electromotive force (emf). To predict the cell potential, standard electrode potential are used. Standard electrode potentials are stated as reduction potentials referenced to the standard hydrogen electrode (SHE).[1] Sometimes, the reactions are reversible and the role of an electrode in a cell changes depend on the relative oxidation/reduction potential of the other electrode. Hence, the oxidation potential for a particular electrode is just the negative of the reduction potential.

A standard cell potential can be obtained from the standard electrode potentials for both electrodes (called half cell potentials). The one that is smaller will be the anode and will undergo oxidation. The cell potential is then calculated as Eq. (1):

$$E^{\circ}_{cell} = E^{\circ}_{red(cathode)} - E^{\circ}_{red(anode)} = E^{\circ}_{red(cathode)} + E^{\circ}_{ox(anode)} \tag{1}$$

During operation of an electrochemical cell, chemical energy is transformed into electrical energy and is expressed mathematically as the product of the cell emf and the electric charge transferred through the external circuit.

$$Electrical\ energy = E_{cell}C_{trans} \tag{2}$$

where E_{cell} is the cell potential measured in volts (V) and C_{trans} is the cell current integrated over time and measured in coulombs (C); C_{trans} can also be determined by multiplying the total number of electrons transferred (measured in moles), times, and Faraday's constant (F).

The emf of the cell at zero current is the maximum possible emf. It is used to calculate the maximum possible electrical energy that could be obtained from a chemical reaction. This energy is referred to as electrical work and is expressed by the following equation:

$$W_{max} = W_{electrical} = -nFE_{cell} \tag{3}$$

where work is defined as positive into the system. Since the free energy is the maximum amount of work that can be extracted from a system, it can be written:

$$\Delta G = -nFE_{cell} \tag{4}$$

A positive cell potential gives a negative change in Gibbs free energy (like galvanic cells). This is consistent with the cell production of an electric current from the cathode to the anode through the external circuit. If the current is driven in the opposite direction by imposing an external potential, then work is done on the cell to drive electrolysis.

The relation between the equilibrium constant, K, and the Gibbs free energy for an electrochemical cell is expressed as follows:

$$\Delta G° = -RT\ lnK = -nFE°cell \tag{5}$$

Rearranging to express the relation between standard potential and equilibrium constant yields:

$$E_{cell}° = \frac{RT}{nF}\ln K \tag{6}$$

The standard potential of an electrochemical cell requires standard conditions ($\Delta G°$) for all of the reactants in standard condition Eq. (6) can be expressed by:

$$E_{cell}° = \frac{0.0591}{n}\log K \tag{7}$$

When reactant concentrations differ from standard conditions, the cell potential will deviate from the standard potential.

$$\Delta G = \Delta G^{\circ} + RT \times lnQ \tag{8}$$

Here ΔG is change in Gibbs free energy, ΔG° is the cell potential when Q is equal to 1, T is absolute temperature (K), R is the gas constant and Q is reaction quotient which can be found by dividing products by reactants using only those products and reactants that are aqueous or gaseous.

In the 20th century German chemist Walther Nernst proposed a mathematical model to determine the effect of reactant concentration on electrochemical cell potential.

Based on Eq. (8), Nernst extended the theory to include the contribution from electric potential on charged species. According to Eq. (5), the change in Gibbs free energy for an electrochemical cell can be related to the cell potential. Thus,

$$nF\Delta E = nF\Delta E^{\circ} - RT\,lnQ \tag{9}$$

Here n is the number of electrons/mole product, F is the Faraday constant (coulombs/mole), and ΔE is cell potential.

Finally, by dividing the Eq. (9) to nF:

$$\Delta E = \Delta E^{\circ} - (RT/nF)\,lnQ \tag{10}$$

Assuming standard conditions (T = 25°C) and R = 8.3145 J/(K mol), the equation above can be stated in base 10 logarithm as follow:

$$\Delta E_{cell} = \Delta E^{\circ}_{cell} - \frac{0.0591}{n} \log Q \tag{11}$$

Electroanalytical methods are a class of techniques in analytical chemistry, which determine an analyte by measuring the potential and/or current in an electrochemical cell containing an analyte.

Electroanalytical method offers a new approach to analysis of various analytes during recent years. In comparison with advanced instrumental methods, they are simple, inexpensive, and sensitive methods, which are able to be portable and detect the analytes without destruction of the sample.

Now days, many environmentally important species can be determined by electrochemical methods. Many electrochemical devices (like sensors and biosensors) are designed, constructed and applied for environmental analysis.

In this chapter, we are going to discuss on a number of electrochemical techniques and their applications in analysis of environmental samples.

The used electroanalytical technique and type of electrodes are determined by nature and amount of the analyte, and the matrix of the sample determine in a measurement.

In general, potentiometric and potentiostatic techniques are two different approaches used in electrochemical measurements. Electroanalytical measurements can be categorized based on which aspects of the cell are controlled and which are measured. The three main categories are (Chart 5.1):

- potentiometry (in potentiometry, electrode potential difference is measured in zero current)
- voltammetry (in voltammetry, current is measured as the potential are changed dynamically)
- amperometry (in amperometry, current is measured in a constant potential)
- coulometry (in coulometry, current is measured over the time and the current should be controlled)

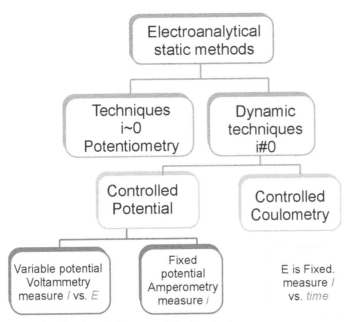

CHART 5.1 Classification of electro-analytical methods.

The most important part of an electroanalytical cell in the measurements is the working electrode. Many studies focus on designing new and various electrodes for selective or even specific analysis of species. This category of analytical electrochemistry has created electrochemical sensors and biosensors.

An electrochemical sensor provides continuous information about its environmental chemical changes, and converts the chemical response into a signal that can be detected by electrochemical methods. An electrochemical biosensor is an analytical device, which converts a biological response into an electronic signal.

5.2 POTENTIOMETRIC METHODS

Potentiometry is an electroanalytical method, which is based on the measurement of a potential under no current flow (or a low current flow). In this method the potential of a solution between two electrodes is measured without destruction of the solution. The potential is then related to the concentration of an analyte [1].

Potentiometric cell generally contains two electrodes. Since practically there is no current in the system, there is no oxidation/reduction process. Thus, instead of cathode and anode, indicator (working) and reference electrodes were used and the changes in the potential difference between the indicator and the reference electrode is measured.

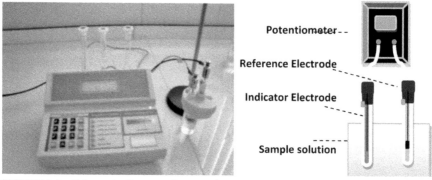

FIGURE 5.1   Schematic diagram of a potentiometric cell and the cell-assembly.

Electrochemical cell potential can be stated as:

$$E^{\circ}_{cell} = E^{\circ}_{indicator} - E^{\circ}_{reference} \qquad (12)$$

A schematic diagram of a potentiometric cell is shown in Fig. 5.1. In a complete potentiometric cell, the indicator or sensing electrode should be used in conjunction with a reference electrode (external reference electrode). The indicator electrode consists of a transducer which is normally an Ag wire coated with AgCl (internal reference electrode) and a sensing element which is placed in the membrane. During the potentiometric measurements, the current is about 10^{-6} μA, which is made by using a high-input-impedance millivolt-meter [2].

An ion-selective electrode and a reference electrode are immersed into the solution of interest and the cell voltage is measured. A dc voltmeter with an input resistance >10 gigaohm can be used as the instrument in all measurements. High-accuracy amplifiers of pH meters with mV readings or Ion-meters are most suitable.

In a normal experiment at least 5–10 mL of analyte solution is required. The required volume of the sample depends on the shape of the electrodes. By recent growth in designing electrodes and microelectrodes, it is possible to do the analysis by smaller amounts of sample [1, 2].

5.2.1 REFERENCE ELECTRODES

To measure the changes in the potential difference of the indicator electrode as the ionic concentration changes, it is necessary to have a stable reference voltage, which acts as a half-cell. Thus, a reference electrode (Fig. 5.2), which should provide a constant potential, is used.

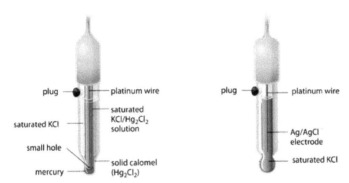

FIGURE 5.2   Reference electrodes.

Silver/Silver chloride reference electrode is the most common reference type used today because it is simple, inexpensive, very stable and nontoxic. The major electrolyte solution is saturated potassium chloride solution, however lower concentrations such as 1 M potassium chloride can also be used. It should be noted that changing the electrolyte concentration changes the potential. Silver chloride is slightly soluble in strong potassium chloride solutions, so, to avoid stripping the silver chloride off the silver wire, it is sometimes recommended the potassium chloride be saturated with silver chloride.

Calomel or mercury/mercurous chloride reference electrodes used to be the most widely used reference electrodes until about 1960, and the majority of older publications on pH refer to this family. Although, its stability is similar to that of the silver/silver chloride, it shows a better thermal hysteresis. Calomel electrodes are less prone to contamination because the mercury/mercurous chloride interface is protected inside a tube, not in direct contact with the electrolyte. The superiorities of a calomel electrode are rivaled by a double junction silver/silver chloride construction. Nowadays, calomel reference electrodes are looked upon with disfavor because of their toxicity, and because they virtually have no performance advantages over a double junction silver/silver chloride, and are also more difficult to construct.

5.2.2 INDICATOR ELECTRODES

These days many indicator electrodes have been commercially available and used in various analyzes. In potentiometric method, the analyte is mostly an ionic species to create a different potential on the electrode interface. Thus, the indicator electrode determines an ionic species. These indicator electrodes are called ion-selective electrodes (ISEs), which are a kind of electrochemical sensors.

The potential of the indicator electrode should be a concentration dependent one.

Based on the structure of the electrodes, potentiometric indicator electrodes are divided into two subdivisions [1]:
- symmetrical ion selective electrodes; and
- asymmetrical ion selective electrodes.

Symmetric electrodes are the classical ion selective electrodes in which the sensing layer (membrane) is placed between two solutions. In asymmetric electrodes, one side of the membrane is in contact with a solid phase, and the other being interfaced to the solution to be tested.

Potentiometric symmetrical or asymmetrical electrodes themselves can be categorizes into following groups:

A. Symmetric Electrodes
- Membrane electrodes
- Liquid membrane electrodes
- PVC membrane electrodes
- ionic materials in organic solvent membrane electrodes (Calcium electrode)
- Solid membrane electrodes
- crystalline membrane electrodes (fluoride electrode based on LaF_3 crystal)
- noncrystalline membrane electrodes (glass electrodes like pH electrode)
- Multiple membrane electrodes (gas sensing electrodes or enzyme-substrate electrodes)

B. Asymmetric Electrodes
- Coated wire electrodes
- Carbon paste electrodes
- All solid state electrode
- Field effect transistors

In conventional ISEs, the ion selective membrane is placed between sample solution and internal reference solution. The drawback of the presence of an internal reference solution, however, is the relatively large electrode dimensions of the ISEs. Based on the composition and materials of the membrane different kinds of ion selective electrodes have been developed.

One of the most common classes of potentiometric ISEs are PVC membrane electrodes. In this kind of electrodes an ion carrier (called selectophore), which is able to interact selectively with cations or anions, is placed in a plasticized PVC membrane. Because of recent theoretical advances in understanding of the response mechanism of this type of electrodes, developments of new ionophores for obtaining highly selective potentiometric membrane sensors is continued.

A typical procedure to prepare a PVC membrane electrode is as follow: mix thoroughly 30–35 mg of powdered PVC, 60–65 mg of a suitable plasticizer, a certain amount of an ionophore (usually 1–10 mg) and 1–5 mg of an ionic additive in 5 mL tetrahydrofuran (THF). The resulting mixture is heated slowly (50°C) until THF evaporates and a concentrated solution forms. Then, a Pyrex tube (3–5 mm o.d.) is dipped into the solution for about 5 s so that a transparent membrane (about 0.3 mm thickness) is formed. The tube is then pulled out from the mixture and kept at room temperature for at least 12 h. The tube was then filled with internal filling solution (usually 10^{-3} M of $M^{n+}Cl_n$). The electrode was finally conditioned for 12–48 h by soaking in a 10^{-2} M, $M^{n+}Cl_n$. A silver/silver chloride electrode is used as an internal reference electrode [1].

Coated-wire electrodes (CWEs) are a type of ISEs in which an electroactive species is incorporated in a thin polymeric support film coated directly on a metallic conductor. The removal of the internal filling solution provides new advantages. The substrate in the wire type electrodes is usually Pt, Cu or Ag wire and graphite rods have also been used. CWEs are manufactured by dipping a metal wire in to a solution of the membrane mixture.

Carbon paste electrodes (CPEs) have attracted attention as ion selective electrodes mainly due to their advantages over membrane electrodes such as renewability, stable response, low ohmic resistance, no need for internal solution. The carbon paste usually consists of graphite powder dispersed in a nonconductive mineral oil. The selectophore also incorporated in the paste. Sometimes nanostructure materials are used to modify the electrode response. The prepared paste packed into a plastic or Teflon tube carefully to avoid possible air gaps, often enhancing the electrode resistance. A copper wire was inserted into the opposite end to establish electrical contact. The external electrode surface was smoothed with soft paper (Fig. 5.3).

In all solid state electrodes, conducting polymers are used. The promising way toward construction of durable ion sensors is covalent binding of ion-recognition sites to the backbone of the conducting polymers. Another approach may be exploiting the conducting polymer as a transducer, which converts ionic data to electric data, in combination with classical ion-selective membranes.

Ion selective field effect transistors (ISFET) work as an extension of CWEs. Field effect transistors incorporate the ion-sensing membrane directly on the gate area of a field effect transistor (FET). A FET is a solid-state device that shows high input impedance and low-output impedance and thus it is capable of monitoring charge buildup on the ion-sensing membrane. In the ISFET two n-type regions, the source and the drain are implanted in the semiconducting p-type bulk silicon.

It should be noted that most of these devices are "ion selective" and not "ion specific"; no practical electrode is exclusively sensitive to one particular ion. This is because the sensing material in the membrane of an ISEs interacts with an ionic specie more strongly than the other ones. However, enzyme biosensors and sensors based on molecular imprinting polymers, which are based on almost specific key-lock mechanism reactions, are examples of specific electrodes. Gas sensing electrodes are also highly selective devices for monitoring gases. The typical gas sensor commonly incorporates a conventional ion selective electrode, immersed in an electrolyte solution, enclosed by a gas permeable membrane. The target gas diffuses through the membrane and reacts with the internal electrolyte, leading to the formation or consumption of an ionic species, which is detectable by the ionic sensor.

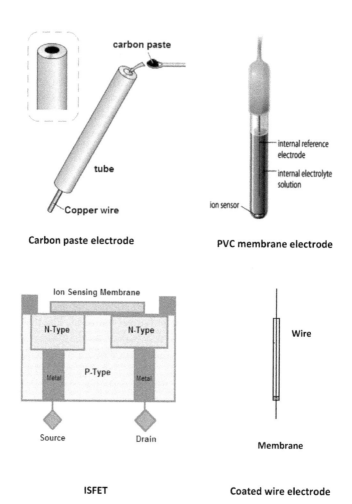

FIGURE 5.3 Some of the indicator electrodes.

Wide range of available sensing materials and electrodes; simple installation; easy and direct measurements, high durability and lifetime, especially with the most recent plastic-bodied; in relatively dilute aqueous solutions interfering ions are dilute enough not to be a major problem, measuring directly the activity of the ion; an effective pH range; accuracy and precision levels of ±2 or 3% for some ions by careful use, frequent calibration, and an awareness of the limitations; the wide operational temperature range for application in aqueous solutions (e.g., that of crystal membranes being in the range 0°C to 80°C and that of plastic membranes being from 0°C to 50°C); unaffected by many common disturbances, like sample color or turbidity, low cost, are the advantages of potentiometric measurements.

Some commercial available indicator electrodes are summarized in Table 5.1.

TABLE 5.1 Some Commercially Available Ion Selective Electrodes for Environmentally Important Ions [1]

Ion	Manufacturer	Model or Art. No.	Type of electrode	Concentration range (ppm)	Working temperature
NO_3^-	Radiometer	ISE25NO_3	PVC membrane	0.2–60000	0–50°C
	ELIT	8021	PVC membrane	3–62000	5–50°C
NO_2^-	ORION	93–46	PVC membrane	0.02–100	0–40°C
	ELIT	8071	PVC membrane	1–460	5–50°C
ClO_4^-	ORION	93–81	PVC membrane	0.7–99500	0–40°C
	ELIT	8061	PVC membrane	2–99000	5–50°C
Br^-	Metrohm	6.0502.100	Solid state	0.4–79900	0–50°C
	Radiometer	ISE25Br	Solid state	0.08–80000	0–60°C
Cl^-	Radiometer	ISE25Cl	Solid state	1.8–35000	0–60°C
	Radiometer	ISE/HS25Cl	Solid state	0.04–35000	0–60°C
	ORION	93–17	PVC membrane	0.18–35500	0–50°C
CN^-	Metrohm	6.0502.130	Solid state	0.2–260	0–80°C
	Radiometer	ISE25CN	Solid state	0.013–260	0–60°C
	ORION	94–06	Solid state	0.2–260	0–80°C
F^-	ORION	94–09	Solid state	0.02-saturated	0–80°C
	Radiometer	ISE25F	Solid state	0.01–20000	0–60°C
SCN^-	ORION	94–58	PVC membrane	0.29–58100	0–50°C
	ELIT	8229	Solid state	3–58000	5–80°C
Cu^{2+}	Metrohm	6.0502.140	Solid state	0.0006–6300	0–80°C
	Radiometer	ISE25Cu	Solid state	0.06–60000	0–60°C
	ELIT	8227	Solid state	1–64000	5–80°C
Cd^{2+}	ORION	94–48	Solid state	0.01–11200	0–80°C
	ELIT	8241	Solid state	1–11000	0–80°C
Pb^{2+}	Radiometer	ISE25Pb	Solid state	0.2–200000	0–60°C
Hg^{2+}	ELIT	8251	Solid state	2–200000	5–80°C
S^{2-}	ELIT	8225	Solid state	0.03–32000	5–80°C

5.2.3 POTENTIAL *VS.* CONCENTRATION

According to the Nernst equation, the measured potential differences of the result-ing electrochemical cell have a linear correlation with the logarithm of the ion activ-ity in the solution.

The essential component of any potentiometric ion-selective electrode is its ion-selective layer, and in fact, the sensing material placed in this layer. If an ion can pass through the boundary between the electrode sensing layer and aqueous phase and senses by the sensing element in an ion-selective layer, an electrochemi-cal equilibrium occurs in the interfaces. And different potentials in the two phases are formed. That is why the sensing material is the source of the selectivity of the sensor, because if it can exchange only one type of ion between the two phases, the resulting potential difference formed between the phases, will then be governed only by the activities of this specific ion in the solution phases, and also the sensing layer of the electrode.

In the case of a membrane separating two solutions of different ionic activities (a_1 being the activity of the ion in phase 1 and a_2, being that in the second phase), the potential difference (E) across the membrane is described by the Nernst equation:

A (phase 1) $\leftrightarrow$ A (phase 2)

$$E = E^0 \mp \frac{RT}{nF} \ln\left(\frac{a_2}{a_1}\right)$$

$$(13)$$

The activity of the target or primary ion in phase 1 being kept constant, then the unknown activity in phase 2 ($a_2 = a_x$) is related to the measured potential (E). In an ideal case, the potential at the electrode membrane depends on a single species of ions. The Nernst equation describes the relation between activity and electrochemi-cal potential:

$$E = E^0 \mp \frac{RT}{nF} \ln\left(\frac{a_x}{a_1}\right) \cong E = Const. + S.Log\ (a_x)$$

$$(14)$$

n_x being the charge of the analyte ion and S = 59.16/n_x (mV) at 298 K. Placing two identical reference electrodes in the two phases (on both sides of the sensing ele-ment), this potential difference can be easily measured. This potential difference, i.e., the electromotive force, is in practice measured between an ion selective elec-trode (containing an internal reference electrode) and a reference electrode that is placed in the sample solution, containing the analyte to be measured. It is notable that this is a measurement at zero current, that is, under equilibrium conditions, wherein equilibrium means that the transfer rate of ions from the solution to the membrane equals that from the membrane to the solution. The resulting signal, which is measured, is the sum of some different potentials formed at all solid-solid, solid liquid and liquid-liquid interfaces.

5.2.4   CHARACTERIZATIONS OF THE POTENTIOMETRIC ELECTRODE

The properties of an ion-selective electrode are characterized by some parameters [2]:

- measuring range
- detection limit
- response time
- selectivity
- lifetime
- accuracy/precision

The measuring range of an ion-selective electrode includes the linear part of the calibration curve as shown in Fig. 5.4. The calibration curve of the electrode is obtained by measuring the potential difference of electrochemical cell when the immersed in a series of ion standard solution (1.0×10^{-8}–1 M).

For many electrodes the measuring range can extend from 1 M down to 10^{-6} or even 10^{-7} M concentrations. The detection limit of an ion-selective electrode is calculated by using the cross-section of the two extrapolated linear parts of the ion-selective calibration curve. Any ion-selective electrode has a lower and an upper detection limit, in which the linear behavior of the potential response *vs.* concentration changes starts to deviate significantly from a Nernstian electrode slope. These points are generally observed in activity ranges, where the electrodes start to lose their sensitivities to the target ion. An ion-selective polymeric membrane sensor is usually filled with an internal reference solution that normally contains relatively high levels of target ion (about 10–100 mM). Under such circumstances, ions start flowing from the inner filling solution, or even from the membrane itself, to the sample side of the membrane.

In practice, detection limits for the most selective electrodes are in the range of 10^{-5}–10^{-6} M. However, recent studies have shown that even sub nM detection limits can be obtained for these devices by different methods such as application of metal buffers to eliminate the contamination of very dilute solutions, using cation-exchange resin in the internal solution of ISEs to keep the primary ion activity at a constant low level, using lipophilic particles such as silica-gel 100 C18-reversed phase into the sensing membrane, using sandwich membranes, and so on [3–6]. Breakdowns in the perm selectivity of the ISE membrane are the reason for the existence of an upper detection limit. Increasing the concentration of analyte ion in the sample solution leads to the consequent increase of the number of complexed carriers in the membrane. The decrease of the free carrier concentration in the membrane causes the membrane to act as an ion-exchanger for the counter ions. When the analyte ion is strongly complexed by the ionophore, and also when highly lipophilic counter ions are present in the sample solution, available free carrier reduces more rapidly.

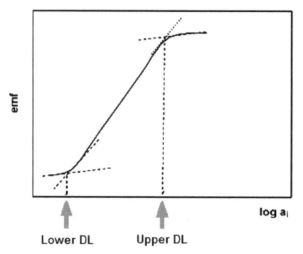

FIGURE 5.4 Typical calibration curve of a potentiometric electrode.

The response time of an electrode is evaluated by measuring the average time required to achieve a potential within ±0.1 mV of the final steady-state potential, upon successive immersion of a series of interested ions, each having a ten-fold difference in concentration. Response of the most of potentiometric electrodes are from 5 s to 2 min. It is notable that the experimental conditions-like the stirring or flow rate, the ionic concentration and composition of the test solution, the concentration and composition of the solution to which the electrode was exposed before experiment measurements were performed, any previous usages or preconditioning of the electrode, and the testing temperature have an effort on the experimental response time of a sensor.

As their names indicate selectivity, which describes an ISE's specificity toward the target ion in the presence of other ions that are called the "interfering ions" is the most important characteristics of these devices. For polymer membrane sensors, interferences by other sample ions are mainly dictated by their competitive extraction into the membrane phase. Therefore, the response of these ISEs can be fully predicted from thermodynamic constants, ionophore and ionic sites concentrations, and, in the case of ionophore based membranes, from the complex formation constants of each ion-ionophore complex in the membrane.

Selectivity is most often expressed as the logarithm of (K_{xy}). Negative values indicate a preference for the target ion relative to the chosen interfering ion. Positive values of log K_{xy} indicate the better preference of an electrode for the interfering ion. The experimental potentiometric selectivity coefficients depend on the activity and a method of their determination.

There are a number of different methods for the determination of potentiometric selectivity coefficients, among which three methods have been, however, much more widely accepted, which are namely [5]:

1. The Separate Solution Method (SSM)
2. The Mixed Solution Method (MSM)
   (a) Fixed Interference Method
   (b) Fixed Primary Method
3. The matched potential method (MPM)
4. The Unbiased Selectivity Coefficients

For explanation of these methods refer to the encyclopedia of the sensors.[1]

The average lifetime for most of the reported ion-selective sensors is in the range of 4–10 weeks. After this time the slope of the sensor will decrease, and the detection limit will increase. The loss of plasticizer, carrier, or ionic additive from the polymeric film because of leaching into the sample is a primary reason for the limited lifetimes of the sensors. However, the lifetime of solid sate electrode (asymmetric electrodes) are longer than PVC membrane electrodes.

In general, accuracy of a measurement shows the closeness of the result to the real value, and precision or reproducibility is a measure of how close a series of measurements, on the same sample, are to each other. Accuracy and precision in the case of ISE measurements, depending on many factors, can be highly variable.

Because the measured potential is directly related to the target ion concentration, any error in potential measurement will cause an error in the concentration. Although these errors are not directly proportional to the concentration, they are dependent on the slope of the calibration curve. For example in the case of a monovalent ion selective electrode with a slope of about 55 mV/decade of concentration, a ±1 mV error in potential measurement will cause ±4% error in the concentration, while for a bivalent ISE showing a slope of around 26 mV/decade of concentration, the same potential reading error will cause an 8% error in the calculated concentration.

It is notable that, because of the decrease in the calibration slope at two ends of the concentration range, where the curve starts to deviate from linearity, the error per mV will become even greater at lower concentrations.

No matter how precise and/or accurate our measuring device, that can be either a simple volt meter or a computer interface and its software, is to achieve the most precise results, one should be able to manipulate the electrode drift and hysteresis (or memory effect), and also restrict the liquid junction potential drifts of the reference electrode, in order for the measured voltage to be reproducible. Different ions and different electrodes exhibit different potential drifts and hysteresis effects.

The main challenges in working with ion-selective electrode are:
- The interference of the other ions in solution

- The influence of the ionic strength of the solution, which can, at high concentrations, reduce the ratio of the measured activity relative to the true concentration.
- The drifts of electrode potential with repeating measurements.

The effect of ionic strength can be fixed by applying a suitable ionic strength adjustment buffer (ISAB) to both the standards and the sample solutions. To gain stable results, and avoiding potential drift, both ISE and reference electrode are recommended to be preconditioned in an appropriate standard solution, by immersion in it, for 10–15 min before being used.

5.2.5 POTENTIOMETRIC MEASUREMENT

To have the precise results, a number of factors should be considered:
- It should be also noted that only responses related to the water soluble components of target element, will be sensed by ISE. The possibility of blocking or contamination of both ISE and reference electrodes, with organic ions is difficulty of working with ISEs. ISEs are generally designed for use in aqueous solutions. If impurities are present in the sample, some pretreatment should be taken before application of the electrodes.
- Sometimes it is needed to gently stirring the solution by a magnetic stirrer (at ~100 rpm) during immersion period of the electrodes can be done. String the solution can be helpful when the potential stabilization rate in systems is low. This will also prevent from the formation of concentration gradients within the solution or around the electrodes. The drawbacks of stirring may be the longer time of analysis, the possibility contamination during the insertion and removal of the magnet, and also the probable heat exchange between the stirrer and the solution. In addition, common methods of stirring require at least 50 mL to preferably 100 mL of the sample, while this is not practical when small samples are available. Although some users prefer to record the results while stirring, others suggest that it is better to switch off the stirrer while reading the results. To ensure good, homogeneous contact between solution and membrane, the sample solution can also be swirled manually, after the immersion of the electrodes. The solution should be left to stand, after the manual stirring. In this way, problems of heat transfer and the inconvenience of putting and taking the magnetic stirrer can be overcome too.
- To reduce the effect of the previous measurements on the next ones, rinse the electrodes with deionized water for 20 or 30 sec, and then dry it between measurements. After rinsing of the electrode, it is soaked in deionized water, and then dried to prevent dilution of sample.

A list of the commonly applied methodology for work with ISEs is as follow:[1]
1. Direct Potentiometry
2. Incremental Methods that include

        (a) Standard Addition
        (b) Sample Addition
        (c) Sample Subtraction
   3.   Potentiometric Titrations

5.2.5.1   DIRECT POTENTIOMETRY

Direct potentiometry simply consists of measurement of the electrode response in an unknown solution. In this way, one can easily read the concentration from the calibration curve, which is plotted by standard solution series either manually, or using special computer software (sometimes the devices directly display the concentration, in the case of self calibrating ion meters.) The major advantage of the direct potentiometric method is its ability to be used in fast measuring of large quantities of samples, even of those having a wide range of concentrations, without needing to alter the ranges, recalibrating or performing sophisticated calculations. In addition, if an ISAB is not used, measuring the volume of the samples or those of the standards will not be necessary. In the case of some elements, the results obtained from this method are quite acceptable, even when the electrodes are directly inserted in a river, or lake samples.

5.2.5.2   STANDARD ADDITION

This method is applied in a relatively large but precisely measured volume of the sample, and measuring its potential (E_x), then addition of a very small, but precisely measured, volume of a standard solution of a known concentration of the target ion to the initial sample, waiting for potential stabilization, and finally measuring the stabilized potential value (E_{x+s}). There are three requirements about the used sample and standard solution:

   1.   The concentration and volume of the added standard must be enough to bring about a significant and measurable change in the measured voltage of the sample solution (ideally, 10 to 20 mV), while the volume of the added standard so small that it does not cause a significant change in the ionic strength of the initial solution.
   2.   The initial volume of the sample solution must be large enough to cover the tips of both the ISE and reference electrodes. The volume should be so large that, after the addition of the standard, the two solutions can be effectively mixed by swirling or stirring, the electrodes being still immersed.
   3.   The volume of the standard should be very small that volumetric errors are not significant. The sample concentration can then be calculated by solving the equation set below:

$$E_x = Const. \pm \frac{RT}{Z_xF} \ln(a_x) \tag{15}$$

$$E_{x+s} = Const. \pm \frac{RT}{Z_xF} \ln\left(\frac{a_x.V_x + a_s.V_s}{V_x + V_s}\right) \tag{16}$$

Which form the general equation:

$$E_{x+s} - E_x = \pm \frac{RT}{Z_xF} \ln\left(\frac{a_x.V_x + a_s.V_s}{V_x.a_x + V_s a_x}\right) \tag{17}$$

where a_x is the unknown activity or concentration of the sample, as is that of the standard and is known, v_x is the sample volume, v_s is the volume of the added standard, Ex and E_{x+s} are described above, R is the universal gas constant, F is the Faraday's constant, and n_x is the absolute value of the charge of the target ion. + is used for cations, while– is used for anions.

5.2.5.3   SAMPLE ADDITION

This method can be regarded as opposite direction of the above method. In this method a relatively large, but precisely measured, volume of a standard solution with a known concentration of the target ion are used and its potential (E_s) was measured. Then, a very small, but precisely measured volume of the analyte is added to the initial standard, wait for potential stabilization, and measures the stabilized potential value (E_{s+x}). It should be noted that all the three requirements mentioned in the above case, apply to this method, too. The sample concentration can then be calculated by solving the similar-to-above equation set below:

$$E_s = Const. \pm \frac{RT}{Z_xF} \ln(a_s) \tag{18}$$

$$E_{s+x} = Const. \pm \frac{RT}{Z_xF} \ln\left(\frac{a_s.V_s + a_x.V_x}{V_x + V_s}\right) \tag{19}$$

Which can be stated as the following general equation:

$$E_{s+x} - E_s = \pm \frac{RT}{Z_xF} \ln\left(\frac{a_s.V_s + a_x.V_x}{V_x + V_s}\right) \tag{20}$$

The major superiorities of these two methods is that both the calibration and sample measurement steps are performed at the same time and in the same solution, eliminating the temperature and ionic strength differences, and making ISAB application unnecessary.

The electrodes are continuously immersed in the solution, throughout the measurement process, minimizing the liquid junction potential of the reference electrode that may even come to several millivolts when the electrodes are removed from one solution and placed in another. This way one can avoid this source of measurement error.

The application of these methods also makes it possible to work even with old or worn electrodes, the response of which may not be completely linear over their whole range, or even over a narrow range of about a decade of concentration. This can be done, as long as their slope is both stable and reproducible over the limited concentration range. The methods can, theoretically yield more precise results than the direct potentiometry, even when the ionic strength of samples under study, is low.

The necessity of thoroughly mixing of the applied, and of course accurately measured, volumes of standard and sample, and the fact that these methods involve more sophisticated calculations, than those involved in the direct potentiometry, are the major drawbacks of these methods.

5.2.5.4   SAMPLE SUBTRACTION

This method involves a chemical reaction. It is performed by the addition of a small and known volume of the sample solution to a standard solution of a second ion with which, the target ion will react stochiometrically, forming a third compound, in the form of complex or precipitate, and as a result, decreasing the concentration and the activity of both ions. The ISEs, used in this method, should be selective toward the second, or the reactive ion in the standard, and not to the target ion. In this way, it is possible to measure an ion, which there is no ISE for its direct measurement. For example, if there is currently no ISE capable of detection of the ion "A^{n+}" or "B^{m-}," and if we know these ions can be removed from solution by being precipitated or complexed by means of ions "C^{n-}" or "D^{m+}" respectively, and also if there is an ISE which is sensitive to "C^{n-}" or "D^{m+}," this method can be used for the indirect determination of "A^{n+}" or "B^{m-}" ions, by a "C^{n-}" or "D^{m+}"-selective electrodes. "A^{n+}" or "B^{m-}" concentrations can, therefore, be measured by first measuring the potential of a pure standard solution of "C^{n-}" or "D^{m+}" salts. The next step is the addition of a known volume of the sample containing "A^{n+}" or "B^{m-}," waiting for the reaction to be completed, and remeasuring the potential by the "C^{n-}" or "D^{m+}"-selective electrode, after its being stabilized. The amount of "$Cn-$" or "$Dm+$" used, to completely react with the "A^{n+}" or "B^{m-}" content of the sample, can be calculated using an equation like the one used for the sample addition method, and the "A^{n+}" or "B^{m-}" content of the added sample can be calculated from these results, using the stochiometric equations.

5.2.5.5   POTENTIOMETRIC TITRATIONS

Potentiometry is usually used to detect the end-point of titrations, at which there is often a very obvious change in the concentrations of the reactants, leading to considerable shifts in the electrode potential. Such end point determinations can often give more precise results compared to the other discussed potentiometric methods, because the precision of the results from all those methods is limited by the accuracy of the, far less accurate, volumetric measurements rather than the, very accurate, measurement of the electrode potential. For instance, in the case of the titration of a Cu^{2+} solution with the common complexing reagent, EDTA, the drop-wise addition of the EDTA solution, causes a gradual decrease in the Cu^{2+} concentration until the end point, when approximately all the Cu^{2+} ions disappear from the solution. The progress of this titration can be monitored using a Cu^{2+}-selective electrode. This method, as well, makes it possible to use an ISE for the measurement of ions, toward which it is not selective, like the sample subtraction method. For example if direct potentiometry cannot be applied for the measurement of "A^{n+}" or "B^{m-}," and if these ions can be titrated by reacting with "C^{n-}" or "D^{m+}," and if the reaction can be monitored by using a "C^{n-}" or "D^{m+}"-selective electrode, the potentiometric titration by "C^{n-}"or "D^{m+}" in the presence of a "C^{n-}" or "D^{m+}"-selective electrode, as an indicator electrode can be used for this purpose. Potentiometric titration can also be used in the determination of elements, the maintenance of the standard solutions of which is hard, or which are toxic, and hence, working with their concentrated standard solutions, is both hazardous and undesirable.

5.3   VOLTAMMETRIC METHODS

Voltammetry is another category of electroanalytical methods used in analysis of various analytes. In voltammetry, information about an analyte is obtained by measuring the current as the potential at an electrode's surface is varied or constant. The analyte should be electroactive, that is, it can oxidize or reduce in a special potential. Or it can be measured by an electroactive species in an indirect manner. In comparison, in potentiometric methods, it is not necessary the analyte to be electroactive. Voltammetry like potentiometry is practically nondestructive method because only a very small amount of the sample is consumed at the electrodes surfaces.

5.3.1   PRINCIPLES OF VOLTAMMETRY

Voltammetry is the study of current as a function of applied potential. The curve of current *vs.* potential $(I = f(E))$ is called voltammograms. The potential is varied step by step or continuously and then the amount of current is measured. To carry out such an experiment at least two electrodes is needed. The first one is a working electrode, which is in contact with the analyte and the desired potential is applied to it. The second electrode acts as the other half cell. It should have a known potential with which the potential of the working electrode can be adjusted, furthermore it can balance the charge added or removed by the working electrode. However, it is very difficult for an electrode to keep a potential constant while the current passing through it. To solve this problem, the roles of providing a reference potential and supplying electrons are divided between the reference electrode and another electrode called auxiliary electrode. The reference electrode is a half cell with a known reduction potential. It just works as reference in measuring and controlling the potential of the working electrode and at no point it doesn't pass any current. On the contrary, the auxiliary electrode passes all the current needed to balance the charged in the system. To achieve this, the auxiliary electrode will often swing to extreme potentials at the edges of the solvent window, where it oxidizes or reduces the solvent or supporting electrolyte. These electrodes, the working, reference, and auxiliary make up the modern three electrode system (Fig. 5.5) [7–9].

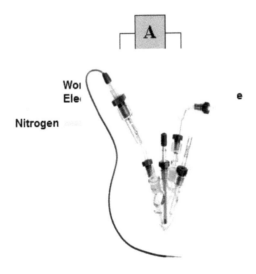

FIGURE 5.5   A typical setup of a voltametric analysis.

To conduct the current in the solution an ionic conductor, which contains ions that can freely move between the electrodes should be used. The electrodes were finally connected to a potentiostat instrument to create a close circuit.

When an analyte oxidizes at the working electrode, the resulting electrons pass through the potentiostat to the auxiliary electrode, and causes reduction of the solvent or some other component of the solution matrix. If an analyte reduces at the working electrode, the current flows from the auxiliary electrode to the cathode. The current from redox reactions at the working and the auxiliary electrodes is called a faradaic current. Since the reaction of interest occurs at the working electrode, the faradaic current using this reaction is considered. A faradaic current due to the analytes reduction is a cathodic current, and its sign is positive. An anodic current is due to an oxidation reaction at the working electrode, and its sign is negative.

Two factors have an important role in the rate of the electrochemical reaction and the current; the rate at which the reactants and products are transported to and from the electrode (called mass transport) and the rate at which electrons pass between the electrode and the reactants and products in solution (called charge transfer).

The concentration of analyte at the electrode surface may not be the same as its concentration in bulk solution. When the potential is applied to the working electrode, the species near the electrode start to react and a faradaic current creates but quickly returns to zero. Thus, the concentrations of species at the electrode surface make different with its concentrations in bulk solution. Because of this difference in concentration, a concentration gradient between the solution at the electrode surface and the bulk solution occurs. This concentration gradient creates a driving force that transports species from the bulk to the electrode surface and vice versa. Thus, the faradaic current continues to flow until there is no difference between the concentrations of species at the electrode and their concentrations in bulk solution.

In addition of diffusion, there are two other mechanisms for mass transfer; convection and migration. Convection occurs when mechanically the solution is mixed. It carries the reactants toward the electrode surfaces and removing products from the electrodes. The most common form of convection is stirring the solution with a stir bar. Other methods that have been used include rotating the electrode and incorporating the electrode into a flow-injection cell. Some times is needed the convection should be eliminated or controlled accurately to provide controlled transport of the analyte to the electrode.

Migration is the movement of a charged ion in the presence of an electric field. Migration occurs when charged species in a solution are attracted to or repelled from the electrode that carries a surface charge. If the electrode carries a positive charge, for example, an anion will move toward the electrode and a cation will move toward the bulk solution. Unlike diffusion and convection, migration only affects the mass transport of charged species. Using a supporting electrolyte at concentrations 100 times more than that of the species being determined eliminates the effect of migration.

The rate of mass transport is one factor influencing the current in voltammetry and the other factor is charge transfer. The ease of electrons move between the electrode and the species reacting at the electrode surface affects the current. When electron transfer kinetics is fast, the redox reaction is at equilibrium. Under these conditions the redox reaction is electrochemically reversible and the Nernst equation works. If the electron transfer kinetics are sufficiently slow, the concentration of reactants and products at the electrode surface and as a result the magnitude of the faradaic current are not what is predicted by the Nernst equation. In this case the system is electrochemically irreversible.

In addition to current resulting from redox reactions (faradaic current), other source of current cane be exist in an electrochemical cell (nonfaradaic current). At beginning of the process, the charge on an electrode is zero. Then, for example a positive potential is applied to it. Now, cations near the electrode surface respond to this positive charge and move around from the electrode. Anions, on the other hand, migrate toward the electrode. This migration of ions continues until the positive charge on the electrode surface equal with the negative charge of the solution near the electrode. Thus, a small and short-life nonfaradaic current (charging current) produces. Every time the electrode potential is changed, a temporary charge current flow produces.

The movement of the ions in response to the electrode surface charge leads to the formation of a structured electrode-solution interface, which is called the electrical double layer (EDL). The charge current is the result of forming EDL.

Even in the absence of the analyte, there is a small, measurable current in an electrochemical cell, which is called residual current. The residual current has two components: a faradaic current due to the oxidation or reduction of trace impurities and the charge current.

Data analysis in voltammetry requires the consideration of kinetics in addition to thermodynamics. In this method, the effects of the applied potential and the behavior of the redox current are described by several well-known laws. Idealized theoretical electrochemical thermodynamic relationships such as the Nernst equation are modeled without a time component. These models alone are insufficient to describe the dynamic aspects of voltammetry. Models like the Tafel equation and Butler-Volmer equation can work for the voltammetry relationships.[9,10]

The applied potential controls the concentrations of the redox species at the electrode surface ($c_O{}^\circ$ and $c_R{}^\circ$) and the rate of the reaction (k°), as described by the Nernst and Butler–Volmer equations, respectively. In the cases where diffusion only plays a controlling part in mass transfer, the current resulting from the redox process (known as the faradaic current) is related to the material flux at the electrode-solution interface and is described by Fick's law.

For a reversible electrochemical reaction (that is, a reaction so fast that equilibrium is always reestablished as changes are made), which can be described by $O + ne^- = R$, the application of a potential E forces the respective concentrations of O and R at the surface of the electrode (that is, ($c_O{}^\circ$ and $c_R{}^\circ$) to a ratio in compliance with the Nernst equation:

$$E = E^0 - \frac{RT}{nF} \ln \frac{c_R^0}{c_O^0} \qquad (21)$$

where R is the molar gas constant (8.3144 J mol^{-1}K^{-1}), T is the absolute temperature (K), n is the number of electrons transferred, F = Faraday constant ($96,485°$C/mol), and E° is the standard reduction potential for the redox couple. If the potential applied to the electrode is changed, the ratio $c_R°/ c_O°$ at the surface will also change. If the potential is made more negative the ratio becomes larger (that is, O is reduced) and, conversely, if the potential is made more positive the ratio becomes smaller (that is, R is oxidized).

For some techniques it is useful to use the relationship that links the variables for current, potential, and concentration, known as the Butler–Volmer equation:

$$\frac{i}{nFA} = k^0 \left\{ c_o° exp\left[-\alpha\right] - c_R° exp\left[\left(1 - \alpha\right)\right] \right\} \tag{22}$$

where θ = nF(E–E°)/RT, k° is the heterogeneous rate constant, a is known as the transfer coefficient, and A is the area of the electrode. This relationship allows to obtain the values of the two analytically important parameters, i and k°.

When new O or R is created at the surface, the increased concentration provides the force for its diffusion toward the bulk of the solution. Likewise, when O or R is destroyed, the decreased concentration promotes the diffusion of new material from the bulk solution. The resulting concentration gradient and mass transport is described by Fick's law, which states that the flux of matter (F) is directly proportional to the concentration gradient:

$$\Phi = -AD_O\left(\partial c_O / \partial x\right) \tag{23}$$

where D_O is the diffusion coefficient of O and x is the distance from the electrode surface. An analogous equation can be written for R. The flux of O or R at the electrode surface controls the rate of reaction, and thus the faradaic current flowing in the cell.

In the bulk solution, concentration gradients are generally small and ionic migration carries most of the current. The current is a quantitative measure of how fast a species is being reduced or oxidized at the electrode surface. The actual value of this current is affected by many additional factors, most importantly the concentration of the redox species, the size, shape, and material of the electrode, the solution resistance, the cell volume, and the number of electrons transferred.

In addition to the general objects above mentioned, each voltammetric technique has its own unique laws and theoretical relationships.

5.3.2 ELECTRODES AND ELECTROLYTES

Practically, it can be very important to have a working electrode with known dimensions and surface characteristics. It is common to clean and polish working electrodes regularly. The earliest voltammetric techniques (polarography) used a mercury working electrode. It has several advantages including a wide cathodic ranges and a renewable surface. Next, several different materials such as platinum, gold, silver, and carbon were also applied as working electrodes. Solid electrodes constructed using platinum, gold, silver, or carbon may be used over a range of potentials, which are negative and positive *vs.* SCE. The electrolyte used has an important effect on the potential window of the working electrode. For example, the potential window for a Pt electrode extends from approximately +1.2 V to –0.2 V *vs.* SCE in acidic solutions, and from +0.7 V to –1 V *vs.* SCE in basic solutions. A solid electrode can replace a mercury electrode for many voltammetric analyzes that require negative potentials, and it is the electrode of choice at more positive potentials. Except for carbon paste electrodes, a solid electrode is fashioned into a disk and sealed into the end of an inert support with an electrical lead. The carbon paste electrode is made by filling the cavity at the end of the inert support with a paste consisting of carbon particles and viscous oil. The problem of solid electrodes is adsorption of a solution species or formation of an oxide layer at their surfaces. For this reason a solid electrode needs frequent reconditioning, by applying an appropriate potential or by polishing.

Todays, to overcome these problems, chemically modified electrodes are widely used for high sensitive electrochemical determination of organic molecules as well as metal ions.

The auxiliary electrode can be almost any well conductive material, which cannot react with the bulk of the analyte solution.

The reference electrode is the most important one among the three electrodes. Saturated calumel electrode (SCE) and Ag-AgCl reference electrode can be used as reference electrodes.

In most voltammetry experiments, a bulk electrolyte (also known as a supporting electrolyte) is used to minimize solution resistance. It is possible to run an experiment without a bulk electrolyte, but the added resistance greatly reduces the accuracy of the results. With room temperature ionic liquids, the solvent can act as the electrolyte.

5.3.3 POTENTIOSTAT

A potentiostat use to control the potential of the working electrode. The potential of the working electrode is measured relative to a constant-potential reference electrode that is connected to the working electrode through a high-impedance potentiometer. To set the working electrode potential the slide wire resistor was adjusted, which is connected to the auxiliary electrode. If the working electrode potential begins to drift, by adjusting the slide wire resistor the potential return to its initial value. The current flowing between the auxiliary electrode and the working electrode is measured by an amperometer. In the larger instruments, the potentiostat package also includes electrometer circuits, A/D and D/A converters, and dedicated microprocessors with memory.

Most voltammetric techniques are dynamic (i.e., they require a potential modulated according to some predefined waveform). Accurate and flexible control of the applied potential is a critical function of the potentiostat. In early analog instruments, a linear scan meant just that, a continuous linear change in potential from one preset value to another. Since the advent of digital electronics almost all potentiostats operate in a digital (incremental) fashion. Thus, the application of a linear scan is actually the application of a "staircase" modulated potential with small enough steps to be equivalent to the analog case. Digital fabrication of the applied potential has opened up a whole new area of pulsed voltammetry, which gives fast experiments and increased sensitivity. Modern potentiostats include waveform generators that let apply a time-dependent potential profile, such as a series of potential pulses to the working electrode. The use of micro and nanometer-size electrodes has made it necessary to build potentiostats with very low current capabilities. Microelectrodes routinely give current responses in the pico to nanoampere range. High-speed scanning techniques such as square-wave voltammetry require very fast response times from the electronics. These diverse and exacting demands have pushed potentiostat manufacturers into providing a wide spectrum of potentiostats tailored to specific applications.

5.3.4   VARIOUS TECHNIQUES OF VOLTAMMETRY

Based on the electrodes and the electrolyte types, how the potential is applied and how the current is measured, different voltammetric techniques were developed.

5.3.4.1   POLAROGRAPHY

Polarography is a voltammetric method in which a mercury electrode is used. Because mercury is a liquid, the working electrode is often a drop suspended from the end of a capillary tube. Hanging mercury drop electrode (HMDE), dropping mercury electrode (DME) or static mercury drop electrode (SMDE) is used as a working electrode. In HMDE, the drop of mercury is extruded by rotating a micrometer screw that pushes the mercury from a reservoir through a narrow capillary tube. In DME, mercury drops form at the end of a capillary tube as a result of gravity.

The mercury drop grows continuously as mercury flows from the tank under and has a short lifetime of several seconds. At the end of its lifetime the mercury drop is replaced by a new drop, either manually or on its own. SMDE uses a solenoid driven needle to control the flow of mercury. Activation of the solenoid momentarily lifts the needle, allowing mercury to flow through the capillary and forming a single, hanging mercury drop. Repeatedly activating the solenoid produces a series of mercury drops. In this way SMDE may be applied as a HMDE or a DME.

Mercury has several advantages as a working electrode. A dropping mercury electrode has advantages of a wide cathodic ranges and a renewable surface. Its high over potential for the reduction of H_3O^+ to H_2 makes accessible potentials as negative as -1 V $vs.$ SCE in acidic solutions and -2 V $vs.$ SCE in basic solutions. A species such as Zn^{2+}, which is difficult to reduce at other electrodes without simultaneously reducing H_3O^+, is easily reduced at a mercury electrode. Other advantage is the ability of metals to dissolve in mercury resulting in the formation of an amalgam. One limitation to using mercury as a working electrode is the ease of oxidizing. Depending on the solvent, a mercury electrode can not be used at potentials more positive than approximately -0.3 V to $+0.4$ V $vs.$ SCE.

There is another type of mercury electrode in which the mercury film electrode coated on a solid electrode. Carbon, platinum, or gold is placed in a solution of Hg^{2+} and held at a potential where the reduction of Hg^{2+} to Hg is occurred and a thin mercury film on the solid electrode surface forms.

In polarography, analytes are transferred from the bulk to the surface of the electrode through diffusion/convection mass transport. Simple principle of polarography is based on an electrolysis with two electrodes, one polarizable (mercury electrode as shown in Fig. 5.6A,B) and one unpolarizable.

A N_2-purge tube for removing dissolved O_2, and a stir bar is can be employed if needed. Electrochemical cells are designed in a various sizes; to analysis the varying solution volumes ranging from more than 100 mL to as even 50 μL.

Polarography is a kind of linear-sweep voltammetry where the applied potential is altered in a linear mode from the initial potential to the final potential (Fig. 5.6C). The current $vs.$ potential response of a polarographic experiment (is called polarograph) has the typical sigmoidal shape (Fig. 5.6D). The curve in a polarography experiment shows the current oscillations, which is due to falling of the mercury drops from the capillary. The plateau on the sigmoid is called limiting current because the electrode surface is saturated and the mercury drop life is short and when each drop of mercury falls, mix with the solution and affects the diffusion mechanism.

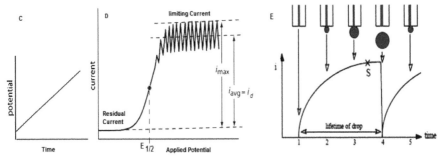

FIGURE 5.6 Polarography A: a polarograph picture; B: schematic diagram of a polarograph; C: applied potential as a function of time in normal polarography; D: A typical polarograph curve; E: current vs. time during drop growth.

There are various limitations in a classical polarography experiment for quantitative analytical measurements.

Because the current is continuously measured during the growth of the mercury drop (Fig. 5.6E) a considerable capacitive current occurs. As the Hg flows from the capillary end, a large increase in the surface area take placed. As a result, the initial current is dominated by capacitive effects as the mercury drop size increases. At the end of the drop life, there is little change in the surface area, which diminishes the contribution of capacitance changes to the total current. At the same time, any redox process, which occurs, will result in faradaic current that decays approximately as the square root of time (due to the increasing dimensions of the Nernst diffusion layer). The exponential decay of the capacitive current is much more rapid than the decay of the faradaic current; hence, the faradaic current is proportionally larger at the end of the drop life. Unfortunately, this process is complicated by the continuously changing potential that is applied to the mercury drop throughout the experiment. Because the potential is changing during the drop lifetime (assuming typical experimental parameters of a 2mV/sec scan rate and a 4 s drop time, the potential can change by 8 mV from the beginning to the end of the drop), the charging of the interface (capacitive current) has a continuous contribution to the total current, even at the end of the drop when the surface area is not rapidly changing.

Typical signal to noise of a polarographic experiment allows detection limits of only approximately 10^{-5} or 10^{-6} M. Dramatically better discrimination against the capacitive current can be obtained using the tast and pulse polarographic techniques. These have been developed with introduction of analog and digital electronic potentiostats. A first major improvement is obtained, if the current is only measured at the end of each drop lifetime. An even greater enhancement has been the introduction of differential pulse polarography. Here, the current is measured before the beginning and before the end of short potential pulses. The latter are superimposed to the linear potential-time-function of the voltammetric scan. Typical amplitudes of these pulses range between 10 and 50 mV, whereas pulse duration is 20 to 50 ms. The difference between both current values is that taken as the analytical signal. This technique results in a 100 to 1000-fold improvement of the detection limit, because the capacitive component is effectively suppressed.

The rate of diffusion to the mercury electrode surface depends on the concentration of the analyte. Ilkovic equation shows this relation as follow (Eq. 24):

$$i_d = 706 \; n \; C \; D^{1/2} \; m^{2/3} \; t^{1/6} \tag{24}$$

where i_d is current in μA, n is the number of electrons transferred, C is the concentration of the analyte, D is the diffusion coefficient, m is the mass of mercury drop, and t is the drop life time. If during the measurements D, m, and t are constant, then simplify the measurement current correlates with concentration (Eq. 22). This can be a base for quantitative analysis. By using a standard series and calibration method, the concentration of an unknown sample can be easily determined.

$$i_d = K C \tag{25}$$

Qualitative information can also be determined from the half-wave potential (called $E_{1/2}$) of the polarogram (Fig. 5.6D). The value of the half-wave potential is related to the standard potential for the redox reaction being studied. $E_{1/2}$ values are listed *vs.* SCE and they are not the same as E° values. Also, they are dependent on the used supporting electrolyte. It is typically varied from −1.9 to +0.2 V. For example for reduction of Pb^{2+} ion in 0.1 M KCl, the $E_{1/2}$ is −0.4 V while in 1M NaOH is about −0.76 V.

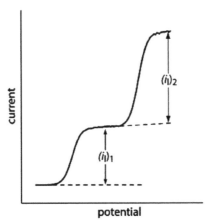

FIGURE 5.7   A typical plarograph for multielectroactive analytes.

One of the most important advantages of polarography and also other voltammetric techniques is that two or even more substances can be determined by a single current potential curve. If more than one electroactive analyte presents in the sample, the curves will add on top of each others (Fig. 5.7).

Polarographic analysis can be directly applied for determination of any substance in solid, liquid or gaseous, organic compounds containing conjugated double or triple bonds including polynuclear aromatic ring systems, as well as compounds like oximes, imines, ketones, aldehydes nitro diazo compounds and halo substituted compounds. Besides, polarography has been used to study hydrolysis, solubility, complex formation, absorption, kinetics of chemical reaction and mechanism of electrode reactions [11–14].

5.3.4.2   LINEAR SWEEP VOLTAMMETRY

The simplest technique that uses this waveform for applying the potential is Linear Sweep Voltammetry (LSV). In this method, the current at a working electrode is

measured while the potential between the working electrode and a reference electrode is swept linearly in time. Oxidation or reduction of species is occurred at the
potential at which the species begins to be oxidized or reduced and. In LSV a fixed
potential range is employed much like potential step measurements. However, in
LSV the voltage is scanned from a lower limit to an upper limit.

The characteristics of the linear sweep voltammogram recorded depend on a
number of factors including, the rate of the electron transfer reaction, the chemical
reactivity of the electroactive species, and the scan rate.

The value of the scan rate may be varied from as low as mV/sec (typical for
polarography experiments) to as high as 1,000,000 V/sec (attainable when ultra-
microelectrodes are used as the working electrode). With a linear potential ramp,
the faradaic current is found to increase at higher scan rates. This is because of the
increased flux of electroactive material to the electrode at the higher scan rates. The
amount of increase in the faradaic current is found to scale with the square root of
the scan rate (see cyclic voltammetry section). This seems to suggest that increasing
the scan rate of a linear sweep voltammetric experiment could lead to increased analytical signal to noise. But, the capacitive contribution to the total measured current
scales directly with the scan rate. As a result, the signal to noise of a linear sweep
voltammetric experiment decreases with increasing scan rate.

Alternating current voltammetry is the application of a sinusoidally oscillating
voltage to LSV. The AC experiment when used in conjunction with a lock in amplifier or frequency analyzer offers considerably increased sensitivity over the early
described techniques and can also reveal important mechanistic and kinetic information not easily available using more tradition voltammetric techniques.

5.3.4.2   CYCLIC VOLTAMMETRY

Cyclic voltammetry (CV) is one the most important voltammetric techniques which
is widely used in many areas of chemistry. Generally it is used for studying the redox processes, for understanding reaction behavior or detecting intermediates, and
for obtaining stability of reaction products, however, it is can be used for quantitative analysis too. In fact, cyclic voltammetry is the electrochemical equivalent of
spectrophotometry. It is the most powerful tool for examining the electrochemical
properties of chemical materials. By CV, information about the rates of electron
transfer between substances and electrodes and also the nature of chemical processes coupled to the electron transfer event can be determined. Properties of reactants
and products can frequently be distinguished from a single voltammogram, or from
a series of voltammograms obtained as a function of scan rate, concentration, pH,
solvent type, temperature, and et cetera.

In this technique, the current is measured after applying a potential to a working electrode in two opposite directions (forward and reverse directions). It means
that a linear sweeping potential is applied to electrode surface. After the potential

reaches a certain maximum value, the potential is reversed and the sign of potential is changed. It is continued by the same scan rate till the applied potential come back to the initial point. The process can then be repeated in a periodically (as shown in Fig. 5.8).

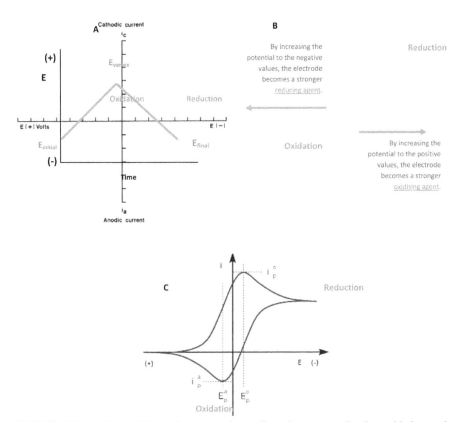

FIGURE 5.8 A: Potential *vs.* time curve in cyclic voltammetry; B: the oxidation and reduction area in an i-E curve; C: A typical cyclic voltamogram

The response obtained from a CV can be very simple, as shown in Fig. 5.8C for a reversible redox system such as:

$$Fe(CN)_6^{-3} + e^- = Fe(CN)_6^{-4}$$

in which the complexed Fe^{3+} is reduced to Fe^{2+}. The important parameters in a cyclic voltammogram are the peak potentials (E_{pc}, E_{pa}) and peak currents (i_{pc}, i_{pa}) of the cathodic and anodic peaks, respectively.

The peak potential of the anodic sweep, E_{pa} and the peak potential for cathodic peak, E_{pc}, can be directly read from the program, and the difference between them, ΔE_p, can be calculated. If the charge transfer process is fast compared to other processes such as diffusion, the reaction is said to be electrochemically reversible, and the peak separation is:

$$\Delta E_p = \left| E_{pa} - E_{pc} \right| = 0.05916/n \qquad (26)$$

And the ratio of the peak currents is equal to one: $\left| i_{pa}/i_{pc} \right| = 1$

Thus, for a reversible redox reaction at 25°C with n electrons ΔE_p should be 0.05916/n V or about 60 mV for one electron. In practice this value is difficult to attain because of such factors as cell resistance.

Irreversibility is occurred when electron transfer rate is slow and causes $\Delta Ep >$ 0.05916/n V, normally more than 70 mV for a one-electron reaction.

The intermediate rates of charge transfer define as quasi reversible reaction.

When the applied potential is equal to E° of the redox couple, the forward and reverse rate constants of redox equilibrium are equal. A standard heterogeneous rate constant is defined for these conditions, expressed as k_s (or $k°$), and having units of cm/s (resulting from concentration of redox active species in mol/cm³ and electron transfer to an electrode of area expressed in cm²).

$k_s > 0.020$ cm/s Reversible

$0.020 > k_s > 5.0 \times 10^{-5}$ cm/s Quasi-reversible

$k_s < 5.0 \times 10^{-5}$ cm/s Irreversible

The formal reduction potential (E°) for a reversible couple is given as follow:

$$E° = \frac{E_{pc} + E_{pa}}{2} \qquad (27)$$

For a reversible reaction, the concentration is related to peak current by the Randles–Sevcik expression (at 25°C):

$$i_p = 2.686 \times 10^5 \, n^{3/2} \, Ac° D^{1/2} v^{1/2} \qquad (28)$$

where i_p is the peak current in ampere, A is the electrode area (cm²), D is the diffusion coefficient (cm² s⁻¹), c° is the concentration in mol cm⁻³, and n is the scan rate in V s⁻¹. As seen, in a reversible reaction, the peak currents are proportional to the square root of the scan rate, thus, by increasing the scan rate the peak current increase too. In case of irreversible reaction, the peak current does not vary as a function of the square root of the scan rate. By analyzing the variation of peak position as a function of scan rate, it is possible to gain an estimate for the electron transfer rate constants and reversibility of the reaction.

Cyclic voltammetry is carried out in quiescent solution to ensure diffusion control. A three-electrode arrangement is used. Mercury film electrodes are used because of their good negative potential range. Other working electrodes include glassy carbon, platinum, gold, graphite, and carbon paste.

Depending on the analysis, a full cycle, a partial of a cycle, or a series of cycles can be used.

5.3.4.3   PULSE METHODS

In order to increase speed and sensitivity, many forms of potential modulation have been developed. The important parameters for pulse techniques are:
- Pulse amplitude, which is the height of the potential pulse. This may or may not be constant depending upon the technique.
- Pulse width, which is the duration of the potential pulse.
- Sampling time, which is the time at the end of the pulse during which the current is measured.
- For some pulse techniques, the pulse period or drop time must also be specified. This parameter defines the total time required for one potential cycle, and is particularly significant for polarography (i.e., pulse experiments using a mercury drop electrode), where this time corresponds to the lifetime of each drop.

a.   Staircase Voltammetry
The potential is varied in a series of steps, with the current sampled at the end of each step. Staircase voltammetry (SCV) is a derivative of linear sweep voltammetry. Though there are similarities between SCV and CV and LSV, SCV allows more control of the waveform that is applied to the working electrode. CV and LSV waveforms are optimized to take full advantage of the resolution of the potentiostat's digital-to-analog converter, while the waveform in SCV is designed for much higher sweep rates than typically encountered in CV or LSV. Sometimes for have a more sensitive signal, staircase is applied to the CV potential scan function too.

b.   Normal Pulse Voltammetry
This technique uses a series of potential pulses, which their amplitudes are increasing. The current measurement is made near the end of each pulse, which allows time for the charge current to decay. It is usually carried out in an unstirred solution at either DME (called normal pulse polarography) or solid electrodes. The potential is pulsed from an initial potential E_i. The duration of the pulse, t, is usually 1 to 100 ms and the interval between pulses typically 0.1 to 5 s.

c.   Differential Pulse Voltammetry
Differential Pulse Voltammetry (DPV) (or Differential Pulse Polarography, DPP) can be considered as a derivative of linear sweep voltammetry. This technique is comparable to normal pulse voltammetry in that the potential is also scanned with

a series of pulses. However, it differs from NPV because each potential pulse has fixed and small amplitude (10 to 100 mV), and is superimposed on a slowly changing base potential. Current is measured at two points for each pulse, the first point (1) just before the application of the pulse and the second (2) at the end of the pulse. These sampling points are selected to allow for the decay of the non-Faradaic current. The difference between current measurements at these points for each pulse is determined and plotted against the base potential.

d. Square Wave Voltammetry

Square wave voltammetry (SWV) is one of the four major voltammetric techniques provided by modern computer-controlled electro-analytical instruments. The excitation signal in SWV consists of a symmetrical square-wave pulse of amplitude E_{sw} superimposed on a staircase waveform of step height ΔE, where the forward pulses of the square wave correspond with the staircase step. The net current, i_{net}, is obtained by taking the difference between the forward and reverse currents ($i_{for} - i_{rev}$) and is centered on the redox potential. The peak height is directly proportional to the concentration of the electroactive species and direct detection limits as low as 10^{-8} M is possible.

Square-wave voltammetry has several advantages. Among these are its excellent sensitivity and the rejection of background currents. Another is the speed (for example, its ability to scan the voltage range over one drop during polarography with the DME). This speed, coupled with computer control and signal averaging, allows for experiments to be performed repetitively and increases the signal to-noise ratio.

Table 5.2 summarized i-E curves and potential *vs.* current for different wave form and pulse form of voltammetric techniques.

TABLE 5.2   Different Wave and Pulse Voltammetric Techniques

Technique	Applied potential vs. time	i-E curve	Normal Concentration range (M)
LSV			$10^{-6} - 10^{-2}$

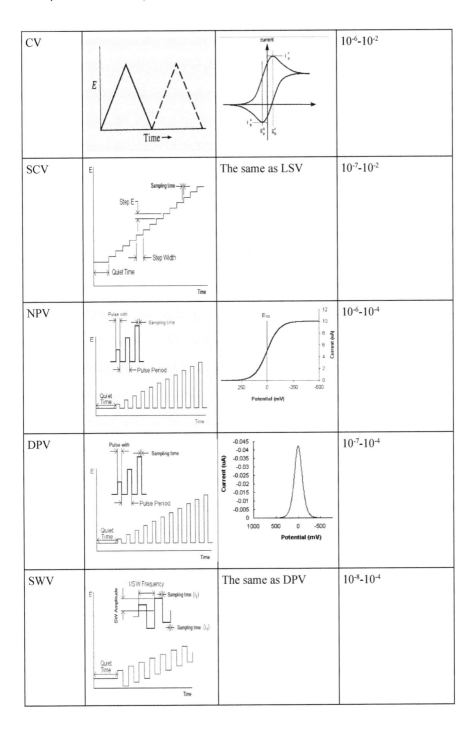

CV			$10^{-6}\text{-}10^{-2}$
SCV		The same as LSV	$10^{-7}\text{-}10^{-2}$
NPV			$10^{-6}\text{-}10^{-4}$
DPV			$10^{-7}\text{-}10^{-4}$
SWV		The same as DPV	$10^{-8}\text{-}10^{-4}$

5.3.4.5   STRIPPING METHODS

The preconcentration techniques have the lowest limits of detection among the commonly used electroanalytical techniques. Sample preparation is minimal and sensitivity and selectivity are excellent. The three most commonly used methods are anodic stripping voltammetry (ASV), cathodic stripping voltammetry (CSV), and adsorptive stripping voltammetry (AdSV).

Even though ASV, CSV, and AdSV each have their own unique features, all have two steps in common. First, the analyte species in the sample solution is concentrated onto or into a working electrode. It is this crucial preconcentration step that results in the exceptional sensitivity that can be achieved. During the second step, the preconcentrated analyte is measured or stripped from the electrode by the application of a potential scan. Any method of applying potential can be used for the stripping step (differential pulse, square wave, linear sweep, or staircase). The most common methods used are differential pulse and square wave due to the discrimination against charge current. However, square wave has the added advantages of faster scan rate and increased sensitivity relative to differential pulse.

The electrode of choice for stripping voltammetry is generally mercury. The species of interest can be either reduced into the mercury, forming amalgams as in anodic stripping voltammetry, or adsorbed to form an insoluble mercury salt layer, as in cathodic stripping voltammetry.

Stripping voltammetry is a very sensitive technique for trace analysis. As with any quantitative technique, care must be taken so that reproducible results are obtainable. Important conditions that should be held constant include the electrode surface, rate of stirring, and deposition time. Every effort should be made to minimize contamination.

Anodic stripping voltammetry–A quantitative, analytical method for trace analysis of metal cations. The analyte is deposited (electroplated) onto the working electrode during a deposition step, and then oxidized during the stripping step. The current is measured during the stripping step.

ASV is most widely used for trace metal determination and has a practical detection limit in the part per-trillion range. This low detection limit is coupled with the ability to determine simultaneously four to six trace metals using relatively inexpensive instrumentation.

Metal ions in the sample solution are concentrated into a mercury electrode during a given time period by application of a sufficient negative potential. These amalgamated metals are then stripped (oxidized) out of the mercury by scanning the applied potential in the positive direction. The resulting peak currents, i_p, are proportional to the concentration of each metal in the sample solution, with the position of the peak potential, E_p, specific to each metal. The use of mercury limits the working range for ASV to between approximately 0 and -1.2 V versus SCE. The use of thin Hg films or Hg microelectrodes along with pulse techniques such as square-wave voltammetry can substantially lower the limits of detection of ASV.

With more than one metal ion in the sample, the ASV signal may sometimes be complicated by formation of intermetallic compounds, such as ZnCu. This may shift or distort the stripping peaks for the metals of interest. These problems can often be avoided by adjusting the deposition time or by changing the deposition potential.

Cathodic stripping voltammetry—A quantitative, analytical method for trace analysis of anions. A positive potential is applied, oxidizing the mercury electrode and forming insoluble precipitates of the anions. A negative potential then reduces (strips) the deposited film into solution.

CSV can be used to determine substances that form insoluble salts with the mercurous ion. Application of a relatively positive potential to a mercury electrode in a solution containing such substances results in the formation of an insoluble film on the surface of the mercury electrode. A potential scan in the negative direction will then reduce (strip) the deposited film into solution.

This method has been used to determine inorganic anions such as halides, selenide, and sulfide, and oxyanions such as MoO_4^{2-} and VO_3^{5-}. In addition, many organic compounds, such as nucleic acid bases, also form insoluble mercury salts and may be determined by CSV.

Adsorptive stripping voltammetry—A quantitative, analytical method for trace analysis. The analyte is deposited simply by adsorption on the electrode surface (i.e., no electrolysis), then electrolyzed to give the analytical signal. Chemically modified electrodes are often used in this method.

AdSV is quite similar to anodic and cathodic stripping methods. The primary difference is that the preconcentration step of the analyte is accomplished by adsorption on the electrode surface or by specific reactions at chemically modified electrodes rather than accumulation by electrolysis.

Many organic species have been determined at micromolar and nanomolar concentration levels using AdSV; inorganic species have also been determined. The adsorbed species is quantified by using a voltammetric technique such as DPV or SWV in either the negative or positive direction to give a peak-shaped voltammetric response with amplitude proportional to concentration.

Molecularly imprinted polymers (MIP) are promising materials, which are currently used as the extraction agent in carbon paste electrodes. The MIP-based solid phase extraction as a high selective preconcentration and separation method and the square wave voltammetry as a high sensitive determination technique lead to low concentration detections.[15–17] However, it suffers from some drawbacks such as long analyzing time, high RSD, complexity of the method.

5.4 AMPEROMETRIC METHOD

Amperometry is the term indicating the whole of electrochemical techniques in which a current is measured as a function of an independent variable that is, typically, time or electrode potential.

5.4.1 SINGLE-POTENTIAL AMPEROMETRY

Any analyte that can be oxidized or reduced is a candidate for amperometric detection. The simplest form of amperometric detection is single-potential, or direct current (DC), amperometry. A voltage (potential) is applied between two electrodes positioned in the column effluent. The measured current changes as an electroactive analyte is oxidized at the anode or reduced at the cathode. Single-potential amperometry has been used to detect weak acid anions, such as cyanide and sulfide, which are problematic by conductometric methods. Another, possibly more important advantage of amperometry over other detection methods for these and other ions, such as iodide, sulfite, and hydrazine, is specificity. The applied potential can be adjusted to maximize the response for the analyte of interest while minimizing the response for interfering analytes.

5.4.2 PULSED AMPEROMETRY (PULSED AMPEROMETRIC DETECTION, PAD)

An extension of single-potential amperometry is pulsed amperometry, most commonly used for analytes that tend to foul electrodes. Analytes that foul electrodes reduce the signal with each analysis and necessitate cleaning of the electrode. In pulsed amperometric detection (PAD), a working potential is applied for a short time (usually a few hundred milliseconds), followed by higher or lower potentials that are used for cleaning the electrode. The current is measured only while the working potential is applied, then sequential current measurements are processed by the detector to produce a smooth output. PAD is most often used for detection of carbohydrates after an anion exchange separation, but further development of related techniques show promise for amines, reduced sulfur species, and other electroactive compounds.

5.4.3 AMPEROMETRIC TITRATION

Amperometric titration refers to a class of titrations in which the equivalence point is determined through measurement of the electric current produced by the titration reaction. It is a form of quantitative analysis.

Consider a solution containing the analyte, A, in the presence of some conductive buffer. If an electrolytic potential is applied to the solution through a working electrode, then the measured current depends (in part) on the concentration of the analyte. Measurement of this current can be used to determine the concentration of the analyte directly; this is a form of amperometry. However, the difficulty is that the measured current depends on several other variables, and it is not always possible to control all of them adequately. This limits the precision of direct amperometry.

The chief advantage over direct amperometry is that the magnitude of the measured current is of interest only as an indicator. Thus, factors that are of critical importance to quantitative amperometry, such as the surface area of the working electrode, completely disappear from amperometric titrations.

5.4.4 CHRONOAMPEROMETRY

It is the technique in which the current is measured, at a fixed potential, at different times since the start of polarization. Chronoamperometry is typically carried out in unstirred solution and at fixed electrode, i.e., under experimental conditions avoiding convection as the mass transfer to the electrode.

Chronoamperometry is an electrochemical technique in which the potential of the working electrode is stepped and the resulting current from faradic processes occurring at the electrode (caused by the potential step) is monitored as a function of time. Limited information about the identity of the electrolyzed species can be obtained from the ratio of the peak oxidation current versus the peak reduction current. However, as with all pulsed techniques, chronoamperometry generates high charging currents, which decay exponentially with time as any RC circuit. The Faradaic current-which is due to electron transfer events and is most often the current component of interest-decays as described in the Cottrell equation. In most electrochemical cells this decay is much slower than the charging decay-cells with no supporting electrolyte are notable exceptions. Most commonly investigated with a three electrode system. Since the current is integrated over relatively longer time intervals, chronoamperometry gives a better signal to noise ratio in comparison to other amperometric technique.

5.5   COULOMETRIC METHODS

Coulometry is the name given to a group of techniques in analytical chemistry that determine the amount of species during an electrolysis reaction by measuring the amount of electricity (in coulombs) consumed or produced.

In this method applied current or potential is used to completely convert an analyte from one oxidation state to another. In these experiments, the total current passed is measured directly or indirectly to determine the number of electrons passed. Knowing the number of electrons passed can indicate the concentration of the analyte or, when the concentration is known, the number of electrons transferred in the redox reaction.

Common forms of coulometry include bulk electrolysis, also known as potentiostatic coulometry or controlled potential coulometry. Potentiostatic coulometry involves holding the electric potential constant during the reaction using a potentiostat. The other, called coulometric titration or amperostatic coulometry, keeps the current (measured in amperes) constant using an amperostat.

An advantage to this kind of analysis over electrogravimetry (in this method analyte solution is electrolyzed. Electrochemical reduction causes the analyte to be deposited on the cathode. The cathode is weighed before and after the experiment, and weighing by difference is used to calculate the amount of analyte in the original solution. Controlling the potential of the electrode is important to ensure that only the metal being analyzed will be deposited on the electrode.) is that it does not require that the product of the reaction be weighed. This is useful for reactions where the product does not deposit as a solid, such as the determination of the amount of arsenic in a sample from the electrolysis of arsenous acid (H_3AsO_3) to arsenic acid (H_3AsO_4).

5.6   ADVANCED ELECTROCHEMICAL TECHNIQUES

Now days to have more sensitive electroanalytical methods and to be able to determine very low concentrations of compounds, some advanced electrochemical techniques are applied in measurement systems. Here, some of the important ones are discussed.

5.6.1   IMPEDANCE/ADMITTANCE TECHNIQUES

Electrical resistance is the ability of a circuit element to resist the flow of electrical current (according to the Ohm's law; $E=IR$). Conductance of a circuit is defined as inverse of the resistance. While this is a well known relationship, its use is limited to only the ideal resistor. In an ideal resistor Ohm's Law can be applied at all current and voltage levels, its resistance value is independent of frequency and AC current and voltage signals though a resistor are in phase with each other. However, the real world contains circuit elements that exhibit much more complex behavior. These elements force us to abandon the simple concept of resistance. Thus, impedance is defined instead of resistance and admittance instead of conductance. Like resistance, impedance is a measure of the ability of a circuit to resist the flow of electrical current, but unlike resistance, it is not limited by the simplifying properties listed above.

Electrochemical impedance is usually measured by applying an AC potential to an electrochemical cell and then measuring the current through the cell. Assume that we apply an altering potential excitation. The response to this potential is an AC current signal. This current signal can be analyzed as a sum of sinusoidal functions (a Fourier series).

Based on this approach, electrochemical Impedance Spectroscopy (EIS) is a relatively new and powerful technique to characterize the electrical properties of materials [18] surface-modified electrodes[19] for study the electrochemical processes.[20] It is used in study of dynamics of bound or mobile charge in the bulk or interfacial regions of any kind of material (solid or liquid). EIS is a sensitive technique based on monitoring the electrical response of a device after application of a periodic small amplitude AC signal in a wide range of frequencies (typically, from 100 kHz to 0.1 Hz). The analysis of the impedance values measured provides information concerning the electric properties of the electrode-sample interface and the underlying reactions.

The Randle modified equivalent circuit is used to fit the EIS data and to determine electrical parameter values for the concentration of analyte and Nyquist plots are used to show the changes in impedance of the electrode surface. As shown in Fig. 5.9 (inset), the circuit includes the electrolyte resistance between working and reference electrodes (R_s); Warburg impedance (Z_w), resulting from the diffusion of ions to the interface from the bulk of the electrolyte; electron-transfer resistance (R_{et}); and electrode/electrolyte interface capacitance (C). The value of Z_w gives information about the diffusion of analyte through the surface layer of the electrode, while the Rs values depend on the solution. In addition, C models the capacitive behavior of the double layer replacing the infrequently ideal capacitance and diffusion behavior. The values of R_{et} increases significantly upon adsorption of analyte on the electrode surface with concentration, reflecting the more hindered charge transfer diffusion.

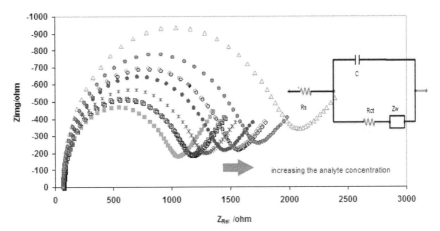

FIGURE 5.9   Nyquist plots of EIS spectra of after interaction with different concentrations of an analyte.

Another different approach in electrochemical measurements is admittometry. This method is fundamentally different approach to SWV measurement, in which the detection limits are improved, while preserving the information content of the

SW voltammogram. In this way, the analyte signal is calculated based on admittance changes related to the changes in electrical double layer.

4.6.2 ELECTROANALYTICAL MEASUREMENTS USING FAST FOURIER TRANSFORM (FFT)

Most electrochemical measurements are performed in the time domain. However, in some cases, we require more information for the obtained data such as knowledge about the frequency content and behavior of the electroanalytical signals and of complete systems. Fortunately, there exists a defined method for transforming data from the time domain into the frequency domain, where information exist about the spectral content of electrochemical data.

The solution of this problem is Fourier Transform algorithm, which has the ability to convert a time domain data to the complex frequency domain. Thus, the spectral data contains information about both the amplitude and phase of the sinusoidal components that make up the signal. In addition, the inverse FT, converts the generated complex frequency-domain signal data back into the time-domain without losing wanted information. Accordingly, it can say that the both the time- and frequency-domain data complement and the two domains can provide a different view of the same electrochemical data.

Application of Fast Fourier transformation algorithm for numerical electrochemical data provides the complex spectrum according to magnitude and phase, which can be used for real time analysis. In this direction, in modern electrochemistry, FFT has been used for digital signal processing to provide a sensitive system when combine by normal electrochemical method in trace analysis of compounds.[21]

Most of the electrochemical signals have a defined value for every possible instant in time. They are called continuous signal like analog voltage and current. To analyze these continuous signals by a computer-based system, the signal should convert to digital signal in which each sample representing a numeric value that is proportional to the measured signal at a specific instant in time. Thus, the sampling process used for electrochemical measurements creates digital signal data spaced on an even interval of time. Fourier Transform is a defined technique for converting or transforming electrochemical signals from the time domain to the frequency domain. Hence, the electrochemical signal does not suffer from environmental noises. The approach used here is designed to separate the voltammetric signal and background signal in frequency domain by using discrete Fast Fourier Transformation (FFT) method. Based on FFT information, the cutoff frequency of the analog filter is set at a certain value. Thus, some of the noises filtrate digitally and decrease the bandwidth of the measurement (Fig. 5.10).[22]

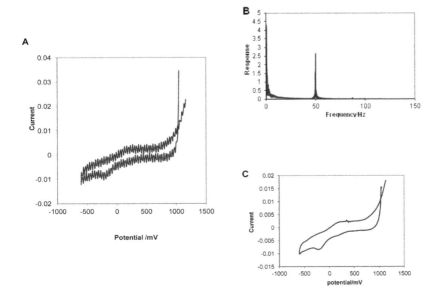

FIGURE 5.10   FFT algorithm is applied into the cyclic voltammogram.

In modern electroanalytical methods, the FFT analysis voltammetry were developed in order to overcome some existing limitations encountered with electronic instrumentations. In fact, the selectivity and precision characteristics of the classic electroanalytical methods depend on the number of filter circuits. Normally, the most potentiostat typically has many filter circuits that are arranged before and after the amplifier. The main problem here is with applying a fast excitation signal for fast electrochemical measurement. Actually, application a fast excitation signal can produce a large charging current due to existing capacitor in the analog filters. On the other hand, by using FFT filtering method, instead of the analog filters, the signal can be measured very quickly. Consequently, the time element per sample is reduced to a matter of less than second (10^{-9} s) rather than several minutes, which happened in the classical analog measurements. Thus, a superior improvement in limit of detections has been occurred.

4.6.3   ELECTROCHEMICAL SENSORS AND BIOSENSORS

Sensors are widely used in daily objects of life. An electrochemical sensor provides continuous information about its environmental chemical changes, and converts the chemical response into an electrical signal that can be detected by modern instrumentations. Electrochemical biosensors are one of the important subdivisions of electrochemical sensors. An electrochemical biosensor is an analytical device, which converts a biological response into a detectable electrical signal.

Each chemical sensor/biosensor composed of a sensing element, a transducer, and a signal processor or a detector. The first part, which is the most significant part of the sensor/biosensor, is a chemical selective material, or the recognition element, which should recognize and differentiate between analyte and its environment and provide a selective or even specific response to the changes in concentration of an analyte. Sensing materials in a sensor can be an organic compound (selectophore), an inorganic complex or even a solid material and in a biosensor can be a microorganism, part of a tissue, a cell, an organelle, a sequence of nucleic acid, an enzyme, or an antibody.

Conversion of a biological or chemical signal into a measurable electrical signal can be done by a physicochemical transducer based on one of the electrochemical techniques.

Signals from the transducer are passed to a processor where they are amplified and analyzed. The data is then converted to concentration units and transferred to a display or/and data storage device.

Using sensors and biosensors in the environmental analysis has the following advantages:

- Easy to use
- Fast response
- Low cost method
- Non-destructive method of analysis
- Can be a portable device by miniaturization of the electrodes and electronic processors

All of the potentiometric ion selective electrodes, which are discussed above, are a kind of ion-sensors.

As an example, a novel atrazine electrochemical biosensor [23] was developed based on antibody combine with Au nanoparticles, multi walled carbon nanotube and ionic liquid on glassy carbon electrode surface. Coulometric fast Fourier transformation square wave voltammetry was used for the electrochemical measurements. In this method, the admittance response of the electrode was integrated in a selected potential range to calculate amount of transferred charge during the adsorption of atrazine. The linear concentrations range of atrazine was from 0.5–100 nM with a detection limit of 0.02 nM.

Another example is a new flow injection enzymatic acetylcholinesterase (AChE) biosensor (Fig. 5.11) [24] was designed for selective determination of monocrotophos (an organophosphate pesticide). The biosensor was constructed by modifying glassy carbon electrode surface with gold nanoparticles (AuNPs) and multiwall carbon nanotubes (MWCNTs) while chitosan microspheres used to immobilize AChE. The measurement method was based on fast Fourier transform continuous cyclic voltammetry (FFTCCV) in which the charge under the peak calculated in a specific potential range. The inhibition of the enzyme activity by monocrotophos was proportional to monocrotophos concentration in the range of 0.1 to 10 μM, with a detection limit of about 10 nM.

Organophosphate and carbamate pesticides exert their toxicity via attacking the hydroxyl moiety of serine in the 'active site' of acetylcholinesterase (AChE). In another paper,[25] a stable AChE biosensor was developed based on self-assembling of AChE to grapheme nanosheet (GN)-gold nanoparticles (AuNPs) nanocomposite electrode for investigation of inhibition, reactivation and aging processes of different pesticides.

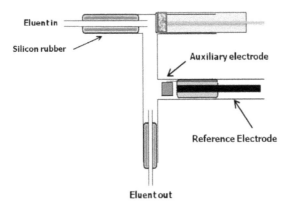

FIGURE 5.11   The diagram of monocrotophos biosensor and the electrochemical cell used in flow injection analysis.

5.6.3.1   GAS SENSORS

The oldest electrochemical sensors date back to the 1950s and were used for oxygen monitoring. Recently, new electrochemical gas sensors have been developed for monitoring of toxic and combustible gases in confined space applications.

An electrochemical gas sensor (Fig. 5.12) basically comprises a working, a counter and sometimes a reference electrode, which are enclosed in the sensor housing and are in contact with a liquid electrolyte. The working electrode is on the inner side of a Teflon membrane, which is permeable to a gas but not to the electrolyte.

The gas sample diffuses through the membrane into the sensor and eventually to the working electrode, leading to an electrochemical reaction (either an oxidation or reduction depending on the type of gas) to occurs. Carbon monoxide is an example of gases that are oxidized, while oxygen and its reduction to water characterized those reduced.

Oxidation reactions lead to the flow of electrons from the working to the counter electrode through the external circuit while reduction reactions result in an opposite flow. These flows, being proportional to the gas concentration, are detected and amplified by the electronic parts of the instrument, providing the data required for drawing a calibration curve. Finally the instrument displays the gas concentration in a predetermined and proper unit (e. g. ppm for toxic gases and percent volume for oxygen sensors).

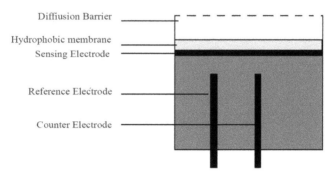

FIGURE 5.12   Structure of a gas sensor.

Due to the importance of gas detection in controlling the air pollutant, and in household security, many such devices have been developed for detecting species like CO_2, CO, SO_2, O_2, O_3, H_2, Ar, N_2, NH_3, H2O and several organic vapors. Different approach of electrochemical techniques can be used for gas sensing. Both potentiometric and voltammetric gas sensors are developed. Different materials such as various metal oxide (such as aluminum oxide, gallium oxide, indium oxide, tin oxide, bismuth oxide, titanium oxide, zirconium oxide, niobium oxide, tantalum oxide, chromium oxide, molybdenum oxide, tungsten oxide, manganese oxide, iron oxide, cobalt oxide, nickel oxide, copper oxide, zinc oxide, cadmium oxide, cerium oxide); composite/multicomponent metal oxide; and conducting polymer.

Up to now some electrochemical gas sensors have been reported a complete list of these gas sensors have been previously reviewed systematicall [2]. For O_2, O_3, H_2, Cl_2, CO, CO_2, NO, NO_2, SO_2, SO_3, H_2S and some Hydrocarbons which are the environmentally important gases, many sensors have been developed during the year.

5.6.4   ARRAY OF SENSORS (ELECTRONIC TONGUE/NOSE)

Sensor arrays provide multicomponent data at onetime without no extra effort for the same sample processing. By designing multichannel electrochemical instrumentation, make it possible to make array of sensors. In 2005, IUPAC established the international nomenclature for potentiometric analysis of liquids. In this case, the electronic tongue is defined as a multisensor system, which consists of a number of low-selective sensors and uses advanced mathematical procedures for signal processing based on pattern recognition and/or multivariate analysis [26]. The second design of electronic tongue is based on sensor arrays where each electrode is selective for one analyte or group of similar analytes and provides information about their quantity. The obtained data are often distorted due to overlapping or interference signals, therefore it is necessary to use multivariate calibration.

Arrays of gas sensors are termed 'electronic noses' while arrays of liquid sensors are referred to as 'electronic tongues'.

Kundu et al. [27] in 2011 developed a new approach for water sample authentication, in real life, using a pulse-voltammetry-method-based electronic tongue instrumentation system. The system is developed as a parallel combination of several neural network classifiers; each dedicated to authenticate a specific category of water sample, and can be extended for more categories of water sample authentication. This proposed system, implemented in a laboratory environment for various water samples available in India, showed encouraging average authentication percentage accuracy, on the order of over 80% for most water categories and even producing accuracy results exceeding 90%, for several categories.

Potentiometry and voltammetry by various modified working electrodes can be applied as detection systems. The challenge in this field is that try to array more sensors in a systems to simultaneous determination of species and the miniaturized the electrodes and processors to have a portable devices.

An automatic electronic tongue for online detection and quantification of organophosphorus and carbamate pesticides using enzymatic screen printed biosensors has been reported [28]. Also, an automatic monitoring electronic tongue based on differential pulse stripping voltammetry (DPSV) was developed for heavy metals analysis. Simultaneous detections of trace Zn(II), Cd(II), Pb(II), Cu(II), Fe(III) and Cr(III) in water samples were performed with three electrochemical sensors. The sensor chip is made of a silicon-based Hg-coated Au microelectrode array (MEA) as the working electrode on one side with an Ag/AgCl reference electrode and a Pt counter electrode on the other side. By a computer controlled multipotentiostat, pumps and valves, the electronic tongue realized in-situ real-time detection of the six above metals at ppb level without manual operation [29].

5.6.5   LAB ON A CHIP

A lab-on-a-chip (LOC) is a device that integrates one or several laboratory functions on a single chip of only millimeters to a few square centimeters in size. LOCs deal with the handling of extremely small fluid volumes down to less than pico liters. Lab-on-a-chip devices are a subset of MEMS devices and often indicated by "Micro Total Analysis Systems" (μTAS) as well. LOC is closely related to microfluidics (the manipulation and study of minute amounts of fluids). However, LOC is defined as the scaling of single or multiple lab processes down to chip-format, whereas "μTAS" is dedicated to the integration of the total sequence of lab processes to perform chemical analysis.

There is a huge potential for development of voltammetric environmental trace-metal analysis, particularly in the field of in situ measurements where other techniques are much more limited for various reasons.

All these developments will be based on both microelectrodes and micro total analytical systems for speciation requirements. Thus, in situ environmental voltammetry will progress, only in close relations with microtechnology, as it is required to develop of reliable procedures for fabrication of microelectrodes and micrototal-analytical systems (μTAS).

Buffle and Tercier-Waeber [30], in an article reviewed application of voltammetry for in-situ analysis of organic and inorganic environmentally important compounds. They described development of new concepts for metal-speciation measurements in association with both voltammetric measurements and biogeochemical processes, new bioanalogical sensors, modern μTAS, and rugged submersible probes based on specifically designed mechanics, electronics and software.

LOCs have advantages such as low fluid volumes consumption (less waste, lower reagents costs and less required sample volumes for diagnostics), faster analysis and response times due to short diffusion distances, fast heating, high surface to volume ratios, small heat capacities, better process control because of a faster response of the system (e.g., thermal control for exothermic chemical reactions), compactness of the systems due to integration of much functionality and small volumes massive parallelization due to compactness, which allows high-throughput analysis lower fabrication costs, allowing cost-effective disposable chips, fabricated in mass production, safer platform for chemical, radioactive or biological studies because of integration of functionality, smaller fluid volumes and stored energies [31].

Some of the disadvantages of LOCs are novel technology and therefore not yet fully developed, physical and chemical effects like capillary forces, surface roughness, chemical interactions of construction materials on reaction processes become more dominant on small-scale. This can sometimes make processes in LOCs more complex than in conventional lab equipment, detection principles may not always scale down in a positive way, leading to low signal-to-noise ratios. As a result, still some advanced works is needed to do analysis by a small chip.

ACKNOWLEDGMENT

The authors are acknowledged the research council of University of Tehran for financial support of this work.

KEYWORDS

- **Advanced Techniques**
- **Amperometry**
- **Coulometry**
- **Electrodes**
- **Potentiometry**
- **Sensors**
- **Voltammetry**

REFERENCES

1. Ganjali, M. R., Norouzi, P., & Rezapour, M. (2006). Encyclopedia of Sensors: Potentiometric Ion Sensors, Potentiometric Ion Sensors, American Science Publisher (ASP), *8*, 197–288.
2. Ganjali, M. R., Norouzi, P., & Faridbod, F. (2010). Research Signpost Transworld Research Network, Electrochemical Sensors, Include 11 Chapters.
3. Püntener, M., Vigassy, T., Baier, E., Ceresa, A., & Pretsch, E. (2004). Improving the Lower Detection limit of Potentiometric Sensors by Covalently binding the Ionophore to a Polymer Backbone, Anal Chim Acta, *503*, 187–194.
4. Qin, W., Zwickl, T., & Pretsch, E. (2000). Improved Detection Limits and Unbiased Selectivity Coefficients Obtained by Using Ion-Exchange Resins in the Inner Reference Solution of Ion-Selective Polymeric Membrane Electrodes, Anal Chem., *72*, 3236–3240.
5. Peper, S., Ceresa, A., & Bakker, E. (2001). Improved Detection Limits and Sensitivities of Potentiometric Titrations, Analysis Chemistry, *73*, 3768–3775.
6. Telting-Diaz, M., & Bakker, E. (2000). Effects of Lipophilic Ion-Exchanger Leaching on the detection Limit of carrier-based Ion-Selective Electrodes, Anal Chem, *73*, 5582–5589.
7. Heineman, P. W. R. (1996). Laboratory Techniques in Electro-Analytical Chemistry, Second Edition, Revised and Expanded (2nd Ed.), CRC ISBN 0–8247–9445–1.
8. Zoski. Cynthia, G. (2007). Hand book of Electrochemistry, Elsevier Science, ISBN: 0–444–51958–0.
9. Bard, Allen J., & Larry, R. Faulkner. (2000). Electrochemical Methods: Fundamentals and Applications (2 Ed.), Wiley ISBN 0–471–04372–9.
10. Nicholson, R. S., & Irving Shain (1964). Theory of Stationary Electrode Polarography, Single Scan and Cyclic Methods Applied to Reversible, Irreversible, and Kinetic Systems, Anal. Chem, *36*, 706–723.
11. Heyrovsky, M. (2011). Polarography-Past, Present, and Future, J. Solid State Electrochem, *15*, 1799–1803.
12. Bruce, D., Kuhn, A., & Sojic, N. (2004). Electrochemical Removal of Metal Cations from Wastewater Monitored by Differential Pulse Polarography, J. Chem. Educ., *81*, 255–258.
13. Barek, J., Fogg, A. G., Muck, A., & Zima, J. (2001). Polarography and Voltammetry at Mercury Electrodes, Crit. Rev. Anal, Chem., *31*, 291–309.

14. Kariuki, S., Morra, M. J., Umiker, K. J., & Cheng, I. F. (2001). Determination of Total Ionic Polysulfides by Differential Pulse Polarography, Anal Chim, Acta, *442*, 277–285.

15. Alizadeh, T., Ganjali, M. R., Zare, M., & Norouzi, P. (2012). Selective Determination of Chloramphenicol at Trace Level in Milk Samples by the Electrode Modified with Molecularly Imprinted Polymer. *Food Chemistry, 130 (4),* 1108–1114.

16. Alizadeh, T., Zare, M., Ganjali, M. R., Norouzi, P., & Tavana, B. (2010). A New Molecularly Imprinted Polymer (MIP)-Based Electrochemical Sensor for Monitoring 2, 4, 6-Trinitrotoluene (TNT) in Natural Waters and Soil Samples, *Biosens Bioelectron, 25(5),* 1166–1172.

17. Alizadeh, T., Ganjali, M. R., Zare, M., & Norouzi, P. (2010). Development of a Voltammetric Sensor Based on a Molecularly Imprinted Polymer (MIP) for Caffeine Measurement, Electrochim, Acta, *55(5),* 1568–1574.

18. Barsoukov, E., & MacDonald, J. R. (2005). Impedance Spectroscopy: Theory, Experiment and Applications (second ed.), Wiley Inter science, New Jersey (Chapter 1).

19. Katz, E., & Willner, I. (2003). Probing Bio molecular Interactions at conductive and Semi-conductive Surfaces by Impedance Spectroscopy: Routes to Impedimetric Immunosensors, DNA-Sensors, and Enzyme Biosensors, Electro analysis, *15*, 913–947.

20. Bard, A. J., Stratmann, M., & Unwin, P. R. (2003). Instrumentation and Electroanalytical Chemistry, *3*, Wiley VCH, NewJersey, (Chapters 1–2).

21. Norouzi, P., Pirali-Hamedani, M., Mirzaei Garakani, T., & Ganjali, M. R. (2011). Application of Fast Fourier Transforms in Some Advanced Electro analytical Methods, Intech, Chapter 15, "Fourier Transforms New Analytical Approaches and FTIR Strategies", Book Edited by Goran Nikolic, ISBN 978–953–307–232–6.

22. Norouzi, P., Garakani, T. M., & Ganjali, M. R. (2012), Using Fast Fourier Transformation Continuous Cyclic Voltammetry Method for New Electrodeposition of Nano-Structured Lead Dioxide. Electrochim Acta, *77*, 97–103.

23. Norouzi, P., Larijani, B., Ganjali, M. R., & Faridbod, F. (2012). Admittometric Electrochemical Determination of Atrazine by Nano-Composite Immune-Biosensor using FFT- Square Wave Voltammetry, Int. J. Electrochem. Sci., *7*, 10414–10426.

24. Norouzi, P., Pirali-Hamedani, M., Ganjali, M. R., & Faridbod, F. (2010). A Novel Acetyl-cholinesterase Biosensor Based on Chitosan-Gold Nanoparticles Film for Determination of Monocrotophos Using FFT Continuous Cyclic Voltammetry, Int. J. Electrochem. Sci. *5*, 1434–1446.

25. Zhang, L., Long, L. J., Zhang, W. Y., Du, D., & Lin, Y. H. (2012). Study of Inhibition, Reactivation and Aging Processes of Pesticides Using Graphene Nanosheets/Gold Nanoparticles-Based Acetylcholinesterase Biosensor, Electro analysis, *24*, 1745–1750.

26. Vlasov, Y., Legin, A., Rudnitskaya, A., Di Natale, C., & Amico, A. D. (2005). Analysis of Liquids (IUPAC Technical Report), Pure Appl Chem., *77*, 1965.

27. Kundu, P. K., Chatterjee, A., & Panchariya, P. C. (2011), Electronic Tongue System for Water Sample Authentication: A Slantlet-Transform-Based Approach, IEEE Transaction, *60*, 1959–1966.

28. Alonso, G. A., Munoz, R., & Marty, J. L. (2013). Automatic Electronic Tongue for On-Line Detection and Quantification of Organophosphorus and Carbamate Pesticides Using Enzymatic Screen Printed Biosensors, Anal. Lett, *46*, 1743–1757.

29. Zou, S. F., Men, H., Li, Y., Wang, Y. P., & Wang, P. (2006). Automatic Monitoring Electronic Tongue with MEAs for Environmental Analysis, Rare Metal Mat Eng., *35*, 381–384.

30. Buffle, J., & Tercier-Waeber, M. L. (2005). Voltammetric Environmental Tracemetal Analysis and Speciation: from Laboratory to in Situ Measurements, Trends in Anal. Chem., *24*, 172–191.

31. Pawell, R. S., Inglis, D. W. Barber, T. J., & Taylor, R. A. (2013). Manufacturing and wetting Low-cost Microfluidic Cell Separation Devices, Biomicrofluidics, *7*, 056501.

PART II

ENVIRONMENTAL APPLICATIONS

CHAPTER 6

ENVIRONMENTAL CHEMISTRY

MAHMOOD M. BARBOOTI

Department of Applied Chemistry, School of Applied Sciences, University of Technology, P.O. Box 35045, Baghdad, Iraq; E-mail: brbt2m@gmail.com

CONTENTS

6.1   INTRODUCTION

This chapter is intended to be a transfer station between the first part of this book, the fundamental of instrumental analysis and the second part, which will deal with the environmental applications. Environmental chemistry is the study of the sources, reactions, transport, effects, and fates of chemical species in water, soil, and air environments, and the effects of technology thereon. Environmental chemists may in more than one group. The first group works on the monitoring, and study the ways of prevention of environmental deterioration. They work together with industrial or production authorities to help in complying with environmental regulations (WHO, EPA, UNEP, etc.). They work with industrial processes to insure that targets for environmental emissions and discharge are achieved. Agricultural participation in environmental pollution must also be considered. Infiltration of residual pesticides and fertilizers need to be monitored as a potential source of soil and water contamination. Even green and traditional practices of manure application are accompanied with the release of nutrients like phosphate into surface waters and cause eutrophication.

The second group of environmental chemists help in the environmental cleanup, or remediation processes to restore the environment by using mechanical, chemical or biological methods to detoxify land or water that has become contaminated.

Another group the scientists are defining environmental issues by doing basic research. Environmental chemists participate in all aspects of environmental studies from collecting the data for basic research, to monitoring environmental quality to developing chemical processes for remediation and environmental cleanup and to provide a healthier environment. Environmental chemists work with biologists, geologists, atmospheric scientists, engineers, lawyers and legislators. Many terms used in environmental chemistry originated with these disciplines. In fact all environmental activities start with environmental analysis to determine the level of pollutants to present the essential data for the legislators, engineers and decision makers.

6.1.1   EARTH SYSTEMS

Earth consists of the following systems:
1.   The hydrosphere: It includes water in all forms, snow, liquid water and water vapor in the atmosphere. It extends from the depths of the sea to the upper reaches of the troposphere where water is found. Ninety-seven percent of the hydrosphere is found in salty oceans, and the remainder is found as vapor or droplets in the atmosphere and as liquid in ground water, lakes, rivers, glaciers and snowfields.
2.   The atmosphere: it includes all of the Earth's air and is divided into troposphere, stratosphere, mesosphere, thermosphere and ionosphere.
3.   The lithosphere, which includes earth's crust, core and mantle.
4.   The biosphere, which includes all life forms within water land and atmosphere.

Each of these systems has its unique identity, but also has substantial interaction with other spheres and affected by events taking place in other spheres. Environmental scientists study the effects of events in one sphere on the other spheres. The events may be natural and human induced like industry and agriculture.

Plants absorb carbon dioxide from atmosphere and produce oxygen by photosynthesis. Plants and animals release carbon dioxide by respiration. Technology uses rocks and minerals to produce metals and nonmetals for the various uses and in the same time dispose polluted water to the hydrosphere, solid waste, and pollutant gases into the atmosphere. Engines combust gases to produce energy requirements and emit many gaseous pollutants to the atmosphere.

Biosphere introduces biomass, nutrients, water, CO_2 and oxygen to the water bodies. Modern technology helped in the improvement of the quantity and quality of plants and animals by genetic engineering and introduces fertilizers and pesticides to improve the agricultural production. Technology uses water bodies for transportation and for hydrometallurgy which results of hazardous wastes like toxic metal ions and cyanide. Technology uses the geosphere as a source of minerals and raw materials for the industry and introduces hazardous wastes and heat energy to the environment.

Chemists play important role in the estimation of the impact of the events like industry and agriculture on environment represented by the earth systems. Environmental science studies the interactions between the physical, chemical, and biological components of the environment, including their effects on all types of organisms. Environmental and earth science study the interactions of four major systems or "spheres" (Fig. 6.1).

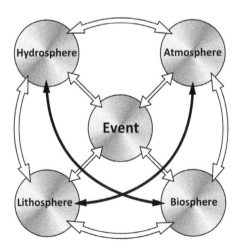

FIGURE 6.1   The interactions of earth systems.

6.1.2 BIOGEOCHEMICAL CYCLES

6.1.2.1   HYDROLOGIC (WATER) CYCLE

The water cycle [1] describes the continuous movement of water on, above and below the surface of the Earth. Although the balance of water on Earth remains fairly constant over time, individual water molecules can come and go, in and out of the atmosphere. The water moves from one reservoir to another, such as from river to ocean, or from the ocean to the atmosphere, by the physical processes of evaporation, condensation, infiltration, runoff, and subsurface flow. In so doing, the water goes through different phases: liquid, solid (ice), and gas (vapor).

The water cycle involves the exchange of energy, which leads to temperature changes. For instance, when water evaporates, it takes up energy from its surroundings and cools the environment. When it condenses, it releases energy and warms the environment. These heat exchanges influence climate. By transferring water from one reservoir to another, the water cycle purifies water, replenishes the land with freshwater, and transports minerals to different parts of the globe. It is also involved in reshaping the geological features of Earth, through such processes as erosion and sedimentation. Finally, the water cycle figures significantly in the maintenance of life and ecosystems.

6.1.2.2   CARBON CYCLE

Flows of Carbon Within the Geobiosphere [2, 3]:

Figure 6.2 shows the reservoirs (green boxes) of carbon in the geobiosphere that participate in the major carbon fluxes. The reservoirs are in gigatons of carbon and the flows are in gigatons of carbon per year. The land based cycle seems on balance to remove carbon dioxide from the atmosphere as expected, but the ocean based cycle seems to have a net outflow of carbon dioxide to atmosphere, contrary to the claims of many sources that photosynthesis in marine organisms reduces atmospheric carbon dioxide, partially offsetting anthropogenic carbon output.

Carbon dioxide does not flow directly from the atmosphere to ocean biota. Carbon dioxide is dissolved in cold ocean surface waters and is emitted by warm ocean surface waters. Ocean biota absorbs carbon dioxide from the ocean as represented by the gross primary production, GPP, arrow and it returns through respiration as well as settling and decay.

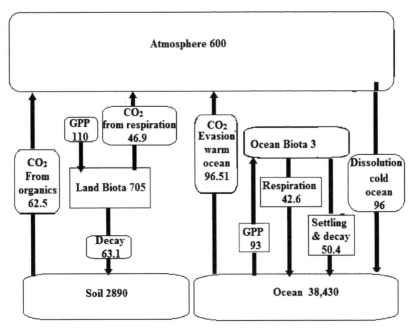

FIGURE 6.2 Reservoirs and flows of carbon on Earth. Reservoirs are in gigatons of carbon and the flows are in gigatons of carbon per year [The data are adapted from Refs. [2, 3].

TABLE 6.1   Distribution of Carbon on the Earth [2].

Source	Moles $C \times 10^{18}$	Relative to atmosphere
Sediments		
carbonate	1530	28,500
Organic carbon	572	10,600
Land		
Organic carbon	0.065	1.22
Ocean		
$CO_2 + H_2CO_3$	0.018	0.3
HCO_3^-	2.6	48.7
CO_3^{2-}	0.33	6.0
Dead organic	0.23	4.4
Living organic	0.0007	0.01
Atmosphere		
CO_2	0.0535	1.0

Carbon is a main component of living things and represents a part of ocean, air, and even rocks. The carbon cycle on the earth is shown in Fig. 6.3. The distribution of carbon on the Earth can be seen in Table 6.1. Because the Earth is a dynamic place, carbon does not stay still.

$$H_2O \qquad + CO_2 \qquad \rightarrow C_6H_{12}O_6 + O_2 \qquad (1)$$

In the atmosphere, carbon is present as carbon dioxide in dry air. In a crowded poorly ventilated room, the P_{CO2} is high. Huge amount of CO_2 is taken from the atmosphere by photosynthesis in land and sea. The gas is returned by respiration and a small amount leaks into the slow, sedimentary part of the geochemical cycle. The carbon becomes part of the plant. Plants that die and are buried may turn into fossil fuels made of carbon like coal and oil over millions of years.

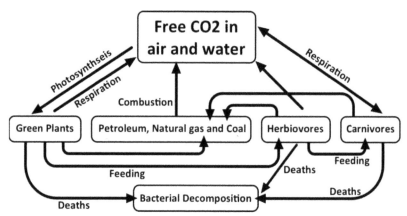

FIGURE 6.3   The carbon cycle Free CO_2 in air and water.

Combustion occurs when any organic material is reacted (burned) in the presence of oxygen to give carbon dioxide, water and energy. If other elements are present they also ultimately combine with oxygen to form a variety of pollutant molecules such as sulfur oxides and nitrogen oxides. About 30–50% of the CO_2 released into the atmosphere by combustion remains there; the remainder enters the hydrosphere and biosphere. The oceans have a large absorptive capacity for CO_2 by virtue of its transformation into bicarbonate and carbonate in a slightly alkaline aqueous medium, and they contain about 60 times as much inorganic carbon as is in the atmosphere. However, efficient transfer takes place only into the topmost (100 m) wind-mixed layer, which contains only about one atmosphere equivalent of CO_2; mixing time into the deeper parts of the ocean is of the order of 1000 years. For this reason, only about 10% of the CO_2 added to the atmosphere is taken up by the oceans [4].

6.1.2.3   NITROGEN CYCLE [5]

Nitrogen is an essential element for plants to grow and produce seeds. Nitrogen is present in the environment in a wide variety of chemical forms including organic nitrogen, ammonium, (NH_4^+), nitrite (NO_2^-), nitrate (NO_3^{-1}), nitrous oxide (N_2O), nitric oxide (NO) or inorganic nitrogen gas (N_2). Organic nitrogen may be in the form of a living organism, humus or in the intermediate products of organic matter decomposition.

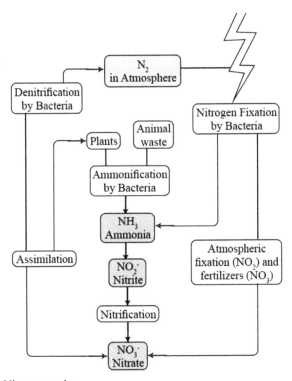

FIGURE 6.4   Nitrogen cycle.

Nitrogen cycle is the process by which nitrogen is transported and converted between its various chemical forms. This transformation can be carried out through both biological and physical processes. The important processes in the nitrogen cycle include fixation, ammonification, nitrification, and denitrification. Although nitrogen represents the majority of atmosphere (78%), it has limited availability for biological use, leading to a scarcity of usable nitrogen in many types of ecosystems. Human activities such as fossil fuel combustion, use of artificial nitrogen fertilizers, and release of nitrogen in wastewater have dramatically altered the global nitrogen cycle. The diagram of nitrogen cycle (Fig. 6.4) shows how these processes fit together to form the nitrogen cycle. A 2011 study found that nitrogen from rocks may also be a significant source of nitrogen.

6.1.2.4   SULFUR CYCLE

Sulfur is present in human and livestock excreta and sulfates are found in most water supplies. Sufficient sulfur is normally available in domestic wastewater in the form of organic sulfides such as mercaptans, and disulfides for the production of odorous gasses by anaerobic and facultative bacteria. The sulfate ion (SO_4^{2-}) is one of the most universal anions occurring in rainfall, especially in air masses that have encountered metropolitan areas. Sulfate concentrations in wastewater can vary from only a few milligrams per liter $(mg.L^{-1})$ to hundreds of milligrams per liter. Generally, for domestic wastewater, the main source of sulfide is sulfate. The main features of sulfur cycle [6] in the environment are shown in Fig. 6.5.

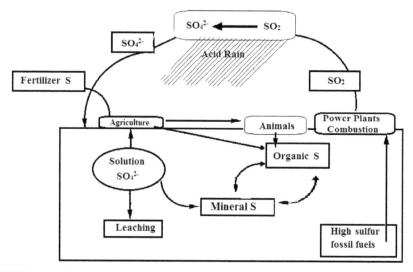

FIGURE 6.5   The main features of sulfur cycle in the environment.

The burning of fossil fuels results in the release of the sulfur contamination in the form of sulfur oxides, SOx, emission to the atmosphere. Rain water washes SOx from air to yield sulfurous and sulfuric acid,

$$2SO_2 + O_2 + 2H2O \rightarrow 2H_2SO_4 \qquad (2)$$

That will return to earth as acid rain.

6.2   WATER

6.2.1   PROPERTIES

Water is the universal solvent and its quality is important to the prevention of diseases. Diseases like cholera, typhoid fever, dysentery and hepatitis are related to impure water supplies. In addition to the removal of suspended materials, water supplies must be disinfected. For disinfection, chlorination is the major process used. Sewer systems are built underground to prevent mixing of sewage with drinking water supplies. Wastewater from industry and hospitals must be treated before disposal to surface water systems. Modern wastewater treatment facilities produce an effluent that is environmentally safe and according to regulations.

Water quality standards include biological, physical, and chemical parameters. The limits are determined by the purpose for which water is to be used. Water quality standards are established for domestic, washing and bathing, aquatic and wildlife uses, and agricultural uses.

6.2.2   IMPORTANT WATER QUALITY FACTORS

Analysts determine water quality by testing for specific chemicals. Each type of water determines what *parameters*, or *analytes*, to be analyzed and the maximum level allowed. For example, chlorine is important in finished drinking water, but not in natural water.

6.2.2.1   PH

The pH of water is an expression for the acidity or basicity of water. It is the Logarithm of hydrogen ion concentration in water. In pure water, $[H^+]$, is in equilibrium with $[OH^-]$, and the pH measures exactly 7. In a lake or pond, the water's pH is affected by its age and the chemicals discharged by communities and industries. Most lakes are basic (alkaline) when they are first formed and become more acidic with time due to the build-up of organic materials. As organic substances decay, carbon dioxide (CO_2) forms and combines with water to produce a weak acid, called "carbonic" acid. The pH of surface waters is important to aquatic life because pH affects the ability of fish and other aquatic organisms to regulate basic life-sustaining processes, primarily the exchanges of respiratory gasses and salts with the water in which they live. Failure to adequately regulate these processes can result in numerous sublethal effects (e.g., diminished growth rates) and even mortality in cases when ambient pH exceeds the range physiologically tolerated by aquatic organisms. Large amounts of carbonic acid lower the pH of water. In summary, fishes can tolerate pH values of about 5.0 to 9.0 [7].

6.2.2.2 ALKALINITY

Alkalinity is the ability of natural water to neutralize acid. It can be determined from a titration. The acid is equivalent to the sum of bicarbonate alkalinity (Moles HCO_3^-), hydroxide Alkalinity (Moles OH), and carbonate alkalinity ($2\times$ Moles CO_3^{2-}). This is total alkalinity. Pure water has zero alkalinity. Alkalinity is a measure of the buffering activity of water against sudden acid or alkali solution disposal. Buffered water, with a pH of 6.0, can have high alkalinity. If you add a small amount of weak acid to both water samples, the pH of the pure water will change instantly (become more acid), but the buffered water's pH won't change easily.

Alkalinity is the measure of the capacity of water to neutralize acids. It's one of the best measures of the sensitivity of the stream to acid inputs of wastewater or acid rains. Alkalinity is important for fish and aquatic life because it protects or buffers against pH changes (keeps the pH fairly constant) and makes water less affected by acid rain. The main sources of natural alkalinity are rocks, which contain carbonate, bicarbonate, and hydroxide compounds. Borates, silicates, and phosphates may also contribute to alkalinity. Waters flowing through limestone regions, that are rich in carbonates, generally have high alkalinity and hence good buffering capacity. Conversely, granite does not have minerals that contribute to alkalinity. Therefore, areas rich in granite have low alkalinity. Alkalinity is determined by titration of 100 mL of sample with standard acid solution 0.01 mol.l^{-1} using (phenolphthalein, methyl orange or mixed indicator). The numerical value of alkalinity as mg.L^{-1} (ppm) $CaCO_3$ is 10 times the number of milliliters of titrant consumed.

EPA issued the method # 310.2 for the determination of Alkalinity based on the automated colorimetric detection of methyl orange change in color due to alkalinity. Methyl orange is used as the indicator in this method because its pH range is in the same range as the equivalence point for total alkalinity, and it has a distinct color change that can be easily measured. The methyl orange is dissolved in a weak buffer at a pH of 3.1, just below the equivalence point, so that any addition of alkalinity causes a loss of color directly proportional to the amount of alkalinity [8].

6.2.2.3 NITRATE AND NITRITE

Nitrite, NO_2^-, and nitrate, NO_3^-, are produced naturally as part of the **nitrogen cycle**, when a bacteria 'production line' breaks down toxic ammonia wastes first into nitrite, and then into nitrate. **Nitrites** are quickly converted to nitrates by bacteria. Nitrite reacts directly with hemoglobin in human blood to produce methemoglobin, which destroys the ability of blood cells to transport oxygen. Water with nitrite levels exceeding 1.0 mg.L^{-1} should not be given to babies. Nitrite concentrations in drinking water seldom exceed a value of 0.1 mg.L^{-1}.

Nitrate washes out from farmland treated with fertilizers by rain water to the water ways. Leaking septic tanks and cesspools also release nitrate to the environment. Manure from farm livestock, animal wastes (including fish and birds) is another important source of nitrate. Finally the discharges from car exhaust contribute in nitrate emission. Nitrate stimulates the growth of plankton and water weeds that provide food for fish. This may increase the fish population. However, if algae grow too wildly, oxygen levels will be reduced and fish will die.

Nitrate can be reduced to toxic nitrites in the human intestine, and many babies have been seriously poisoned by well water containing high levels of nitrate-nitrogen. Healthy drinking water must have a maximum of 10 mg/L of nitrate-nitrogen.

EPA issued a method for the determination of nitrate-nitrate using automated colorimetry. A filtered sample is passed through a column containing granulated copper- cadmium to reduce nitrate to nitrite. The nitrite (that was originally present plus reduced nitrate) is determined by diazotizing with sulfanilamide and coupling with N-(1-naphthyl)-ethylenediamine dihydrochloride to form a highly colored azo dye, which is measured colorimetrically. Separate, rather than combined nitrate-nitrite, values are readily obtained by carrying out the procedure first with, and then without, the Cu-Cd reduction step [9].

6.2.2.4   OXYGEN DISSOLVED IN WATER

Dissolved oxygen, DO, is essential for aquatic life. Oxygen levels in water can be reduced through over fertilization of water plants by run-off from farm fields containing phosphates and nitrates (the ingredients in fertilizers). Under these conditions, the numbers and size of water plants increase a great deal. With limited sun shine (cloudy weather) for several days, respiring plants will use much of the available DO. When these plants die, they become food for bacteria, which in turn multiply and use large amounts of oxygen. The minimum amount of DO is 4–5 ppm to support life of large number of fish types. The DO level in good fishing waters generally averages about 9.0 ppm.

Oxygen is measured in its dissolved form as DO. If more oxygen is consumed than is produced, dissolved oxygen levels decline and some sensitive animals may move away, weaken, or die. DO levels fluctuate seasonally and over a 24-hour period. They vary with water temperature and altitude. Cold water holds more oxygen than warm water and water holds less oxygen at higher altitudes. Thermal discharges, such as water used to cool machinery in a manufacturing plant or a power plant, raise the temperature of water and lower its oxygen content. Aquatic animals are most vulnerable to lowered DO levels in the early morning on hot summer days when stream flows are low, water temperatures are high, and aquatic plants have not been producing oxygen since sunset [10].

The Winkler method involves filling a sample bottle completely with water (no air is left to bias the test). An excess of Mn(II) salt, iodide (I^-) and hydroxide (OH^-)

ions is added to a water sample causing a white precipitate of $Mn(OH)_2$ to form. This precipitate is then oxidized by the dissolved oxygen in the water sample into a brown manganese precipitate. In the next step, a strong acid (either hydrochloric acid or sulfuric acid) is added to acidify the solution. The brown precipitate then converts the iodide ion (I^-) to iodine. The amount of dissolved oxygen is directly proportional to the titration of iodine with a thiosulfate solution [11]. Today, the method is effectively used as its colorimetric modification, where the trivalent manganese produced on acidifying the brown suspension is directly reacted with EDTA to give a pink color. As manganese is the only common metal giving a color reaction with EDTA, it has the added effect of masking other metals as colorless complexes.

EPA issued method 360.1 for DO determination on basis of membrane electrode. In summary, DO in a sample is measured using a probe. The most common probes depend on electrochemical reactions. This method is recommended as an alternative to the modified Winkler procedure when interferences are present, a nondestructive method is desired, or when continuous monitoring is desired [10].

6.2.2.5   BIOCHEMICAL OXYGEN DEMAND (BOD)

Dissolved oxygen is used by microorganisms for biochemical oxidation of organic matter, which is their source of carbon. The BOD is used as an approximate measure of the amount of biochemically degradable organic matter present in a sample. The DO content of the liquid is determined before and after incubation for 5 days at 20°C. The difference gives the BOD of the sample after allowance has been made for the dilution, if any, of the sample. BOD_5 is expressed in mg.L^{-1}, where 5 indicates the number of days in incubation. As determined experimentally by incubation in the dark, BOD includes oxygen consumed by the respiration of algae. A further complication in the BOD test is that much of the oxygen- consuming capacity of samples may be due to ammonia and organically bound nitrogen, which will eventually be oxidized to nitrite and nitrate if nitrifying bacteria are present.

The BOD test is thus useful for determining the relative waste loadings to treatment plants and the degree of oxygen demand removal provided by primary treatment. Compounds constitutionally resistant to breakdown will not exert an oxygen demand on the receiving waters, but substances amenable to breakdown will generally contribute to the pollution load.

6.2.2.6   CHEMICAL OXYGEN DEMAND

The chemical oxygen demand (COD) is the amount of oxygen consumed by organic matter from boiling acid potassium dichromate solution. It provides a measure of the oxygen equivalent of that portion of the organic matter in a water sample that is susceptible to oxidation under the conditions of the test. It is an important and rapidly measured variable for characterizing water bodies, sewage, industrial wastes and treatment plant effluents.

The sample is boiled under reflux with potassium dichromate and silver sulfate catalyst in strong sulfuric acid. Part of the dichromate is reduced by organic matter and the remainder is titrated with ferrous ammonium sulfate. Depending on the aim of the analysis, COD can be determined on unfiltered and/or filtered samples. When both determinations are carried out, the difference gives the COD of the particulate matter. Samples containing settleable solids should be homogenized sufficiently by means of a blender to permit representative sampling for the COD determination in unfiltered samples. For the analysis of filtrate, the original (not homogenized) sample is used.

6.2.3   WATER POLLUTION

The ability of water to dissolve large number of chemicals including toxic materials made the danger of water-borne toxic chemicals to be high regarding the safety of water supplies in industrialized nations. This is particularly true in arid (dry) regions where the volume of ground water is more than the volume of surface water. Ground water is contaminated by infiltration of irrigation water, carrying any excess amounts of pesticides and fertilizers. Industrial plants discharges and the drainage from both operational and defunct mines are considered important sources of hazardous inorganic pollutants for ground water.

The industrial revolution impact on river water quality is significant and numerous chemical spills and heavy metal deposition from air emissions finds their way to the water streams. However, much effort is directed for the restoration of surface water quality. In the 1970s, the water quality in the Rhine River dropped to its lowest levels resulting in the implementation of various Rhine restoration programs, including the reintroduction of endangered and/or temporarily extinct fish species. One of the commitments was the installation of the world's largest industrial sewage treatment plant (approx. 110 million m^3 waste water per year, equivalent to a city of approx. 3 million inhabitants) located at the BASF SE site in Ludwigshafen am Rhine (Rhine-Neckar Metropolis Region, Germany). Along with this project, electrofishing was carried out by BASF SE at regular intervals in order to investigate potential trends in fish populations close to the industrial site in Ludwigshafen. Starting in 1976, the species caught have been identified, their relative abundance determined and their overall health status observed [12].

The 1986 fire at a Sandoz Ltd. storehouse at Schweizerhalle, an industrial area near Basel, Switzerland, resulted in chemical contamination of the environment. The storehouse, which was completely destroyed by the fire, contains pesticides, solvents, dyes, and various raw and intermediate materials. The majority of the approximately 1250 t of stored chemicals was destroyed in the fire, but large quantities were introduced into the atmosphere, into the Rhine River through runoff of the fire-fighting water, and into the soil and groundwater at the site. The chemicals discharged into the Rhine caused massive kills of benthic organisms and fish, particularly eels and salmonids. Public and private reaction to the fire and subsequent chemical spill was very strong. This happened only a few months after the Chernobyl accident and it destroyed the myth of immunity of Switzerland regarding such catastrophes [13].

The 1988 average for heavy metals releases in the Rhine River was: cadmium, 2.8 tons/yr., lead, 0.6 kilo-tons/yr., and zinc, 3.8 kilo-tons/yr. The good news is that these values were down significantly from 1970, when the values were: cadmium, 207 tons/yr., lead, 1.8 kilo-tons/yr., and, zinc, 12.6 kilo-tons/yr.

All the environmental pollutants in fresh water, and the air for that matter, ultimately end up in the sea. About 80% of ocean pollution comes from land. Nonpoint pollution occurs as a result of runoff. Nonpoint source pollution includes many small sources, like septic tanks, cars, trucks, and boats, plus larger sources, such as farms, ranches, and forest areas. Millions of motor vehicle engines drop small amounts of oil each day onto roads and parking lots. Much of this, too, makes its way to the sea [14]. Shipping accidents account for about 10% of ocean pollution. Ocean dumping is another major source of pollution for oceans with 10% contribution. Off-shore oil activities have a 1% share in the pollution but disastrous accidents like that occurred in the Gulf of Mexico from the shell operations may have a long-run environmental effects.

This pollution is important and affects the fish production, a major source of protein (27%) for many countries. People living in the South Pacific and the Far East will be affected the most by reduced sea productivity. Selected WHO guidelines relevant to drinking water, are shown in Table 6.2.

TABLE 6.2   WHO Drinking Water Quality Guidelines [15]

Parameter	WHO Guideline value, mg. L^{-1}
Fecal coliform or E. coli	0.0*
Aluminum	0.2 **
Ammonia	1.5 **
Cadmium	0.003
Chloride	250 **
Color	15TCU**
Copper	2
Hydrogen Sulfide	0.005 **
Iron	0.3**
Lead	0.01
Nitrate	10
Sodium	200**

Sulfate	250* *
Turbidity	5 NTU**
Total dissolved solids	1000**
Zinc	3* *
Arsenic	0.01 (P)
Barium	0.7
Boron	0.5 (T)
Chromium	0.05 (P)***
Fluoride	1.5****
Manganese	0.4 (C)
Molybdenum	0.07
Selenium	0.01
Uranium	0.015 (P, T)*****

*Not detectable in a 100 mL sample.
**May not be toxic but could result in consumer complaints.
***For total chromium.
****Volume of water consumed and intake from other sources should be considered when setting national standards.
*****Only chemical aspects.
P = provisional guideline value, as there is evidence of a hazard, but the available information on health effects is limited;
T = provisional guideline value because calculated guideline value is below the level that can be achieved through practical treatment methods, source protection, etc.;
C = concentrations of the substance at or below the health-based guideline value may affect the appearance, taste or odor of the water, resulting in consumer complaints.

6.2.4 WATER CHEMISTRY

Water is essential for 99.9% of all life on Earth. It is the medium that allows necessary biological reactions to occur. Water is the medium with which the nutrients and minerals are carried to aquatic life. Water also carries waste away. Chemical parameters play an important role in health, abundance, diversity and the life within the stream. Changes in chemical parameters can affect other parameters. Change in temperature can affect the amount of dissolved gas in water, rate of plant growth, rate of photosynthesis, and metabolic rate of organisms. Change in the pH of water impacts aquatic life, where extreme values >11.0 or <4.5 are lethal and extremely acidic water results in metals (e.g., Aluminum, Zinc, Copper) becoming bio—available. More metals and other compounds in soil and rock, which are toxic to aquatic life, will be suspended in the water column. Phosphorus is mostly available to plants as orthophosphate, PO_4^{3-}. It occurs naturally in rocks and enters the water column through the weathering of rock, runoff from animal production (especially from poultry litter), wastewater from treatment plants, poorly functioning septic systems, and breakdown of organic matter. Very small amounts (0.01 mg/L) can cause large algal blooms.

World Health Organization, Geneva, published in 2008 the recommendations for analytical achievability for inorganic chemicals for which guidelines values have been established (Table 6.3) [16].

TABLE 6.3 Analytical Achievability for Inorganic Chemicals Determination in Water [16]

Parameter	Field methods		Laboratory methods				
	Col	Absor	IC	FAAS	EAAS	ICP	ICP/MS
Naturally occurring chemicals							
Arsenic		#		+(H)	+++++(H)	++(H)	+++
Barium				+	+++	+++	+++
Boron		++				++	+++
Chromium			#	+	+++	+++	+++
Fluoride	#	+	++				
Manganese	+	++		++	+++	+++	+++
Molybdenum					+	+++	+++
Selenium		#		#	+++(H)	++(H)	+
Uranium						+	+++
Chemicals from industrial sources and human dwellings							
Cadmium		#			++	++	+++
Cyanide	#	+	+				
Mercury					+		
Chemicals from agricultural activities							
Nitrate/nitrite	+++	+++	#				
Chemicals used in water treatment or materials in contact with drinking-water							
Antimony				#	++(H)	++(H)	+++
Copper	#	+++		+++	+++	+++	+++
Lead		#			+	+	++
Nickel		+		#	+	+++	++

Thus, the study of chemical equilibrium applies to natural systems. Specifically, carbonate equilibrium and other equilibria are important in of Natural systems. Carbonic acid, H_2CO_3, is the most important weak acid in nature. Mathematical relationship known as the "Mass Action Equation" is obeyed for all equilibrium systems. The equilibrium and mass action equation can be written for aqueous solutions carbonic acid has the following equilibrium:

$$H_2CO_3 \leftrightarrows H^+ + HCO_3^- \quad K_{a1} = \frac{\left[H^+\right]\left[HCO_3\right]}{[H_2CO_3]} \tag{3}$$

And

$$HCO_3^- \leftrightarrows H^+ + CO_3^{2-} \quad K_{a2} = \frac{\left[H^+\right]\left[CO_3^-\right]}{[HCO_3^-} \tag{4}$$

Carbon dioxide in air is in equilibrium between air and water.

$$CO_{2(air)} \leftrightarrows \tag{5}$$

$$CO_{2(aqueous)} + H2O \leftrightarrows H_2CO_3 \tag{6}$$

When calcium is added to an aqueous solution of carbonate, calcium carbonate (a sparingly soluble salt) is formed.

6.2.4.1   WATER SOLUBILITY OF MATERIALS

Ionic inorganic substances dissolve in water. Hydration occurs when ions are surrounded by water molecules. Oil (a saturated hydrocarbon, CH_3-CH_2-CH_2-CH_2-CH_2-CH_2-CH_2-CH_3) does not dissolve in water. Plastic (Polystyrene for example) does not dissolve in water. It is possible to modify the structure of some molecules to make them soluble in water. This is accomplished by making part of the molecule ionic (or at least giving it a dipole moment).

Some organic molecules are soluble with water. Examples of organic molecules that dissolve in water are acetic acid, ethyl alcohol and ethylene glycol. Sparingly soluble compounds are in equilibrium between an ionic form and a nonionic form. Calcium carbonate is an example of such a substance.

$$CaCO_{3(s)} \quad \overset{K_{sp}}{\leftrightarrows} \quad Ca^{2+} + CO_3^{2-} \tag{7}$$

Insoluble form   Soluble form

The solubility product or K_{sp} is an equilibrium expression that governs such reactions. K_{sp} is a characteristic of inorganic solid from which the solubility of that substance in water can be calculated. For Calcium Carbonate:

$$K_{sp} = [Ca^{2+}][CO_3^{2-}] = 4.47 \times 10^{-9} \tag{8}$$

The solubility of calcium carbonate in pure water (in a system not exposed to air and ignoring reactions with water) can be calculated as the square root of the K_{sp} (Eq. 6). The solubility of calcium carbonate in water-if not open to the air, is 6.69×10^{-5} moles $CaCO_3/L$.

Mixing water with air is an important factor and would change the concentration of CO_3^{2-}. This will reflect the CO_3^{2-} from dissolved calcium carbonate and from CO_3^{2-} from dissolved carbon dioxide. The calculation of the solubility of calcium carbonate for a system open to the air is more difficult than in closed system. Calcium carbonate in water, open to the air, is an important system to understand because it represents the weathering process when limestone $(CaCO_3)$ encounters fresh water. We will deal with this in a future lecture.

6.2.4.2   SOLUBILITY VERSUS INTRINSIC SOLUBILITY

Solubility calculations may lead to erroneous results of as much as 100% if salt water is used instead of pure water. Since most natural waters contain at least a few dissolved salts, we should try to understand what causes this error. The errors encountered for solubility in salt solutions results from an effect called "activity." Salts modify the properties of water by reducing an ion's activity in solution. This reduced activity results in greater than expected solubility.

6.2.4.4   DISTRIBUTION OF SPECIES IN A SYSTEM AT EQUILIBRIUM

In the carbonate equilibrium, Fig. 6.6 shows the dependence of the carbonate species occurrence on the pH of water. It is clear that below a pH value of 4.5, carbonic acid is the dominant species and no bicarbonate can exist. Below a pH value of 8.3, all carbonate disappear and the bicarbonate is the dominant species, which is the case of natural waters. During the titration, the addition of acid to a natural water sample will cause strongest base to be neutralized first followed by the next strongest base neutralized second and so on.

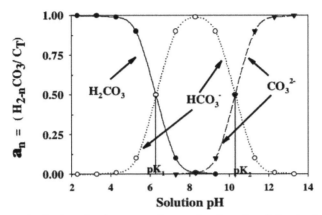

FIGURE 6.6   Distribution of carbonate species as a fraction of total dissolved carbonate in relation to pH [17].

Example 1

About 100 mL of a water sample required 20.0 mL of 0.0100 M HCl to reach an end point of pH = 4.2 express the alkalinity of this sample as mg $CaCO_3/L$

Since 1 Mole of HCl = 1 Equivalent of HCl

   0.0100 M HCl = 0.0100 Eq /L of HCl

   20.0 mL of this acid = 0.200 mEq of HCl

   1 mEq of HCl = 1 mEq of $CaCO_3$

   Therefore,

   0.200 mEq of HCl = 0.200 mEq of $CaCO_3$

   0.200 mEq of $CaCO_3$ * 50.045 mg CaCO3/mEq = 10.1 mg of $CaCO_3$ in 100 mL of water samples. Thus, the amount per liter would be 100.1 mg $CaCO_3/L$.

6.2.5   OXIDATION/REDUCTION IN NATURAL ENVIRONMENT (18)

Inorganic species in water exist in various oxidation states depending on the pH of water. Nitrogen, sulfur, and carbon are present as N_2, NH_3, NO_3^-, NO_2, NO, NO_2^-, H_2S, S, SO_2, SO_3^{2-}, SO_4^{2-}, CH_4, CH2O, CO, and CO_2. In these forms the elements exist in various oxidation states. Mercury can be found as Hg, $HgCH_3^+$, and $Hg(CH_3)_2$. Methyl Mercury (both forms) is much more toxic than mercury metal. For the same element, the toxicity is highly dependent on the oxidation state. Chromium VI is more toxic than other forms of chromium (i.e., Cr^{3+}). Reduction is defined as the gain of electrons and oxidation is defined as the loss of electrons. The hydrogen oxidation-reduction of is defined to be zero and everything is relative to this standard. The more positive the potential, the greater will be the tendency for the reaction to proceed as written.

$$Fe^2 + 2\ e^- \rightarrow Fe \ \text{.............................} -0.44\ V \tag{9}$$

$$2H^+ + 2\ e^- \rightarrow H_2 \ \text{.............................} 0.00\ V \tag{10}$$

$$Cu^{2+} + 2\ e^- \rightarrow Cu \ \text{.............................} 0.34\ V \tag{11}$$

$$NO_3^- + 4\ H^+ + 3\ e^- \rightarrow NO + 2\ H_2O \ \text{.........} 0.96\ V \tag{12}$$

The oxidation states of nitrogen in the common forms in the aqueous environment are: (+5) in nitrate ion, (+3) in nitrite ion, and (–3) in ammonium ion. The transformations between these nitrogen oxidation states are governed by the following equations:

$$1/2NO_3^- + H^+ + e^- \rightarrow 1/2NO_2 + 1/2\ H_2O \ \text{......} E^\circ = +14.15\ V \tag{13}$$

$$1/6NO_2^- + 4/3H^+ + e^- \rightarrow 1/6NH_4 + 1/3\ H_2O \ \text{......} E^\circ = +15.14\ V \tag{14}$$

$$1/8NO_3^- + 5/4H^+ + e^- \rightarrow 1/8NH_4 + 3/8\ H_2O \ \text{......} E^\circ = +14.90\ V \tag{15}$$

At various pH values, the dominant species of C, N and S are shown in Fig. 6.7.

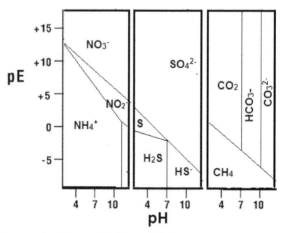

FIGURE 6.7   Diagram for Nitrogen, Sulfur and Carbon.

6.2.5.1   DISINFECTING TREATED WASTEWATER

Disinfection can be done with either chlorine, ozone or chlorine dioxide. Chlorine has the undesirable aspect of forming trace level of chlorinated hydrocarbons as a by-product. The active disinfecting agent in chlorination is the hypochlorite ion, OCl^-.

K

$$Cl_2 + H2O \leftrightharpoons H^+ + Cl^- + HOCl \tag{16}$$

$$K = 4.5 \times 10^{-4}$$

K_a

$$HOCl \leftrightharpoons H^+ + OCl^-$$
(17)

$$K_a = 2.7 \times 10^{-6}$$

Chlorine dioxide, ClO_2, is a substitute for chlorine gas that does not produce trihalomethanes as a byproduct.

Reductant

$$ClO_2 + 4H^+ + 5e^- \rightarrow Cl^- + 2\,H_2O \tag{18}$$

Reductant

$$ClO_2 + e^- \rightarrow ClO_2^- \tag{19}$$

Ozone, O_3, is the third disinfecting method employed in wastewater treatment. Water disinfected with ozone is often treated with a low level of chlorine as well, because ozone decomposes fairly rapidly to O_2 and does not remain in the water much time after treatment.

$$O_{3(g)} + 2H^+ + 2e \rightarrow O_{2(g)} + H_2O \tag{20}$$

6.2.6   PHASE INTERACTIONS, SOLUBILITY OF GASES IN WATER

Water is present, in all three phases solid, liquid and gas. Phase interactions are the basis of much of the interesting chemistry that shapes our planet. Dissolved oxygen in water makes it possible for fish and insects to live underwater. The oxygen in the atmosphere originated from microorganisms that lived in the sea. The transfer of oxygen from the sea to the atmosphere made it possible of life to develop on the land.

Phase interactions are responsible for the transport of nutrients (and pollutants) in soil, and the formation of all the sedimentary minerals. The earth is unique among bodies of the solar system, in that it has oceans of liquid water. The phase interactions that result from the oceans stabilize the earth's temperature and move water to the land where it condenses as fresh water to support both plant and animal life forms. An understanding of phase interactions is essential to understanding the chemical processes of the environment.

To examine the role that dissolved gases play in the environment, let's reexamine the solubility of calcium carbonate in water. In a sealed bottle containing only pure water and pure calcium carbonate, we find the following:

Equilibrium conditions for $CaCO_3$ in water, closed to the air:

pH = 9.95 and the $[Ca^{2+}] = 1.26 \times 10^{-4}$

Equilibrium conditions for $CaCO_3$ in water open to the air:

pH = 8.40 and the $[Ca^{2+}] = 3.98 \times 10^{-4}$

The pH drops over 2.5 pH units and the calcium concentration increases by over three times by exposing calcium carbonate to atmospheric carbon dioxide. This occurs because of the following reactions:

$$CO_{2(g)} + H2O \rightarrow H_2CO_3 \tag{21}$$

$$H_2CO_3 + CaCO_3 \rightarrow 2HCO_3^- + Ca^{2+} \tag{22}$$

Note that the solubility of gases is greater at lower temperature. This is because the enthalpy of dissolution for gases in water is negative.

6.2.6.1   NATURE OF COLLOIDS

Colloids are substances, which remain suspended in water for extended time periods and range in diameter 0.001 to 1 μm. A colloid is a type of mixture intermediate between a solution and a *heterogeneous mixture* with properties also intermediate between the two. The particles are approximately 10 to 10,000 angstroms in size and generally cannot be filtered, or settled out in an easy manner. Colloids may be colored or translucent because of the Tyndall effect, which is the scattering of light by particles in the colloid. Colloid particles may be seen in a beam of light such as dust in air in a "shaft" of sunlight. There are three types of colloids 1, hydrophilic colloids, 2, hydrophobic colloids, 3, and association colloids. The common element between the three classes of colloids is the fact that they are held in suspension by electrostatic interactions with water molecules. The three classes of colloids differ from each other in their chemical composition.

Hydrophilic colloids are large molecules that contain functional groups that can make hydrogen bonding with water. Proteins and some synthetic polymers are examples of hydrophilic colloids.

Hydrophobic colloids are substances that have charged surfaces in water, and form an electrical "double layer" that holds them in suspension. Clays form a negative charge on their surface when placed in water, and remain in suspension by the electrostatic interaction between the negative surface charge and positive charges from cations in the water. Figure 6.8 illustrates colloidal clay particles suspended in solution by these electrostatic interactions.

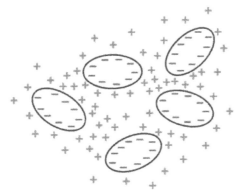

FIGURE 6.8   Hydrophobic colloids suspended in solution.

Association colloids are molecules that are hydrophobic at one end and hydrophilic at the other end. Soaps and detergents form association colloids in water. Their molecular structure is similar to Fig. 6.9 below, where the carboxylic acid group is the hydrophilic portion and the hydrocarbon chain is the hydrophobic part of the molecule.

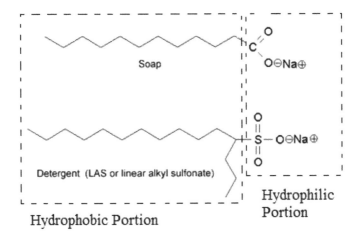

FIGURE 6.9   Compounds that form association colloids when placed in water.

When placed in water, these molecules form a structure where the hydrocarbon "tails" collect into tiny oil droplets and the charged carboxylate groups interact with the water through an electrostatic interaction to keep the droplet suspended. Figure 6.10 illustrates association colloids in suspension.

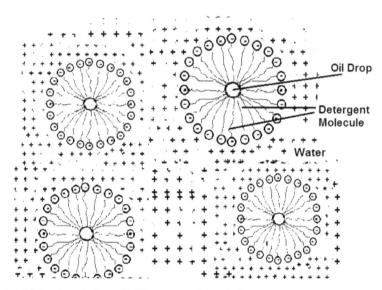

FIGURE 6.10   Association colloids suspended in solution.

6.2.6.2   IMPORTANCE OF COLLOIDS TO ENVIRONMENTAL CHEMIST

Colloids are of interest to the environmental chemist for several reasons:
- both organic and inorganic substances are transported as colloids;
- the analytical measurement for the quantity of colloids in water is called "suspended solids;"
- although most pesticides and herbicides are not water soluble, they aggregate with some natural materials.

Clays are the most common hydrophobic colloids in natural waters. Clay comes from the weathering of sedimentary rocks.

The composition and structure of clays give them the properties that result in colloidal behavior in water. Clays are flat sheets of alternating layers of silicon oxides and aluminum oxides, held together by ionic attraction for cations sandwiched between the sheets. Figure 6.11 is an example of kaolinite clay, illustrating the alternating sheets of silicon oxides and aluminum oxides. The negative surface charge of clay particles results when an aluminum (+3) or silicon (+4) is replaced with a sodium (+1), potassium (+1) or ammonium (+1) ion, giving an overall negative charge to the particle.

Clays tend to sorb chemical species from water. Substances that are not transported by dissolving in water will be transported if they sorb to the surface of clay. In addition, both organic and inorganic materials can be trapped between the sheets of aluminum and silicon oxides, providing an effective mechanism for transporting these materials in the aqueous environment.

These properties of clays also make them useful agents for cleaning contaminated water. Clay liners are placed at the bottom of landfills because of their ability to retain pollutants, preventing them from getting into the groundwater [19].

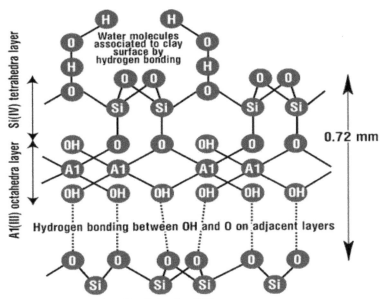

FIGURE 6.11   An example of kaolinite clay [20].

6.3   NATURE AND TYPES OF ENVIRONMENTAL POLLUTANTS

Two important definitions must be clarified in the very beginning of the environmental pollution section; contaminant is a chemical substance at greater than background levels that has no detrimental effect; and pollutant is a chemical substance at greater than background levels that has a detrimental effect. Environment suffers from various types of pollutants coming from different sources. Table 6.4 summarizes the types of environmental pollutants.

TABLE 6.4   Types of Aquatic Pollutants

Chemical		Physical	Biological
Inorganic	**Organic**		
Nutrients (Nitrate, Phosphate)	Trichloroethylene, Chloroform, Carbon Tetrachloride	Sediments	Oxygen Depleting Substances (Sewage)
Heavy Metals (Hg, Cd, Pb)	Herbicides, Pesticides	Thermal Pollution	Pathogens
Radionuclides (Th, U)	Oil and Grease, PAHs, Hydrocarbons		

Synthetic organic compounds represent a wide range of environmental pollutants. They may accumulate in the food chain and degrade very slowly in the environment. In general, carbon is the basic structural component of these compounds. Organic compounds are characterized by low solubility in water, reasonable solubility in fatty tissue and may be volatile. Molecules that contain only carbon, hydrogen and chlorine are not soluble in water. Molecules that contain only carbon, hydrogen and chlorine are "fat" soluble. Molecules that contain only carbon, hydrogen and chlorine are volatile. Organic molecules that contain oxygen are soluble in water than the corresponding hydrocarbon. Organic molecules that contain oxygen are less volatile than the corresponding hydrocarbon.

Regarding the reactivity, saturated molecules that contain only carbon, hydrogen and chlorine are nonreactive. Branched hydrocarbons are less reactive than "straight chain" hydrocarbons. Hydrocarbons that contain a double bond, like ethylene, are more reactive than the corresponding saturated hydrocarbon. Hydrocarbons that contain a triple bond, like acetylene, are more reactive than the corresponding hydrocarbon with a double bond. Hydrocarbons that contain oxygen, like alcohols and carboxylic acids, are more reactive than the corresponding hydrocarbon.

In photochemical smog formation, the peroxyacids that contain C=O and C–O of a peroxide group (-OOH). Peracetic acid is an example:

$$\text{H}_3\text{C} - \text{C}(\!=\!\text{O}) - \text{O} - \text{O} - \text{H}$$

The major organic pollutants are the pesticides, herbicides, PCBs, PAHs and by-products of manufacturing processes, soaps, like sodium oleate, surfactants like alkylsulfate and alkylbenzene sulfonate; Organochlorine insecticides like DDT, Methoxychlor, Dieldrin, Aldrin and Endrin, Nitroaniline Herbicides, 3-(4-Chlorophenyl)-1, 1-dimethylurea; Dipyridilium Herbicides; Diaquat, Paraquat; Additional Common Herbicides like Alachlor, Atrazine, 2,4-Dichloro-phenoxyacetic acid, Propanil. The Carbamate Insecticides: Carbaryl, Carbofuran. Organophosphate Insecticides, methyl parathion, Ethyl parathion, Polychloronated Biphenyls; Botanical Insecticides like Allethrin, Nicotine, Rotenone, Dioxins and Furans

6.4   THE ATMOSPHERE

6.4.1   COMPOSITION OF AIR

In addition to the main constituents, N_2 and O_2 clean air contains the gases presented in Table 6.5. Some of the gases are very stable and has long residence time like N_2, CO_2 and O_2 and others have very short residence time like CO, NH_3, SO_2 and NO. The average composition of the atmosphere up to an altitude of 25 km is given in Table 6.5 [21].

TABLE 6.5   Average Composition of the Atmosphere up to an Altitude of 25 km

Gas Name	Chemical Formula	Percent Volume	Cycle
Nitrogen	N_2	78.08%	Biological and Microbial
Oxygen	O_2	20.95%	Biological and Microbial
*Water	H_2O	0 to 4%	Physic-chemical
Argon	Ar	0.93%	No Cycle
*Carbon Dioxide	CO_2	0.0360%	Anthropogenic and bio-genic
Neon	Ne	0.0018%	No Cycle
Helium	He	0.0005%	Physio-chemical
*Methane	CH_4	0.00017%	Biogenic and chemical
Hydrogen	H_2	0.00005%	Biogenic and chemical
*Nitrous Oxide	N_2O	0.00003%	Biogenic and chemical
*Ozone	O_3	0.000004%	Chemical

6.4.2   AIR POLLUTION

Atmospheric air is polluted with variety of materials released from natural events and human activities. The major inorganic gaseous pollutants are carbon monoxide (CO), sulfur dioxide (SO_2), nitrogen oxides (NO, NO_2), NOx and ozone (O_3). The organic substances represent primary ingredients of photochemical smog, and some mountains are blue in color due to the presence of monoterpenes. Monoterpenes are hydrocarbons with the basic formula of $C_{10}H_{22}$, and are produced in quantities by pine trees. The air conditioning fluids are the main pollutants for air: chlorofluorocarbons (CFC's) CCl_3F, CCl_2F_2, CFC-11, CFC-12. Compounds that contain bromine, in addition to carbon, fluorine and chlorine are called Halons like $CBrF_3$, Halon-1211 and $CBrClF_2$ Halon-1301.

Air particulates are another important category of air pollutants. The are represented by:

- Aerosol: A suspension of solid or liquid particles in a gas;
- TSP: Total suspended particulate matter;
- RSP: Respirable suspended particulate matter;
- PM10: Particulate matter of less than 10 um; equivalent to RSP;
- PM2.5: Particulate matter of less than 2.5 um.

Particulates transport toxic substances into the respiratory tract. Toxic compounds (lead, cadmium, beryllium, Poly aromatic hydrocarbons, PAHs, etc.) can then be adsorbed into the body where they affect biological processes as if they had been ingested by any other means. If the particles are taken with breathing they will remain in the respiratory tract. The larger particles are deposited in the respiratory tract with high efficiency. PM enters the atmosphere from windblown dust, salt particle from wind blowing over the oceans, and combustion processes (soot, smoke, fly ash and gases).

Meteorological conditions significantly affect the gases and particulate matter emitted into the atmosphere. Pollutant transport and transformation (gas-to-aerosol conversion) depend on wind speed and atmospheric stability, as well as solar radiation.

AEI is determined as a 3-year running annual mean PM2.5 concentration averaged over the selected monitoring stations in agglomerations and larger urban areas, set in urban background locations to best assess the PM2.5 exposure to the general population.

Humans can be adversely affected by exposure to air pollutants in ambient air. In response, the European Union has developed an extensive body of legislation which establishes health based standards and objectives for a number of pollutants in air. These standards and objectives are summarized in Table 6.6. These apply over differing periods of time because the observed health impacts associated with the various pollutants occur over different exposure times [22].

TABLE 6.6   European Union Health-Based Standards and Objectives for Air Pollutants

Pollutant	Concentration	Permitted exceedences each year
Fine particles (PM2.5)	25 $\mu g/m^3$	n/a
Sulfur dioxide (SO_2)	350 $\mu g/m^3$	24
	125 $\mu g/m^3$	3
Nitrogen dioxide (NO_2)	200 $\mu g/m^3$	18
	40 $\mu g/m^3$	n/a

PM10	50 µg/m³	35
	40 µg/m³	n/a
Lead (Pb)	0.5 µg/m³	n/a
Carbon monoxide (CO)	10 mg/m³	n/a
Benzene	5 µg/m³	n/a
Ozone	120 µg/m³	25 days averaged over 3 years
Arsenic (As)	6 ng/m³	n/a
Cadmium (Cd)	5 ng/m³	n/a
Nickel (Ni)	20 ng/m³	n/a
Polycyclic Aromatic Hydrocarbons	1 ng/m³ (as Benzo(a)pyrene)	

6.4.2.1 CARBON MONOXIDE: HEALTH EFFECT

Carbon monoxide, CO, enters the blood stream and binds preferentially to hemo-globin to form carboxyhemoglobin, thereby replacing oxygen. The common sources of carbon monoxide are the incomplete combustion (internal combustion engine), biomass burning, methane oxidation, oxidation of nonmethane hydrocarbon, and decay of plant matter. The reaction with OH radical may be a sink for the gas.

$$.OH + CO \rightarrow CO_2 + H. \tag{23}$$

$$H. + O_2 + M \rightarrow HO_2. + M \tag{24}$$

Soil microorganism may remove the gas from air stream.

Carbon monoxide-Atmospheric chemistry

$$OH + CO + O_2 \rightarrow CO_2 + HO_2. \tag{25}$$

$$HO_2. + NO \rightarrow NO_2 + OH \tag{26}$$

$$NO_2 + hv \rightarrow NO + O. \tag{27}$$

$$O. + O_2 \rightarrow O_3 + M \tag{28}$$

And the net reaction is

$$CO + 2O_2 + hv \rightarrow CO_2 + O_3. \tag{29}$$

The net reaction can be viewed as a catalytic oxidation of CO to CO_2. Net formation of O_3 occurs.

The control strategies on the automobile source include the use of a leaner air/fuel mixture (higher air/fuel ratio); and the use of catalytic exhaust reactors. The latter involves pumping of excess air into the exhaust pipe and passing air-exhaust mixture through a catalytic converter to oxidize CO to CO_2. The addition of oxygenates to gasoline is another promising alternative including methanol, ethanol, Methyl tertiary butyl ether, MTBE.

6.4.2.2   SULFUR DIOXIDE

Sulfur dioxide has health effects like causing irritation and increasing resistance in the respiratory tract. In sensitive individuals, the lung function changes may be accompanied by perceptible symptoms such as wheezing, shortness of breath, and coughing. The gas may also lead to increased mortality, especially if elevated levels of suspended particles are also present.

Sources:

- Combustion of S-containing fuel in electric power plants, vehicles.

$$S \text{ (organic S + FeS}_2 \text{ pyrite)} + O_2 \rightarrow SO_2 \tag{30}$$

$$\text{Oxidation of H}_2S: 2H_2S + 3\,O_2 \rightarrow 2\,SO_2 + 2\,H_2O \tag{31}$$

– H_2S is produced as an end product of the anaerobic decomposition of S-containing compounds by micro organisms.

- Oxidation of DMS

SO_2 is converted into sulfuric acid in either gas or liquid phase. Sulfur dioxide can be removed from smokestacks before entering the atmosphere by washing the smoke by absorption in alkaline solution. Recently, the gas was successfully absorbed from air streams in dilute urea solutions [23].

6.4.2.3   NITROGEN OXIDES

6.4.2.3.1   HEALTH EFFECTS

The NO causes cellular inflammation at very high concentrations and may be taken into hemoglobin in the blood and effect the transport of oxygen around the body. NO_2 causes irritation of the lungs and lowering resistance to respiratory infection such as influenza.

Sources:
Fuel combustion in power plants and automobiles are the artificial sources of nitrogen oxides.

$$N_2 + O_2 \rightarrow NO \tag{32}$$

$$2NO + O_2 \rightarrow 2NO_2 \tag{33}$$

They may result from natural sources like electrical storms and bacterial decomposition of nitrogen-containing organic matter. Inter-conversion of NO and NO_2 in the atmosphere may be induced by light of certain wavelength. It involves the formation of ozone and oxygen radicals as intermediate species. In another mechanism, the interconversion of nitrogen oxides may result in ozone formation.

$$NO_2 + hv \rightarrow NO + O \tag{34}$$

$$O + O_2 + M \rightarrow O_3 + \tag{35}$$

$$NO + O_3 \rightarrow NO_2 + O_2 \tag{36}$$

or

$$NO_2 + hv \rightarrow NO + O \tag{37}$$

$$O + O_2 + M \rightarrow O_3 + M \tag{38}$$

$$HO_2\cdot + NO \rightarrow NO_2 + OH \tag{39}$$

$$RO_2\cdot + NO \rightarrow NO_2 + RO. \tag{40}$$

Nitric acid may be formed by a gas-phase reaction

$$NO_2 + OH \rightarrow HNO_3 \quad \text{daytime (dominate pathway)} \tag{41}$$

Nitrate salts also form

$$HNO_3 + NH_3 \rightarrow NH_4NO_3 \tag{42}$$

$$HNO_3 + NaCl \rightarrow NaNO_3 + HCl \tag{43}$$

The emission of nitrogen oxides can be controlled by:
1. Lowering the combustion temperature of the furnace in electric power plants
2. Install catalytic converters: catalytic converters in cars can remove 76% of NOx from tailpipes. The converter is a three-way catalytic converter for automobile exhaust (Remove CO, NO and Hydrocarbons). A catalytic converter uses the basic redox chemical reactions to help reduce the danger

of pollutants emission from a car. Almost 98% of the harmful fumes are converted into safer gases. Exhaust gases are directed into a metal housing containing a ceramic interior with thin wall channels that are coated with a washcoat of aluminum oxide. Precious metals (catalysts) are contained in this part of the converter. The overall reactions involved in the converter are shown in Fig. 6.12.

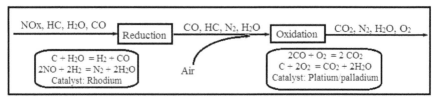

FIGURE 6.12   Chemical changes in the automobile catalytic convertor system.

6.4.3   OZONE DEPLETION

Ozone is an essential component of the atmosphere that absorbs ultraviolet radiation from the sun, heating the gases of the stratosphere. Ozone is measured in Dobson. Dobson took the Ozone in the column of air over Labrador, Canada, if compressed it would form a slab approximately 3 mm thick. One Dobson Unit (DU) is defined to be 0.01 mm thickness at STP; the ozone layer over Labrador then is ~300 DU.

Ozone depletion: process where ozone is concentration reduced by manmade chemicals, primarily chlorinated fluorocarbons or CFCs. Additional UV radiation reaching the earth's surface will lead to additional cancer to the human population. Atmospheric methyl chloroform concentration in air is decreasing since 1991 after the production and use of CFC's decreased. The VOCs react with nitrogen oxides to produce ozone. Many VOCs are toxic. Some VOCs, benzene, xylenes and toluene are carcinogens and formaldehyde and ethylene are known to harm plants. VOCs' most significant environmental effect is the role they play in ozone formation.

6.4.4   CARBON DIOXIDE EMISSIONS AND GLOBAL WARMING

Global warming [24] is an environmental threat. Satellite data and computer advances permitted mathematical models to be used to study the earth's climate. The basic idea behind global warming goes back to the energy balance of incoming solar radiation. Earth exists in a state of thermal equilibrium with space, (energy of incoming radiation = energy reradiated back into space). The incoming radiation has short wavelength (UV light). Absorption, scattering, and emission processes change the wavelength of the incoming light to low energy (longer wavelength).

The energy reradiated back into space mostly long wavelength (Visible and infrared light). Water vapor and carbon dioxide absorb visible and infrared light and decrease the energy released to space. This leads to higher temperatures at the earth's surface. "green house" gases. Nitrogen oxides, CFCs and HCFCs are also important greenhouse gases.

Most experts agree that the earth's temperature has increased between 0.3 and 0.6°C since the beginning of 20th century. In the meantime, the carbon dioxide concentration changed from 280 ppm to 360 ppm. Night time temperatures have increased more than daytime temperatures. The Solution: is to reduce the amount of greenhouse gases (water and carbon dioxide). The international concern with the relation of green house gases with ozone depletion resulted in Kyoto 1997 Kyoto Protocol. This protocol aims at the reduction of the emissions of CO_2 to 5.2% below 1990 levels. The industrialized countries would be permitted to exceed their targets if they "purchased" emission rights from countries.

6.5   LAND AND SOIL ENVIRONMENT

6.5.1   SOIL CHEMISTRY

Soil is a complex system holding all the activities of human being in addition to land plants and land animals. Soil is the result of the erosion of rock matter by natural agents such as wind, water and glaciers. For a real soil chemistry study, two important definitions must be given. Mineral is an inorganic solid, naturally occurring with definite crystal structure and rock is a solid, cohesive mass or aggregate of two or more minerals. Twenty five minerals make up most of the earth's crust. Because most of the mass of the earth's crust consists of oxygen (49.5%) and silicon (25.7%), it is reasonable, then that the primary mineral composition of the earth's crust would be of these two elements. The most abundant minerals of the earth's crust are silicate, SiO_2, and orthoclase, $KAlSi_3O_8$. The major mineral groups of the earth's crust are given in Table 6.7.

TABLE 6.7   Major Mineral Groups Clays

Mineral Groups	Examples	Formula
Silicates	Olivine	$(Mg,Fe)_2SiO_4$
Oxides	Magnetite	Fe_3O_4
Carbonates	Dolomite	$CaCO_3. MgCO_3$
Sulfides	Pyrite	$FeSn_2$
Sulfates	Gypsum	$CaSO_4. 2H_2O$
Halides	Fluorite	CaF_2
Native Elements	Sulfur	S

6.5.1.1   MINERALS AND THE WEATHERING PROCESS

Weathering of aluminum-silicate rocks produces clays. The weathering process breaks the parent minerals down into microcrystalline, secondary minerals of aluminum silicate hydroses. There are three major clay groups: Montmorillonite, $Al_2(OH)_2Si_4O_{10}$, Illite, $K_{0-2}Al_4(Si_{8-6}Al_{0-2})O_{20}(OH)_4$, and Kalolinite, $Al_2Si_2O_5(OH)_4$. Clays are flat sheets of silicon dioxide (SiO_2) and alumina (Al_2O_3) molecules, bound together by ionic attraction to cations sandwiched between the sheets.

Soil is a variable mixture of minerals, organic matter and water capable of supporting plant life. A typical soil contains approximately ~5% organic matter and ~95% inorganic matter. The inorganic components of soil come from the breakdown of rocks and minerals. The key to understanding soil types is to know the parent minerals from which the soil was formed. The organic components of soil consist of decaying plant material, viable plants and microbes. As we discussed earlier in this course, microbes living in the soil mediate many important biogeochemical transformations.

Soil Horizons result from weathering of bedrock. There are typically three horizons in any soil. The horizon at the surface is called the "A" Horizon and consists of topsoil. This is the region where plant growth occurs and most microbes live. The middle region of soil is called the "B" Horizon, which consists of sub soil. The deepest soil layer is the "C" Horizon, which consists of weathered parent rocks. Bedrock is the material under the "C" Horizon (Fig. 6.13).

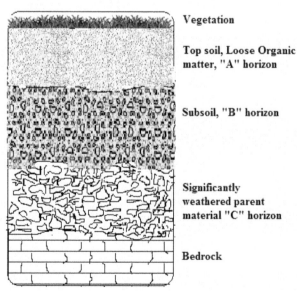

FIGURE 6.13   Soil horizons.

Soil is a precious resource to life as air and water-and just as susceptible to pollution. Soil must be conserved by reducing soil loss from cultivated land through the use of modern agricultural practices. The agricultural system depends on healthy soils to grow food. No nation has ever been successful without first being able to provide food for its citizens. Plants extract the elements like carbon, hydrogen, oxygen, phosphorous, nitrogen, potassium, sodium and calcium from the soil.

A healthy soil has:
- physical properties that allow roots to easily penetrate to the smallest particles,
- a high exchange capacity to allow nutrients to be used as needed, and
- Correct chemicals needed for optimum growth rates.

6.5.2   SOIL NUTRIENTS

In the same method, pollutants are transported and moved into growing plants. Nutrients are stored in soil on "exchange sites" of the organic and clay components. Calcium, magnesium, ammonium, potassium and most of the micronutrients are present as cations under most soil pH values.

The cation exchange capacity of soils is an important measure of its ability to store these nutrients and provide them to growing plants as needed. Organic substances (humic and fulvic acids, humus) contain exchange sights because of the presence of carboxylic acids. An organic component is an essential ingredient of all healthy soils.

The carboxylic acids exchange protons in soil in the same manner that protons are exchanged in aqueous solution. Soil pH is therefore an important parameter, which affects the ability of a soil sample to exchange cations and the ease with which nutrients move through the soil. Ideally, a soil sample should have a pH near 6.5 to provide for optimum nutrient storage capacity and ease of movement to the plant roots. Phosphorous is present in soil as orthophosphate (PO_4^{3-}). The two forms of phosphate that are present in soil under most conditions, HPO_4^{2-} and $H_2PO_4^-$, are anions. Phosphate complexes of most metals have very low solubility, and phosphate is therefore relatively immobile in soil.

Nitrogen, phosphorous and potassium are used in large quantities by all growing organisms. Since they are typically present in soil at the percentage level, they are called macronutrients (Table 6.8). Deficiency of any of these three elements reduces plant growth and lowers crop production. Fertilizers are substances used to replenish these nutrients. Chemical fertilizers and manure are used for this purpose. Compost work just as well and permit organic wastes to be put to good use.

TABLE 6.8   The macronutrients in Soil

Element	Symbol	Chemical Form in Soil
Calcium	Ca	Ca^{2+}
Carbon	C	HCO_3^-, CO_3^{2-}
Hydrogen	H	H^+
Magnesium	Mg	Mg^{2+}
Nitrogen	N	NO_3^-, NH_4^+
Oxygen	O	HO^-
Phosphorous	P	H_2PO^{4+}, HPO_4^{2-}
Potassium	K	K^+
Sodium	Na	Na^+

Micronutrients are classified as "essential" and "nonessential" (Table 6.9). An essential plant nutrient is one that is required for life, whereas a nonessential plant nutrient (present in soil a very low levels) will increase crop yield-but its absence will not cause the organism to die.

TABLE 6.9   The Essential Micronutrients

Element	Symbol	Chemical Form in Soil
Boron	B	H_3BO_3
Chlorine	Cl	Cl^-
Copper	Cu	Cu^{2+}
Iron	Fe	Fe^{2+}, Fe^{3+}
Manganese	Mn	Mn^{2+}
Molybdenum	Mo	MoO_4^{2-}
Sulfur	S	SO_4^{2-}
Zinc	Zn	Zn^{2+}

The nonessential micronutrients: Aluminum, Cadmium, Cobalt, Lead, Mercury, Nickel, Selenium and Silicon.

6.5.3   PESTICIDES IN SOIL

Pesticides are used in all forms of agriculture to control unwanted insects and plants. Paris green, a form of arsenic, was used to control the potato beetle. Modern synthetic pesticides are used since 1950. Fields that were treated with Paris green 50 years ago still contain unacceptable levels of arsenic, since the arsenic cannot be degraded into another substance and the only natural removal mechanism is leaching. Chlorinated hydrocarbons became popular as broad-spectrum pesticides after World War II, the most widely used being DDT (dichlorodiphenyltrichloro-ethane). Other pesticides that are classified as chlorinated hydrocarbons include:

Methoxychlor, Dieldrin, Endrin, Chlordane, Aldrin, Endrin, Heptachlor, Toxaphene and Lindane.

- They do not last as long as nondegradable substances like arsenic, but they do have unacceptably long half-lives in the soil.
- The very long time needed to degrade chlorinated hydrocarbons,
- They are concentrated in the food chain,

This led to legislation banning their use. Unfortunately, these compounds are still in common use in underdeveloped countries. Modern pesticides have half-lives of weeks or days, and if used properly, do not build up in soil like their predecessors.

6.5.4   CHEMICAL WASTES IN SOIL

It was a common practice to deposit industrial wastes in landfills or simply bury containers of waste chemical in the soil. Waste materials are not easily degraded and have the potential to pollute the groundwater. Once in the soil, these substances can enter the food chain by being incorporated into plant tissue, which is eaten by livestock. The most infamous example of such a case occurred in Michigan, where dairy cattle were exposed to PCBs. PCB contaminated milk was consumed by people who drank the milk and the cattle had to be destroyed.

Lead can be a problem. Lead washes off buildings that have been painted with lead based paint (common with old buildings). Lead additives to gasoline caused significant increases to lead soil levels near major highways. Legislation to remove the lead additives stopped this source of lead, but the legacy of lead based gasoline will remain in the soils near major streets for years to come.

KEYWORDS

- **Atmosphere**
- **Earth Systems**
- **Global Warming**
- **Pollution**
- **Soil**
- **Water Environment**

REFERENCES

1. Source: English Wikipedia, (2005). Water cycle http://ga.water.usgs.gov/edu/watercycleprint. html

2. Mackenzie, F. T., & Lerman, A. (2006). Geobiology, *25*, Carbon in the Geobiosphere Earth's Outer Shell, Springer, Abraham, QH344, M33.

3. Raven, P., Evert, R., & Eichhorn, S. (1992). Biology of Plants, 5th Ed., Worth.

4. Lower, S. K. (1999). Chem. Environmental Chemistry, Carbonate Equilibria in Natural Waters, Simon Fraser University, 26 p, June.

5. Morford, S. L., Houlton, B. Z., & Dahlgren, R. A. (2011). "Increased Forest Ecosystem Carbon and Nitrogen Storage from Nitrogen Rich Bedrock", Nature, *477 (7362)*, 78–81.

6. Barbooti, M. M. (2013). Chemistry and Technology of Sulfur and Sulfuric Acid, Iraqi Academy Press, Baghdad.

7. Robertson Bryan, R. (2004). PH Requirements of freshwater Aquatic Life, Technical Memorandum, Robertson Bryan, Inc.

8. EPA method #: 310.2 Alkalinity (Colorimetric, Automated, Methyl Orange), Auto analyzer, http://www.caslab.com/EPA-Methods/PDF/EPA-Method-3102.pdf.

9. O'Dell, J. W. (1993). Determination of Nitrate-Nitrite Nitrogen by Automated Colorimetry, Method 353.2, Revision 2.0, August, Environ, Monitor, Systems Lab., R&D Office, U.S Environmental Protection Agency.

10. EPA, Dissolved Oxygen and Biochemical Oxygen Demand, http://www.caslab.com/EPA-Method-360_1.

11. Montgomery, H. A., Thom, C. N. S., & Cockburn, A. (1964). Determination of Dissolved Oxygen by the Winkler Method and the Solubility of Oxygen in Pure Water and Sea Water, *J. Appl. Chem., 14(7)*, 280–296.

12. Pawlowski, S., Jatzek, J., Brauer, T., Hempel, K., & Maisch, R. (2012). 34 Years of Investigation in the Rhine River at Ludwigshafen, Germany Trends in Rhine Fish Populations Environ. Sci. Europe 24:28.

13. Giger, W. (2009). The Rhine Red, the Fish Dead-the 1986 Schweizerhalle Disaster, a Retrospect and Long-Term Impact Assessment, Environ. Sci. Pollut. Res. Int. 1: S98–111.

14. National Oceanic and Atmospheric Administration | Department of Commerce, USA. Gov, Revised January 11, 2013, http://oceanservice.noaa.gov/facts/pollution.html.

15. World Health Organization (WHO) Drinking Water Quality Guidelines.

16. World Health Organization (WHO) Guidelines for Drinking-Water Quality, 3rd Ed., 1 Recommendations, Geneva (2008).

17. http://inside.mines.edu/~epoeter/_GW/19WaterChem3/WaterChem3pdf.pdf

18. Manahan, S. E. (2007). Environmental Chemistry, 9th Ed., CRC, Boca Raton.
19. Barbooti, M. M., Su, H., Punamiya, P., & Sarkar, D. (2014). *Int J. Environ. Sci. Technol.,* *11(1).*
20. http://www.soils.wisc.edu/virtual_museum/kaolinite/index.html
21. Pidwirny, M. (2006). "Atmospheric Composition", *Fundamentals of Physical Geography, 2nd Ed.,* Date Viewed. http://www.physicalgeography.net/fundamentals/7a.html
22. European Commission, http://ec.europa.eu/environment/air/quality/standards.htm.
23. Barbooti, M. M., Ibraheem, N. K., & Ankosh, A. (2011). Removal of Nitrogen Dioxide and Sulfur Dioxide from Air Streams by Absorption in Urea Solution, *J. Environ. Protect, 2,* 175–185.
24. Tucker, M. (1995). Carbon Dioxide Emissions and Global GDP, Ecological Economics, *15(3),* 215–223.

CHAPTER 7

ENVIRONMENTAL SAMPLING AND SAMPLE PREPARATION

ASLI BAYSAL[1*] and MUSTAFA OZCAN[2]

[1]T.C. Istanbul Aydin University, Health Services Vocational School of Higher Education, 34295 Sefakoy Kucukcekmece - Istanbul, Turkey;
*E-mail: baysalas@itu.edu.tr

[2]Istanbul Technical University, Science and Letters Faculty, Chemistry Department 34469 Maslak Istanbul-Turkey

CONTENTS

7.1 INTRODUCTION

The purpose of an analytical study is to obtain information about an object or substance [1–13]. The substance could be a solid, a liquid, a gas or a biological material. The information to be obtained can be varied. This information about analytical results is valid only if the sample represents the material of interest. To get varied information, we need instrumentation processes. Although there is much interest in such noninvasive devices, most analysis is still done by taking a part or portion of the object under study and analyzing it in the laboratory. Some basic steps are shown in Fig. 7.1. The other important tasks in environmental sampling also include filling in the chain-of-custody form, use of GPS to fix sample locations for a second time visit, soil sampling designs including grids (square and hexagonal) and stockpile sampling. Specific sampling plans for each type of pollutant should be considered. Measurement procedures, which include sampling and sample preparation, are closely related to the accuracy, representativeness and sensitivity of the results for an object or substance.

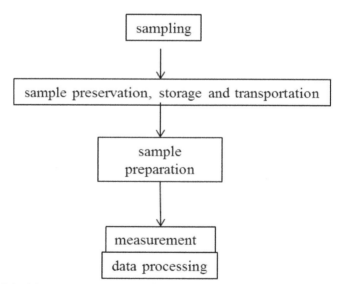

FIGURE 7.1 Measurement procedure of sample.

The measurement procedure is mainly divided into three basic steps, which are:
i)   **Sampling:** It where the sample is obtained from the object to be analyzed. This is collected such that it represents the original object or target population. Sampling is done with variability within the object in mind.
ii) **Sample preservation, storage and transportation:** This is another important step; there is usually a delay between sample collection and analy-

sis. Sample preservation ensures that the sample retains its physical and chemical characteristics so that the analysis truly represents the object under study.

iii) **Sample preparation:** Most samples are not ready for direct introduction into instruments. There might be several processes within sample preparation itself. Homogenization/size reduction, extraction, concentration and clean up might be possible steps within sample preparation.

Once the sample is prepared, analysis is carried out by an instrument of choice. A variety of instruments are used for different types of analysis, depending on the information to be acquired. Common types of analytical instrumentation and the sample preparation associated with them are listed in Table 7.1. The sample preparation depends on the analytical techniques to be employed and their capabilities.

TABLE 7.1   Common Analytical Instrumentation and the Sample Preparation

Analytes	Sample preparation technique	Instrumentation
Organics (volatile/non-volatile)	Extraction	Gas chromatography
	Derivatization	Liquid chromatography
	Clean up	Gas chromatography/Mass spectroscopy
	Transfer to vapor phase	
	Concentration	Liquid Chromatography /Mass spectroscopy
Inorganic (heavy/trace metal)	Digestion	Atomic absorption spectroscopy
	Extraction	Inductively coupled plasma
	Derivatization	
	Speciation	
	Concentration	
Ions	Extraction	Ion chromatography
	Concentration	UV-VIS spectroscopy
	Derivatization	

The purpose of environmental analysis is mainly background monitoring of chemical substances and pollution monitoring of hazardous materials in the environment. These two subjects include regulatory enforcement, regulatory compliance, routine monitoring, emergency responses and scientific research. The importance or effect of the determination or analysis of this chemical analysis depends on the parameters mentioned above which are sampling, storage and sample preparation (treatment).

Chemical analysis may also be categorized with respect to the type of substance being analyzed. Inorganic analysis is concerned with the determination of atoms and inorganic compounds, whereas organic analysis involves the determination of organic compounds. According to the type of substance under analysis, the sample preparation can be varied as a sample preparation of inorganic substances and organic substances.

Another critical point in environmental analysis is to conduct all determination under the guidance of a carefully designed quality assurance (QA) program. In many cases, environmental analyzes must also be conducted following strictly designed regulatory requirements, dictating such things as sampling and analysis methodology, quality control procedures and complete documentation policies. Specific legally mandated QA requirements can vary significantly depending upon many parameters, including the country, the regulating agency and the sample type.

7.2 SAMPLING FUNDAMENTALS

Fundamentals of the measurement procedure for the environmental samples [1–13] are explained in this Section.

7.2.1 SAMPLING

Sampling is an important and critical process for data quality and success of the analysis. Sampling includes some important rules. The rules of the sampling procedure are summarized in Fig. 7.2.

The goal of the environmental sampling, *the sampling frame* [1], is that the whole sampling procedure can be planned answering the question of where, when, what, how, and how many. A frame ensures a representative sample collection. It must support the goals of an investigation.

Another important point about sampling is *the representativeness* [2] of the whole. Representative sampling is made in various ways. Sampling should also be obtained to represent all matrices in the whole object and the target population's heterogeneity should not cause any sampling error. For the purpose of the target populations' or whole objects' homogeneity or representativeness, different sampling approaches can be used for environmental samples, for example;

   i) *Judgment sampling*: the sample is collected from the target population using the available information about the analytes distribution within the population. Its bias is higher (or more) than the bias of the random sampling and it is also encountered in many protocols.

   ii) *Random sampling*: the sample is collected at random from the target population or the whole object. Obtaining correct random sampling is difficult. A convenient method to ensure the collection of a random sample is to divide the target into equal units. Random sampling does not make any assumptions about the population, on the other hand it requires more time and expense because of a higher number of samples is needed to characterize the target.

iii) *Stratified sampling*: the target population is divided into units and random samples are collected from each unit. Each unit is analyzed separately and the mean value gives the target population. Analysis of each unit represents a more homogeneous population than the target population.

iv) *Systematic sampling*: the object or target population is sampled at regular time or intervals of time or space. If the object or population is heterogeneous at regular temporal or special intervals, systematic sampling causes sampling errors.

v) *Composite sampling*: is the combination of several single samples of the target population. It is the most adaptable of the sampling schemes and it saves time and costs.

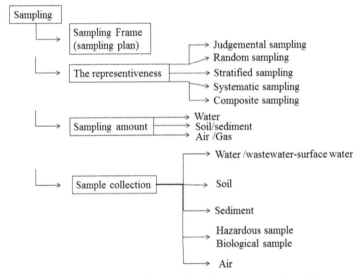

FIGURE 7.2 Some important rules for sampling of environmental analysis.

We briefly explained about the sampling frame or plan and ways of representing the sample. Now we are going to explain another point of the sampling procedure, which is the *sample amount* [3]. The sample amount should be sufficient to perform all required analysis with an additional amount for extra analysis, double check and other quality analysis. The importance of the sample amount of the sampling is about representativeness and satisfying the performance criteria and constraints. The sample amount depends on the sample homogeneity and sample type such as air, soil, water, etc. If the sample is heterogeneous, a large amount of the sample should be used for correct sampling and then a larger portion of the sample is homogenized for subsampling. If the sample is homogenous, a minimal sample could be collected. Taking too many or too large samples can cause storage, transportation and cost problems. Some environmental sample types and their sample amounts are explained below;

   i)   Water-wastewater samples; the minimum required sample amount is approximately 5 mL for TPHs, 100 mL for total metals, 1 L for trace organics according to EPA (1995) rules.

   ii) Soil-sediment samples; the minimum required sample amount for the total characterization is approximately 200 g of soil, if the analysis includes detection of low solubility organic contaminants, more soil samples are needed according to EPA (1995) rules.

   iii) Air samples; the air sample amount depends on the chemical concentration of the detection parameter and the sensitivity of the measurement. More than 10 m^3 samples may be required for the determination of analytes.

Table 7.2 shows the minimum required sample amount.

TABLE 7.2   Sample Type and Required Sample Amount

Sample type	Required sample amount
Water Ground/surface water Waste water	5 mL for TPHs 100 mL for total metals 1.0 L for trace organics
Soil Sediment	200 g of soil for total characterization
Air	>10 m^3

The next step is *sample collection* [4]. There is a wide range of sampling techniques which means sampling equipment or/and tools for the collection which depend on the type of environment, the material being sampled and the subsequent analysis of the sample, the ease of the use and contamination risks and the location of the sampling.

The sampling tools are mostly divided according to the type of environment.

   i)   **Water (sea water/surface water/ground water)–wastewater**—In this part we will detail specific sampling and equipment considerations for each of the most common water quality parameters. There are, however, two general tasks that are accomplished any time water samples are taken. Firstly, sampling containers should be prepared according to EPA guidance. Re-usable sample containers and glassware must be cleaned and rinsed before the first sampling run and after each run by following either the general method for determination of conductivity, total solids, turbidity, pH, and total alkalinity or the acid wash procedure for the determination of nitrates and phosphorus.

After the preparation of sample containers, water sampling can be divided into two sections:

- Surface water/wastewater–three common sampling tools are depicted:
  ✓ the pond or grab sampler–it is appropriate for near shore sampling and when there is limited direct access by contaminants. It is also used for sampling from lagoons, pit banks and disposal ponds.
  ✓ the weighted bottle sampler–it is used for the collection of samples from a predetermined water body.
  ✓ the Kemmerer bottle–it is a Teflon, acrylic or stainless steel tube attached to a rope and has a stopper. It is used for sampling from a boat or other similar structures.
- Groundwater–similar to surface water/wastewater sampling tools, three sampling tools can be used which are shown in Fig. 7.3:
  ✓ The bailer pump–it is a pipe with an open valve and a check valve at the bottom. It is easy to use and is a commonly used sampling container (Fig. 17.3a).
  ✓ Peristaltic pump–it has ball bearing rotors and tubing is attached to the end of the rotor. It is suitable for small diameters and has a depth limitation.
  ✓ Bladder pump–it has a Teflon or stainless steel bladder and compressed gas source. It is suitable for collecting volatile compounds (Fig. 7.3b).

(a) Bailer   (b) Bladder pump

FIGURE 7.3   Basic diagram of tools for ground water sampling.

ii) **Soil**—Accurate sampling requires the use of the correct tool. Soils are usually treated as two- or three-dimensional materials. The sampling location is divided as a layer and is horizontal. A classical approach for sampling soils is to use a spoon, shovel, scoop or other scraping tool. Depending upon the amount of material needed for the analysis and the purpose of the sampling effort, increments can be randomly taken from each horizon or layer in each of the cuts. These are combined into a single sample representing the layer at the pit location.

iii) **Sediment**—Mostly solid sampling tools can be used for sediments, especially surface sediment. In addition to classical solid sampling tools, four different tools can be used which are;

✓ Ekman dredge–small and light devices to collect bottom materials. The materials can be easy or not to collect, including gravel, rocks etc. (Fig. 7.4a)

✓ Petersen grab and Ponar grab–if the sediment is large debris or includes rocks, gravel etc., the Petersen and Ponar grab should be used for collection (Fig. 7.4b).

✓ Kajak-Brinktrurst corer–it is used to compare time-dependent deeper sediment or to collect undisturbed sediment samples (Fig. 7.4).

Tools used for sediment sampling shown are in Fig. 7.4.

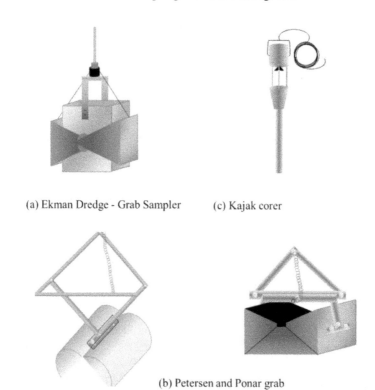

(a) Ekman Dredge - Grab Sampler   (c) Kajak corer

(b) Petersen and Ponar grab

FIGURE 7.4   A diagram of tools for sediment sampling

iv) Hazardous samples or biological samples—Both hazardous sampling and biological sampling need special attention compared to other environmental sampling tools.

v) Air—The most straightforward method of measuring components in air is to capture a sample and return it to the laboratory for analysis. More complex instruments are also available for deployment to enable on-site measurements. Air sampling collectors are based on particulate matter (PM). Mostly all tools are PM10 (<10μm largest particles), PM2.5 (<2.5μm particles) and PM1.0 (<1 μm smallest particles). Air samples have different types, like particulates and aerosols, gas. The components of sampling tools used for the collection of aerosols and gaseous materials differ from each other and they are so diverse. In recent years mostly high volume samplers have been used for the collection of different kinds of air samples (Fig. 7.5). This sampler is useful for particles <100 mm in diameter and requires a protection guard. The filtration head of the sampler is variable depending on the PM. The air samples are collected on a filter using different mechanical methods and are then analyzed.

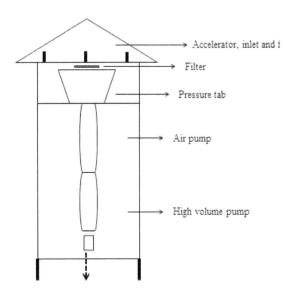

FIGURE 7.5   Basic air sampling tool.

Sampling tools for different sample types for the collection of samples are summarized in Table 7.3.

TABLE 7.3 Sample Tools Use for the Collection of Environmental Sample According to Sample Type

Sample type		Collectors
Water	Surface/Waste water	The pond or grab sampler
		The weighted bottle sampler
		The Kemmerer bottle
	Groundwater	The bailer pump
		Peristaltic pump
		Bladder pump
Soil		Spoon
		Shovel
		Scoop
		Other scraping
Sediment		Ekman dredge
		Petersen grab/ Ponar grab
		Kajak-Brinktrurst corer
Hazardous sample		Consultation is needed
Biological sample		Special attention
Air		High volume sampling

7.2.2   SAMPLE PRESERVATION, STORAGE AND TRANSPORTATION

Once the sample has been correctly and properly collected according to analyzes, the second step before sample preparation is sample preservation, storage and transportation. Careful handling (storage, preservation and transportation) is essential and representative sample collection of the whole substance until analysis is resulted. Once the sample has been correctly collected in a sample vessel, some interactions can happen between the sample and sampling vessel, or a chemical and biological reaction can form in the sample while it is waiting for analyzes. These reactions and interactions may change, destroy, increase or decrease the analyte/ analytes concentration or type in the sample via microbial degradation, chemical reaction, volatilization, adsorption, absorption, diffusion, photoreaction, etc. Also the sample may be contaminated in the time before the analysis. Therefore, sample storage and transportation must be performed carefully and properly so as not to allow any reactions, interactions or contamination.

To maintain the original conditions and to avoid any changes between the sampling and sample preparation, some precautions must be taken. Table 7.4 shows those precautions in storage, preservation and transportation.

TABLE 7.4   Some Important Parameters and Their Precautions

Parameter	Preservation	Maximum holding time
pH	-	ASAP
Salinity	-	ASAP
Temperature	-	ASAP
Metals	Use plastic bottle Add HNO_2 to pH<2	6 months
Organic	Low pH and temperature Add $HgCl_2$ to prevent from bacteria Add $Na_2S_2O_3$ to remove Cl_2	1–3 months
Ammonia	Low temperature Add H_2SO_4 to pH<2	1 month
PAH	Low pH and temperature Use amber glass bottle or glass teflon septum cap	15 days
Inorganic anions (Br, Cl, F)	Use plastic or glass bottle	1 month
Nitrate, Nitrite	Low temperature Use plastic or glass bottle	2 days
Sulfide	Low temperature Add zinc acetate and NaOH ~ pH 9 Use plastic or glass bottle	7 days
Tissue	Freeze Use plastic bottle or foil	ASAP
Biochemical oxygen demand	Low temperature Use plastic or glass bottle	48 hrs
Chemical oxygen demand	Low temperature Use plastic or glass bottle	28 days
Biological materials (DNA- RNA)	Freeze at –80°C	Years

To avoid those problems some precautions can be taken, such as,
- Choice of correct sample collection vessel: the material type is critical for the selection of vessels to avoid problems.
  ✓ Glass–if adsorption to glass wall is not important. The glass vessels are used for BOD, COD, dissolved oxygen, phenols, and volatile-suspended- total solids analysis. Especially Pyrex glass vessels are more suitable for organic analysis.
  ✓ Plastic–it is suitable for inorganic analysis to avoid adsorption to glass wall.
  ✓ cap/septum
  ✓ amber bottle–to preserve a photochemical reaction, for example PAH analysis.
- find the correct place for waiting
  ✓ room temperature
  ✓ cold storage
- add chemicals for the duration of time
  ✓ acid–to prevent metal oxide or metal hydroxides, volatilization of ammonia, or bacteria degradation of total phenolics
  ✓ base–to prevent volatilization of cyanide, to protect phenols
  ✓ other chemicals–zinc acetate, ascorbic acid, Na2S2O3, etc.

Proper preservation may not be enough for the data quality. The preservation type is not the only way to avoid any changes, destroy, increases or decreases in the analyte/analytes concentration or type. Each of the samples has its own storage time between analyzes and sampling. Some analyzes or analytes have some storage time after sampling. These time changes range from as soon as possible to months.

7.2.3 BASIS OF THE SAMPLE PREPARATION FOR ENVIRONMENTAL SAMPLES

Sample preparation [1–9] is an important step not in only environmental analysis but also in all analytical procedures. Usually, this stage is a more time consuming step for the sampling process than collection, analysis or data management. Another important point of this stage is that it is open to errors. Because of being open to errors and time consuming, reducing error sources, speeding up or automating sample preparation is a necessity. Ideal sample preparation should involve the minimum number of steps, be environmentally friendly, economical and cause fewer errors.

Generally, to reduce these disadvantages or to achieve accurate sample preparation, the sample component of interest must be,
i)   in a solution;
ii) free from interfering matrix elements;
iii) at an appropriate concentration for detection and measurement.

Accurate sample preparation can be achieved by ensuring these three objectives. Different sample preparation approaches can be used for this purpose:

    i)   sample matrices
        a. organic analysis
        b. inorganic analysis
    ii) sample type
        a. solid
        b. liquid
        c. gas

In this part of this chapter, we are going to discuss sample preparation according to sample type. Samples can be in various forms; such as solid, liquid, gas or other (semisolid, colloids, etc.). Most samples are found in a solid or liquid form. A gas form sample is collected on solid phase or detected directly using appropriate techniques.

Sample preparation steps are explained in Fig. 7.6 according to sample type. Preparation of the sample for the determination of analyte is not easy. Some problems often occur and further or extra stages are needed. Some of them are summarized in Table 7.5.

TABLE 7.5   Some Problem and Solution Methods for the Sample Preparation

Problem	Solution
Concentrated Sample	Dilution of sample with a solvent which compatible with method
	Usage of smaller initial sample
Diluted Sample	Increase of injection volume
	Concentration of sample
	Liquid-liquid extraction
	Solid-phase extraction
	Evaporation
	Lyophilization
Reactive/Thermally/Hydrolytically unstable Sample	Derivatization of sample to stabilize
	Cooling down the sample
	Protecting sample from light/air
Sample contain unwanted high-molecular weight components	Separation of high molecular weight substances
	Size exclusion chromatography
	Dialysis
	Ultrafiltration
	Precipitation
	Supported-liquid membrane

TABLE 7.5   *(Continued)*

Problem	Solution
Sample contain particles	Removal of particles
	Filtration
	Centrifugation
	Sedimentation
Applied solvent not compatible with method	Removal of solvent
	Evaporation
	Lyophilization
	Distillation
	Application of another solvent which is compatible with method
Recovery of method is not good	Controlling sample preparation steps
	Application of internal standard

Some main sample preparation methods are explained according to sample matrices.

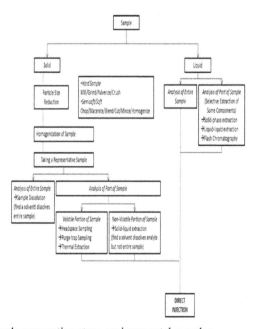

FIGURE 7.6   Sample preparation steps environmental samples.

7.2.3.1   FOR INORGANIC ANALYSIS

Most of the routine sample preparation methods for the determination of inorganic analysis include (or use) digestion or dissolving of the sample, especially soil, sediment and air samples. First, the sample is dried to remove water content, and then if it is needed, further homogenization of the portion occurs and then digestion can be made. Between 0.1 and 0.5 g of sample portions is digested by various digestion methods. The aim of digestion is to dissolve the analyte into aqueous solution. Digestion procedures minimize interference and provide optimal precision and accuracy. Digestion methods are explained below;

i)   *dry digestion:* dry digestion generally consists of heating the sample at 400–600°C for 12–24 hrs. in a furnace. The procedure results in the formation of ash and the sample can be used directly or can be dissolved by wet digestion methods. Inaccuracies can arise both from the volatilization of metals and the retention of metals in an insoluble form in the crucible.

ii)   *wet digestion:* typical wet digestion consists of heating or mixing the sample with a reagent to break down organic matter. Reagents can be an oxidizing acid or a combination of several acids. Some matrixes and commonly used reagents are shown in Table 7.6. Wet digestion has some advantages; less loss due to low temperature and liquid conditions and more homogenization of the sample. However, wet digestion containing acids needs great attention or care and gives high blank values. In addition, wet digestion is carried out in an open vessel so the temperature cannot be controlled exactly during the procedure.

iii) *microwave digestion:* microwave digestion is a kind of wet digestion but it occurs in a closed vessel. This closed vessel controls the temperature and pressure by highly resistant polymers. In addition, the vessels prevent the contamination or adsorption of the contaminants. Generally, an acid or acid combination, which is mentioned in wet digestion, is used for the digestion process. This digestion method is faster and sample loss is much lower than in the wet digestion method.

TABLE 7.6   Some Matrix and Related Reagent for Dissolving of the Environmental Samples

Sample matrix	Reagent
Clean/easy oxidized material (all type environmental sample: water, sea water, surface water,)	HNO_3
Readily oxidizable material (soil, sediment, waste, air)	$HNO_3 + HCl$ $HNO_3 + H_2SO_4$
Difficult to oxidize organic material (soil, sediment, waste)	
Containing any silicates (air, filters contain silicates)	$HNO_3 + HClO_4 + HF$

7.2.3.2   FOR THE ORGANIC ANALYSIS

The primary sample preparation method for organic analysis is the extraction procedure. Extraction procedures can be applied for organic analysis types.

- *Non-volatile organic (NVO) or semi volatile organic (SVO) analysis:* The extraction procedures for the nonvolatile or semi volatile organic matrix are
  - ✓ Liquid–liquid extraction; passing the substance between water phases to organic phase using a separator funnel.
  - ✓ Solid phase extraction; retains the analyte from the liquid sample to the solid sorbent.
  - ✓ Soxhlet extraction; dry solid sample is converted and extracted in a solvent.

This procedure occurs in an open vessel and is available only for nonvolatile and semi volatile organic substance determination. A general explanation is given above but there are many extraction methods, which can be used for NVO and SVO analysis (see Table 7.7).

TABLE 7.7   Methods for the Sample Preparation of SVO and NVO Determination

Extraction method	Application	Information
Liquid- liquid	SVO NVO (for liquid samples)	Low cost Relatively fast and simple procedure Solvent usage approx. 500 mL
Solid phase	SVO NVO (for liquid samples)	Medium cost Relatively fast and simple procedure Solvent usage approx. 100 mL
Solid phase micro-extraction	SVO NVO (for liquid samples)	Low cost Relatively fast and simple procedure Non solvent usage
Soxhlet	SVO NVO (for solid samples)	Low cost 4–24 hours procedure duration and simple procedure Solvent usage approx. 200–500 mL
Ultrasonic	SVO NVO (for solid samples)	Medium cost Too fast and simple procedure Solvent usage approx. 200 mL
Supercritical fluid	SVO NVO (for liquid and solid samples)	High cost Fast and not simple procedure Solvent usage approx. 20 mL
Pressured fluid/ accelerated solvent	SVO NVO (for solid samples)	High cost Too fast and not simple procedure Solvent usage approx. 20 mL

7.2.3.3   FOR THE VOLATILE ORGANIC ANALYSIS

- Volatile organic (VO) analysis: the extraction methods except solid phase micro extraction are not available as sample treatment methods for volatile compound determination. Sample treatments for the VO compound are listed below:
  - ✓ *Purge and trap:* this is purging the sample while heating with an inert gas and trapping VOs in sorbent. This method is relatively fast and there is no need for solvent consumption but it has high costs and it is complicated.
  - ✓ *Headspace:* it can be explained as the equilibrium between the sample and gas phase. The sample phase is introduced and diffused into the gas phase until the headspace has reached the equilibrium. The advantages of this method are that is has low costs, it is fast and simple and that it uses no solvent.

7.3   SAMPLING PROCESS

This section provides the fundamentals of the main procedures for the sampling of common environmental matrices for a wide range of chemical contaminants. The use of these procedures as basic information will enable researchers or analysts to create their own sampling procedures. Between the beginning of the sampling procedure for sample preparation and the handling and analysis section of environmental analysis, some common steps must be done correctly [14–17]. This procedure includes seven main steps:

1. **Reading and understanding of the sampling and analysis plan.** This should include a field sampling project with sufficient detail for proper sampling, sample handling, and completing field documentation.
2. **Identifying the types and quantities of the necessary sampling equipment and supplies.** The project team should make a list of all the needed equipment and supplies according to the volume of the work.
3. **Arranging the sampling points.** Identify the sampling points and record coordinates with the assistance of a Global Positioning System (GPS). The maps, sampling grids, treatment system process diagrams, GPS, etc. are needed for the identification of the sampling points.
4. **Collecting the field and quality assurance/quality control sample.** After the sampling points have been identified, we can start the collection of field and quality assurance/quality control samples according to the procedures described in the first and second step. All samples should be prepared according to the appropriate sample preservation method until the collected sample reaches the laboratory and analysis should not lose any analytes and use the proper containers for any contamination.

5. **Completing field documentation.** Completing forms, field logs and sampling forms is a separate and distinctive step in the sampling process. Field documentation establishes the basis for informed data interpretation and efficient and accurate report preparation.

- **Chain-of-Custody (COC).** The COC form is usually the only written means of communications with the analytical laboratory and legal team who are responsible for sample integrity. All COC forms must include at least some of the information listed below:
  ➤ project name and point of contact
  ➤ signature of a sampling team member
  ➤ field sample ID no
  ➤ date and time of sampling
  ➤ grab or composite sample designation
  ➤ signatures of individuals involved in sample transfer
  ➤ transfer bills (air, shipping, etc. number)
  ➤ analytical requirements

The form includes extra information, such as:
  ➤ the description of sample matrices
  ➤ the description of sampling points
  ➤ the number of containers for each sample
  ➤ preservation procedure, like used chemicals
  ➤ special requirements or instructions
  ➤ a record of sample disposal

- **Sample numbering and labeling.** Each sample must have its own unique identification number (ID no) to match the sample and the requested analysis. The ID no must be simple, short and consecutive. The sample ID should include such items as:
  ➤ project name
  ➤ sampling point ID
  ➤ date and time of sampling
  ➤ sampler's name
  ➤ requested analysis
  ➤ chemical preservation method

- **Sample tracking.** This brings together the following field and laboratory information that is vital for data review, data management and report preparation. Sample tracking records provide some information according to the explanation below:
  ➤ field sample ID
  ➤ laboratory sample ID
  ➤ sample collection date
  ➤ sample location (soil boring, depth, monitoring well, grid unit, etc.)
  ➤ sampler's name

> ➢ sample matrix
> ➢ requested analysis
> ➢ laboratory name (if several laboratories are used)
> ➢ sample type (field sample, field duplicate, equipment blank)

6. **Packaging the collected samples for transfer**. Packaging should be done properly and correctly according to the correct holding time, temperature, and safe packing for the transfer so as not to cause any leaks, breakages, etc. of the samples.

7. **Transferal to the laboratory.** Some sampling details are given below in sections for each sample type, such as air, soil and water.

7.3.1 AIR SAMPLING PROCESS

Firstly, site or sampling area information is defined for the sampling. A general description is obtained of the region, state or area where it is located, the kind of sample (dust, indoor, outdoor etc.), and what kind of analysis is required for the project. The project organization is planned according to this information. All supplies, equipment and project plans are prepared according to the above sections.

A general view of the air sample collection should be as follows [18, 19]:

✓ The sample should be representative of the whole.

✓ The technique employed, particularly the volume sampled and the analytical measurement method, is suited to the pollutant and the application, (e.g., range, analytical limit of detection, linearity, response speed, and measurement uncertainty).

✓ The sample/measurement system is leak-tight.

✓ The material and condition (e.g., temperature, wind, season) of the sample/measurement systems is such that there is neither loss of pollutants nor addition of interfering contaminants.

✓ Any supporting measurements that are required such as volumetric flow rate, oxygen and moisture are conducted using suitable techniques and are simultaneous with the sampling/measurement process.

✓ The volume of air sampled is accurately measured and corrected to standard conditions of temperature and pressure (e.g., using a cumulative gas meter, or a flow rate meter)

The overall project view includes:

• where to monitor/sample;
• when to monitor.

8. The time period in which the sample/measurement is taken.
9. The duration of sampling.
10. The frequency of monitoring.

The important units for air measurements are:

- Mass Concentration: The mass of pollutants per unit volume of waste gas emitted (e.g.
- mg/Nm³).
- Volume concentration: The volume of pollutants per unit volume of waste gas emitted
- (e.g., ppm, ppb)
- Volumetric flow rate: The volume of waste gas emitted per unit time (e.g., Nm³/hr.)
- Mass flow rate: The mass of pollutants emitted per unit time (e.g., kg/hr.)

7.3.2   SOIL SAMPLING PROCESS

This section briefly explains soil sampling design. Soil sampling can be planned in different ways depending on the purpose of the project. Soil sampling varies in terms of systematic sampling on a grid system, sample compositing, and sampling for specific chemical parameters, such as heavy metal determination or organic carbon determination.

7.3.2.1   GRID SAMPLING

An important point about soil sampling is how to take or arrange the sampling points. The grid is used to identify the sampling points and is mostly used for surface soil, the bottom and sidewalls of excavation pits and trenches. The grid properties are explained in the project plan (size of the sampling grid, number of sample points and the locations of sampling points). Sampling grids can be square, rectangular or hexagonal (triangular) with the grab samples collected from each grid nodule [14].

The most commonly used grid sampling are:

- **Square grid.** A sample landmark is divided using squares (20 ft. each border according to field size), and each square is properly marked. There are different options for sampling using a grid system. For example, collecting samples in the grid nodules (a) or taking these border points (b). Square grid sampling is shown in Fig. 7.7.

- **Hexagonal grid.** Another name is triangular sampling. This type of the grid is best achieved using a simple and effective device, like a rope. Firstly, the landmark is identified, and then the size of the grid is determined and remarked as hexagonal or triangular. After grid has been pointed and laid out, we may collect grab samples from grid nodules or the points of triangular corners.

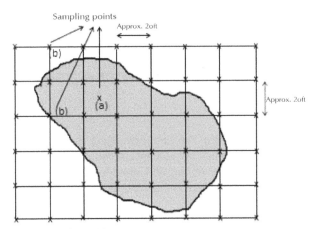

FIGURE 7.7   Square sampling grid.

7.3.2.2   STOCKPILE SAMPLING

Sampling of stockpiled soil is used for the remediation process. An important point about the stockpile sampling is the size of the stockpile. Small and medium size stockpiles are usually sampled using a three-dimensional simple random sampling strategy. A large stockpile may be characterized using a systematic sampling approach. Also, stockpile volume is an important point for this sampling. Prior to sampling, the stockpile is divided into three-dimensional segments from which samples will be collected. Then sampling is performed as grab.

7.3.3   WATER SAMPLING PROCESS

Water sampling includes surface sources (such as, sea, lakes, rivers, lagoons, and ponds), groundwater, drinking water and tap water, etc. Sampling varies widely for water samples, therefore no universal or exact plan for can be recommended. Water samples can be taken easily according to explanation section of sampling, sample preservation, sample tools and related EPA or other sampling standards.

7.4   SAMPLING AND ANALYSIS PLAN

We summarized the whole sampling plan using information explained previously. Sampling and analysis plans should contain these steps [15–17, 20, 22]:
1. scope and application;
2. method summary;
3. sample preservation, containers, handling, and storage;

4. interference and potential problems: interference or potential problems associated with the sample are explained in the plan;
5. equipment/apparatus: equipment needed for the collection of samples may be included in this section;
6. reagents: all reagents used in the whole procedure are listed;
7. sampling procedures:

Preparation
A. Determine the extent of the sampling effort, the sampling methods to be employed, and the types and amounts of equipment and supplies needed.
B. Obtain the necessary sampling and monitoring equipment.
C. Decontaminate or preclean equipment, and ensure that it is in working order.
D. Prepare scheduling and coordinate with staff, clients, and regulatory agency, if appropriate.
E. Perform a general site survey prior to site entry, in accordance with health and safety requirements specific to the parent organization for the field collection crew.
F. Use a GPS unit to identify and record sample location coordinates. If required, the proposed locations may be adjusted based on site access, property boundaries, and obstructions.

Sample Collection
Samples are collected using proper sample collection tools and it is necessary to explain exactly which ones, how many samples are to be taken, etc.

Sample Preservation, Containers, Handling, and Storage Calculations
Once samples have been collected, the following procedures should be followed:
A. Transfer the sample(s) into suitable, labeled sample containers specifically for the analyzes to be performed.
B. Preserve the sample, if appropriate or use pre-preserved sample bottles. Do not overfill bottles if they are pre-preserved.
C. Cap the container securely, place in a resealable plastic bag, and cool to 4°C.
D. Record all pertinent data in the site logbook and/or on field data sheets.
E. Complete the Chain-of-Custody record.
F. Attach custody seals to the cooler prior to shipment.
G. Decontaminate all nondedicated sampling equipment prior to the collection of additional samples.

Decontamination
For collection techniques, all sampling equipment must be decontaminated prior to reuse. Equipment decontamination will consist of the following five steps:

1) detergent wash;
2) tap water rinse;
3) acetone rinse;
4) deionized water rinse; and
5) air dry.
6) quality assurance/quality control: there are no specific quality assurance activities, which apply to the implementation of these procedures. However, the following QA/QC procedures apply:
   - all data must be documented on field data sheets or within site logbooks.
   - all instrumentation must be operated in accordance with the operating instructions as supplied by the manufacturer, unless otherwise specified in the work plan. Equipment checkout and calibration activities must occur prior to sampling/operation, and they must be documented.
7) data validation: if it is required, this section may be prepared using related standards.
8) health and safety: follow U.S. EPA and Occupational Health and Safety (OSHA) health and safety procedures.

KEYWORDS

- **Air sampling**
- **Handling**
- **Preservation**
- **Sample preparation**
- **Soil sampling**
- **Storage**
- **Storage calculations**
- **Transportation**
- **Water sampling**

REFERENCES

1. Stoeppler, M. (1997). Sampling and Sample Preparation, Practical Guide for Analytical Chemists, Springer: Berlin.
2. Mitra, S. (2003). Sample Preparation Techniques in Analytical Chemistry. Wiley-Interscience: New Jersey.
3. Meyers, R. (2013). Encyclopedia of Analytical Chemistry, Wiley, 1999.
4. Worsfold, P., Townshend, A., & Poole, C. (2005). Encyclopedia of Analytical Science, Elsevier.
5. Zahang, C. (2007). Fundamentals of Environmental Sampling and Analysis, Wiley-Inter sciences: New Jersey.

6. Radojevic, M., & Bashkin, N. V. (2006). Practical Environmental Analysis, 2nd Ed. RSC Publishing: Cambridge.
7. Lopez-Avila, V. (1999). Sample Preparation for Environmental Analysis, Critical Reviews in Analytical Chemistry, 29.
8. Csuros, M., & Csuros C. (2010). Environmental Sampling and Analysis for Metals, Taylor Francis.
9. Namieśnik, J., & Górecki, T. (2001). Preparation of Environmental Samples for the Determination of Trace Constituents, *Polish Journal of Environmental Studies, 10*, 77–84
10. Sava, R. (1994). Guide to Sampling Air, Water, Soil, and Vegetation for Chemical Analysis, Environment Hazard Assessment Program, Environmental Protection Agency, California.
11. Water EPA, United States Environmental Protection Agency (2013). http://water.epa.gov/ (accessed July 27, 2013)
12. Air Science, EPA, United States Environmental Protection Agency, (2013). http://www2.epa.gov/science-and-technology/air-science (accessed Aug 8, 2013)
13. Guidance on Choosing a Sampling Design for Environmental Data Collection, EPA, December 2002.
14. Popek, E. P. (2003). Sampling & Analysis of Environmental Chemical Pollutants, A Complete Guide, Academic Press.
15. EPA, Preparation of Soil Sampling Protocols: Sampling Techniques and Strategies, EPA/600/R-92/128 July 1992.
16. U. S. EPA Environmental Response Team Standard Operating Procedures, (2012).
17. U.S. EPA, Sampling and Analysis Plan, Soil Vapor Extraction System Big Mo and Former Benzene Pipeline Areas, (2012).
18. Environment Protection Agency, Office of Environmental Enforcement (OEE), Air Emissions Monitoring Guidance Note: 2 (AG2).
19. U.S EPA, High Volume Indoor Dust Sampling at Residences for Determination of Risk-Based Exposure To Metals, Sop No: Src-Dust-01 (2004).
20. U.S. EPA, Field Sampling Guidance Document #1205, 1999.
21. Technical Standard Operating Procedure, Surface Water Sampling, Adapted from Ert/Reac Sop 2013 Rev 1.0.
22. Technical Standard Operating Procedure, Sediment Sampling, Sop # Src-Ogden-04.

CHAPTER 8

PRINCIPLES AND APPLICATIONS OF ION CHROMATOGRAPHY IN ENVIRONMENTAL ANALYSIS

MOHAMMAD REZA GANJALI[1*], MORTEZA REZAPOUR[2], PARVIZ NOROUZI[1], and FARNOUSH FARIDBOD[1]

[1]Center of Excellence in Electrochemistry, Faculty of Chemistry, University of Tehran; *E-mail: Ganjali@khayam.ut.ac.ir

[2]IP Department, Research Institute of Petroleum Industry (RIPI), P.O. Box 14665-137, Tehran, Iran

CONTENTS

8.1   INTRODUCTION

Information on the nature and concentration of ions in samples can be of great importance in different aspects of human life. Such qualitative and quantitative information can be acquired through myriads of methods one of the most important of which is ion chromatography (IC), which is a subdivision of chromatographic techniques.

The term "chromatography," which is formed from the combination of two Greek words χρῶμα (Chroma) meaning color and γράφειν (graphein) meaning writing, is an umbrella term for a group of analytical methods used for the separation of the components of a mixture and the consequent determination of each. The name of the method stems from the initial applications where the separated substances were recognized based on the color they left (which resembled some sort of color writing) on the stationary phase. To use these methods for the separation of mixtures, the samples are dissolved in a so-called mobile phase constituted of a liquid, which carries the analytes through another so-called stationary phase. Due to the fact that the different components of the sample will move at different speeds, the difference in the nature of their affinity to the mobile and stationary phases, leads to different partition coefficients, and hence they are separated.

IC is a member of this group of techniques, and is commonly applied to the determination of single, double and triple charged cations and anions. Due to the versatility of the method, as well as its reasonable costs (at least these days) and other analytical figures of merit, the technique has turned to a common and easy to use laboratory tool in many analytical chemistry labs.

8.2   MECHANISMS OF SEPARATION IC

Ion-chromatography, per se, is an umbrella term, which covers the three subdivisions of ion exchange, ion pair formation and ion exclusion techniques. Although according to IUPAC's definition of ion exchange chromatography, the separation of the species in this method happens due to the differences in the ion-exchange affinities of different species and only in the case of separating inorganic ions and detecting them through conductivity or indirect UV detections, the method is classified as ion chromatography [1–3], one might riotously argue that neither should the detection technique be regarded as a definition criteria, nor should there be an emphasis on the ion's being inorganic in nature [3]. More generally, and perhaps to overcome the shortcomings of the mentioned definition, IC is also described as any liquid chromatographic separation of ionic species in columns in conjunction with a flow-through detector [4].

It should be noted that, although ion chromatography is considered to cover all such methods used for the separation and detection of ionic species, specifically speaking, it refers only to ion exchange chromatography (IEC). In this sense, ion exclusion chromatography (IEC) and ion pair chromatography (IPC) although leading to the same overall end with only different mechanisms, are regarded as specialized and independent applications of the mother method.

In ion exchange chromatography, a reaction between the ionic species, a mobile phase and a stationary phase with functional groups (e.g., sulfonic acid or quaternary ammonium groups), is the basis of separation. For the sake of the electroneutrality of the stationary phase, the functional groups should be accompanied by counter ions originating from the mobile phase, commonly referred to as the "eluent ion." Due to the fact that ions with the same charge can reversibly exchange across the mobile and stationary phases, the exchange process starts once the two phases meet in the column and can reach an equilibrium state over time (Fig. 8.1).

If ions A$^-$ and A$'^-$ are present in the mixture to be separated by IC, the ions randomly exchange with the eluent ion E and vice versa. For the examples shown in Fig. 8.1, we have:

$$\text{Stationary phase}-N^+R_3\ E\cdot + A\cdot \leftrightarrow \text{Stationary phase}-N^+R_3\ A\cdot + E\cdot \tag{1}$$

$$\text{Stationary phase}-N^+R_3\ A\cdot + E\cdot \leftrightarrow \text{Stationary phase}-N^+R_3\ E\cdot + A'\cdot \tag{2}$$

The equilibrium constant K for the interaction between each ion and the stationary phase can be written as:

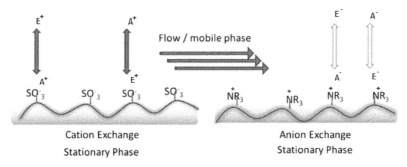

FIGURE 8.1   Schematic diagram of an ion exchange process, where the analyte and eluent ions, A and E respectively, compete in the exchange process.

$$K_A = \frac{stationary\ phase - N^+R_3A^-}{stationary\ phase - N^+R_3E^-} = \frac{[A^-]_s\,[E^-]_M}{[E^-]_s\,[A^-]_M} \tag{3}$$

$$K_{A'} = \frac{[stationary\ phase - N^+R_3A'^-].[E^-]}{\left[stationary\ phase - N^+R_3E^-\right].[A'^-]} = \frac{[A'^-]_s\,[E^-]_M}{[E^-]_s\,[A'^-]_M} \tag{4}$$

In case the interactions between each ion and the functional groups are different, the ions shall be separated as they go along the column. These different interactions can be reflected by the differences in the equilibrium constants of each ion's reaction, which are in this case also referred to as the selectivity coefficient for each ion.

As mentioned above, the separation mechanism in ion pair chromatography, although leading to the identical result of separation and detection of the analytes, is very different. One major difference is that ion-pair chromatography is reversed phased in nature (i.e., the mobile phase is hydrophilic and the stationary phase is hydrophobic). Another difference is that, to make the ionic analyte capable of interacting with the phydrophobic stationary phase, surfactants (such as hexadecyltrimethyl ammonium salts or dodecylsulfonic acid) are added to act the mobile phase to play a so-called ion pairing. This way neutral ion pairs of the surfactants and the analyte ions are formed. These ion-pairs will be capable of having hydrophobic interactions with the stationary phase (Fig. 8.2), and hence the separation due to the different interactions of the various ion pairs formed with the stationary phase, will start.

Another rather specialized branch of IC is ion exclusion chromatography (IEC), which is commonly applied to the separation of weakly acidic or basic compounds like carboxylic and amino acids or carbohydrates. This relatively limited range of applications is due to the limited number of compounds that can participate in this special mechanism. An illustration of the ion exclusion mechanism for acetic acid is shown in Fig. 8.3.

In Fig. 8.3, it can be seen that an ion exchange stationary phase grafted with $-SO_3^-$ (sulfonate) functional groups, in which the sulfonic groups are neutralized with H^+ counter ions, interact with the acetic acid molecules in the mobile phase. The solvated (hydrated) surface of the stationary phase, which is known as the Donnan membrane is separated from the mobile phase with a so-called Donnan membrane represented by the dotted line in Fig. 8.3. The Donnan membrane has an overall negative charge, and only nondissociated species can pass through it. If the mobile phase is a strong acid, which hinders the dissociation of weaker acid such as carboxylic acids, these can also pass through the membrane making further interaction with the stationary phase possible. The molecular form of the weak acids that passes through this hypothetical membrane is hence excluded from the stationary phase and other accompanying ionic species, which cannot take part in this interaction.

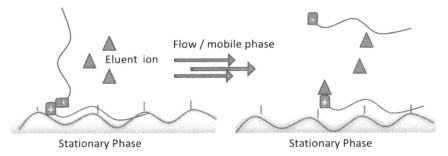

FIGURE 8.2 Illustration of the ion-pair formation mechanism and the subsequent interactions of the ion-pairs with the stationary phase.

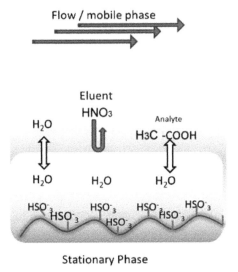

FIGURE 8.3 Illustration of the ion exclusion phenomenon at a sulfunated stationary phase, for an acetic acid sample in a nitric acid containing mobile phase; it can be seen that only neutral species can pass through the Donnan membrane.

Based on this principle, different acids of different acidity constants (K_a) can be separated. There is no need to mention that since very strong acids readily dissociate, and can therefore not diffuse the Donnan membrane, they are the first species to be washed out of the column. Oppositely, the weaker an acid is in the mobile phase, the more the species is delayed in the column. Figure 8.4 shows how the elution of the mentioned acids is dependent on their acidity constants (pK_a).

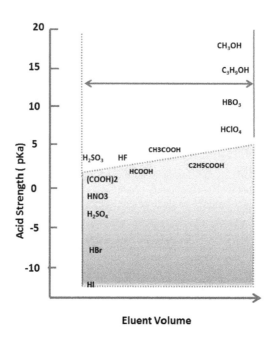

FIGURE 8.4 Elution volume *vs.* the pK$_a$ of the acids ion exchange chromatography.

Chelation (complexation) ion chromatography (CIC), is the name given to a mode of IC in which the stationary phase that is used is one not only interacting with the solute ions through a simple ion-exchange mechanism but it also has the potentials to form coordinate bonds (complexes) based on multipoint interactions (chelate formation) with the ions present in the stationary phase. The solutes used in this case are commonly inorganic cations. The coordination capacity stems from the chelating functional groups that are present on the stationary phases used in CIC. Examples of these chelating functional groups are iminodiacetic acid (IDA) $HN(CH_2CO_2H)_2$ or aminophosphonic acid (APA) [5, 6]. The strength of these interactions is reflected by the stability constants for each metal cation and the specific immobilized polydentate ligand, and as the ligands applied in the method are typically weak acids, retention is heavily dependent on the eluent pH. It should be emphasize that the retention of the species in this method results from a combination of ion-exchange and coordination interactions, and to minimize the former, which can lead to complexation interactions, that can be very specific for special analyte leading to the selectivity of the method and its better separation, are sometimes minimized through the use of a relatively high ionic strength eluent. Such reagents might be 0.5 to 1 M of an inorganic salt, such as KNO_3 which reduces the role of ion exchange interactions making coordination interactions the dominant phenomenon.

8.3   INSTRUMENTATION OF CHROMATOGRAPHY

A schematic illustration of an IC instrument including schematic figures of its components is shown in Fig. 8.5. In general, the device constitutes an eluent reservoir, a pump, an injector, at least one column (an analytical column which is commonly used together with a guard column), the suppressor and the detector. Below, there is a very brief description of the different sections, as well as some descriptive information about each. However, the manuscript does not intend to focus very much on details.

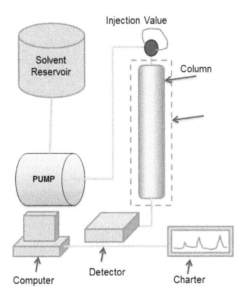

FIGURE 8.5   The schematic illustration of the components of an IC.

8.3.1   THE ELUENT RESERVOIR

The eluent (mobile phase) reservoir is a container in which the eluents are stored. To begin the measurement the eluent should be pumped into the system. The eluents are prepared by dissolving buffers, acids or bases in aqueous solvents or in a mixture of such with organic solvents. The mobile phase should not contain any solid particles, since such pollution can not only perturb the pumping process but also damage the seals or valves or gather on the frits or on the inlet of the column, leading to blockages and increase the system's pressure. Therefore, the eluents are normally removed of such particles using a filter. In some cases, the eluents can alternatively be prepared by an electrolytic eluent generation unit.

8.3.1.1   IC ELUENTS

IC mobile phases for anions are typically aqueous solutions. In specific cases, however, small amounts of an organic solvent are added to prevent undesirable lipophilic interactions. The mobile phases can further contain dilute electrolytes or complex buffer solutions like sodium or potassium hydroxides or carbonates, light linear or aromatic carboxylic acids, aromatic or aliphatic sulfonic acids or other suitable inorganic salts. The concentration of these eluents, as well as their affinity towards the stationary phase, influences their elution strength and can also affect the choice of the detector.

For cations in general, either dilute organic or inorganic acids or, protonated organic bases are among the main eluents used. The most popular eluents for the standard IC separation of inorganic monovalent and divalent alkali and alkaline earth metal cations include light amines together with dilute nitric or methane sulfonic acids. In these cases the retention of the analyte is affected the pH of the mobile phase. On the other hand, in the case of transition and heavy metal cations, the selectivity of the procedure can be controlled by adding a chelating agent to the eluent [6].

8.3.2   THE PUMP

The first step in IC is to precisely pump the eluents throughout the system at a controlled rate. No need to mention that, the pumps should not only be inert and not to have reactions with the mobile phase, it should also function with very high accuracy and precision. As mentioned the major role of a pump is producing accurate and precise pulse-free flows. The pumps should be equipped with vacuum degassing means, a proportioning valve for gradient elusions and also be able to produce a range of pressures on the columns.

The most conventional pumping systems used in ion chromatographic instruments, are reciprocating piston pumps (Fig. 8.6), which satisfy all of the basic requirements mentioned above.

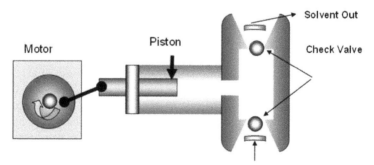

FIGURE 8.6   A reciprocating piston pump.

Pneumatic and displacement pumps may also be used in IC systems. Pneumatic pumps work based on a highly pressurized flexible container and is capable of creating constant pressure pulse free flows. They are much cheaper than the reciprocating piston pumps, but can produce limited pressure ranges and they have problems during gradient elution programs. Another drawback is the fact that refilling of these pumps is rather problematic (which is of course simplified in the newer versions). Displacement type pumps, on the other hand, work through a syringe-like mechanism, can have varying volumes as compared to the reciprocating type. Further, due to their syringe like mechanism the eluents flow created by these pumps is pulse free.

8.3.3 THE INJECTOR

The role of the sample injection system is to reproducibly inject desired amounts of the sample into the ion chromatograph for the purpose of separation. In addition to good accuracy, precision and being rapid, the injection of the sample should not disrupt the flow of the mobile phase in the system. It is interesting to know that, the injection per se, is the largest source of imprecision in all chromatographic analyzes, including IC [7–9]. Commonly, manual or automated injections in ion chromatography are carried out through an injection valve. The most common type of injection valves used are the ones with six ports, which have the capability of s witching to two positions (Fig. 8.7). Ion chromatographic systems are commonly equipped with one injection valve, but for the sake of more complicated analysis like 2D techniques additional valves can be installed. Currently, ion chromatographs with management modules have been constructed to make this time-consuming procedure easier.

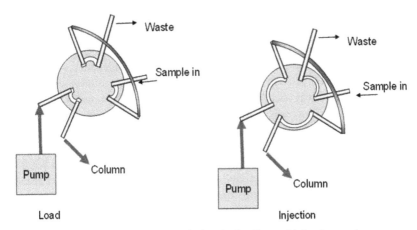

FIGURE 8.7 A six port injection valve during the loading and injection modes.

8.3.4 THE COLUMN

The IC columns (Fig. 8.8) look very much like HPLC columns. The column body, as well as the frits that are used for holding the stationary phase, are made of polymers. The most common type of polymer used for the construction of IC columns is polyether ether ketone (PEEK), and the columns are normally 3–30 cm long and have an inner diameter of 1 to 7.8 mm [10].

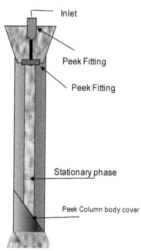

Inlet

Peek Fitting

Peek Fitting

Stationary phase

Peek Column body cover

FIGURE 8.8   A typical IC column and its components.

Since the impurities of the samples may damage the analytical columns used in IC separations, sample pretreatment is necessary. The common sample pretreatment techniques include sample centrifuging or filtration, but this may not be enough since some components cannot be eliminated through these techniques. So sometimes, application of a so-called guard column for the maintenance of the analytical column is advised. A guard column is usually a smaller column with the same packing as the analytical one, placed before the analytical column to trap potential impurities and hence increase the life time of analytical column, especially in the case of the rather expensive analytical columns.

Apart from the role of the column and its structure, the most important function performed by this part is actually the result of the packing material used within the columns. As mentioned, the selectivity of the column depends on a series of factors ranging from the functional groups of the stationary phase, their surface density (ion-exchange capacity of the packing material) as well as lack of hindrance, and hydrophobicity of the packing material or stationary phase. Other important characteristics of the stationary phase are its porous structure, specific surface area, and hydrophobicity in the case of organic polymers. The functional groups present in the stationary phase have various charges and sizes and their potential for polarization, and their lipophilicities differ. Actually the nature of these groups, not only defines the type of the interaction between the analyte ions and the stationary phase, but it also influences the type of suitable eluent and hence the selectivity of the column [6].

A few inorganic materials like silica, alumina, titania, zirconia, and porous graphitic have been used as IC stationary phases. But they either suffer being instable at alkaline pH (e.g., silica), or are not inert in the presence of phosphate and carboxylate ions (e.g., titania, zirconia). So, organic stationary phases have become more common. Mainly, there are four types of resins used as the packing material of IC columns, regarding the nature of functional groups. These include strong acidic and basic resins, which are typically grafted with sulfonic acid groups and quaternary amino groups respectively and weak acidic and basic resins commonly containing carboxylic acid and amino groups. There are also specially tailored chelating resins that are used in the so-called chelation ion chromatography (CIC) in which case, the functional groups is capable of forming multiple coordinate bonds with the analytes. Typical chelating functional groups are iminodiacetic acid (IDA) or aminophosphonic acid (APA) [7].

The most common polymer backbone for ion exchange resins are cross-linked polystyrene polymers, in which styrene and small amounts of divinyl benzene are copolymerized to form a cross-linked polymer network (reaction 1). It is noteworthy that cross-linking increases the lifetime of the resins by decreasing their solubility in solvents. Functionalized polymethacrylate and poly vinyl alcohol backbones are also used for the same purpose (More-efficient columns packed with silica-based ion-exchangers are often preferable for the separation of cations in acidic media).

REACTION 1: Styrene and small amounts of divinyl benzene are copolymerized to form a cross-linked polymer network.

The functional groups of the resins are grafted on the polymeric backbone through a chemical reaction but there are also examples in which the functionalization is performed on the monomers prior to synthesizing the resin.

Ion exchangers are created by chemically introducing suitable functional groups into the organic or polymeric matrix. In the most common types of the stationary phase the surface of an inorganic oxide or the organic polymer is modified through performing reactions on its surface, which creates a monolayer of functional groups on the surface. Examples of this type are sulfonated poly(styrene-divinylbenzene). Sometimes, especially in the case of inorganic material, where the surface of the matrix does not have the binding sites required for functionalization or is not resistant to aqueous media, it is covered with a layer that can satisfy these requirements (silica particle covered with a layer of poly(butadiene-maleic) acid (PBDMA) [11]. A variation of the packing materials is prepared through coating a layer of charged nanoparticles on the surface of a larger supporting matrix, which modifies surface area of the support material and increases the ion-exchange capacity [6], yet as another strategy, columns packed with a hydrophobic adsorbent can be saturated with hydrophobic ionic molecules, which forms a stable coating with aqueous eluents, and help obtain ion exchangers with complex functional groups [6, 12].

Resins are also classified based on the size distribution of their porosities. In this regard, resins can be divided into macroporous or microporous groups, among which microporous resins, have found much more applications in the field. Both types are synthesized through a so-called suspension polymerization technique and their physicochemical properties can be adjusted through varying the reactions conditions during the synthesis procedure. Further, column efficiency and peak resolution are dependent on the particle size of the adsorbent. Hence there is a clear tendency to decrease particle sizes of the stationary phase used in anion exchange columns. Modern IC columns are generally packed with 3–5 micron ion-exchangers. The other path to improve column efficiency is to further develop monolithic porous ion exchange columns for IC. However, only silica monolithic columns have, to date, demonstrated impressive efficiencies for small ions, but these are not stable hydrolytically [13].

8.3.5   THE SUPPRESSOR

IC systems equipped with conductivity detectors can work under two distinctive modes referred to as suppressed and nonsuppressed (indirect) modes. Through the application of a suppressor, conductivity detection becomes a rather solute specific detector. The different commercially available suppressors included electrolytically regenerated, micro membrane, capillary electrolytic, suppressors etc. The technique is however viewed differently for anion and cation separations.

8.3.5.1   ANION SUPPRESSION

Very often a so-called suppressor is used between the column and the conductivity detector to increase the detection sensitivity for anions, while decreasing the background conductivity of the eluent. These suppressors are actually cation exchangers, which replace the cations in the mobile phase with proton. For example, if sodium carbonate is present in the mobile phase and the species to be detected is NaCl, the eluent conductivity of about 800 μS reduces to around 20 μS since sodium is exchanged with H^+ and now the species in the eluent is carbonic acid. This is while the analyte (i.e., NaCl) with a conductivity of 125 μS, now changes to HCl (425 μS). This greatly increases the sensitivity by both reducing the background signal while doing the opposite to the analyte signal.

8.3.5.2   CATION SUPPRESSION

In the case of cation chromatography, however, there are different viewpoints and arguments about the application of suppression, due to problems such as nonlinear calibration behavior observed in the case using suppression of cations that are not completely dissociated from their counter ions, or the fact that in the absence of suppression in cation analysis, transition and main group metals can be analyzed in the same run, eliminating the necessity of a second and independent-unsuppressed for them.

So, one thing is for sure, and that is the fact that cation suppression is not, contrary to its counterpart procedure for anions, always recommended. This is the case, especially when the analyte concentrations are high, and there is actually no need to try to increase the sensitivity of the procedure. However, there have also been arguments that, in the absence of suppression, in some cases the response factors are higher, but high background noise lowers the sensitivity [14].

8.3.6   ION CHROMATOGRAPHY DETECTORS

Different detection systems can be used in ion chromatography, and the detection method is chosen based on the nature of the analyte. Like any analytical method, a suitable detector should have a high sensitivity and short response times, the produced signal should be a linear function of the concentration of the analyte over a wide concentration range, it should produce minor baseline drifts and have as small a noise as possible. The detection methods used for IC fall into the two broad categories of electrochemical and spectrometric.

8.3.6.1   ELECTROCHEMICAL DETECTION

The primary analytical advantage of electrochemical detection (ECD) in flowing systems, such as IC, is the high sensitivity contrary to relative simplicity and low cost. Owing to the inherent features of electroanalytical techniques, and the versatility of electrochemical detectors, ECD meets most of the requirements of ion chromatographic trace analyzes, including linearity of signal over a reasonable concentration range of the analyte, precision and accuracy of electrode potential control, sufficiently fast response, long-term stability of the detector, high sensitivity and low detection limits, as well as, reproducibility of the results. The ECD fall into the classes mentioned in the following sub-sections.

8.3.6.1.1   CONDUCTOMETRIC DETECTORS

Since the majority of common ionic species have very weak or even no UV absorption, conductivity detectors have changed to the most common detection system in IC. The technique is nonselective and functions based on the device's sensitivity to small alterations in the conductivity of the mobile phase, caused by the presence of the analytes based on their elution times. To increase the sensitivity of these detectors, ion chromatographs are equipped with suppression systems, which as mentioned earlier, increase their sensitivity. It is ineluctable to state the detection system can also be used under nonsuppressed modes, in case factors like the sample matrix or its concentration, per se, can yield ample sensitivity.

8.3.6.1.2   AMPEROMETRIC DETECTION

Another powerful technique used as the detector of an IC instrument is voltammetry. Voltammetric methods are applicable to ionic species that are electroactive or have electroactive functional groups (i.e., which can be reduced or oxidized in the presence of the mobile phase). Amperometric detection is the most important and also a very sensitive voltammetric-based detection system used in ion chromatography. In principle, the current caused by the oxidation or reduction of an electroactive species, which passes between a working and a reference electrode having s a certain potential difference, is detected. There is no need to mention that the current will be observed only if the redox potential of the species allows its electrochemical reaction at the applied potential. The detector is common for the analysis of some cations such as Fe^{3+}, Co^{2+} but the main application is for anions like nitrite, nitrate, thiosulfate, halogens and pseudohalogens [6]. Amperometric detections can be performed under single potential (SP), and the very common pulsed amperometric detection (PAD) modes.

In the PAD mode, the detection is performed under a set of potentials. Initially a potential is applied and after a period required for the establishment of an equilibrium current, the current is measured. Next a positive potential pulse is applied to remove the reaction products formed, from the electrode surface through oxidation, before the potential of the working electrode is returned to its initial value. This is cyclically repeated with each cycle lasting about <1 s.

8.4.6.1.3   ION SELECTIVE ELECTRODES (ISES)

Another electrochemical detection method used in ion chromatography is the application of ion sensitive electrodes. This detection system is not very common due to some technical limitations, but successful results have been reported on specific applications of the technique in the ion chromatography of environmental samples. The ion selective electrodes have been discussed in "Principles of Electroanalytical Methods" chapter.

8.3.6.1.4   NEW ELECTROCHEMICAL DETECTORS

Recently advanced methods tailored by electrochemists have also been used as the detection module of ion chromatographs. Fast scan cyclic voltammetry (CV) measurements, which were introduced as detection methods for LC and flow injection system in the 1980s and the detection potential of which for the determination of electroactive compounds, are examples of such methods. Further, the new method of using fast voltammetric detection using Fast Fourier Transformation filtering (FFTF) method to make signal measurements quick is a new approach [15–20]. The techniques have been discussed in "Principles of Electroanalytical Methods" chapter.

8.3.6.2   SPECTROSCOPIC DETECTION

Another very powerful detection system used in ion chromatography is based on the interactions of the analytes with light, or their spectral emissions upon excitation. These methods also include the application of mass spectroscopic techniques for the detection of ionic species.

8.3.6.2.1   UV/VIS DETECTORS

The major prerequisite for the applicability of spectrometric detection is that the eluent have no absorbance in the wavelengths applied for the detection, while the analytes do, or their concentration can somehow be related to an absorbance signal. The most common mode is the direct spectrometry at the maximum absorption wavelength of the analyte(s). In case the analyte either has a limited or no absorption at all in the applicable wavelength range, indirect measurements, where the increase or decrease of the absorbance of a species upon the presence of the analyte is used as an indicator of the analytes concentration, could be a resort. Due to the inherent properties of inorganic ionic species UV/Vis detectors have a limited range of application for them. Since many inorganic ions do not have absorption in the applicable wavelength ranges, in the case of multivalent and transition metals, post column derivatization techniques are applied. In these techniques, the cation is chelated with an organic ligand having a proper absorption behavior in the applicable wavelength, even in the complexed (chelated) form. An examples of such ligands is 4-(2-pyridylazo)-resorcinol also known as PAR. Another measure is also taken to induce the capability of absorption to nonabsorbing species, which can include postcolumn reactions. In general, such reactions are designed to induce UV/Vis absorbance, fluorescence, or electrochemical activity to the analyte through converting it to a new species or complex.

8.3.6.2.2   FLUORESCE DETECTORS

In some especial cases in IC, where organic compounds with conjugated π-electrons (especially aromatic components) can be excited to show fluoresce properties, this property is used for the detection of the analyte. Such detectors are typical in clinical and organic analyzes, where some analytes have fluorescent properties. For species that are incapable of fluorescence, derivatization might help. This detection technique is very sensitive to the composition of the eluents and interference of the contaminants. Another limitation of the method could be mentioned as its rather narrow linearity range, which is due to the fact that fluorescence signals are prone to the self-absorption phenomenon, where a fraction of the fluorescence of the material is absorbed by the material itself.

8.3.6.2.3   REFRACTOMETRIC DETECTORS

Differential refractometry is another optical detection technique that can be used as an ion chromatographic detector. This detector is also referred to as RI (Refractive Index), is a nonspecific detection technique suitable for universal use as the quantity measured is the change in the refractive index of the pure eluent caused by the presence of the analyte.

However, the great temperature sensitivity of the refractive index means that the method is very much susceptible to ambient interference as well as those of the operating conditions. Provided that the temperature is absolutely stable, the method has a linear range of about three orders of magnitude. As simple inorganic ions have an extremely low refractive index, they can only be determined indirectly through using eluents, either having high refraction or to which very strong refracting compounds have been added.

8.3.6.2.4 MASS SPECTROMETRIC DETECTORS

Contrary to HPLC-MS systems, where a range of ionization sources including electrospray ionization sources (ESI), atmospheric pressure chemical ionization (APCI), and atmospheric pressure photochemical ionization (APPI) are applicable in the event of using a mass-spectrometric detector, the case is different with IC-MS. The application of these different sources is based on the polarity and mass of the analytes, and the flow rate of the mobile phase used in HPLC, however due to the nature of the analytes, and mobile phases in IC only electrospray ionization sources (ESI), which is a soft ionization source capable of transferring multivalent ions into gas phase, is used in IC–MS systems. The importance of these ionization sources in systems like HPLC and IC is due to the necessity of having very low pressures in the MS detector while the ions leaving the chromatograph are under rather high pressures induced on the mobile phase by the pumps. So, sources like ESI have the role of converting of the mobile phase and the analyte to a gas phase under atmospheric pressure.

Mass spectrometric detection is applicable to thermally stable analytes. Further it is important that the gasification of the mobile phase not leave salts behind. Sometimes, methanol or acetonitrile or other suitable organic solvents are added to the mobile phase to improve the vaporization.

Application of IC–MS, not only allows qualitative and quantitative analysis of the samples, but it also makes it possible to acquire structural information about the samples. Mass spectrometric detection is used in selected ion monitoring or scans modes, based on the nature of the analysis. In the former mode, the molar mass of the analyte is acquired for quantitative analyzes, while the latter can yield information on the retention time, as well as, mass spectra and mass distribution for qualitative analysis.

8.3.6.2.5 ICP-MS DETECTORS

ICP-MS is another very powerful combined detector for the determination of different organic and inorganic compounds, which if coupled with ion-chromatography will constitute a technique for the separation and determination of ionic species enjoying the strength of both methods that are very robust and powerful analytical techniques per se.

The detection method, however, suffers drawbacks. One example is the speciation of arsenic and chromium where species such as $^{40}Ar^{35}Cl^+$ and $^{40}Ar^{12}C^+$ are reported to form in the plasma, as a result of the presence of chlorides or carbon in the matrix, which interfere with the accurate determination of ^{75}As or ^{52}Cr at m/z=75 and m/z=52, respectively. Naturally IC can be used to eliminate interfering species like chloride or the suppression of the interfering species formed in plasma through the so-called collision-reaction cell techniques can be applied to rid of the interferences [21].

8.3.7   DATA ACQUISITION AND CALCULATION OF RESULTS

Chromatographic data acquisition has come a long way from earlier days when the output of the detector was simply connected to a pen-and-ink recorder moving at a fixed chart speed so that the resulting chromatogram could be recorded. Now, the results of the chromatographic separation are almost always stored and displayed on a computer. The computer uses an A/D (analog to digital) board to convert the analog signal from the detector to digital. The digital information is stored and manipulated to report the results to the user. The scale of both the vertical axis (detector signal) and the horizontal axis (elution time) can be adjusted to give a record of the separation.

8.4   PREPARATION OF ENVIRONMENTAL SAMPLES FOR IC ANALYSIS

Determination of ionic species like inorganic anions, cations, transition metal ions and light organic acids and bases can be achieved through the application of a variety of analytical methods among which ion chromatography is the commonest. The other methods, including colorimetry, gravimetric analysis, turbidimetry, titrimetry, and electrochemical techniques such the use of selective electrode (ISE) and amperometric titrations, all suffer from some limitations. The major shortcoming of most of these methods,, that is, interference from other species, is very easily overcome by the predetection separation of the ionic species through ion-chromatography, which naturally eliminates the interference problem. Furthermore, IC offers lower detection limits in the order of sub-ppb to upper detection limits of ppm for most samples. The regulatory applications of the method are more for mixtures of anions, while cation determination methods are most based on spectroscopic techniques. An irreplaceable application of IC is its ability to be used in the speciation of cations.

8.4.1   SAMPLE PREPARATION

The general issues of focus in the sampling process are that the sample represents its source well, and that its storage does not adversely affect this. The nature of the sample also influences the procedure required for the preparation of the sample. It is evident that, water samples are easier to take and prepare. In case more complicated samples like soils or sludge are to be analyzed, further measures should be taken since such samples can evidently not be introduced to the ion chromatograph and require additional pretreatments, like extraction, digestion, alkali fusion, etc. In the lines below a brief account of the methods used for the preparation of each of the mentioned samples is given.

8.4.1.1   WATER SAMPLES REQUIRING NO PREPARATION

An advantage of ion chromatographic analysis of water samples is that many of such, like drinking water samples, require very little or no pretreatment and preparation measures to be taken prior to the analysis. The only measures required before starting the analysis, in many cases, would be filtration through 0.45 μm filters to remove particulates. For other more concentrated samples, like wastewater, which have a rather high ionic strength, only a dilution step would be required before or after filtration.

8.4.1.2   EXTRACTION

Since most ionic species can readily dissolve in water, water extraction can be used to transfer such species to an aqueous phase before injection to IC. To do this, a known amount of the sample is added to a known volume of an extracting solvent, suitable for the solute ions, and mixed. Mixtures of water and miscible solvents like light alcohols or dilute solutions acids, bases, and salts like potassium chloride or phosphate buffers and even the mobile phase itself can be used in the extraction step. However, in case application of pure water is possible, it would be preferred since this could help avoid unwanted signals, which might interfere with those of the analytes.

8.4.1.3   ACID DIGESTION AND ALKALI FUSION

It is clear that mere extraction of the samples might be inefficient for rather solid samples, due to the low recovery of the analyte ions achieved. Under such circumstances, or whenever the analyst finds it suitable considering the nature and matrix of the sample, acid digestion or alkali fusion techniques can be of use. In the case of acid digestion of IC samples, use of concentrated acids or their mixtures can lead to large interfering peak or column overloading, because of the high concentration of the acid colon. So, analysts either try to avoid such procedures as reflux distillation or to reduce such adverse effects through further pretreatment techniques.

Another alternative method, for the preparation such samples is alkali fusion. In general the solid sample and a proper alkaline compound (e.g., Na_2CO_3, Na_2O_2, lithium tetraborate, or $NaOH$) are mixed and heated, until they are molten. Next the melt is cooled before being dissolved in a digestion solution. The resulting solution might require further treatments like filtration or dilution before being injected to the ion chromatograph.

8.4.1.4   SAMPLE COMBUSTION

Another conventional method for preparing solid samples for analyzes is sample combustion, which involves complete combustion of the sample in the presence of excess oxygen. As a result the metallic elements remain in the ash, while the nonmetallic ones are converted to gaseous compounds, which can be absorbed by a scrubbing solution used later on. The choice of the scrubbing solution, like choosing an extraction solvent, depends on the nature of the gasses produced. Solutions such as dilute alkaline solution (e.g., sodium hydroxide), which might further contain a reducing agent like a hydrazine salt, or the solution of an oxidizing agent for sulfate and phosphate oxides are examples that can be mentioned. This solution, after necessary treatments, is injected to the ion chromatograph.

8.4.2   SAMPLE TREATMENT

Usually after preparing the sample solutions, we require to perform some treatment thereon before injecting them into the ion chromatograph. This is done to remove solid particles which might contaminate or block the column, concentrate the sample or eliminate interferences from matrix and can range from a simple filtering step as mentioned in the case of water samples above to rather complex measures required for matrix elimination.

8.4.2.1   FILTERING

To remove the solid particles, which might be present in the sample the solution and damage the different parts of the IC, it should be passed through a 0.45 µm ≥ filter. Some auto-injection systems already include filtering steps that eliminate this need; however, in case such devices are not being used one can use commercial disposable filter for this purpose. The step is mainly performed to avoid the blockage or contamination f the different parts of the equipment and is very necessary in many samples.

8.4.2.2 MATRIX ELIMINATION

Samples with more complex matrixes may also require chemical, as well as physical, treatment to reduce the effects caused by interfering species present in their matrix, on the overall analysis results. This can be achieved through solid phase extraction (SPE). Solid phases might be chosen from the packing materials used in the chromatographic columns including silica, alumina, C18, anion-exchange resins, cation exchange resins, neutral polymer, amino, and activated carbon [22, 23]. As in the case of filters used for the elimination of solid particles, there are commercially available cartridges that can be used for removing interfering species before the analysis.

Solid phase extraction with different material can be used for eliminating neutral organic contaminants (which might decrease the lifetime of IC columns), adjust sample pH and ionic strength of the sample, remove carbonate and cationic species which can precipitate in the presence of alkaline eluents [22, 25], selectively remove halides and sulfate from samples [24] and allow the determination of very low concentrations of anions, such as bromate in ozonated water samples with chloride interfering ions, which helps avoid the masking of the analyte by the excess chloride.[22,26]

Sample treatment techniques, including sample preconcentration or matrix elimination are also automated to be performed by the IC instrument itself. In case of such automatic preconcentration, known volumes of the sample initially pass a concentrator column packed with the ion exchange material to eliminate the analyte ions. In a next step the sample concentrator column is put in line with the mobile phase, for the column to enter the analysis circuit. An interesting fact is that since the analyte ions are trapped and the solution passes the concentrator column, the matrix is also eliminated and hence this method can also be regarded as a matrix elimination measure. This can also be achieved through using chelating resins as the packing material and is not necessarily limited to the application of ion exchange resins.

8.5 ENVIRONMENTAL SAMPLES ANALYSIS BASED ON DETECTION TECHNIQUES

As mentioned above, ion chromatography is a very strong technique that can find important roles in the analysis of environmental samples, which can be reflected by the wealth of research that is focused on applications of the method. After a general and rather brief overview of the separation principles and instrumental considerations of the technique, we tend to provide the reader with an overview of the research performed on the IC analyzes of environmental sample. We have classified the methods based on the detection systems used in each and also provide some detail about some specific cases. Further, the text provides you with some selection of regulatory procedures as well as research procedure conducted on the analyzes of heavy metal ions. There is no need to mention that, there are certainly much more studies performed on the IC analysis in the literature, and the focus of this text is merely on the reports on environmental samples.

8.5.1   IC WITH CONDUCTOMETRIC DETECTORS

Different water samples ranging from rain, surface, and drinking waters to sea, dam, pond, hot spring, underground and river waters and even power plant, oil field waters and brines, as well as polar ice core, snow, fog samples, atmospheric aerosols, flue and stack gasses, ambient air, plant sap, landfill leachates, edible vegetable oils, cola beverages, ozonized water, oyster tissue, river sediments, estuary water, marine sediment, acid rain have been analyzed through ion chromatographs with conductometric detection systems in either suppressed or unsuppressed modes as well as in combination with UV-Vis detectors, as shown in Table 8.1. The variety and quantity of the work performed using this detection system makes it the most commonly used detection system used in ion chromatographic analysis.

TABLE 8.1   Some Reports on Environmental Sample Analysis Using Ion Chromatography

No.	Ions	Sample matrix	Detection technique	Ref.
1	NO_3^-, NO_2^-, F^-, Br^-, SO_4^{2-}, PO_4^{3-}	Sea water	Conductivity	27
2	NO_3^-, NO_2^-, or NH_4^+	Water samples	Conductivity	28
3	NO_3^-, NO_2^-, or NH_4^+	Soil samples	Conductivity	29
4	NO_3^-, NO_2^-, or NH_4^+	Polar ice core	Conductivity	30
5	NO_3^-, NO_2^-, or NH_4^+	Natural water	Conductivity	31
6	NO_3^-, NO_2^-, or NH_4^+	Rain water	Conductivity	32
7	NO_3^-, NO_2^-, or NH_4^+	Mineral water	Conductivity	33
8	NO_3^-, NO_2^-, or NH_4^+	Surface water	Conductivity	34
9	NO_3^-, NO_2^-, or NH_4^+	Drinking water	Conductivity	35
10	NO_3^-, NO_2^-, or NH_4^+	Snow	Conductivity	36
11	NO_3^-, NO_2^-, or NH_4^+	Drinking water	Conductivity	37
12	NO_3^-, NO_2^-, or NH_4^+	Water sample	Conductivity	38
13	NO_3^-, NO_2^-, or NH_4^+	Drinking water	Conductivity	39
14	NO_3^-, NO_2^-, or NH_4^+	Water sample	Conductivity	40

TABLE 8.1 *(Continued)*

No.	Ions	Sample matrix	Detection technique	Ref.
15	NO_3^-, NO_2^-, or NH_4^+	Dam water, river water	Conductivity	41
16	NO_3^-, NO_2^-, or NH_4^+	Power plant water	Conductivity	42
17	NO_3^-, NO_2^-, or NH_4^+	Rain water	Conductivity	43
18	NO_3^-, NO_2^-, or NH_4^+	Reference materials	Conductivity	44
19	NO_3^-, NO_2^-, or NH_4^+	Fog samples	Conductivity	45
20	NO_3^-, NO_2^-, or NH_4^+	Rain water	Conductivity	46
21	NO_3^-, NO_2^-, or NH_4^+	Waters from peat lands	Conductivity	47
22	NO_3^-, NO_2^-, or NH_4^+	Atmospheric aerosols	Conductivity	48
23	NO_3^-, NO_2^-, or NH_4^+	Rain water	Conductivity	49
24	NO_3^-, NO_2^-, or NH_4^+	Drinking water	Conductivity	50
25	NO_3^-, NO_2^-, or NH_4^+	Fog samples	Conductivity	51
26	NO_3^-, NO_2^-, or NH_4^+	Rain and snow	Conductivity	52
27	NO_3^-, NO_2^-, or NH_4^+	Natural waters	Conductivity	53
28	NO_3^-, NO_2^-, or NH_4^+	Roof runoff waters	Conductivity	54
29	NO_3^-, NO_2^-, or NH_4^+	Rainwater	Conductivity	55
30	NO_3^-, NO_2^-, or NH_4^+	Drinking water	Conductivity	56
31	NO_3^-, NO_2^-, or NH_4^+	Atmospheric aerosols	Conductivity	57
32	NO_3^-, NO_2^-, or NH_4^+	Sea water	Conductivity	58
33	NO_3^-, NO_2^-, or NH_4^+	Flue gas	Conductivity	59
34	NO_3^-, NO_2^-, or NH_4^+	Stack gases	Conductivity	60
35	NO_3^-, NO_2^-, or NH_4^+	Ambient air	Conductivity	61
36	NO_3^-, NO_2^-, or NH_4^+	Ambient air	Conductivity	62
37	NO_3^-, NO_2^-, or NH_4^+	Atmospheric aerosols	Conductivity	63
38	NO_3^-, NO_2^-, or NH_4^+	Ambient air	Conductivity	64
39	NO_3^-, NO_2^-, or NH_4^+	Atmospheric air	Conductivity	65
40	NO_3^-, NO_2^-, or NH_4^+	Atmospheric air	Conductivity	66
41	NO_3^-, NO_2^-, or NH_4^+	Atmospheric air	Conductivity	67
42	Total nitrogen and phosphorus	Wastewaters	Conductivity	68

TABLE 8.1   *(Continued)*

No.	Ions	Sample matrix	Detection technique	Ref.
43	SO_3^{2-}, SO_4^{2-}, SO_5^{2-}, $S_2O_3^{2-}$ $S_2O_4^{2-}$ $S_2O_5^{2-}$, $S_2O_6^{2-}$, $S_4O_6^{2-}$	Hot-spring waters	Conductivity	69
44	Cl^-, NO_2^-, SO_4^{2-}, NO_3^-	Plant sap	Conductivity	70
45	Cl^-, NO_2^-, SO_4^{2-}, NO_3^-	Oil field waters	Conductivity	71
46	Cl^-, SO_4^{2-}, PO_4^{3-}	Landfill leachates	Conductivity	72
47	ClO_4^-	Power plant waters	Conductivity	73
48	Cl^-, NO_2^-, SO_4^{2-}, PO_4^{3-}, SCN^-	Natural water	Conductivity	74
49	Cl^-, NO_3^-, SO_4^{2-}, PO_4^{3-}	Impurity Anions in High Pure Fluoride Reagents	Suppressed Conductivity	75
50	F^-, Cl^-, Br^-, I^-, NO_3^-, NO_2^-, SO_4^{2-}, $H_2PO_4^-$, HCO_2^-, $CH_3CO_2^-$	Rainwater at Maracaibo, Venezuela	Conductivity	76
51	F^-, Cl^-, Br^-, I^-, NO_3^-, NO_2^-, SO_4^{2-}, $H_2PO_4^-$	Synthetic samples	Conductivity	77
52	Cl^-, NO_2^-, SO_4^{2-}, SO_3^{2-}	Synthetic samples	Conductivity	78
53	Cl^-, Br^-, NO_3^-, NO_2^-	Synthetic samples	Conductivity	79
54	ClO_4^-	Drinking water and groundwater	Conductivity	80
55	Cl^-, NO_3^-, SO_4^{2-}, SO_3^{2-}	Edible vegetable oils	Conductivity, or Photometric at 520 nm	81
56	PO_4^{3-}	Water samples	Conductivity	82
57	NO_3^-, PO_{4-}^3	Sea water	Conductivity	83
58	PO_4^{3-}	Cola beverages	Conductivity	84
59	BrO_3^-	Ozonized water	Conductivity	85
60	BrO_3^-	Drinking water	Conductivity	86
61	BrO_3^-	Drinking water	Conductivity	87
62	I_2	Ground water and soil	Conductivity	88
63	NO_3^-, PO_4^{3-}, SO_4^{2-}	Reference materials (Oyster tissue Buffalo River sediments)	Conductivity	89

TABLE 8.1 *(Continued)*

No.	Ions	Sample matrix	Detection technique	Ref.
64	F^-, Cl^-, Br^-, SO_4^{2-}, I^-, Na^+, K^+, Mg^{2+}, Ca^{2+}	Mineral waters	Suppressed conductivity	90
65	Cl^-, NO_3^-, SO_4^{2-}, Na^+, K^+, Mg^{2+}, Ca^{2+}, Sr^{2+}, Ba^{2+}	Oil field water	Non-suppressed conductivity	91
66	F^-, Cl^-, Br^-, NO_3^-, SO_4^{2-}, Na^+, K^+, Mg^{2+}, Ca^{2+}, Sr^{2+}	Sea and estuary water	Suppressed conductivity	92
67	Na^+, K^+, Mg^{2+}, Ca^{2+}, Li^+, NH_4^+	Brine water	Suppressed conductivity	93
68	Mg^{2+}, Ca^{2+}, Sr^{2+}, Mn^{2+}	Sea water, fish otoliths	Suppressed conductivity	94
69	I^-, Br^-, SO_4^{2-}	Oilfield waters	Suppressed conductivity	95
70	NO_3^-, NO_2^-, PO_4^{3-}	Estuary water (untreated)	Suppressed conductivity	96
71	NO_3^-, NO_2^-, PO_4^{3-}	Estuary water (untreated)	Suppressed conductivity	97
72	NO_3^-, PO_4^{3-}	Sea water	Suppressed conductivity	98
73	NO_3^-, PO_4^{3-}	Marine sediment	Suppressed conductivity	99
74	Mg^{2+}, Ca^{2+}, Sr^{2+}	Brine, Sea water	Suppressed conductivity	100
75	Sr^{2+}	Sea water	Non-suppressed conductivity	101
76	NH_4^+	Sea water	Suppressed conductivity	102
77	NH_4^+	Sea water	Suppressed conductivity	103
78	MoO_4^{2-}	Water samples	Conductivity	104
79	SCN^-	Human Semen Samples	Conductivity	105
80	haloacetic acids	Drinking water	Suppressed Conductivity	106
81	SO_4^{2-}	Coastal sediments	Conductivity	107
82	ClO_4^-	Soil samples	Conductivity	108

TABLE 8.1   *(Continued)*

No.	Ions	Sample matrix	Detection technique	Ref.
83	phosphite	Eutrophic Fresh-water Lake	Suppressed Conductivity	109
84	Br, NO_3^-, SO_4^{2-}, PO_4^{3-}	wakame samples (a kind of seaweed)	Conductivity	110
85	Se(IV), Se(VI)	Drinking water	Conductivity	111
86	Phosphoric Acid	Air of work place	Conductivity	112
87	F-, Cl-, Br	Household Products	Suppressed Conductivity	113

Rantakokko, et al. [37] used direct ion chromatography for the simultaneous determination of fluoride, chloride, nitrite, bromide, nitrate, sulfate, and phosphate and iodide as well as some organic acids (were lactate, acetate, propionate, formate, oxalate, and citrate) in raw and drinking water samples using a selective high capacity column AS11-HC. The detection limit of the method was reported to be less than 1 µg/L for the ions analyzed without suppression and ranged from 3 to 15 µg/L for the others which were analyzed under suppressed mode [37]. Another study was performed by Bruno, et al. [38] to deliver the formidable task of analysis of inorganic anions (nitrate, nitrite and phosphate) in concentrated salt waters through a simple column-switching method, using two in-line guard columns, (Dionex AG9-HC 4 mm) which were connected to the analytical column (Dionex AS9-HC 4 mm) using a four way pneumatic valve. This made it possible to elute chloride off into the waste and further the analytes were detected by a conductometeric and/or a UV spectrophotometry detector, after separation, without the need for sample pretreatment and dilution, which was the result of the setup, used by the researchers. The detection limits for nitrate, nitrite and phosphate were reported to be 100, 300, 1000 ppb, respectively, through spiking a synthetic sample containing 20,000 ppm of Cl- and 3000 ppm of SO_4^{2-}. The upper detection limit for the analytes was 60 ppm and the calibration curve was linear between the two concentration ranges [38].

Stefanovic et al. [29] reported nonsuppressed IC equipped with conductometric detection for the simultaneous determination of fluoride, chloride, nitrite, bromide, nitrate and sulfate, with a low-capacity anion-exchange column SuperSep (Metrohm), and phtalic acid dissolved in high-purity water, 2-amino-2-hydroxy-methyl-1,3-propendiol and acetonitrile as the mobile phase. Computer optimization procedures were applied to achieve the desired separation parameters. The method was reported to possess various advantages as compared to other common nonsuppressed IC methods including high selectivity, fast analysis, low quantification and detection limits [39].

Schminke et al. [40] reported the application of an IC procedure for the simultaneous determination of disinfection by-products bromate, chlorite, chlorate, as well as, the so-called seven standard anions, fluoride, chloride, nitrite, sulfate, bromide, nitrate and orthophosphate. The separation was performed using a homemade anion-exchanger of high capacity, which allowed the direct injection of large amounts of the sample without any sample pretreatment. Suppressed conductometric detection mode was used for the determination of fluoride, chloride, nitrite, sulfate, bromide, nitrate, orthophosphate and chlorate, while chlorite and bromate were determined after postcolumn reaction, based on chlorpromazine. The detection limits for bromate and chlorite in deionized water were 0.100 ppb and 0.7 ppb, respectively, while they equaled 0.7 ppb for bromate and 3.5 ppm for chlorite in hard drinking water. The rest of the detection limits method's ranged between those of nitrite and chlorate which were 100 ppb and 1.6 ppm, respectively [40]. Mouli et al. [57] used ion chromatography for the analysis of atmospheric aerosol samples from Tirupati, South India, for F^-, Cl^-, NO_3^- SO_4^{2-}, Na^+, K^+, Mg^{2+}, Ca^{2+} and NH_4^+ [57].

Zhu [63] used combination of ion-pair reagents and zwitter ions as mobile phase in fused silica C_{18} column to separate F^-, Cl^-, Br^-, NO_2^-, NO_3^-, pyruvate and the detection was a suppressed conductivity mode. The eluent was a mixture of 1 mmol/L tetrabutyl-ammonium hydroxide (as ion-pair reagent), 5 mmol/L 3-(N-morpholine)-propane-sulfonic acid (zwitterion) and 0.5 mmol/L Na_2CO_3 (inorganic additive). The LODs for F^-, Cl^-, Br^-, NO_2^-, NO_3^-, pyruvate were reported to be 0.017, 0.014, 0.0048, 0.036, 0.16 and 0.017 mg/L, respectively [63].

Mirani, et al. [94] applied IC with conductometric detection for the determination of alkaline and alkaline earth ions in fish otoliths. They performed the measurements on coastal, off-shore and sediment waters and fish otoliths (Engraulis encrasicholus, Mullus barbatus, Umbrina cirrhosa, Sciaena umbra, Pagellus erythrinus) in the Adriatic Sea and the Canal of Sicily using an IONPAC CS12A chromatographic column and 18 mm and methanesulfonic acid as the eluent. The detection limit for one E. encrasicholus fish otolith, weighing 2.6 mg, were reported to be equal to or less than 0.1 ppb for Li^+, 59 ppb for Na^+, 46 ppb for NH_4^+, 23 ppb for K^+, 13 ppb for Mg^{2+}, 88 ppb for Mn^{2+}, 2.567 ppb for Ca^{2+}, and 13 ppb for Sr^+. The method was reported to separate sodium and potassium, and ammonium's overlapping peaks in seawater and also that of calcium, magnesium and strontium peaks in fish otolith [94]. In another experiment Wilson, et al. [97] reported the low level determination of inorganic nitrogen and related anions nitrite-N, nitrate-N, phosphorous-P, sulfate, bromide, chloride, sulfide, fluoride, ammonia, calcium, and magnesium in fish and shrimp samples in a single run using a combination of UV and conductivity detectors. Wang et al. [103] reported the application of a purge-and-trap (P&T) preconcentration method for the analysis of volatile organic compounds (VOCs) in aqueous samples. Trace amounts of ammonium ion in high-salinity water samples were measured through IC. The optimum purge and trap conditions were determined and the concentration of ammonium over the range of 1.2–5.9 μM could be measured with the method. Low levels of ammonium in matrices with high concentrations of sodium were reported to be easily analyzed and the detection limit was down to 1.35 ppb. Nikashima reported a simple and sensitive suppressed IC method with conductivity detection for the determination of molybdate in environmental waters in highly saline water samples after extraction and preconcentration with a chelating resin immobilized with carboxymethylated polyethylenimine (Presep® PolyChelate). They reported the determination of as low as 0.6 ppb of molybdate, in highly saline water with a 500-μL injection [104].

8.5.2 IC WITH AMPEROMETRY, ION SELECTIVE AND ADVANCED ELECTROCHEMICAL DETECTORS

The ion chromatographic separation and determination of NO_2^-, I^-, S^{2-}, CN^-, Cl^-, Br^-, SCN^-, SO_4^{2-}, HCO_2^-, As^{3+}, As^{5+} in environmental samples have been reported in some works as summarized in Table 8.2. The detectors used in these cases are amperometric detection systems.

TABLE 8.2   Some Examples of IC Analysis of Environmental Samples Using Amperometric Detection Systems

No.	Ion	Matrix	Detection technique	Ref.
1.	NO_2^-	Rain, well water, snow	Amperometry	114
2.	I^-	Seawater	Amperometry	115
3.	CN^-, Cl^-, Br^-, I^-	Lake water	Amperometry	116
4.	Br^-, I^-, NO_3^-, SCN^-	Drinking water	Amperometry	117
5.	CN^-	Drinking water	Amperometry	118
6.	Br^-, NO_3^-	Sea waters	Amperometry	119
7.	As^{3+}, As^{5+}	Natural water	Amperometry	120

8.5.3 IC WITH UV/VIS DETECTORS

Apart from the instances on the combination of UV/Vis and conductometric detectors the method has been independently used in numerous environmental application of IC (Table 8.3), for the detection of PO_4^{3-}, NO_3^-, NO_2^-, CN^-, SCN^-, F^-, Cl^-, Br^-, I^-, IO_3^-, S_2^-, SO_4^{2-}, $S_2O_3^{2-}$, Ni^{2+}, Co^{2+}, Cu^{2+}, Mn^{2+}, Zn^{2+}, Fe^{2+}, Cu^{2+}, Pb^{2+}, Cd^{2+}, Cr^{6+}, Cr^{3+}, Na^+, K^+, NH_4^+, Ca^{2+}, Mg^{2+}, Hg^{2+}, NH_4^+ ions in many different samples.

TABLE 8.3 IC Analyzes of Some Environmental Samples Using Ultraviolet-Visible Spectrometry Detection

No.	Ions	Sample matrix	Detection Technique	Ref.
1	Cr(VI)	Colla corii asini one of the well-known traditional Chinese plant	UV/Vis (at 545 nm)	121
2	Total organic carbon	Deionized, mineral, tap and river water	Indirect UV (at 272 nm)	122
3	Cl^-, Br	Sedimentary and igneous rocks	UV (at 210 nm)	123
4	NO_2^-, NO_3^-	Meat products	UV (at 225 nm)	124
5	I^-	Sea water	UV (at 226 nm)	125
6	I^-	Soil and water	UV (at 230 nm)	126
7	I^-, IO_3^-	Mineral and drinking waters	UV/Vis (at 288 nm)	127
8	S^{2-}, SO_4^{2-}, $S_2O_3^{2-}$, SCN^-	Hot spring waters	UV/Vis (at 350 nm)	128
9	BrO_3^-	Drinking water	Vis (at 530 nm)	129
10	Cl^-, NO_2^-, NO_3^-, SO_4^{2-}	Environmental water Polar ice core samples	Indirect UV (at 270 nm)	130
11	NO_2^-, NO_3^-, Br, I^-	Sea water	UV (at 225 nm)	131
12	BrO_3^-	Drinking water	UV (at 352 nm)	132
13	BrO_3^-	Drinking water	UV (at 352 nm)	133
14	NO_3^-, NO_2^-, NH_4^+	Sea water	UV (at 225 nm)	134
15	NO_3^-, NO_2^-	Drinking water	UV (at 214 nm)	135
16	NO_3^-	Snow samples	UV (at 215 nm)	136
17	I^-	Natural water	UV (at 226 nm)	137
18	NO_2^-	Drinking and	UV (at 265 nm)	138
19	Cr^{3+} and Cr^{6+}	Waste water	UV/ Vis (at 530 nm)	139
20	Cd^{2+}, Co^{2+}, Mn^{2+}	Spiked water sample	UV/Vis (at 520 nm)	140

TABLE 8.3   *(Continued)*

No.	Ions	Sample matrix	Detection Technique	Ref.
21	NO_3^-, NO_2^-, NH_4^+	Lake and River-Water	Vis (at 520 nm)	141
22	Ca^{2+}, Mg^{2+}	River water	Vis (at 590 nm)	142
23	I^-	Sea water (untreated)	UV (at 226 nm)	143
24	I^-	Sea water (untreated)	UV (at 226 nm)	144
25	I^-	Sea water (untreated)	UV (at 210 nm)	145
26	I^-, Br^-, NO_3^-	Sea water	UV (at 210 nm)	146
27	Br^-	Sea water (untreated)	UV (at 220 nm) and Non-suppressed conductivity	147
28	NO_3^-, Cl^-	soil extracts (diluted)	UV (at 210 nm)	148
29	NO_3^-, NO_2^-	Sea water (untreated)	UV (at 225 nm)	149
30	Cu^{2+}, Ni^{2+}, Zn^{2+}, Co^{2+}, Cd^{2+}, Mn^{2+}, Hg^{2+}	Environmental samples	UV/Vis	150
31	SiO_3^{2-}, PO_4^{3-}	Environmental waters	UV/Vis	151

Hu, et al., reported an electrostatic ion chromatographic (IC) method for fast and direct analysis of I^- in seawater using a reversed-phase ODS packed column (250×4.6 mm I.D.) modified by coating with Zwittergent-3–14 micelles. The eluent was an aqueous solution containing 0.2 mM $NaClO_4$ and 0.3 mM Zwittergent-3–14 and UV detection at 210 nm was evaluated [145]. In a study conducted by Raessler et al. [148] low concentrations of nitrate were analyzed in the presence of very high concentrations of chloride ion (>30,000 ppm) and a hundred-fold excess of bromide ions through IC equipped with a UV detection set at λ = 210 nm and it was reported that although chloride ions have no interference with the detection at the mentioned wavelength, bromide concentrations >1500 ppb had severe interferes with nitrate determination at the trace level. When the nitrate concentrations were above 500 ppb, however, the interference was reported to be negligible. Hence the authors spiked all samples with known amounts of nitrate standards so as to overcome the Br^- interference by increasing the overall nitrate concentration up to the critical level of 500 ppb. This way they achieved LODs as low as 25 ppb. Ito et al. [149] introduced a quick and sensitive IC procedure with a UV detector set at 225nm for the measurement of NO_2^- and NO_3^- in seawater. They used two monolithic ODS columns with lengths of 50 mm and 100 mm and inner diameters of 4.6 mm in series. The columns were reported to be coated using and equilibrated with 5 mM aqueous solutions of cetyltrimethyl ammonium chloride (CTAC). The mobile phase was a 0.5 M NaCl solution, and in 3 minutes, good separation of the ions was achieved without considerable interferences. The detection limits at S/N=3 were reported to be as low as 0.8 ppb (for nitrite) and 1.6 ppb (for nitrate). Srijaranai et al. [150] reported the application of 1-(2-pyridylazo)-2-naphthol (PCR) as a post column reagent during the IC analysis of low concentrations of Cu^{2+}, Ni^{2+}, Zn^{2+}, Co^{2+}, Cd^{2+}, Mn^{2+} and Hg^{2+}, using a UV detector. This so-called ion exchange chromatography PCR separation, took about 14 min. The mobile phase was a 3 mM solution of 1 2,6-pyridinedicarboxylic acid (PDCA) and 3 mM oxalate with pH of 12.5. The method was reported to have detection limits as low as 4.5 ppb for all cations except for Hg^{2+}in which case he LOD reached 6 ppb.

8.5.4 IC WITH FLUORESCENCE AND CHEMILUMINESCENCE DETECTORS

There have also been a few reports on the application of fluorescence and chemiluminescence detectors in the analysis of environmental samples through IC. Fluorescence detection has been used in the detection of Se^{4+}, Se^{6+} in river and drinking waters, and Al^{3+} in tap water, and chemiluminescence detectors have been used in the detection of ionic species in water samples (Table 8.4).

TABLE 8.4 IC Analysis of Some Water Samples with Chemiluminiscence Detection Systems

No.	Ion	Matrix	Detector	Ref.
1	Cl⁻, Br⁻, NO_3^-, NO_2^-, SO_4^{2-}	Water samples	Chemiluminiscence	154
2	$C_6H_6O_2$, $C_6H_6O_3$	Water samples	Chemiluminiscence	155
3	Iron (II) and (III)	Water samples	Chemiluminiscence	156

For the determination of the Se^{4+}, Se^{6+} ion-exchange chromatography equipped with a spectrofluorimetry detector was used [152]. 2,3-diaminonaphthalene (DAN) was reported as a derivatizing agent. Since postcolumn derivatization led to peak broadening, an air segmented continuous flow auto analyzer was applied. DAN interacts with Se^{4+} but not with Se^{6+} so the authors used an on-line reduction system to change Se^{6+} to Se^{4+}.

Aluminum ion was also determined through high-performance ion chromatography and postcolumn derivatization with 8-hydroxyquinoline-5-sulphonate. The derivatization procedure made it possible to use fluorescence detection at 512 nm (excitation at 360 nm). The linear range of the method was 5 to 10,000 pg L^{-1}, and the detection limit equaled 1 pg L^{-1} [153].

The chemiluminescence of the neutralization reaction of nitric acid and potassium hydroxide, which was enhanced by the addition of $iron^{3+}$ to the acid, is suppressed in the presence of anions such as Cl⁻, Br⁻, NO_3^-, NO_2^-, SO_4^{2-} to the base. This was used as a new postcolumn chemiluminescence detection method for the determination of the anions after separation with anion-exchange chromatography using a potassium hydroxide solution as eluent. The linear calibration ranges for each anion extended from 100 ng ml^{-1} to 100 pg ml^{-1} [154].

In another instance resorcinol and phloroglucinol were determined in environmental water samples by IC equipped with chemiluminescence detection. Using 50 mM NaOH as the mobile phase, the analytes were separated with an IonPac AS19 column and determined based on the chemiluminescence reaction of luminol and K$_4$Fe(CN)$_6$ in alkaline pH. Under the optimal conditions, the detection limit of resorcinol and phloroglucinol were 4.0 and 4.3 µg L^{-1} and the responses were linear in the range of 0.05–1.0 mg L^{-1} [155].

8.5.5 IC WITH MS-SPECTROMETRIC AND ICP-MS DETECTORS

The use of MS-MS detectors has been reported in the detection ion chromatography of a water sample [157]. An electrospray ion chromatography-tandem mass spectrometer (IC-MS/MS) was used for the determination of bromate ions in water samples. The method included solid phase extraction of the samples with an ion exchange column and their elution with water/methanol ammonium sulfate eluent. The limit of detection for bromate ions was reported to have improved by a factor of 10. A requirement of the SPE was to remove any major ions, which can displace bromate (i.e., SO$_4^{2-}$, Cl$^-$, and HCO$_3^-$), with barium-form, silver-form, and acid (H$^+$-form) exchange resins. A methanolic sulfate eluent was used which permitted IC-MS coupling via an electrospray interface and the limit of quantitation was 0.1 µg/L.

There are, however, more reports on the application of ICP-MS detection in some environmental samples (Table 8.5). Inductively coupled plasma mass spectrometery (ICP-MS) was used as a detector for sulfide, sulfite, sulfate, and thiosulfate after the IC separation with a Dionex IonPac(R) AS12A column [158], the analysis was performed under cool and normal plasma conditions.

TABLE 8.5 Summary of the Work on the IC-ICP-MS Analysis of Environmental Sample

No.	Ion	Matrix	Detector	Ref.
1	S^{2-}, SO$_3^{2-}$, SO$_4^{2-}$, SCN$^-$	Samples with biological matrices	ICP-MS	158
2	BrO$_3^-$	Drinking waters	ICP-MS	159
3	BrO$_3^-$	Drinking water	ICP-MS	160
4	BrO$_3^-$, IO$_3^-$	Ozonized water	ICP-MS	161
5	PO$_4^{3-}$	Sea water (untreated)	ICP-MS	162
6	I$^-$, IO$_3^-$	Sea water (untreated)	ICP-MS	163
7	Br$^-$, BrO$_3^-$	Sea water (diluted)	ICP-MS	164
8	ClO$_4^-$	Antarctic snow and ice	MS	165

Also, a method for ultra-trace determination of bromate in drinking water, using inductively coupled plasma mass spectrometry (ICP-MS) was reported [159]. The method uses a microbore column in combination with a tailored anion exchanger with high-capacity and performance. The authors reported that the high capacity of the separation column and the optimized elution system based on NH_4NO_3 allows for the direct analysis of almost every water sample without matrix elimination. The method also uses large injection volumes and does not require trace enrichment because of the sensitivity of the detection system (i.e., ICP-MS). The detection limits for bromate in the drinking and mineral water samples were in the range of 50–65 µg/L, corresponding to absolute detection limits of 44–58 pg.

In another method [160] inductively coupled plasma mass spectrometery was used for the analysis of bromate in drinking waters. The method evaluated three chromatographic columns in terms of detection limits, analysis time and tolerance to potentially interfering inorganic anions and found the LODs for all columns to be in the range of 1–2 µg/L in direct analysis of bromate. The analysis time was reported to be 5 minutes when using a Dionex AG10 column and 100 mM NaOH as the eluent.

In another work bromate, iodate and other halogen anions were determined in drinking water through IC using either ICP-MS or postcolumn derivatization [161]. The samples were directly injected into the IC, and halogen anions were separated. The eluates were directly introduced into ICP-MS and detected at 79 and 127 u. The detection limit for bromate and iodate upon the injection of 0.5 mL were 0.45 µg/L and 0.034 µg/L, respectively. IC combined with ICP-MS was also applied to the simultaneous determination of bromate, bromide and other halogen anions in raw and ozonized water. Good agreement was obtained for the determined values by IC-ICP MS and postcolumn derivatization. Furthermore, several bromine species different from bromate or bromide were detected by IC-ICP-MS set-up.

Phosphate levels in sea-water were also determined through ion-exclusion chromatography in combination with ICP-Ms [230]. The method was rapid (about 6 min for each run), simple and accurate and required no pretreatment steps. The detection limit was 2.0 ppb of P and the injection of only 100 µL of sea-water samples was required.

Iodide and iodate content of sea water was also measured [163] through IC-ICP-Ms with a G3154A/101anion-exchange column (provided by Agilent), with an eluent containing 20 mM NH_4NO_3 at pH 5.6. This way the build-up of salts on the sampler and skimmer cones was reported to be reduced. Through the proposed method, no specific pretreatment measures except for dilution of the sample were required. The research also focused on the effects of competing ion (NO_3^-) in the eluent on the retention time and detection sensitivity of the method. The method showed linear responses over a range of 5.0–500 µg/L and the detection limits for iodate and iodide were 1.5 µg/L and 2.0 µg/L, respectively.

Another study [164] was performed on the determination of bromate and bromide in water samples, like sea water, containing high concentrations of chloride, through IC-ICP-Ms. Reducing the interference of chloride on the analyzes results, as well as avoiding clogging were evaluated through the use of ammonium salts like $NH_4H_2PO_4$, $(NH_4)_2HPO_4$, $(NH_4)_2CO_3$, and NH_4NO_3 as mobile-phase components and a mobile phases containing 20 mM NH_4NO_3 at pH of 5.80 was found to be compatible with the anion-exchange column and enabled good analysis of bromate and bromide in a rather short time (about 7 min) with detection limits ranging from 2.0 to 3.0 µg L^{-1} for bromate and bromide, upon the direct injection of 50 µL^{-1} samples without matrix elimination.

8.5.6   IC WITH ATOMIC EMISSION AND ATOMIC ABSORPTION SPECTROSCOPIC DETECTORS

Determination of inorganic and amino acid forms of selenium (SeO_3^{2-}, SeO_4^{2-}, $C_5H_{11}NO_2S$, $C_6H_{12}N_2O_4S_2$) by ion chromatography coupled with inductively coupled plasma (ICP) atomic emission spectroscopic (AES) detection has also been reported.[166] Three chromatographic systems were compared and the effects of sample matrices on the analyzes was investigated. Elution of seleno-cystine and seleno-cysteine was strongly suppressed in bacterial cell extract sample matrices analyzed with the Dionex AS10 column. The results showed that the interferences could be resolved with a Dionex AS11 column.

Also a method using direct current plasma atomic emission spectrometry detection for ion chromatographic determination of Cr^{3+} and Cr^{4+} species was proposed. The chromium-containing species were detected on the basis of the atomic emission of chromium. Both anion and cation separator columns gave similar results when used with varying sample matrices. The detection limit of the method, for chromium species, was less than 1.0 ppb and the method was applied to the determination of chromium species in human serum, natural water, and industrial process stream samples [167].

In the case of using atomic absorption spectrometry AAS there have been some reports, too. An on-line method for the speciation of organic and inorganic selenium (SeO_3^{2-}, SeO_4^{2-}, $C_5H_{11}NO_2S$, $C_6H_{12}N_2O_4S_2$) was proposed.[168] The method consisted of liquid chromatography, UV irradiation-hydride generation-quartz cell atomic absorption spectrometry. The method showed a good resolution between the four species (selenite, selenate, selenocystine and selenomethionine) in a short time (<15 min).

8.6 STANDARD PROCEDURES FOR IC ANALYZES

8.6.1 ANALYSIS OF ANIONS IN WATER SAMPLES BY IC (EPA 300.1)

A variety of standard methods have been proposed for the determination of anionic species in different aqueous samples. These include ASTM D4327; 6581; 15061; D5257; D5827; WK 652; and the United States Environment Protection Agency (EPA) 300.1; 317; 321.8; 326; 214.1;214.2; 218.8; 6860 and UOP 953–97.

The method proposed in EPA 300.1 divides anions into the two groups of A $(Br^-, NO_2^-, NO_3^-, Cl^-,$ ortho-Phosphate, $F^-, SO_4^{2-})$ and B $(BrO_3^-, ClO_2^-, Br^-$ and $ClO_3^-)$ and mentions that the volume of the samples analyzed for the determination of group B should be 5 times more than that of group A, while applying the method.

8.6.1.1 INSTRUMENTATION AND OPERATING CONDITIONS

The method uses Dionex AG9-HC column, 2 mm as the guard and Dionex AS9-HC column, 2 mm, as the analysis column. A suppressed conductivity detector Dionex CD20, and Dionex Anion Self Regenerating Suppressor (ASRS) are also used in the method and the ASRS suppressor is set to work at a current setting of 100 mA using an external source DI water mode. The eluent is a 9.0 mM Na_2CO_3 used with a flow rate of 0.40 mL/min. The sample loop is 10 µL for group A and 50 µL for group B ions and the system backpressure and the background conductivity are 2800 psi and 22 µS, respectively. A consideration of the method is that in case 4 mm columns are used, the injection volume of the samples should be four times that of the case of 2mm columns, for the detection limits to be comparable.

8.6.1.2 REAGENTS

Distilled or deionized water, free of the analyte anions and filtered, so as not to contain particles larger than 0.20 microns, are used. The eluent solution is a 9.0 mM sodium carbonate solution and stock standard solutions, 1000 mg/L (1 mg/mL) of the anions are either purchased or prepared from their sodium or potassium salts. The method describes the stock standards of most anions to be stable for at least 6 months when stored at 40°C, except for the chlorite standard which is only stable for two weeks, and nitrite and phosphate solutions that are only stable for 1 month under equivalent conditions and hence ethylenediamine (EDA) is prescribed as the preservation solution, which is described to serves a dual purpose both as a preservative for chlorite by chelating iron as well as any other catalytically destructive metal cations and removing hypochlorous acid/hypochlorite ion by forming an organochloroamine. EDA is also mentioned to preserve the bromate concentrations through binding with hypobromous acid/hypobromite, which is a by-product of the reaction of ozone or hypochlorous acid/hypochlorite with bromide ion and if is not removed from the matrix may form bromate ion.

EPA 300.1 also prescribes those samples to be analyzed for ortho-phosphate, not be held at room temperature for more than 12 hours, due to the instability of the analyte.

Further 100 mg/mL, and 0.50 mg/mL dichloroacetate (DCA) solutions are mentioned as a surrogate solution and the method suggests that in case sample are collected from treatment plant employing chlorine dioxide, they be purged with an inert gas (helium, argon, nitrogen) prior to addition of EDA.

8.6.1.3 PROCEDURE

In the case of refrigerated samples the operator is advised to ascertain that the samples have reached thermal equilibrium with the environment before that sample analysis, by allowing them to warm on the bench for at least 1 hour. Next 10.0 mL of the surrogate and sample is prepared for either manual injection or filling autosampler vials. About 20 µL of the surrogate solution is added to a 20 mL disposable plastic micro beaker and exactly 10.0 mL of the sample is transferred into the micro beaker and mixed.

Next, using a 10 mL syringe, the sample is taken from the micro beaker and a 0.45 µm particulate filter is directly attached to the syringe, so as to filter the sample before its entrance into the autosampler vial (some vials are designed to automatically filter the sample, in which case the last step is not necessary). The sample can also be manually loaded into the injection loop. In the case of manual injection, the loop should be thoroughly washed between sample analyzes using each new sample matrix.

As mentioned earlier for 2 mm columns 10 µL of group A and 50 µL of group B anion samples are injected while in the case of 4 mm columns these volumes are increased by a factor of 4, reaching 40 µL and 200 µL, respectively, both for the standard and sample.

EPA 300.1 advises that if the response exceeds the calibration range, the sample be diluted with or the harder route of new calibration concentrations be employed.

8.6.2 ANALYSIS OF CATIONS IN WATER SAMPLES BY IC (UOP 959–98)

Determination of cationic species (ammonium) in water samples has also been described by ASTM D6910; D6919; UOP 959–98. UOP 959–98 describes the method for the ion chromatographic determination of ammonium ion water samples.

8.6.2.1   INSTRUMENTATION AND OPERATING CONDITIONS

Cation self-regenerating suppressor, conductivity suppressor device, Model CSRS-II (4-mm), Dionex, Cat. No. 46079, Chromatographic column, Model IonPac CS12A Analytical, dimensions 250-mm length by 4-mm ID, Dionex, Cat. No. 46073, Conductivity cell, with shield, Dionex, Cat. No. 44132, ion chromatography system, equipped with a PeakNet Chromatography Workstation (includes the integrator), LC10 chromatography organizer with injection valve, GP40 gradient pump with degas, CD20 electrochemical detector and E01 eluent/solvent organizer, Dionex, Model DX 500, Mobile phase reservoirs, 2-L, Dionex, Cat. No. 39163, 2 required, are among the IC instruments used in this method. The sample injection volume is mentioned to be 25 μL and the column operates at ambient temperature. A gradient elution program is also mentioned in the method and the total analysis time is 30 minutes.

8.6.2.2   REAGENTS

Ionically pure, 18 megohm-cm, organic-free, water, ammonium chloride, 99.998% pure, Aldrich Chemical, Cat. No. 25,413–4, 1.0 M Methanesulfonic acid solution, 3.0 mM methanesulfonic acid solution as mobile Phase A, 20.0 mM Methanesulfonic acid solution as mobile phase B and methanesulfonic acid, 99% pure, Aldrich Chemical, Cat. No. M860–6 are used in the method.

8.6.2.3   PROCEDURE

The ion chromatograph is assembled according to the manufacturer's guidelines. Mobile Phase A and B reservoirs are filled with two liters of the corresponding solutions. The mobile is pumped at 1.0 mL/min and the composition of the mobile phase is set at 83 vol % of mobile phase A and 17 vol % of mobile phase B and the suppressor is turned on. Next the conductivity detector's sensitivity is set at 10 μS and the detector is given about 20 min to stabilize, which is identified by the background conductivity reaching a steady level less than 3 μS, according to the method.

Fresh standards are prepared for the calibration stage and the calibration curve is obtained through duplicate measurements. Ammonium samples are kept at pH of about 5 to avoid the evaporation of ammonia and are stored in a refrigerator prior to analysis, and are analyzed within 24 h. About 25 μL of the sample is injected into the IC and immediately the mobile phase gradient and integrator are started. The ammonium peak is obtained and in case it falls within the calibration range the next steps are followed. The method also suggests daily analysis of less than 0.00005 mass-% ammonium as a blank sample.

8.6.3   DETERMINATION OF PERCHLORATE IN SOILS AND SOLID WASTES USING IC (E6860)

Determination of per-chlorate in water, soils and solid wastes using IC-Ms has also been described by EPA 6860. The method is divided into two distinct sections regarding the sample preparation techniques. Here we would like to mention the section dealing with soil and solid waste samples, due to the nature of the sample preparation techniques used.

8.6.3.1   INSTRUMENTATION AND OPERATING CONDITIONS

An ion chromatograph for separation of the sample components, coupled with an electrospray ionization source and mass spectrometer for fragmentation and detection of sample components is used. The application of single stage MS or MS/MS instruments is also mentioned as less preferred arrangements.

The IC system is required to be equipped with a programmable solvent delivery system and all necessary accessories including injection loop, analytical columns, chromatography pump, purging gases, etc. The method also requires a conductivity suppressor for the removal of mobile phase ions before introduction into the MS detection system.

The choice of the mobile phase is dependent on the column. The method suggests that Dionex IonPac® AS20 column, 2.0 mm × 50 mm be used with 45 mM KOH, or Dionex IonPac® AG16 column, 2.0 mm ×50 mm; Dionex IonPac® AS16 column, 2.0 mm × 250 mm or Metrohm Metrosep ASUPP5–150 column, 4.6 ×100 mm be used with 2.5 mM NH4OH, 35, 45 or 55 mM KOH, 12.8 mM Na2CO3, 4 mM NaHCO3, 8.6 mM CH3CN, respectively. C_{18} (2000 mg) or equivalent columns are also used for the extract solution cleanup. Details on the systems appropriate for each combination are expressed in detail in EPA 6860.

Matrix diversion valves like Rheodyne® or equivalent 6-port valves; a conductivity suppressors like Dionex Anion Self Regenerating Suppressor ASRS® ULTRA II, Metrohm Advanced IC Liquid Handling Suppressor Unit or equivalent; optionally a conductivity detector (for measuring the output and effectiveness of the conductivity suppressor) including the Dionex CD25A Conductivity Detector, Metrohm Advanced IC Detector or an equivalent; an auxiliary pump for introducing organic additives into the eluent stream prior to introduction into the mass spectrometer for improved electrospray efficiency like Dionex AXP-MS or equivalent; a mixing tee for the introduction of organic additives from the auxiliary pump into the eluent stream like the Upchurch Scientific Micro Static Mixing Tee or equivalent; a chromatographic oven like Dionex LC30 Chromatography Oven, Metrohm Advanced IC Separation Center or equivalent; electrospray ionization (ESI) source and IC/MS and IC/MS/MS instrument are used in the method.

8.6.3.2 REAGENTS

All of the chemicals are reagent or HPLC-grade according to the instructions of the method. Reagent water, acetonitrile, ammonium hydroxide, potassium hydroxide, ammonium acetate, sodium carbonate, sodium bicarbonate, 100 mM NH4OH, 2.5 mM NH4OH (mobile phase), 35 mM KOH (mobile phase), 45 mM KOH (mobile phase), 55 mM KOH (mobile phase), 12.8 mM Na2CO3, 90% CH_3CN (postcolumn eluent additive), 50% CH_3CN (postcolumn eluent additive), 0.01 M $CH_3CO_2NH_4$ (postcolumn eluent additive), sodium chloride, sodium sulfate, sodium perchlorate, stock standard solution (1000 mg/L ClO4-) are used in the method.

8.6.3.3. PROCEDURE

8.6.3.3.1 SAMPLE COLLECTION, PRESERVATION AND STORAGE

Solids samples are collected in 4-oz amber glass bottles and subjected to extraction within 28 days of sampling. All samples and extracts are advised to be stored with headspace to reduce potential anaerobic biodegradation.

The solid sample preparation procedure includes weighing 1 g of the sample, and transferring it to a 15-mL centrifuge tube. Next enough reagent water is added to the tube, so that the final volume reaches 10 mL. About 50 μL of internal recovery and calibration standard spiking solution is added to the sample tube. It is required that the final concentration of IRCS in the sample be exactly the same as that in the calibration standards. At the next stage the mixture is vortexed, and then sonicated for at least 10 min, and further vortexed afterwards. Next the sample is centrifuged for 5 min, if necessary, under a speed leading to a visibly adequate separation. The supernatant extract is filtered using a plastic syringe fitted with an 0.45-μm or 0.2-μm PTFE membrane filter and is dispensed into an autosampler vial for analysis.

The method suggests that in case high levels of organic contaminants are not expected to exist in the solid sample extract, which can be indicated by the fact that the supernatant extract is relatively clear and not highly colored, the analysis can be started. Otherwise the sample should go under a cleanup stage using a C18 column or other equivalents including (Supelclean™ ENVI Carb-II column from Supelco or graphitized carbon by United Chemical Technologies, Inc) for removing organic contaminants. The detail of this stage is mentioned in the standard procedure document.

8.6.3.3.2 SAMPLE ANALYSIS

The steps for the analysis of liquid and extracted samples are the same and include setting up the IC/MS instrument as mentioned in the operating conditions and then establishing a stable baseline which is mentioned to take about 15–30 min. Next an initial calibration is established according to the instructions of EPA 6860.

As the next step a volume of the sample, suitable for the analytical column and instruments used, is injected into the ion chromatograph (the method prescribes injecting 100 µL) and the results are recorded and further processed.

8.7   EXPERIMENTAL PROCEDURES FOR HEAVY METALS DETERMINATION

Further to the above mentioned regulatory procedures, several studies have been directed to the determination of different species through ion chromatography. Some samples of the details of the actual procedures are included here to elaborate on the methods used for the determination of heavy metal ions, in research studies.

Huang et al.[169] reported the determination of vanadium, molybdenum and tungsten in complex matrices through chelation IC-ICP-MS. They used a Dionex ion chromatography 4000i equipped with a 20 µL loop with a separation column packed with a tailored chelation resin (bis(2-aminoethylthio)methylated resin and γ–aminobutyrohydroxamate [170, 171]. The IC was equipped with a PE SCIEX Elan 6000 (Perkin-Elmer, Norwalk, CT, USA) ICP-MS and a cross-flow nebulizer with a Scott type double pass spray chamber was used.

The authors reported using reagents and standards prepared from analytical reagent grade chemicals (Merck), and pure water (18 MΩcm, Milli-Q water purification system, Millipore, Bedford, MA, USA). Ammonium molybdate, sodium tungstate, vanadyl sulfate, nitric acid, potassium bromide, acetic acid, sodium acetate, sodium hydrogen carbonate, sodium carbonate, sodium hydroxide, sodium chloride, potassium chloride, calcium chloride, magnesium chloride, sodium sulfate, oxalic acid, ethylenediaminetetraacetic acid (EDTA), ethylene diamine, sodium nitrate, hexamethylenetetraamine, paraformaldehyde, zinc chloride, and sodium iodide (Merck, Darmstadt, Germany), thiourea, sodium azide, and silver nitrate (Wako, Japan), the hydro chlorides of XAD-4 (polystyrene-divinylbenzene copolymer) and 2-mercaptoethylamine (Sigma), were the chemicals used.

They evaluated the different parameters affecting the separation as well as detection by ICP-MS. Since, the samples were all aqueous solutions no specific preparation was required for the samples.

Amm, et al. [172] reported the high-performance chelation IC for the determination of trace Zn^{2+}, Pb^{2+}, Ni^{2+}, and Cu^{2+} in sea water through Xylenol orange-impregnated resins. The ion chromatographic set-up was identical with that used by Challenger et al. [173]. The eluent and postcolumn reagent were pumped through two Constametric Model III pumps (Laboratory Data Control, Riviera Beach, FL, USA). The injection of the samples was performed using a steel six-port injector (Rheodyne, Cotati, CA, USA) used in series with a titanium six-port injector (Valco, Schenkon, Switzerland). Through this arrangement, the researchers were able to use both direct injection and preconcentration modes and an Eldex Laboratories (Menlo Park, CA, USA) was reported to be used for preconcentration. The authors reported mixing the eluting metals with the PCR at a zero dead volume T-piece that precedes a 1.4 m poly(tetrafluoroethylene) reaction coil. The detections were made using a spectral array detector (Dionex, Sunnyvale, CA, USA).

Polystyrene-divinylbenzene neutral PLPRS resin (Polymer Laboratories, Church Stratton, UK) (10 μm particle size, 100 A0 pore size) was coated with xylenol orange (BDH) and next packed into a 10×46 cm² i.d. PEEK columns (Alltech Chromatography, Camforth, UK). Xylenol orange coated Amberlite neutral XAD-2 (BDH) packed in a 20×2 cm² i.d. glass column for eluent cleanup. The absorbance measurements were performed between 490 and 540 nm and the flow rates for both the eluent and PCR were 1 cm³ min⁻¹.

The authors reported using analytical-reagent grade reagents obtained from BDH, except for PAR, which was supplied by Fluka. They degassed the eluent and PCR with helium before use and prepared all solutions using distilled, deionized water. The eluent used was 0.5 mol dm⁻³ potassium nitrate containing 0.05 mol dm-3 lactic acid, the pH of which was adjusted using dilute ammonia or nitric acid solutions. The PCR was reported to be a 1.2×10^{-3} mol dm⁻³ PAR buffered to pH 10.2 using ammonium nitrate.

Between the analyzes, the metal species possibly remaining in the column from the previous run, were cleaned using a 0.5 mol dm⁻³ KNO_3 eluent (pH =1.2) with nitric acid. The next step was described to be the preconcentration procedure in which the column was switched to the preconcentration pump, used for delivering 10 cm³ of the ammonium acetate buffer (pH=6.0), and then a measured volume of the sample with identical pH. Next a small volume of DDW was pumped through the column, to remove the excess NaCl and the alkaline earth metals retained by the column much less than the analytes, under the operating conditions. Finally the column was switched back to the eluent pump and the analysis is performed.

It was reported that step-gradient elution increased the sharpness of the peaks, especially in the case of a short column. It was further mentioned that pH values of the individual steps could be altered depending on the analyte ion to be metals being determined. According to this procedure the relative retention times of analyte ions was adjusted by changing the gradient program, and hence which allowed riding of large interfering peaks. It was also stated that due to the presence of weak acid groups in the structure of xylenol orange, the column had a rather large buffering capacity, which was overcome by the sharp pH changes caused by step gradients. If the pH value was fixed at 6.0 before preconcentration, Cu^{2+} and Mg^{2+} left the column. This was while transition metals and lead were retained in the column.

S. Motellier and H. Pitsch reported on-line preconcentration and simultaneous analysis of ultra-trace amounts of transition metal ions [174]. They used Millex-HA type, 0.45/zm pore diameter (Millipore) filters for filters used for the precipitation of calcium which were rinsed with 1% HNO_3 (3 mL) and water (10 mL) each time and the first 10 mL of the filtered solutions was also were discarded. The IC system used was an Action Analyzer (Waters), equipped with an inert 625 pump driven by a 600E controller unit and a Reagent Delivery Module (RDM, Waters) connected to the flow stream at the column outlet via a T-shaped connector, and a diode-array detector (Waters 990), which was set at 520 nm for the purpose of the analysis. The delivery of the postcolumn reagent was accomplished by nitrogen pressure.

The injection could be performed directly or indirectly through a built-in injector (Waters 125) equipped with a 100/μL loop. For on-line preconcentration, the sample was reported to be taken with a DQP-1 pump (Dionex) directly from the laminar flow exhaust hood onto the concentrator column located in the loop of an automated switching valve (9010, Rheodyne). The columns used were a CG5, which acted as the guard together with a CS5 analytical column (Dionex). The filling resin contained both quaternary ammonium and sulfonate functional groups, in a pellicular layer located on the core of the beads [175, 176]. The cation- and anion-exchange capacities were reported to be 0.071 and 0.033 mequiv/mL of the resin. An MFC-1 column (Dionex) was used for the purification of the eluent and the concentrator column was a MetPac-CC1 column (Dionex) packed with a macroporous iminodiacetate chelating resin (0.45 mequiv). Yridine-2,6-dicarboxylic acid (PDCA) and pyridyl-azo-resorcinol (PAR) were from Fluka and Merk. The metal solutions were prepared by diluting of Specpure solutions (Johnson and Matthey) (1 g L^{-1}) and later their pH was adjusted at 3 with HCl. The other chemicals were reported as Suprapur grade chemicals from Merck and the water was purified with a MilliQ system (Millipore).

The optimized mobile phase was PDCA, 6×10^{-3} M; CH$_3$COOH, 5×10^{-2} M; CH$_3$COONa, 5×10^{-2} M, pH 4.5, with a flow rate of 1 mL min^{-1} and the postcolumn reagent (PAR, 4×10^{-4} M in NH$_3$, 3 M; CH$_3$COOH, 1 M (pH 9.7)) had a flow rate of 0.4 mL min^{-1}.

KEYWORDS

- **Data acquisition**
- **Detectors**
- **Experimental procedures**
- **Instrumentation**
- **Mechanisms of separation**
- **Samples preparation**
- **Standard procedures**

REFERENCES

1. Ettre, L. S. (1993). Nomenclature for Chromatography (IUPAC Recommendations 1993), Pure & Appl. Chem., *65*, 819–872.
2. Engelhardt, H., & Rohrschneider, L. (1998). Deutsche Chromatograpische Grundbegriffe Zur IUPAC Nomenklatur, Universität Saarbrücken.
3. Kolb, M., Viehweger, K. H. (Ed.). Practical Ion Chromatography, an Introduction, 2nd ed., Printed by Metrohm Ltd., CH-9101 Herisau, Switzerland 8.792.5003–2001–02.
4. Schwedt, G. (1985). Separation and Preconcentration Methods in Inorganic Anion Analysis (Review), Fresenius Z. Anal, Chem., *320*, 423–428.
5. Nesterenko, P. N., Jones, P., & Paull, B. (2011). High Performance Chelation Ion Chromatography, Cambridge, UK: RSC Publishing, 303.
6. Fanali, S., Haddad, P. R., Poole, C. F., Schoenmakers, P., & Lloyd, D. K. (2013). Liquid Chromatography: Fundamentals and Instrumentation, Chapter 8-Ion Chromatography, Elsevier.
7. Cook, H. A., Dicinoski, G. W., & Haddad, P. R. (2003). Mechanistic Studies on the Separation of Cations in Zwitterionic Ion Chromatography, *J. Chromatog, A, 997*, 13–20.
8. Paull, V. B., & Nesterenko, P. N. (2005). New Possibilities in Ion Chromatography Using Porous Monolithic Stationary phase Media, Trends Anal Chem, *24*, 295–303.
9. Nesterenko, P. N., & Haddad, P. R. (2000). Zwitterionic Ion-Exchangers in Liquid Chromatography, Anal Sci., *16*, 565–574.
10. Fritz, J. S., & Gjerde, D. T. (2009). Ion Chromatography, 4th Ed. WILEY-VCH Verlag GmbH & Co. KGaA, Weinheim.
11. Kolla, P., Koehler, J., & Schomburg, G. (1987). Polymer-Coated Cation-Exchange Stationary Phases Based on Silica, Chromatographia, *23*, 465–472
12. 13-Nesterenko, E. P., Nesterenko, P. N., & Paull, B. (2008). Anion-Exchange Chromatography on Short Reversed-Phase Columns Modified with Amphoteric (N-dodecyl-N, N-dimethylammonio) Alcanoates. *J. Chromatog, A, 1178*, 60–70.
13. Paull, V. B., & Nesterenko, P. N. (2005). New Possibilities in Ion Chromatography using Porous Monolithic Stationary phase Media, Trends Anal Chem., *24*, 295–303.
14. Small, H. (1989). Ion Chromatography, Springer.
15. Pourjavid, M. R., Norouzi, P., Rashedi, H., & Ganjali, M. R. (2010). Separation and Direct Detection of Heavy Lanthanides using New Ion-Exchange Chromatography: fast Fourier Transform Continuous Cyclic Voltammetry system. *J. Appl. Electrochem, 40*, 1593–1603.
16. Norouzi, P., Pirali-Hamedani, M., Faridbod, F., & Ganjali, M. R. (2010). Flow Injection Phosphate Biosensor Based on PyOx-MWCNTs Film on a Glassy Carbon Electrode using FFT Continuous Cyclic Voltammetry. Int. J. Electrochem. Sci., *5*, 1225–1235.
17. Norouzi, P., Larijani, B., & Ganjali, M. R. (2012). Ochratoxin a Sensor Based on Nanocomposite Hybrid Film of Ionic Liquid-Graphene Nano-Sheets using Coulometric FFT Cyclic Voltammetry. *Int. J. Electrochem.* Sci., *7*, 7313–7324.
18. Pourjavid, M. R., Norouzi, P., Ganjali, M. R., Nemati, A., Zamani, H. A., & Javaheri, M. (2009). Separation and Determination of Medium Lanthanides: A New Experiment with use of Ion-Exchange Separation and Fast Fourier Transform Continuous Cyclic Voltammetry, *Int. J. Electrochem.* Sci., *4*, 1650–1671.
19. Ganjali, M. R., Norouzi, P., Dinarvand, R., Farrokhi, R., & Moosavi-Movahedi, A. A. (2008). Development of Fast Fourier Transformations with Continuous Cyclic Voltammetry at an Au Microelectrode and its Application for the Sub Nano Molar Monitoring of Methyl Morphine Trace Amounts, Mater Sci. Eng. C, S, *28*, 1311–1318.

20. Pourjavid, M. R., Norouzi, P., & Ganjali, M. R. (2009). Light Lanthanides Determination by Fast Fourier Transform Continuous Cyclic Voltammetry after Separation by Ion-Exchange Chromatography, *Int. J. Electrochem.* Sci., *4*, 923–942.

21. Chen, Z., Khan, N. I., Owens, G., & Naidu, R. (2007). Elimination of Chloride Interference on Arsenic Speciation in Ion Chromatography Inductively Coupled Mass Spectrometry using an Octopole Collision/Reaction System, *Microchem J.*, *87*, 87–90.

22. Jackson, P. E. (2000). Ion Chromatography in Environmental Analysis, Encyclopedia of Analytical Chemistry, Meyers, R. A. (Ed.), 2779–2801, John Wiley & Sons Ltd., Chichester.

23. Haddad, P. R., & Jackson, P. E. (1990). Ion Chromatography: Principles, and Applications, in *J. Chromatog,* Library, Elsevier, Amsterdam, *46*.

24. Henderson, I. K., Saari-Nordhaus, R., & Anderson, J. M. (1991). Sample Preparation for Ion Chromatography by Solid Phase Extraction, *J. Chromatog.* A, *546*, 61–71.

25. Lucy, C. A. (1996). Practical Aspects of Ion Chromatographic Determinations, LC/GC., *14*, 406–415.

26. Joyce, R. J., & Dhillon, H. S. (1994). Trace Level Determination of Bromate in Ozonated Drinking Water Using Ion Chromatography, *J. Chromatog.* A, *671*, 165–171.

27. Wang, R. Q., Wang, N. N., Ye, M. L., & Zhu, Y. (2012). Determination of Low-Level Anions in Seawater by Ion Chromatography with Cycling-Column-Switching, *J. Chromatog* A., *1265*, 186–190.

28. Romano, J. P., & Kro, J. (1992). Regulated Method for Ion-Analysis, *J. Chromatog*, *602*, 205–211.

29. Pontes, F. V. M., Carneiro, M. C., Vaitsman, D. S., Monteiro, M. I. C., Da Silva, L. I. D., De Souza, E. D. F., & Neto, A. A. (2012). Fast and Simultaneous Ultrasound-Assisted Extraction of Exchangeable-NH_4^+, NO_3^- and NO_2^- Species from Soils Followed by Ion Chromatography Determination, Chem. Speciation and Bioavailability, *24*, 227–233.

30. Ivsak, J., & Pentchuk, J. (1997). Analysis of Ionsin Polar Ice Core Samples by Use of Large Injection Volumes in Ion Chromatography, *J. Chromatog*, *770*, 125–127.

31. Gros, N., & Gorenc, B. (1997). Performance of Ion Chromatography in the Determination of Anions and Cations in Various Natural Waters with Elevated Mineralization, *J. Chromatog.,* *770*, 119–124.

32. Michalski, R. (2004). Simultaneous Determination of Nitrite and Sulfite in Rainwater, Arch. Environ, Prot., *30*, 61–63 (in Polish).

33. Michalski, R. (2003). Selected Anions and Cations in Mineral Waters, Arch. Environ. Prot., *29*, 8–18.

34. Krawczyk, W., Lefauconnier, B., & Pettersson, L. E. (2003). Chemical Denutiation Rates in the Bayelva Catchment, Svalbard, in the fall of 2000. Phys. Chem. Earth, *28*, 1257–1271.

35. Michalski, R., & Olsińska, U. (1996). The Determination of Bromates in Water by means of Ion Chromatography, Acta Chromatog, *6*, 127–133.

36. Michalski, R. (1993). Determination of Inorganic Anions and Cations in Snow by using Ion Chromatography, Arch. Environ Prot, *3*, 232–241 (in Polish).

37. Rantakokko, P., Yritys, M., & Vartiainen, T. (2004). Ion Chromatographic Method for the Determination of Selected Inorganic Anions and Organic Acids from Raw and Drinking Waters using Suppressor Current Switching to Reduce the Background Noise. *J. Liq. Chromatog*, *27*, 829–842.

38. Bruno, P., Caselli, M., de Gennaro, G., De Tommaso, B., Lastella, G., & Mastrolitti, S. (2003). Determination of Nutrients in the Presence of High Chloride Concentrations by Column-Switching Ion Chromatography, *J. Chromatog.*, *1003*, 133–141.

39. Stefanovic, S. C., Bolanca, T., & Curkovic, L. (2001). Simultaneous Determination of Six Inorganic Anions in Drinking Water by Non-suppressed Ion Chromatography, *J. Chromatogr*, *918*, 325–334.

40. Schminke, G., & Seubert, A. (2000). Simultaneous Determination of Inorganic Disinfection by-Products and the Seven Standard Anions by Ion Chromatography, *J. Chromatogr.*, *890*, 295–301.
41. Morales, J. A., de Graterol L. S., & Mesa, J. (2000). Determination of Chloride, Sulfate and Nitrate in Groundwater Samples by Ion Chromatography, *J. Chromatogr.*, 884, 185–190.
42. Toofan, M., Pohl, C. A., Stillian, J. R., & Jackson, P. E. (1997). Preconcentration Determination of Inorganic Anions and Organic Acids in Power Plant Waters Separation Optimization through Control of Column Capacity and Selectivity, *J. Chromatogr, 761*, 163–168.
43. Morales, J. A., de Graterol, L. S., Velasquez, H., De Nava, M. G., & De Borrego, B. S. (1998). Determination by Ion Chromatography of Selected Organic and Inorganic Acids in Rainwater at Maracaibo, Venezuela, *J. Chromatogr, 804*, 289–294.
44. Lee, J. H., Kim, J. S., Min, B. H., Kim, S. T., & Kim, J. H. (1998). Determination of Anions in Certified Reference Material by Ion Chromatography, *J. Chromatogr*, *813*, 85–90.
45. Achilli, M., Romele, L., & Martinotti, W. (1995). Sommariva, G. Ion Chromatographic Determination of Major Ions in Fog Samples. *J. Chromatogr, 706*, 241–247.
46. Oikawa, K., Muranob, K., Enomotoc, Y., Wadad, K., & Inomatae, T. (1994). Automatic Monitoring System for Acid Rain and Snow Based on Ion Chromatography, *J. Chromatogram*, *671*, 211–215
47. Shotyk, W. (1993). Ion Chromatography of Organic-Rich Natural Waters from Peatlands, Part I: Cl^-, NO_2^-, Br^-, NO_3^-, HPO_4^{2-}, SO_4^{2-} and Oxalate. *J. Chromatogr. A, 640*, 309–316.
48. Dabek-Zlotorzynska, E., & Dlouhy, J. F. (1993). Automatic Simultaneous Determination of Anions and Cations in Atmospheric Aerosols by Ion Chromatography, *J. Chromatogr.*, *640*, 217–226.
49. Schumann, H., & Ernst, M. (1993). Monitoring of Ionic Concentrations in Airborne Particles and Rain Water in An Urban Area of Central Germany, *J. Chromatogr, 640*, 241–249.
50. Umile, C., & Huber, J. F. K. (1993). Determination of Inorganic and Organic Anions in One run by Ion Chromatography with Column Switching, *J. Chromatogr.*, *640*, 27–31.
51. Zhang, Q., & Anastesio, C. (2001). Chemistry of a Fog Waters in California's Central Valley Part 3: Concentrations and Speciation of Organic and Inorganic Nitrogen. Atm. Environ., *35*, 5629–5643.
52. Hoffmann, P., Karandashev, V. K., Sinner, T., & Ortner, H. M. (1997). Chemical Analysis of Rain and Snow Samples from Chernogolovka/Russia by, IC, TXRF and ICP-MS, Fresenius *J. Anal, Chem.*, *357*, 1142–1148.
53. Rocha, F. R., & Reis, B. F. (2000). A Flow System Exploiting Multicommunication for Speciation of Inorganic Nitrogen in Waters, Anal, Chem. Acta., *409*, 227–235.
54. Polkowska, Z., Górecki, T., & Namieśnik, J. (2002). Quality of Roof Runoff Waters from an Urban Region (Gdańsk, Poland), Chemosphere, *49*, 1275–1283.
55. Astel, A., Mazerski, J., Polkowska, Ż., & Namieśnik, J. (2004). Application of PCA and Time Series Analysis in Studies of Precipitation in Tricity (Poland), Adv. Environ. Res., *8*, 337–349.
56. Jackson, P. E., Weigert, C., Pohl, C. A., & Saini, C. (2000). Determination of Inorganic Anions in Environmental Waters with a Hydroxide-Selective Column, *J. Chromatogr, 884*, 175–184.
57. Mouli, P. C., Mohan, S. V., & Reddy, S. J. (2003). A Study of Major Inorganic Ion Composition of Atmospheric Aerosols at Tirupati, *J. Hazard, Mat., 96*, 217–228.
58. Dahllof, I., Svensson, O., & Torstensson, C. (1997). Optimizing the Determination of Nitrate and Phosphate in Sea Water with Ion Chromatography using Experimental Design, *J. Chromatogr, 771*, 163–168.

59. Nonomura, M., & Hobo, T. (1998). Simultaneous Determination of Sulfur Oxides, Nitrogen Oxides and Hydrogen Chloride in Flue Gas by Means of an Automated Ion Chromatographic System, *J. Chromatogr.*, *804*, 151–155.

60. Fujimura, K., & Tsuchiya, M. (1988). Determination of Nitrogen Oxides and Sulfur Oxides in Nitrate Melts by Ion Chromatography. Bunseki Kagaku, *37*, 59–63.

61. Michalski, R. (2002). Simultaneous Determination of Some Inorganic Compounds in Air by Means Ion Chromatography, Chem. Anal., *47*, 855–866.

62. Bari, A., Ferraro, V., Wilson, L. R., Luttinger, D., & Husain, L. (2003). Measurements of Gaseous HONO, HNO$_3$, SO$_2$, HCl, NH$_3$ Particulate Sulfate and PM2.5 in New York, Atmos. Environ., *20*, 2825–2835.

63. Zhu, Y., Ling, Y. Y., & Chen, J. F. (2004). Determination of Inorganic Anions and Organic Acids in Atmospheric Aerosols by Mobile Phase Ion Chromatography Using Zwitterion as Eluent, Chin. *J. Anal. Chem.*, *32*, 79–82.

64. Krochmal, D., & Kalina, A. (1997). A Method of Nitrogen Dioxide and Sulfur Dioxide Determination in Ambient Air by Use of Passive Samplers and Ion Chromatography, Atm. Environ., *31*, 3473–3479.

65. Nonomura, M., Hobo, T., Kobayashi, E., Murayama, T., & Satoda, M. (1996). Ion Chromatographic Determination of Nitrogen Monoxide and Nitrogen Dioxide after Collection in Absorption Bottles, *J. Chromatogr.*, *739*, 301–306.

66. Zellweger, C., Ammann, M., Hofer, P., & Baltensperger, U. (1999). NOx Speciation with a Combined Wet Effluent Diffusion Denuder-Aerosol Collector Coupled to Ion Chromatography. Atm. Environ, *33*, 1131–1140.

67. Ali-Mohamed, A. Y., & Hussain, A. N. (2001). Estimation of Atmospheric Inorganic Water-soluble Particulate Matter in Muharraq Island, Bahrain, (Arabian Gulf), by Ion Chromatography, Atm. Environ., *35*, 761–768.

68. Colombini, S., Polesello, S., & Valsecchi, S. J. (1998). Use of Column-Switching Ion Chromatography for the Simultaneous Determination of Total Nitrogen and Phosphorus after Microwave Assisted Persulfate Digestion, Chromatogr A, *822*, 162–166.

69. Miura, Y., Saitoh, A., & Koh, T. J. (1997). Determination of Sulfur Oxyanions by ion Chromatography on a Silica ODS Column with Tetrapropylammonium Salt as an Ion-Pairing Reagent, Chromatogr. A, *770*, 157–164.

70. Madden J. E., Avdalovic, N., Kackson, P. E., & Haddad, P. R. (1999). Critical Comparison of Retention Models for Optimization of the separation of Anions in Ion ChromatographyIII. Anion Chromatography Using Hydroxide Eluents on a Dionex AS11 Stationary phase, *J. Chromatogr. A*, *837*, 65–74.

71. Liu, X., Jiang, S. X., Chen, L. R., Xu, Y. Q., & Ma, P. (1997). Determination of Inorganic Ions in Oil Field Waters by Single Column Ion Chromatography, *J. Chromatogr A*, *789*, 569–573.

72. Manning, D. A. C., & Bewsher, A. (1997). Determination of Anions in Landfill Leachates by Ion Chromatography, *J. Chromatogr A*, *770*, 203–210.

73. Toofan, M., Christopher, J. R., Pohl, A., & Jackson, E. (1997). Pre-concentration Determination of Inorganic Anions and Organic Acids in Power Plant Waters separation Optimization through Control of Column Capacity and Selectivity, *J. Chromatogr. A*, *761*, 163–168.

74. Gros, N., & Gorene, B. (1997). Performance of Ion Chromatography in the Determination of Anions and Cations in various Natural Waters with Elevated Mineralization, *J. Chromatogr A*, *770*, 119–124.

75. Hu, Z. Y., Ye, M. L., Wu, S. C., Pan, G. W., Zhang, T. T., & Liu, L. Y. (2012). Determination of Impurity Anions in High Pure Fluoride Reagents by Ion Chromatography with Column-Switching, Chinese, *J. Anal Chem.*, *40*, 703–708.

76. Morales, J. A., Graterol, L. S., Velasquez, H., Nava, M. G., & Borrego, B. S. (1998). Determination by Ion Chromatography of Selected Organic and Inorganic Acids in Rainwater at Maracaibo, Venezuela, *J. Chromatogr*. A, *804*, 289–294.

77. Umemura, T., Kamiya, S., & Haraguchi, H. (1999). Characteristic Conversion of Ion Pairs among Anions and Cations for Determination of Anions in Electrostatic Ion Chromatography using Water as a Mobile Phase, Anal Chim Acta, *379*, 23–32.

78. Huang, Y., Mou, S., & Liu, K. (1999). Conductimetric Detection of Anions of very Weak Acids by Incomplete Suppressed Ion Chromatography, *J Chromatogr* A, 141–148.

79. Ohta, K., & Tanaka, K. S. (1998). Uppressed Ion Chromatography of Inorganic Anions and Divalent Metal Cations with Pyromellitic Acid as Eluent, *J Chromatogr* A, *804*, 87–93.

80. Wirt, K., Laikhtman, M., Rohrer, J., & Jackson, P. E. (1998). Low Level Perchlorate Analysis in Drinking Water and Ground Water by Ion Chromatography, Am. Environ. Lab., *10*, 1–5.

81. Buldini, P. L., Ferri, D., & Sharma, J. L. (1997). Determination of Some Inorganic Species in Edible Vegetable Oils and Fats by Ion Chromatography, *J. Chromatogr* A, *789*, 549–555.

82. Mattusch, J., & Wennrich, R. (1996). Elimination of Sulfate Interferences in the Chromatographic Determination of O-Phosphate Using Liquid-Liquid Extraction, Anal Bioanal Chem., *356*, 335–338.

83. Dahllof, I., Svensson, O., & Torstensson, C. (1997). Optimizing the determination of Nitrate and Phosphate in Sea Water with Ion-Chromatography using Experimental Design, *J. Chromatogr* A, *771*, 163–168.

84. Bello, M. A., & Gonzalez, A. G. (1996). Determination of Phosphate in Cola Beverages using Nonsuppressed Ion Chromatography: An Experiment Introducing Ion Chromatography for Quantitative Analysis. *J. Chem. Educ.*, *73*, 1174–1175.

85. Inoue, Y., Sakai, T., Kumagai, H., & Hanaoka, Y. (1997). High Selective Determination of Bromate in Ozonized Water by Using Ion Chromatography with Post column Derivatization Equipped with Reagent Preparation Device, Anal. Chim Acta, *346*, 299–305.

86. Kackson, L. K., Joyce, R. J., Laikhtman, M., & Kackson, P. E. (1998). Determination of Trace Level Bromate in Drinking Water by Direct Injection Ion Chromatography, *J. Chromatogr* A, *829*, 187–192.

87. Colombini, S., Polesello, S., Valsecchi, S., & Cavalli, S. (1999). Matrix Effects in the Determination of Bromate in Drinking Water by Ion Chromatography, *J Chromatogr* A, *847*, 279–284.

88. Tucker, H. L., & Flack, R. W. (1998). Determination of Iodide in Ground Water and Soil by Ion Chromatography, *J Chromatogr* A, *804*, 131–135.

89. Colina, M., & Gardiner, P. H. E. (1999). Simultaneous Determination of Total Nitrogen, Phosphorus and Sulfur by Means of Microwave Digestion and Ion Chromatography, *J. Chromatogr*, A., *847*, 285–290.

90. Gros, N., & Gorenc, B. (1997). Performance of Ion Chromatography in the Determination of Anions and Cations in Various Natural Waters with Elevated Mineralization, *J. Chromatogr* A, *770*, 119–124.

91. Liu, X., Jiang, S. X., Chen, L. R., Xu, Y. Q., & Ma, P. (1997). Determination of Inorganic Ions in Oil Field Waters by Single-Column Ion Chromatography, *J. Chromatogr* A, *789*, 569–573.

92. Gros, N., Camoes, M. F., Oliveira, C., & Silva, M. C. R. (2008). Ionic Composition of Seawaters and Derived Saline Solutions determined by Ion Chromatography and Its Relation to other Water Quality Parameters, *J. Chromatogr* A, *1210*, 92–98.

93. Hodge, E. M., Martinez, P., & Sweetin, D. (2000). Determination of Inorganic Cations in Brine Solutions by Ion Chromatography, *J. Chromatogr* A, *884*, 223–227.

94. Marini, M., Campanelli, A., & Abballe, F. (2006). Measurement of Alkaline and Earthy Ions in Fish Otolith and Sea Water using a High Performance Ion Chromatography, Mar. Chem., *99*, 24–30.

95. Kadnar, R., & Rieder, J. (1995). Determination of Anions in Oil-Field Waters by Ion Chromatography, *J. Chromatogr* A, *706*, 301–305.

96. Bruno, P., Caselli, M., de Gennaro, G., de Tommaso, B., Lastella, G., & Mastrolitti, S. (2003). Determination of Nutrients in the presence of High Chloride Concentrations by Column-Switching Ion Chromatography, *J. Chromatogr* A, *1003*, 133–141.

97. Wilson, B., Gandhi, J., & Zhang, C. L. (2011). Analysis of Inorganic Nitrogen and Related Anions in High Salinity Water using Ion Chromatography with Tandem UV and Conductivity Detectors, *J. Chromatogr* Sci., *49*, 596–602.

98. Carrozzino, S., & Righini, F. (1995). Ion-Chromatographic Determination of Nutrients in Seawater, *J. Chromatogr* A., *706*, 277–280.

99. Dahllof, I., Svensson, O., & Torstensson, C. (1997). Optimizing the Determination of Nitrate and Phosphate in Seawater with Ion Chromatography using Experimental Design, *J. Chromatogr.*, A, *771*, 163–168.

100. Singh, R. P., Pambid, E. R., Debayle, P., & Abbas, N. M. (1991). Ethylenediamine Hydrochloric-Acid Zinc(II) Eluent for the Suppressed Ion Chromatographic-Separation of Strontium(II) from a Large Amount of Calcium(II)-Application of the Method to the Simultaneous Determination of Magnesium(II), Calcium(II) and Strontium(II) in High Salinity Subsurface Waters, Analyst, *116*, 409–414.

101. Butt, S. B., Farhat, W., Jan, S., Ahmed, S., Mohammad, B., & Akram, N. (2004). Optimization of a Mobile Phase for Monitoring Strontium in Seawater using Non-Suppressed Ion Chromatography, *J. Liq. Chromatogr.* Relat Technol, *27*, 1729–1742.

102. Huang, Y., Mou, S., & Riviello, J. M. (2000). Determination of Ammonium in Seawater by Column-Switching Ion Chromatography, *J Chromatogr* A, *868*, 209–216.

103. Wang, P. Y., Wu, J. Y., Chen, H. J., Lin, T. Y., & Wu, C. H. (2008). Purge and Trap Ion Chromatography for the Determination of Trace Ammonium Ion in High Salinity Water Samples, *J Chromatogr* A, *1188*, 69–74.

104. Nakashima, Y., Inoue, Y., Yamamoto, T., Kamichatani, W., Kagaya, S., & Yamamoto, A. (2012). Determination of Molybdate in Environmental Water by Ion Chromatography Coupled with a Preconcentration Method Employing a Selective Chelating Resin, Anal. Sci., *28*, 1113–1116.

105. Demkowska, I., Polkowska, Z., Kielbratowska, B., & Namiesnik, J. (2010). Application of Ion Chromatography for the Determination of Inorganic Ions, Especially Thiocyanates, in Human Semen Samples as Biomarkers of Environmental Tobacco Smoke Exposure, *J. Anal. Toxicolo, 34*, 533–538.

106. Verrey, D., Louyer, M. V., Thomas, O., & Baures, E. (2013). Determination of Trace-Level Haloacetic Acids in Drinking Water by Two-Dimensional Ion Chromatography with Suppressed Conductivity, *Michrochem J.*, *110*, 608–613.

107. Yang, Y., Chen, Q. Q., & Zhang, G. R. (2013). Determination of Sulfate in Coastal Salt Marsh Sediment with High Chloride Concentration by Ion Chromatography: A Revised Method. Instrumentation Sci. Technol, *41*, 37–47.

108. MacMillan, D. K., Dalton, S. R., Bednar, A. J., Waisner, S. A., & Arora, P. N. (2007). Influence of Soil Type and Extraction Conditions on Perchlorate Analysis by Ion Chromatography, Chemosphere, *67*, 344–350.

109. Han, C., Geng, J. J., Xie, X. C., Wang, X. R., Ren, H. Q., & Gao, S. X. (2012). Determination of Phosphite in a Eutrophic Freshwater Lake by Suppressed Conductivity Ion Chromatography, Environment, Sci. Technol., *46*, 10667–10674.

110. Iiyama, T., & Fukushi, K. (2012). Determination of Bromide, Nitrate, Phosphate, and Sulfate in Water Extract of Wakame (Undaria pinnatifida) using Ion Chromatography, Bunseki Kagaku, *61*, 869–875.

111. Xu, S. X., Zheng, M. L., Zhang, X. F., Zhang, J. L., & Lee, Y. I. (2012). Nano TiO_2-Based Preconcentration for the Speciation Analysis of Inorganic Selenium by using Ion Chromatography with Conductivity Detection, *Michrochem J.*, *101*, 70–74.

112. Ning, X., & Zhang, J. H. (2013). Determination of Phosphoric Acid in the Air of Workplace by Ion Chromatography Asian *J. Chem.*, *25*, 1684–1688.

113. Zhang, S., Zhao, T. B., Wang, J., Qu, X. L., Chen, W., & Han, Y. (2013). Determination of Fluorine, Chlorine and Bromine in Household Products by Means of Oxygen Bomb Combustion and Ion Chromatography, *J Chromatogr* Sci., *51*, 65–69.

114. Okutani, T., & Yugeta, Y. (1985). Determination of a Micro Amount of Nitrite Ion by Reversed Phase Ion-Pair Chromatography with an Amperometric Detector, Bunseki Kagaku, *34*, 777–780.

115. Ito, K., & Sunahara, H. (1988). Ion Chromatography of Iodide Ion in Seawater Using Concentrated Sodium Chloride Solution as Eluent, Bunseki Kagaku, *37*, 292–295.

116. Mehra, H. C., & Jr. Frankenberger, W. T. (1990). Simultaneous Determination of Cyanide, Chloride, Bromide, and Iodide in Environmental Samples using Ion Chromatography with Amperometric Detection, *Microchem. J.*, *41*, 93–97.

117. Schwedt, G., & Roessner, B. (1987). Ion-Chromatography Spurenanalyse Amperometrisch Detektierbarer Anionen in Wässern, Fresenius *J Anal Chem.*, *327*, 499–502.

118. Christison, T. T., & Rohrer, J. S. (2006). Direct Determination of Free Cyanide in Drinking Water by Ion Chromatography with Pulsed Amperometric Detection, *J. Chromatgr.* A, *1155*, 31–39.

119. Tirumalesh, K. (2008). Simultaneous Determination of Bromide and Nitrate in Contaminated Waters by Ion Chromatography using Amperometry and Absorbance Detectors, Talanta, *74*, 1428–1434.

120. Butler, E. C. V. (1988). Determination of Inorganic Arsenic Species in Aqueous Samples by Ion-Exclusion Chromatography with Electro-Chemical Detector, *J. Chromatogr.*, *450*, 353–360.

121. Chinaka, S., Takayama, N., Michigami, Y., & Ueda, K. (2013). Analysis of Hexavalent Chromium in Colla Corii Asini with Online Sample, *J. Chromatogr* A, *1305*, 171–175.

122. Fung, Y. S., Wu, Z., & Dao, K. L. (1996). Determination of Total Organic Carbon in Water by Thermal Combustion-Ion Chromatography Anal Chem., *68*, 2186–2190.

123. Blackwell, P. A., M. R., Cave, A. E., & Davis, S. A. (1997). Malik, Determination of Chlorine and Bromine in Rocks by Alkaline Fusion with Ion Chromatography Detection. *J. Chromatogr.* A, *770*, 93–98.

124. Siu, D. C., & Henshall, A. (1998). Ion Chromatographic Determination of Nitrate and Nitrite in Meat Products *J. Chromatogr* A, *804*, 157–160.

125. Ito, K. (1997). Determination of Iodide in Seawater by Ion Chromatography Anal Chem., *69*, 3628–3632.

126. Papadoyannis, I. N., Samanidou, V. F., & Moutsis, K. V. (1998). Determination of Silver Iodide by High Pressure Ion Chromatography in Soil and Water Matrices after Solid Phase Extraction, *J. Liq. Chromatogr Rel. Technol.*, *21*, 361–379.

127. Bichsel, Y., & Gunten, U. (1999). Determination of Iodide and Iodate by Ion Chromatography with Postcolumn Reaction and UV/Visible Detection, Anal Chem., *71*, 34–38.

128. Miura, Y., Fukasawa, K., & Koh, T. (1998). Determination of Sulfur Anions at the Ppb Level by Ion Chromatography Utilizing their Catalytic Effects on the Post Column Reaction of Iodine with Azide, *J. Chromatogr* A, *804*, 143–150.

129. Achilli, M., & Romele, L. (1999). Ion Chromatographic Determination of Bromate in Drinking Water by Post-Column Reaction with Fuchsin, *J. Chromatogr* A, *847*, 271–277.

130. Ohta, K., & Tanaka, K. (1998). Simultaneous Determination of Common Inorganic Anions, Magnesium and Calcium Ions in Various Environmental Waters by Indirect UV-Photometric Detection Ion Chromatography using Trimellitic Acid–EDTA as Eluent Anal. Chim Acta, *373*, 189–195.

131. Ito, K., Nomura, R., Fujii, T., Tanaka, M., Tsumura, T., Shibata, H., & Hirokawa, T. (2012). Determination of Nitrite, Nitrate, Bromide, and Iodide in Seawater by Ion Chromatography with UV Detection using Dilauryldimethylammonium-Coated Monolithic ODS Columns and Sodium Chloride as an Eluent. Analysis Bioanal Chem., *404*, 2513–2517.

132. Wang, N. N., He, S. W., & Zhu, Y. (2012). Low-Level Bromate Analysis by Ion Chromatography on a Polymethacrylate-based Monolithic Column Followed by a Post-Column Reaction, European Food Technology, *235*, 685–692.

133. Cordeiro, F., Robouch, P., De la Calle, M. B., & Schmitz, F. (2011). Determination of Dissolved Bromate in Drinking Water by Ion Chromatography and Post Column Reaction: Inter laboratory Study, *J. AOAC Int., 94*, 355–358.

134. Ohguni, H., Tanaka, M., Kanesada, A., Fujimoto, S., Hirokawa, T., & Ito, K. (2012). Simultaneous and Selective Determination of Nitrite, Nitrate, and Ammonium Ions in Seawater Samples by Ion Chromatography, Bunseki Kagaku, *61*, 685–690.

135. Kok, S. H., Buckle, K. A., & Wootton, M. (1983). Determination of Nitrate and Nitrite in Water using High-Performance Liquid Chromatography, *J Chromatog, 260*, 189–192.

136. Neubauer, J., & Heumann, K. G. (1988). Determination of Nitrate at the Ng/g Level in Antarctic Snow Samples with Ion Chromatography and Isotope Dilution Mass Spectrometry, Fresenius' Z. Anal. Chem., *331*, 170–173.

137. Ubom, G. A., & Tsuchiya, Y. (1988). Determination of Iodide in Natural Water by Ion Chromatography, Water Res., *22*, 1455–1458.

138. Kim, H. J., & Kim, Y. K. (1989). Determination of Nitrite in drinking Water and Environmental Samples by Ion Exclusion Chromatography with Electrochemical Detection, Anal, Chem., *61*, 1485–1489.

139. Chen, S. J., Zhang, X. S., Yu, L. Y., Wang, L., & Li, H. (2012). Simultaneous Determination of Cr(III) and Cr(VI) in Tannery Wastewater using Low Pressure Ion Chromatography Combined with Flow Injection Spectrophotometry, Spectrochim. Acta, *88*, 49–55.

140. Cheam, V., & Li, E. (1988). Ion Chromatographic Determination of Low Level Cadmium (II), Cobalt (II) and Manganese (II) in Water, *J Chromatogr* A, *450*, 361–371.

141. Tanaka, M., Ohguni, H., Adachi, K., Tsumura, T., Hirokawa, T., & Ito, K. (2010). Simultaneous Determination of Nitrite, Nitrate, and Ammonium Ions in Lake and River Water Samples by Ion Chromatography, Bunseki Kagaku, *59*, 879–884.

142. Smith, D. L., & Fritz, J. S. (1988). Rapid Determination of Magnesium and Calcium Hardness in Water by Ion Chromatography, Anal Chim Acta, *204*, 87–93.

143. Chandramouleeswaran, S., Vijayalakshmi, B., Karthikeyan, S., Rao, T. P., & Iyer, C. S. P. (1998). Ion Chromatographic Determination of Iodide in Sea Water with UV Detection, Mikrochim Acta, *128*, 75–77.

144. Ito, K. (1999). Semi-Micro Ion Chromatography of Iodide in Seawater, *J Chromatogr* A, *830*, 211–217.

145. Hu, W. Z., Yang, P. J., Hasebe, K., Haddad, P. R., & Tanaka, K. (2002). Rapid and Direct Determination of Iodide in Seawater by Electrostatic Ion Chromatography, *J. Chromatogr* A, *956*, 103–107.

146. Hu, W. Z., Haddad, P. R., Hasebe, K., Tanaka, K., Tong, P., & Khoo, C. (1999). Direct Determination of Bromide, Nitrate, and Iodide in Saline Matrixes Using Electrostatic Ion Chromatography with an Electrolyte as Eluent, Anal Chem., *71*, 1617–1620.

147. Hu, W. Z., Cao, S. A., Tominaga, M., & Miyazaki, A. (1996). Direct Determination of Bromide Ions in Sea Water by Ion-Chromatography using Water as the Mobile Phase, Anal. Chim Acta, *322*, 43–47.

148. Raessler, M., & Hilke, I. (2006). Ion-Chromatographic Determination of Low Concentrations of Nitrate in Solutions of High Salinity, Microchim Acta, *154*, 27–29.

149. Ito, K., Takayama, Y., Makabe, N., Mitsui, R., & Hirokawa, T. (2005). Ion Chromatography for Determination of Nitrite and Nitrate in Seawater Using Monolithic ODS Columns, *J. Chromatogr.* A, *1083*, 63–67.

150. Srijaranai, S., Autsawaputtanakul, W., Santaladchaiyakit, Y., Khameng, T., Siriraks, A., & Deming, R. L. (2011). Use of 1-(2-pyridylazo)-2-Naphthol as the Post Column Reagent for Ion Exchange Chromatography of Heavy Metals in Environmental Samples, *Microchem J.*, *99*, 152–158.

151. Yokoyama, Y., Danno, T., Haginoya, M., Yaso, Y., & Sato, H. (2009). Simultaneous Determination of Silicate and Phosphate in Environmental Waters using Pre-Column Derivatization Ion-Pair Liquid Chromatography, Talanta, *79*, 308–313.

152. Shibata, Y., Morita, M., & Fuwa, K. (1985). Determination of Ultra-Trace Levels of Selenite and Selenate in Water using High-Performance Liquid Chromatography with Automated Fluorimetric Detection and an On-Line Reduction System, Analyst (London), *110*, 1269–1270.

153. Jones, P., Ebdon, L., & Williams, T. (1988). Determination of Trace Amounts of Aluminum by Ion Chromatography with Fluorescence Detection, Analyst (London), *113*, 641–644.

154. Sakai, H., Fujiwara, T., & Kumamaru, T. (1996). Determination of Inorganic Anions in Water Samples by Ion-Exchange Chromatography with Chemiluminescence Detection Based on the Neutralization Reaction of Nitric Acid and Potassium Hydroxide, Anal. Chim Acta, *331*, 239–244.

155. Wu, H. W., Chen, M. L., Shou, D., & Zhu, Y. (2012). Determination of Resorcinol and Phloroglucinol in Environmental Water Samples using Ion Chromatography with Chemiluminescence Detection, Chin *J. Anal. Chem.*, *40*, 1747–1751.

156. Chen, Y. C., Jian, Y. L., Chiu, K. H., & Yak, H. K. (2012). Simultaneous Speciation of Iron (II) and Iron(III) by Ion Chromatography with Chemiluminescence Detection, Anal. Sci., *28*, 795–799.

157. Charles, L., Pepin, D., & Casetta, B. (1996). Electrospray Ion Chromatography–Tandem Mass Spectrometry of Bromate at Sub-ppb Levels in Water, Anal. Chem., *68*, 2554–2558.

158. Divjak, B., & Goessler, W. (1999). Ion Chromatographic Separation of Sulfur-Containing Inorganic Anions with an ICP–MS as Element-Specific Detector, *J. Chromatogr* A, *844*, 161–169.

159. Nowak, M., & Seubert, A. (1998). Ultra-Trace Determination of Bromate in Drinking Waters by Means of Microbore Column ion Chromatography and On-Line Coupling with Inductively Coupled Plasma Mass Spectrometry, Anal. Chim Acta, *359*, 193–204.

160. Creed, J. T., Magnuson, M. L., Pfaff, J. D., & Brockhoff, C. (1996). Determination of Bromate in Drinking Waters by Ion Chromatography with Inductively Coupled Plasma Mass Spectrometric Detection. *J. Chromatogr.* A, *753*, 261–267.

161. Yamanaka, M., Sakai, T., Kumagai, H., & Inoue, Y. (1997). Specific Determination of Bromate and Iodate in Ozonized Water by Ion Chromatography with Post Column Derivatization and Inductively-Coupled Plasma Mass Spectrometry, *J. Chromatogr* A, *789*, 259–265.

162. Yang, L., Sturgeon, R. E., & Lam, J. W. H. (2001). On-Line Determination of Dissolved Phosphate in Sea-Water by Ion-Exclusion Chromatography Inductively Coupled Plasma Mass Spectrometry, *J. Anal At Spectrom*, *16*, 1302–1306.

163. Chen, Z. L., Megharaj, M., & Naidu, R. (2007). Speciation of Iodate and Iodide in Seawater by Non-Suppressed Ion Chromatography with Inductively Coupled Plasma Mass Spectrometry. Talanta, *72*, 1842–1846.
164. Chen, Z. L., Megharaj, M., & Naidu, R. (2007). Determination of Bromate and Bromide in Seawater by Ion Chromatography, with an Ammonium Salt Solution as Mobile Phase, and Inductively Coupled Plasma Mass Spectrometry, Chromatographia, *65*, 115–118.
165. Jiang, S., Li, Y. S., & Sun, B. (2013). Determination of Trace Level of Perchlorate in Antarctic Snow and Ice by Ion Chromatography Coupled with Tandem Mass Spectrometry Using an Automated Sample On-Line Preconcentration Method, Chinese, Chem. Lett., *24*, 311–314.
166. Harwood, J. J., & Su, W. (1997). Analysis of Organic and Inorganic Selenium Anions by Ion Chromatography-Inductively Coupled Plasma Atomic Emission Spectroscopy, *J. Chromatogr.* A, *788*, 105–111.
167. Urasa, I. T., & Nam, S. H. (1989). Direct Determination of Chromium (III) and Chromium (VI) with Ion Chromatography Using Direct Current Plasma Emission as Element-Selective Detect. *J. Chromatogr, Sci.*, *27*, 30–37.
168. Vilano, M., Padro, A., Rubio, R., & Raudet, G. (1998). Organic and Inorganic Selenium Speciation Using High-Performance Liquid Chromatography With UV Irradiation and Hydride Generation-Quartz Cell Atomic Absorption Spectrometric Detection, *J. Chromatogr* A, *819*, 211–220.
169. Huang, C. Y., Lee, N. M., Lin, S. Y., & Liu, C. Y. (2002). Determination of Vanadium, Molybdenum and Tungsten in Complex Matrix Samples by Chelation Ion Chromatography and On-Line Detection with Inductively Coupled Plasma Mass Spectrometry, Anal Chim Acta, *466*, 161–174.
170. Sutton, R. M. C., Hill, S. J., Jones, P., Sanz-Medel, A., & Garcia-Alonso, J. I. (1998). Comparison of the Retention behavior of Uranium and Thorium on High-Efficiency Resin Substrates Impregnated or Dynamically Coated with Metal Chelating Compounds, *J. Chromatogr* A, *816*, 286–291.
171. Chen, M. J., & Liu, C. Y. (1999). Preparation and Properties of a New Chelating Resin Containing Sulfur and Nitrogen Donor Atoms, *J. Chin. Chem.* Soc., *46*, 833–840.
172. Amm, A., Emara, K. M., & Khodari, M. (1994). Quantification of Tiaprofenic Acid using Voltammetric and Spectrophotometric Techniques, *Analyst*, *119*, 1071–1074.
173. Challenger, J., Hill, S. J., & Jones, P. (1993). Separation and Determination of Trace Metals in Concentrated Salt Solutions using Chelation Ion Chromatography, *J. Chromatogr*, *639*, 197–205.
174. Motellier, S., & Pitsch, H. (1996). Simultaneous Analysis of Some Transition Metals at Ultra Trace Level by Ion-Exchange Chromatography with Online Preconcentration, *J. Chromatogr.* A, *739*, 119–130.
175. Smith, R. E. (1988). Ion Chromatography Applications, CRC Press, Boca Raton, FL., 78–82.
176. Haddad, P. R., & Jackson, P. E. (1990). Ion Chromatography Principles and Applications, *Journal of Chromatography Library*, Elsevier, Amsterdam, *46*, 56–66.

CHAPTER 9

ENVIRONMENTAL APPLICATIONS OF CAPILLARY ZONE ELECTROPHORESIS

OSCAR NÚÑEZ

Department of Analytical Chemistry, University of Barcelona.
Martí i Franquès, 1-11, 08028, Barcelona, Spain.
E-mail: oscar.nunez@ub.edu

CONTENTS

9.1 INTRODUCTION

In capillary zone electrophoresis (CZE), cationic and anionic analytes may be separated based on differences in their charge-to-size ratio, and subsequently measured on-column by UV detection. Compared with the traditional chromatographic techniques, CZE provides an alternative separation principle characterized by high separation efficiency, rapid separations, and by a low consumption of reagents as well as solvents. Based on these advantages, CZE has been implemented in a broad of applications areas including pharmaceuticals, proteins, peptides, agrochemicals, raw materials, water, DNA, surfactants and fine chemicals [1]. Today, CZE is one of the most promising separation techniques in bio-analysis [2–5], food control and safety [6, 7], and even environmental applications [5, 8], where low analyte concentrations are expected. In this chapter, examples of some CZE environmental applications will be presented and discussed.

Unfortunately, CZE suffers from relatively high concentration detection limits because the sample volumes introduced into the capillary tubes under standard conditions are limited to the low nanoliter (nL) level, and because UV detection is accomplished directly on the capillary with a short optical path length. Thus, for trace analysis applications, the amount of analyte injected into the capillary or the detector sensitivity must be increased. The latter aspect may be accomplished by using light paths in connection with UV detection, or alternatively by using more sensitive detectors such as laser-induced fluorescence (LIF) detection [9–11] or even capillary zone electrophoresis coupled to mass spectrometry (CE-MS) techniques [12–14]. Both bubble cells and z-shaped cells have been used as extended light paths for UV detection, which typically provides an enhancement of the signal-to-noise by a factor of 3–6 [15], which is in general not enough for ultra-trace analysis of contaminants in environmental samples. High mass sensitivity has been reported with LIF detection but one of its handicaps is that is only applicable for some analytes (those with fluorescence properties) and the number of wavelengths available with commercial LIF detectors is limited. Regarding MS, CE-MS techniques are frequently used to increase sensitivity but this will require specialized interfaces and will not be addressed in this chapter.

The second approach, to increase the amount of analyte injected into the capillary, is then one of the most frequently proposed to improve sensitivity in CE when UV detection us used. This may be accomplished by analyte enrichment during the sample preparation step, by using off-line extraction and/or preconcentration methods such as solid phase extraction (SPE) [16, 17] or liquid-liquid extraction (LLE) [18]. Another approach involves increasing the amount of analyte introduced into the capillary tube by on-column electrophoretic-based preconcentration methods based on sample stacking, for instance when the sample is of lower conductivity than the background electrolyte (BGE), procedure known as field amplified sample injection (FASI). In this chapter, several examples of CE environmental applications and the way addressed to increase sensitivity by means of on-column electrophoretic-based preconcentration methods will be discussed.

But when dealing with environmental applications, all the steps on the analytical method development including sample preparation, preconcentration method, detection and quantitation are important. For this reason, in order to present a complete picture of the diversity of environmental applications dealing with capillary zone electrophoresis, some selected CZE environmental applications will be presented and discussed at the end of this chapter.

9.2   CZE ENVIRONMENTAL APPLICATIONS

As previously commented, CZE is the most widely used mode in CE due to its simplicity of operation and its versatility, and today is appearing as an alternative to other separation techniques, such as liquid chromatography, in multiple application fields. Despite the low sensitivity characteristic of this technique, CZE has been proposed for multiple environmental applications. However, target analyte concentrations in environmental samples are usually very low, so several CZE applications will require of off-line and/or on-line preconcentration methods.

Some examples of CZE environmental applications are summarized in Table 9.1 [17, 19–30]. As can be seen, most of the applications focused in the analysis of a variety of compounds in environmental water samples. Although direct CZE analysis of water samples is sometimes proposed (depending on the matrix and concentration level of analytes) extraction and preconcentration methods are frequently required. For instance, SPE using C18 cartridges has been described for the CZE analysis of acidic herbicides in surface water samples achieving very low limits of detection (0.2–0.6 µg/L) [17], and Komarova and Kartsova [25] used Diapak C-16 cartridges for the preconcentration of chlorophenoxycarboxylic acid herbicides in natural and potable waters.

TABLE 9.1 Environmental Applications of Capillary Zone Electrophoresis

Compounds	Matrix	Capillary	BGE	Sample Introduction / preconcentration	Detection	LODs	Ref.
Low molecular weight organic acids	Water	Fused-silica capillaries 83 cm (76 cm effective length) x 75 μm I.D.	5 mM TRIS (Tris(hydroxymethyl)aminomethane), 2 mM trimellitic acid, 0.6 mM tetradecyltrimethylammonium bromide and 0.6 mM calcium hydroxide (pH 8.5)	Electrokinetic injection: 45 s at -5 kV; Separation voltage: -30 kV	UV 254 nm	0.5–5 μg/L	[19]
17 Chlorophenolic pollutants	Industrial waste waters	Fused-silica capillaries of 50 μm I.D.	10 mM phosphate buffer (pH 8.23) with 35% acetone (v/v)	Separation voltage: +30 kV	UV 214 nm	100–270 μg/L	[20]
Sulfonated azo dyes	River water	Fused-silica capillaries 57 cm (50 cm effective length) x 75 μm I.D.	1:5 dilution of 10 mM phosphoric acid and tetrabutylammonium hydroxide buffer (pH 11.5), and 25 mM of triethylamine, final pH 11.55.	Hydrodynamic injection: 4 s; Separation voltage: +15 kV; SPE preconcentration: Oasis HLB 1 mL cartridges	UV 460 nm	0.1–4.53 mg/L	[21]
Acidic herbicides	Surface water	Fused-silica capillaries 41.5 cm (32 cm effective length) x 50 μm I.D.	5 mM sodium tetraborate in water-acetonitrile mixture (70:30 v/v), pH 9.0	Hydrodynamic injection: 10 s at 0.3 psi; Separation voltage: +12 kV; SPE preconcentration: C18 500 mg cartridges	UV 320 nm	0.2–0.6 μg/L	[17]

TABLE 9.1 *(Continued)*

Compounds	Matrix	Capillary	BGE	Sample Introduction / preconcentration	Detection	LODs	Ref.
Haloacetic acids	Water	Fused-silica capillaries 41.5 cm (32 cm effective length) x 50 μm I.D.	Phosphate, citrate and borate buffers at several conditions.	Hydrodynamic injection: 3-10 s at 50 mbar Electrokinetic injection: 3-10 s at -5 kV Separation voltage: -25 kV	Contactless conductivity detection UV 200 nm	0.1 mg/L	[22]
Phenolic compounds	Sea sand and soil	Fused-silica capillaries 58.5 cm (50 cm effective length) x 50 μm I.D.	30 mM 2-(N-cyclohexyl-amino) -ethanesulphonic acid (pH 9.7 adjusted with sodium hydroxide)	Hydrodynamic injection: 3 s at 5 kPa Separation voltage: +20 kV	UV 220 nm	270–410 μg/L	[23]
Ammonium ion	River water and sewage samples	Fused-silica capillaries 72 cm (50 cm effective length) x 100 μm I.D.	20 mM borate adjusted at pH 20 with 1 M sodium hydroxde	Hydrodynamic injection: 5 s Separation voltage: +10 kV	UV 190 nm	0.24 mg/L	[24]
Chlorophe-noxy-car-boxylic acid herbicides	Natural and Potable waters	Fused-silica capillaries 70 cm (60 cm effective length) x 75 μm I.D.	10 mM sodium tetraborate pH 9.2	Hydrodynamic injection: 30 s at 30 mbar Separation voltage: +25 kV SPE preconcentration: Diapak C16 cartridges	UV 228.8 nm	0.5–1 μg/L	[25]
Perfluorinat-ed carboxylic acids (C6-C12)	Environmental waters	Fused-silica capillaries 60 cm (50 cm effective length) x 75 μm I.D.	50 mM TRIS solution at pH 9.0 and 50% methanol	Hydrodynamic injection: 10 s Separation voltage: +25 kV	Indirect photometric detection (chromophor 2,4-dinitro-benzoic acid)	0.6–2.4 mg/L	[26]

TABLE 1 (*Continued*)

Compounds	Matrix	Capillary	BGE	Sample Introduction / preconcentration	Detection	LODs	Ref.
Microcystins	Lake waters and water bloom samples	Fused-silica capillaries 64.5 cm (56 cm effective length) x 50 μm I.D.	25 mM tetraborate buffer	Hydrodynamic injection: 2 s at 50 mbar Separation voltage: +25 kV	UV 238 nm	50 mg/L	[27]
Mercury(II)	Water samples	Fused-silica capillaries 60.2 cm (50 cm effective length) x 75 μm I.D.	100 mM boric acid with 10% (v/v) methanol at pH 8.5	Hydrodynamic injection: 5 s at 0.5 psi Separation voltage: +22 kV Dispersive liquid-liquid microextraction and back extraction	UV 200 nm	0.62 μg/L	[28]
Bisphenol A and naphtols	River water	Fused-silica capillaries 60.2 cm (50 cm effective length) x 75 μm I.D.	50 mM sodium tetraborate (pH 9.5) with 30% (v/v) methanol	Hydrodynamic injection: 5 s at 0.5 psi Separation voltage: +25 kV Cloud point extraction	UV 214 nm	0.2–0.5 μg/L	[29]
Glyphosate and aminomethyl-phosphonic acid	Surface water	Fused-silica capillaries 60 cm (50 cm effective length) x 75 μm I.D.	50 mM sodium tetraborate	Hydrodynamic injection: 10 s at 5 mbar Separation voltage: +25 kV	UV 210 nm	5–20 μg/L	[30]

Other extraction strategies have also been used for the preconcentration of analytes to allow their analysis by CZE. For instance, Zhong et al. [29] developed a cloud point extraction (CPE) method to preconcentrate bisphenol A (BPA), α-naphthol and β-naphthol prior to performing CZE analysis. Parameters influencing the CPE extraction, such as triton X-114 concentrations, pH value, extraction time and temperature were systematically evaluated. The application of CPE to a 10 mL water sample allowed increasing capillary zone electrophoretic signal in comparison to the direct water analysis by CZE as can be seen in Fig. 9.1.

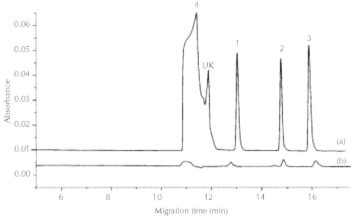

FIGURE 9.1   Electropherograms of the standard solution obtained by (a) CPE-CZE analysis and (b) direct CZE analysis (1, BPA; 2, α-naphthol; 3, β-naphthol, 4, Triton X-114). CPE conditions: 0.1% (v/v) Triton X-114; pH 3; cloud point temperature 30°C. CZE conditions as described in Table 9.1. Reproduced from Ref. [29] with permission of Elsevier.

Several authors have studied the applicability of CZE methods for environmental analysis in comparison to other separation techniques. An interesting study can be found in the work published by Professor Riekkola's group where phenolic compounds were analyzed by both gas chromatography-mass spectrometry (GC-MS) and CZE to compare the techniques and to find out if CZE was a suitable tool for analysis of phenols extracted from environmental matrices [23]. For that purpose, a self-constructed pressurized hot water extraction equipment was used in dynamic mode to extract spiked phenolic compounds (phenol, 3-methylphenol, 4-chloro-3-methylphenol and 3,4-dichlorphenol) from sea sand and soil. When using CZE, phenols extracted from the soil at 300°C were separated with good resolution at pH 9.7, as can be seen in Fig. 9.2, and coextracted compounds (e.g., polyaromatic hydrocarbons) did not interfere with the analysis.

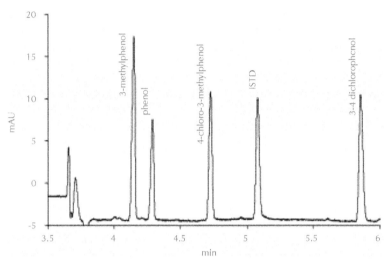

FIGURE 9.2 Electropherogram of phenols spiked in a soil matrix and extracted by pressurized hot water extraction (T=300°C, t=20 min). CZE conditions as described in Table 1. Reproduced from Ref. [23] with permission of Elsevier.

Moreover, the analytical performance values obtained by GC-MS and CZE were generally of similar magnitude, showing that CZE was a suitable technique for the analysis of this family of compounds in environmental matrices.

Fukushi et al. [24] developed a method for the CZE detection and determination of ammonium ion in environmental water samples, achieving a limit of detection of 0.24 mg/L, and with similar results to those usually obtained with ion chromatography (IC) showing that CZE is a suitable technique for the analysis of inorganic ions in environmental matrices. An interesting review can be found in the literature addressing the quantitative aspects of CZE for the analysis of inorganic ions [31]. Innovative applications of CZE to the determination of inorganic ions in environmental aquatic samples of high salinity and, particularly, in seawater have also been reviewed [32].

But nowadays the use of on-line preconcentration methods in CZE is increasing, and also in the environmental field, because no special requirement but a CE instrument is necessary for their application. Most of these methods are based in electrophoretic principles and play with the sample injection in order to increase the amount of analyte introduced into the capillary without losing separation efficiency. The number of on-line electrophoretic-based preconcentration methods is huge, so this chapter will focus only on some of them based in the stacking phenomena such as large volume sample stacking (LVSS), field amplified sample injection (FASI), and electrokinetic supercharging (EKS). Some examples of their application to the analysis of environmental samples will be addressed in the next sections.

9.3 ON-LINE PRECONCENTRATION METHODS

9.3.1 LARGE VOLUME SAMPLE STACKING (LVSS)

The mechanism for analyte preconcentration in the electrophoretic-based methods, which can also be called on-column preconcentration methods, is based on the principle of stacking analytes in a narrow band between two separate zones in the capillary where the compounds have different velocities [33]. One of the most common ways of on-column preconcentration based in sample stacking is large volume sample stacking (LSVV). Their fundamentals were described in Chapter 4.

Some examples of LVSS-CZE applications in the environmental field are presented in Table 9.2 [34–39]. As can be seen, LVSS is an in-line preconcentration method able to achieve sensitivity enhancements in the analysis of environmental water samples up to 40-fold when compared with conventional hydrodynamic injection in CZE. Additionally, it should be pointed out that in most cases extended path length capillaries were used [34–37].

Because sensitivity enhancements achieved by LVSS are not too high, the application of this methodology to the environmental field requires its combination with other preconcentration procedures. For instance, Quesada-Molina et al. [36] developed a LVSS-CZE method for the analysis of metribuzin degradation products in soil samples after the application of pressurized liquid extraction (PLE). This PLE procedure was followed by an off-line preconcentration and sample cleanup procedure by SPE using a LiChrolut EN sorbent column. These last two procedures were also suitable for the direct treatment of groundwater samples before CE analysis. The combination of both off-line and on-line preconcentration procedures provided a significant improvement in sensitivity (500-fold for water samples). This method will be described in depth in Section 9.4.1.

An interesting work is the one described by Herrera-Herrera et al. for the analysis of quinolone antibiotics in water samples [38]. The authors proposed for the first time the use of oxidized multiwalled carbon nanotubes (o-MWCNT) for the dispersive solid phase extraction (dSPE) of 11 quinolone antibiotics in water samples.

TABLE 9.2 Large Volume Sample Stacking In-line Preconcentration Methods for the CZE Analysis of Environmental Samples

Compounds	Matrix	Capillary	BGE	Sample Introduction / preconcentration	Detection	Sensitivity enhancement	LODs	Ref.
Naphthalenes and benzenesulfonates	Waters	Fused-silica capillaries 64.5 cm (50 cm effective length) x 50 μm I.D. with extended pathlength of 150 μm (buble cell)	20 mM sodium tetraborate buffer	Large volume sample injection: 200 s at 50 mbar Separation voltage: +30 kV	UV 225 nm	40	5-10 μg/L	(34)
Sulfonamides	Ground waters	Fused-silica capillaries 64.5 cm (56 cm effective length) x 75 μm I.D. with extended pathlength of 200 μm (buble cell)	45 mM sodium phosphate buffer with 10% (v/v) methanol at pH 7.3	Large volume sample injection: 30 s at 7 bar Sample matrix: imidazole 10 mM with 10% methanol Separation voltage: +25 kV	UV 265 nm	15	2.6-23 μg/L	(35)
Metribuzin degradation products	Soil and ground waters	Fused-silica capillaries 48.5 cm (40 cm effective length) x 75 μm I.D. with extended pathlength of 200 μm (buble cell)	40 mM sodium tetraborate buffer pH 9.5	Large volume sample injection: 200 s at 50 mbar Separation voltage: +15 kV	UV 220 nm	4-36	10-20 μg/L	(36)
Sulfonylurea herbicides	Ground waters	Fused-silica capillaries 48.5 cm (40 cm effective length) x 50 μm I.D. with extended pathlength of 200 μm (buble cell)	90 mM ammonium acetate buffer pH 4.8	Large volume sample injection: 1 min at 7 bar Sample matrix: methanol:water 1:9 (v/v) Separation voltage: +20 kV	UV 240 nm	--	45-116 ng/L	(37)

TABLE 9.2 *(Continued)*

Compounds	Matrix	Capillary	BGE	Sample Introduction / preconcentration	Detection	Sensitivity enhancement	LODs	Ref.
Quinolone antibiotics	Water	Fused-silica capillaries 67 cm (60 cm effective length) x 75 μm I.D.	65 mM phosphate buffer at pH 8.5	Large volume sample injection: 3 s at 20 psi Sample matrix: water with 10% (v/v) acetonitrile Separation voltage: +15 kV	UV 280 nm	20	28-94 ng/L	(38)
Haloacetic acids	Water	Fused-silica capillaries 57 cm (50 cm effective length) x 50 μm I.D.	20 mM acetic acid-ammonium acetate buffer pH 5.5 containing 20% (v/v) acetonitrile	Large volume sample injection: 15 s at 20 psi Separation voltage: +25 kV	UV 200 nm	25	49-200 μg/L	(39)

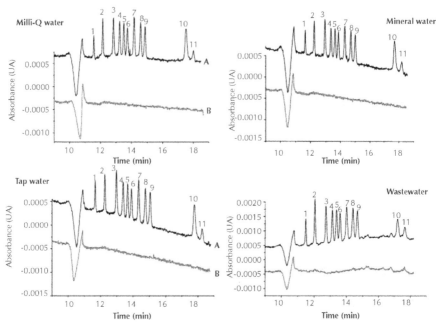

FIGURE 9.3 LVSS-CZE electropherograms of spiked water samples (Milli-Q, mineral, tap and wastewater) containing 1 µg/L of each antibiotic (A) and nonspiked water samples (B) after optimum o-MWCNTs-dSPE conditions. CZE conditions as described in Table 9.2. Reproduced from Ref. [38] with permission of Elsevier.

Several dSPE parameters such as volume sample and pH, o-MWCNT amount, volume and type of eluent were optimized. The application of the developed method to the analysis of spiked Milli-Q, mineral, tap, and wastewater samples (see an example in Fig. 9.3) resulted in good recoveries values ranging from 62.3 to 116% with relative standard deviations values lower than 7.7% in all cases.

9.3.2   FIELD AMPLIFIED SAMPLE INJECTION (FASI)

Among in-line enrichment procedures, field amplified sample injection (FASI) is very popular since it is quite simple only requiring the electrokinetic injection of the sample after the introduction of a short plug of a high-resistivity solvent such as methanol or water. Fundamentals of FASI were presented in Chapter 8.

Several environmental applications of FASI-CZE can be found in the literature [39–41]. As an example, Bernad et al. compared both LVSS and FASI for the in-line preconcentration CZE analysis of haloacetic acids in water samples [39]. When negatively charged analytes are being analyzed by FASI, electroosmotic flow (EOF) must be taken into account in order to prevent removal of low electrophoretic mobility compounds from the capillary when the enhanced sample electrokinetic injection is performed. The authors prevented that by working at low pH using a 200 mM formic acid-ammonium formate buffer solution at pH 3. Additionally, injection times for both the plug of water (hydrodynamic mode) and sample (electrokinetic mode) were simultaneously optimized. Under optimal conditions, 25 s for both water plug hydrodynamic injection (3.5 kPa) and sample electrokinetic injection (−10 kV), sensitivity enhancements up to 300-fold for some haloacetic acids such as dibromoacetic acid were obtained. These enhancements were 10-fold higher than the ones obtained by LVSS for the same compounds. Although the important decrease in LODs (values between 4–50 µg/L), again the sensitivity achieved was not enough for the analysis of this family of compounds in real water samples, so the combination of the proposed method with an off-line SPE step was mandatory. By using Oasis WAX SPE cartridges, specifically proposed for preconcentration of acidic species, sample salinity was considerably removed, and with the combination of both SPE and FASI sensitivity enhancements between 6250 and 26,000 were obtained. The method was applied to the analysis of tap water being able to quantify seven haloacetic acids at concentrations below 13 µg/L (see Fig. 9.4).

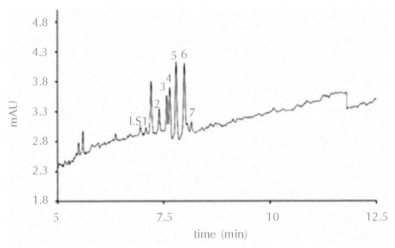

FIGURE 9.4 Analysis of Barcelona (Spain) tap water by SPE-FASI-CZE. 1, dichloroacetic acid; 2, bromochloroacetic acid; 3, trichloroacetic acid; 4, dibromoacetic acid; 5, bromodichloroacetic acid; 6, chlorodibromoacetic acid; and 7, tribromoacetic acid. Reproduced from Ref. [39] with permission of Wiley-VCH.

9.3.3 ELECTROKINETIC SUPERCHARGING (EKS)

As previously commented the development of new in-line preconcentration strategies to achieve CE sensitive methods able to analyze environmental samples at low concentration levels is needed. A recent in-line preconcentration method for CE that has great potential is that of electrokinetic supercharging (EKS). This method is the combination of electrokinetic injection under field-amplified conditions (FASI) and transient isotachophoresis (tITP) and was first described for the analysis of rare-earth ions by the group of Professor Hirokawa (42,43). Their fundamentals were described in Chapter 4.

Some environmental applications of EKS are summarized in Table 9.3 [44–48]. As can be seen this kind of methodology has been employed for the analysis of nonsteroidal anti-inflammatory drugs (NSAIDs) or hypolipidaemic drugs in water samples. For instance, professor Haddad's group proposed the use of electrokinetic supercharging in-line preconcentration method for the analysis of seven NSAIDs in wastewater samples [44]. They examined the application of FASI and found an improvement in detection limits by 200-fold providing LODs down to 0.6–2.0 µg/L, which were insufficient for the determination of NSAIDs as environmental pollutants in water samples. Sensitivity was then improved by EKS. The optimum EKS method involved the hydrodynamic injection of a leading electrolyte (LE, 100 mM of sodium chloride, 30s, 50 mbar), the electrokinetic injection of the sample for a long time (200 s, –10 kV), and finally the hydrodynamic injection of a terminating electrolyte (TE, 100 mM of 2-(cyclohexylamino) ethanesulphonic acid, 40s, 50 mbar). With this method they were able to improve sensitivity by 2400-fold obtaining LODs of 50–180 ng/L. The proposed method was validated and applied to the analysis of wastewater samples. Although with this kind of water matrices LODs increased by approximately 10-fold, the values were lower than values previously found in wastewater samples from European and Mediterranean cities. As an example, Fig. 9.5 shows the electropherogram obtained from EKS of a wastewater sample spiked with 20 µg/L of the evaluated NSAIDs, as well as the one from a wastewater blank sample.

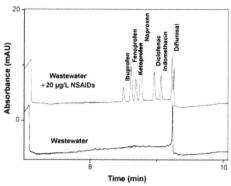

FIGURE 9.5   Electrophrerogram obtained from EKS of (A) wastewater sample spiked with 20 µg/L of the NSAIDs, and (B) blank wastewater sample. CE conditions as described in Table 9.3. Reproduced from Ref. [44] with permission of Elsevier.

TABLE 9.3 Electrokinetic Supercharging In-line Preconcentration Methods for the CZE Analysis of Environmental Samples

Compounds	Matrix	Capillary	BGE	Sample Introduction / preconcentration	Detection	Sensitivity enhancement	LODs	Ref.
Non-steroidal anti-inflammatory drugs	Wastewater	Fused-silica capillaries 85 cm (76.6 cm effective length) x 50 μm I.D.	15 mM sodium tetraborate (pH 9.2) with 0.1% (w/v) hexadimethrine bromide (HDMB) and 10% (v/v) methanol	LE: 100 mM sodium chloride, 30 s at 50 mbar Sample EK injection: 200 s, -10 kV TE: 100 mM 2-(cyclohexylamino)ethanesulphonic acid, 40 s at 50 mbar Separation voltage: -28 kV	UV 214 nm	2400	50-180 ng/L	(44)
Non-steroidal anti-inflammatory drugs	Wastewater	Fused-silica capillaries 85 cm (76.6 cm effective length) x 50 μm I.D.	15 mM sodium tetraborate (pH 9.2) with 0.1% (w/v) hexadimethrine bromide (HDMB) and 10% (v/v) methanol	*Counter-flow EKS* LE: BGE+water Sample EK injection: 220 s, -16 kV combined with negative hydrodynamic pressure of 50 mbar to counter-balance EOF TE: 100 mM 2-(cyclohexylamino)ethanesulphonic acid, 48 s at 50 mbar Separation voltage: -28 kV	UV 214 nm	11800	10-47 ng/L	(45)
Non-steroidal anti-inflammatory drugs	River water and human plasma	Fused-silica capillaries 88.5 cm (80 cm effective length) x 50 μm I.D.	10 mM sodium tetraborate (pH 8 adjusted with NaOH) + 50 mM sodium chloride, with 10% v/v methanol	LE: BGE + methanol (3s, 50 mbar) Sample EK injection: 700 s, 2 kV TE: 50 mM 2-(cyclohexylamino)ethanesulphonic acid, 12 s at 50 mbar Separation voltage: -30 kV	UV 214 nm	2000	0.9-2 μg/L	(46)
Non-steroidal anti-inflammatory drugs	Water	Fused-silica capillaries 80 cm (71.5 cm effective length) x 50 μm I.D.	50 mM ammonium hydrogen carbonate (pH 9.2) with 10% methanol	*Pressure-assited EKS* LE: BGE+water (3 s, 50 mbar) Sample EK injection: 20 min, -14 kV combined with positive hydrodynamic pressure of 50 mbar TE: 8 mM 3-(cyclohexylamino)1-propanesulphonic acid, 20 s at 50 mbar Separation voltage: -28 kV	UV 214 nm	50000	6.7-18.7 ng/L	(47)
Hypolipidaemic drugs	Water	Fused-silica capillaries 88 cm (88 cm effective length) x 50 μm I.D.	60 mM ammonium hydrogen carbonate (pH 9.0) with 60% methanol	LE: BGE + water (5s, 40 mbar) Sample EK injection: 170 s, -10 kV TE: 1 mM 3-(cyclohexylamino)1-propanesulphonic acid, 10 s at 50 mbar Separation voltage: -25 kV	ESI-MS (ion trap)	1000	180 ng/L	(48)

Lately some modifications of the method by combining the application of an additional pressure during the electrokinetic injection of the sample were proposed. For instance, counter flow electrokinetic supercharging (CF-EKS) carried out by applying a negative hydrodynamic pressure of 50 mbar to counterbalance the EOF was also evaluated for the analysis of NSAIDs in wastewater, achieving a 11800-fold sensitivity enhancement and LODs in the range 10.7–47.0 ng/L [45]. Pressure-assisted electrokinetic supercharging (PA-EKS) was also evaluated for the analysis of this family of compounds [47]. In this case, a positive hydrodynamic pressure of 50 mbar during sample injection to improve stacking of NSAIDs was used. Sensitivity enhancements up to 50000-fold were observed with LODs down to 6.7 ng/L.

Similar methodology was proposed for Professor Haddad's group for the analysis of hypolipidaemic drugs in water samples by CE-MS using electrospray (ESI) as ionization source and an ion trap as mass analyzer [48]. The electrophoretic separation was carried out in this case by counter EOF conditions by reversing EOF with hexadimethrine bromide. Using EKS, the sensitivity of the method was improved 1000-fold in comparison to conventional injection under FASI conditions, obtaining LODs down to 180 ng/L.

Other environmental applications of EKS can be found in the literature but using nonaqueous capillary electrophoresis (NACE) as separation technique [49–51]. For instance, Lu and Breadmore [50] proposed the use of EKS-NACE for the analysis of phenolic acids in water by using a 30 mM ammonium acetate methanolic solution as

NACE BGE. With the application of EKS conditions, a sensitivity enhancement of 300- to 440-fold was observed, with LODs in the range of 1.0 to 2.5 ng/mL.

9.4 SELECTED ENVIRONMENTAL APPLICATIONS OF CZE

As presented above, CZE is becoming a good approach for the analysis of contaminants in environmental samples. However, all steps included in the analytical method, including sample preparation (extraction and/or cleanup procedure), sample preconcentration (off-line or on-line), sample analysis (for instance by CZE), detection and quantitation must be optimized and considered simultaneously, because any of them could directly affect the others. For instance, when dealing with on-column electrophoretic-based preconcentration methods in CZE, sample salinity is one of the most important factors to achieve a good sensitivity enhancement, and this parameter will strongly depend on the sample treatment (extraction, cleanup, preconcentration, etc.) carried-out.

In this section, some selected CZE environmental applications from the literature will be presented as examples by addressing the full method development in order to provide a complete picture of the diversity of CZE method development for environmental analysis.

9.4.1 DETERMINATION OF METRIBUZIN DEGRADATION PRODUCTS IN SOIL AND GROUND WATER SAMPLES

Metribuzin is a selective systemic herbicide, belonging to the group of triazinone herbicides, used for pre- and postemergence control of many grasses and broad-leaved weeds in soy beans, potatoes, tomatoes, maize and cereals, among others. Usually, the analysis of metribuzin and its degradation products has been accomplished by different chromatographic methods such as reversed-phase thin-layer chromatography with UV detection [52], and high-performance liquid chromatography (HPLC) with diode array detection [53] or mass spectrometry detection [54]. Quesada-Molina et al. [36] developed a CZE method with UV-detection for the simultaneous monitoring of the major degradation products of metribuzin, that is, deaminometribuzin (DA), deaminodiketometribuzin (DADK) and diketometribuzin (DK) in soil and ground water samples, following the procedure described in this section.

9.4.1.1 SAMPLE TREATMENT

Pressurized liquid extraction (PLE) is one of most frequently used sample treatments for the analysis of environmental soil and sediment samples [55].

In this example, extraction of target analytes from environmental soils was carried out by PLE following the next procedure (36): a portion of 5.00 g of spiked soil sample (previously dried, mixed and sieved through a 2-mm sieve before fortification) was transferred into the extraction cell of an accelerated solvent extraction (ASE) system together with 10 g of sodium sulfate (used as drying agent). Methanol was used for extraction at 103°C for 5 min in static time, with two extraction cycles. Other extraction conditions were as follows: pressure 1500 psi, preheat 3 min, flush volume 100% and purge time 60 s. Total final volume of the methanolic extract was 55–60 mL (depending on the soil moisture). Solution was concentrated to 1 mL in a rotary evaporator at 37°C, diluted with 25 mL of deionized water and preconcentrated by SPE before being analyzed by CZE.

9.4.1.2   OFF-LINE PRECONCENTRATION

SPE is a very common sample treatment procedure for the preconcentration and cleanup of environmental water samples, and the selection of the correct SPE stationary phase good selectivity and high recoveries can be achieved [16]. SPE usually involves several steps: preconditioning (for activation), washing, elution, evaporation and reconstitution.

In this example, SPE using LiChrolut EN cartridges was proposed for the off-line preconcentration and cleanup of ground water samples, as well as water diluted extracts coming from PLE treatment of soil samples, following the next procedure [36]: SPE LiChrolut EN cartridges were preconditioned with 6 mL of acetone, 6 mL of methanol and 6 mL of deionized water. In the case of groundwater samples a volume of 100 mL was treated by SPE. Both, water samples or soil extract were loaded through the SPE cartridge at 2 mL/min by using a vacuum operated pumping system. Cartridges were then washed with 5 mL of water and 5 mL of methanol:water (45:55 v/v) and air-dried for 15 min. Elution of target analytes was achieved with 4 mL of methanol, and extracts were evaporated to dryness under a nitrogen current at 20°C. Finally, 500 µL (for ground water samples) or 2 mL (in the case of soil samples) of methanol:water (10:90, v/v) were used to reconstitute the extracts, which were directly analyzed by the proposed CZE method.

9.4.1.3   ANALYSIS BY LVSS-CZE

Because of the requirement to analyze low concentration levels of metribuzin and its degradation products, after sample treatment and off-line SPE preconcentration of soil and ground water samples, Quesada-Molina et al. [36] proposed the analysis of these compounds by applying also an on-column electrophoretic-based preconcentration method in CZE. The combination of several preconcentration methods,, that is, off-line SPE together with an on-column electrophoretic-based CZE preconcentration method, is very common when contaminants at low concentration levels must be analyzed, especially with CE techniques, in environmental applications.

LVSS procedure was carried out as follows: samples were loaded with a pressure of 50 mbar for 200 s (this way, 80% of the capillary column was filled with the sample solution). After sample injection, a negative voltage (–25 kV) was applied. Sample matrix removal from the capillary was controlled by monitoring the electric current, which progressively increased to its normal value as the low-conductivity sample injected zone was eliminated from the capillary tube. When current reached 95–99% of the normal value, the high voltage was switched from negative to positive (15 kV). At this stage, stacking process could be considered complete, and CZE separation was carried out.

CZE separation was performed using a 40 mM sodium tetraborate buffer, pH 9.5, by applying a voltage of 15 kV (normal mode) and using fused silica capillaries of 48.5 cm × 75 μm i.d. (internal diameter). Capillary temperature was kept at 25°C.

9.4.1.4   DETECTION AND QUANTITATION

UV-detection by monitoring two wavelengths: 220 nm for DA, and 260 nm for DK and DADK, was used. Quantitation was carried out by external calibration using *p*-aminobenzoic acid (PABA) as internal standard (I.S.). The use of internal standards external calibration methods is very common when UV-detection is carried out in CZE in order to correct small shift variations in the UV signal.

9.4.1.5   RESULTS

Very good results were obtained by the method proposed by Quesada-Molina et al. [36] for the analysis of metribuzin major degradation products. As an example, Fig. 9.6 shows the degradation pathway of metribuzin as well as the electropherograms obtained for a nonspiked soil sample (A) and the same soil sample spiked with target compounds.

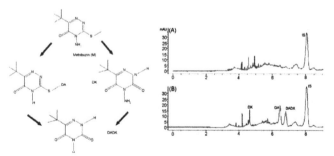

FIGURE 9.6 Degradation pathway of metribuzin and electropherograms of (A) blank soil sample and (B) soil sample spiked with 200 μg/Kg of DK, DA, DADK and 0.5 μg/mL of I.S. Reproduced from Ref. [36] with permission of Elsevier.

The combination of PLE and both off-line and on-line preconcentration procedures provided a significant improvement in sensitivity. LVSS provided preconcentration factors of 4, 36 and 28 for DK, DA and DADK, respectively, and with SPE a preconcentration of 500-fold for the case of water samples and of 2.5-fold in the case of soil samples was obtained. The method proposed by Quesada-Molina et al. [36] was suitable for the monitoring of these residues in environmental samples with high sensitivity (LODs down to ng/L), precision and satisfactory recoveries (ranging from 62 to 102%).

9.4.2 DETERMINATION OF MERCURY(II) IN WATER SAMPLES

Mercury is one of the prevalent toxic heavy metals with significant concern because of its persistent accumulation, high toxicity, wide use and large distribution [56]. Its measurement is a challenging task by its trace level presence in complicated matrices and the interference of other elements. Therefore, in order to determine the trace level of mercury ions and to improve the detection sensitivities, high-efficiency preconcentration methods and high selective and sensitive analysis techniques are required. Li et al. [28] developed a method for the determination of mercury in water samples that combined dispersive liquid-liquid microextraction (DLLME) with back-extraction (BE) and analysis by CZE, following the procedure described in this section.

9.4.2.1 SAMPLE TREATMENT

In this example mercury was extracted from water samples by using DLLME, which was firstly introduced in 2006 by Rezaee et al. [57] and successfully applied for the extraction and the preconcentration of organic and inorganic compounds from water samples [58, 59].

The authors combined DLLME with a back-extraction procedure (by changing the mercury complexes formed) following the scheme shown in Fig. 9.7. Briefly, 10 mL of water sample containing mercury(II) and 1-(2-Pyridylazo)-2-naphthol (PAN) used as chelating reagent were placed in a 15 mL conical centrifuge tube and 1 mL of 0.1 mol/L boric acid buffer solution (pH 6.5) was added.

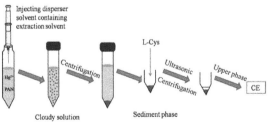

FIGURE 9.7 Scheme of a DLLME-BE procedure. Reproduced from Ref. [28] with permission of Springer.

Then, 800 μL of ethanol (used as disperser solvent) containing 30 μL of chlorobenzene (used as extraction solvent) were placed in a 1.5 mL centrifuge tube and were shaken vigorously. Then the mixture was rapidly (within 3 s) injected into the sample solution by using a 1.00 mL syringe, and the solution was gently shaken. A cloudy solution (water/ethanol/chlorobenzene) was then formed in the tube. After centrifugation for 3 min at 3000 rpm, the dispersed fine droplets of chlorobenzene were sedimented in the bottom of the conical tube. Then a BE procedure was carried out. For that purpose, 10 μL of 0.1% (w/v) L-Cys was injected into the tube and then ultrasonicated for 5 min. In this step, L-Cys displaced PAN to form more hydrophilic and stable complexes with L-Cys over their PAN counterparts (Hg-PAN) [60]. After centrifugation, the hydrophilic Hg-L-Cys complexes were extracted into the upper aqueous phase, which was directly analyzed by CZE.

9.4.2.2   ANALYSIS BY CZE

CZE separation was performed using a 100 mM boric acid and 10% (v/v) methanol (pH 8.5), by applying a voltage of 22 kV (normal mode) and using fused silica capillaries of 60.2 cm (50 cm effective length) x 75 μm i.d. Capillary temperature was kept at 25°C. Samples were hydrodynamically injected at 0.5 psi for 5 s.

9.4.2.3   DETECTION AND QUANTITATION

UV-detection was carried out at 200 nm. Quantitation was performed by external calibration and the method showed to be linear in the range between 1 and 1000 μg/L (r^2 0.9991).

9.4.2.4   RESULTS

To evaluate the feasibility of the proposed DLLME-BE-CZE method, the authors analyzed tap water and seawater samples. The electropherograms from the extracts of tap and seawater samples are shown in Fig. 9.8.

Under optimal conditions, an enrichment factor of 625 was achieved, with a LOD of 0.62 μg/L, and the method showed a good precision with relative standard deviations (RSD, n=6) of 4.1%. Recoveries were determined with tap water and seawater spiked at concentration levels of 10 and 100 μg/L, respectively, and ranged from 86.6% to 95.1%, with corresponding RSDs of 3.95–5.90%. An interesting aspect of this method is that besides preconcentrating traces of Hg^{2+}, this procedure could significantly eliminate the interference from foreign ions and matrix. Additionally, the present DLLME-BE-CE with simple UV detection obtained similar or higher detection sensitivity to/than some hyphenation methods with a very simple instrumental setup, low costs, and being a rapid, simple, accurate and environmental friendly analytical method.

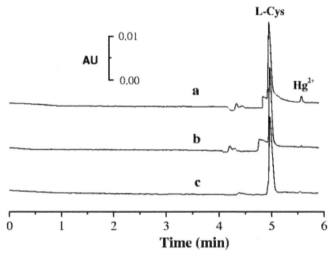

FIGURE 9.8 Electropherograms of mercury(II) of the extract obtained by DLLME-BE-CZE from (a) seawater sample spiked with 2 μg/L standard mercury(II) solution, (b) nonspiked seawater sample and (c) nonspiked tap water. Reproduced from Ref. [28] with permission of Springer.

9.4.3 DETERMINATION OF SULFONAMIDES IN GROUND WATER

Sulfonamides are antibacterial and anti-infective drugs commonly used for the treatment of diseases in medicine and veterinary practice such as gastrointestinal and respiratory infections. Because of their broad spectrum of activity and their low costs, they are widely used nowadays, but the problem is that sometimes they are applied without respecting safety recommendations, which results in undesirable residues in animal tissues, meat or bio-fluids such as milk. Moreover, sulfonamides can also be used against human disease; they spread to the surface water through urban waste water because actual procedures for waste water treatment cannot completely remove these compounds [61]. Also, because of their use in veterinary care, they are found in soils, ground and surface water due to the use of animal excretions as manure. Soto-Chinchilla et al. [35] established and validated a LVSS-CZE method for the analysis of nine sulfonamides (see structures in Fig. 9.9) in ground water samples, following the procedure described in this section.

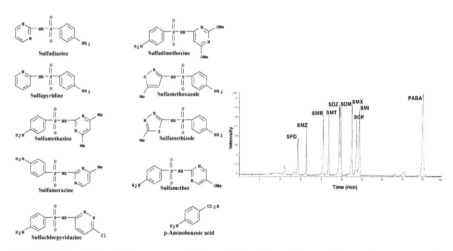

FIGURE 9.9   Chemical structures of the studied sulfonamides and electropherograms of a ground water sample spiked with 50 μg/L of each sulfonamide. Reproduced from Ref. [35] with permission of John Wiley & Sons, Inc.

9.4.3.1   SAMPLE TREATMENT

As previously described, the necessity to analyze contaminants in environmental water samples at low concentration levels requires the combination of several pre-concentration methods previous to CZE analysis. In this example, the authors proposed the use of HLB-SPE cartridges for the cleanup and off-line preconcentration of water samples previous to the analysis by LVSS-CZE. Briefly, cartridges were preconditioned with 3 mL of methanol and 3 mL of acetic acid (50%, pH 3). Then, 10 mL of water samples were loaded through the cartridge, and analytes were eluted by using 3 mL of methanol. The methanolic extract was then evaporated to dryness under nitrogen gas at 50°C and reconstituted with an imidazole solution (10 mM imidazole with 10% methanol) used as sample matrix and containing 150 μg/L PABA as an I.S.

9.4.3.2   ANALYSIS BY LVSS-CZE

LVSS procedure was carried out as follows: samples were loaded with a pressure of 7 bar for 30 s (this way the whole capillary column was filled with the sample solution). After sample injection, a negative voltage (–28 kV) was applied. As commonly done with this on-column preconcentration procedures, sample matrix removal from the capillary was indicated by monitoring the electric current, which progressively increased to its normal value as the low-conductivity injected zone was eliminated from the capillary. When current reached 95–99% of the normal value, the high voltage was switched from negative to positive (25 kV), and separation was carried out by CZE.

CZE separation was performed using a 45 mM sodium phosphate buffer (pH 7.3) with 10% methanol by applying a voltage of 25 kV (normal mode) and using fused silica capillaries of 64.5 cm (56 cm effective length) x 75 μm i.d. (internal diameter), with an optical path length of 200 μm. Capillary temperature was kept at 25°C.

9.4.3.3   DETECTION AND QUANTITATION

UV detection by monitoring all sulfonamides at 265 nm was proposed. It should be pointed out that in order to improve method sensitivity, in this example, the authors used a bubble cell capillary with an extended optical path length of 200 μm. Quantitation was carried out by external calibration with using PABA as I.S.

9.4.3.4   RESULTS

The method proposed by Soto-Chinchilla et al. [35] by combining an SPE off-line preconcentration and cleanup procedure with an on-column electrophoretic-based preconcentration method (LVSS), together with the use of an extended path length capillary, provided a significant improvement in sulfonamide LODs, with values ranging from 2.59 to 22.95 μg/L. As an example, the electropherogram of a ground water sample, spiked with all sulfonamides at 50 μg/L, is given in Fig. 9.9. Satisfactory recoveries, with values higher than 89.2% for most of the sulfonamides, were obtained. These results showed the usefulness of the proposed method for the detection and analysis of sulfonamides in environmental water samples.

9.4.4   DETERMINATION OF NON-STEROIDEAL ANTI-INFLAMATORY DRUGS IN RIVER WATER SAMPLES

Non-steroidal anti-inflammatory drugs (NSAIDs) have been use widely as pain relievers and, due to their antipyretic effect, for the treatment of different diseases. The continuous environmental input of such drugs may lead to a relatively long-term concentration and thereby promotes continuous but unnoticed adverse effects on aquatic and terrestrial organisms. Moreover, elimination of acidic pharmaceutical in sewage treatment plants was found to be rather low and consequently sewage effluents are one of the main sources of these compounds and their metabolites [62]. Botello et al. [46] developed an electrokinetic supercharging EKS-CZE method for the analysis of NSAIDs in River water samples, following the procedure described in this section.

9.4.4.1. SAMPLE TREATMENT

In this example, river water samples were processed by LLE. For that purpose, 1 mL of river water (previously filtered through a 0.2 μm nylon membrane filter to eliminate particulate matter) was diluted to a final volume of 10 mL in Milli-Q water, and then 10 mL of dichloromethane were added. The mixture was vortexed and centrifuged at 3000 rpm for 3 min. The organic phase was collected and evaporated to dryness under a gentle stream of nitrogen. The residue was then reconstituted in 1 mL of Milli-Q water adjusted to pH 8.5 with 0.1 M NaOH and transferred to a glass vial for EKS-CZE analysis.

9.4.4.2   ANALYSIS BY EKS-CZE

EKS on-column preconcentration procedure was carried out as follows: first, capillary was rinsed with the LE solution (10 mM sodium tetraborate with 50 mM NaCl and 10% methanol, also used as BGE for CZE), followed by the introduction of a solvent plug of methanol into the capillary at 50 mbar for 3 s. Then, sample injection was carried out at –2 kV for 700 s. Finally, a TE solution (50 mM CHES) was hydrodynamically injected at 50 mbar for 12 s, and then CZE separation was carried out.

CZE separation was performed using a 10 mM sodium tetraborate buffer (pH 8 adjusted with NaOH) with 50 mM sodium chloride and 10% (v/v) methanol solution as BGE, by applying a capillary voltage of –30 kV (reverse polarity), and using fused silica capillaries of 88.5 cm (80 cm effective length) × 50 μm I.D. Capillary temperature was kept at 25°C.

9.4.4.3   DETECTION AND QUANTITATION

UV detection by monitoring all NSAIDs at 214 nm was proposed. Quantitation was carried out by external calibration showing a linear range from 0.3–50 μg/L.

9.4.4.4   RESULTS

Very good results were obtained with the EKS-CZE method proposed by Botello et al. [46] for the analysis of NSAIDs in river water samples. This strategy enhanced detection sensitivity 2000-fold compared with normal hydrodynamic injection, providing LODs down to 0.9 μg/L in river water samples, with satisfactory precision in terms of repeatability (RSD values lower than 7.4%) and reproducibility (RSD values lower than 7.8%). As an example, Fig. 9.10 shows the electropherograms obtained when a river water sample was analyzed by EKS-CZE.

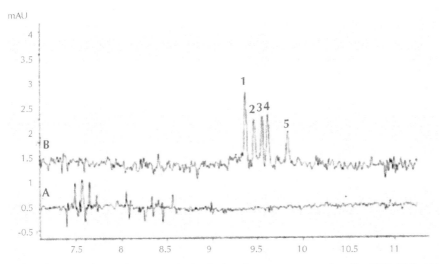

FIGURE 9.10   Electropherogram obtained when river water was analyzed by EKS-CZE. (A) blank river water sample. (B) River water sample spiked at a concentration of 2 μg/L for the five NSAIDs. Peak identification: 1, naproxene; 2, fenoprofen; 3, diclofenac; 4, ketoprofen; 5, piroxicam. Reproduced from Ref. [46] with permission of John Wiley & Sons, Inc.

9.5   SUMMARY AND CONCLUSIONS

Today, CZE is becoming very popular in multiple application fields, such as bioanalysis, food control and safety, and also in environmental applications, because of its high separation efficiency, rapid separations, and low consumption of reagents as well as solvents. In this chapter, an overview of CZE environmental applications has been presented.

Although CZE suffers from relatively high concentration detection limits when working under conventional conditions, and the expected concentration levels of contaminants in environmental samples are very low, CZE is still a good alternative to other separation techniques for the analysis of environmental samples as detection can be improved by combining preconcentration methods (off-line and/or online). Many electrophoretic-based in-line preconcentration methods that play with the sample injection in order to increase the amount of analyte introduced into the capillary without losing separation efficiency are available. Several CZE environmental applications using electrophoretic-based in-line preconcentration methods based on stacking phenomena, such as LVSS, FASI and EKS, have been presented and discussed. Huge sensitivity enhancements (for instance up to 50,000-fold by using EKS) can be achieved with CZE for environmental applications without any especial instrumental requirement.

Today, analysts are playing an important role in exploring multiple possibilities of in-line electrophoretic-based preconcentration methods in CZE by combining existing procedures with new ones, which is making CZE a very promising technique for environmental applications and the number of publications in this field will be increasing in the future.

Finally, some selected CZE environmental application examples have been described from the point of view of sample treatment (extraction and preconcentration procedures), analysis (with or without on-column preconcentration), detection and quantitation, in order to give a complete picture of the diversity of environmental applications dealing with CZE.

KEYWORDS

- **Drugs**
- **Electrokinetic supercharging**
- **Field amplified sample injection**
- **Inorganics**
- **Large volume sample stacking**
- **On-line preconcentration**
- **Organics**
- **Selected applications for soil and water**

REFERENCES

1. Altria, K. D., & Bryant, S. M. (1997). Reader Survey: The Current Status of Capillary Electrophoresis. LC-GC, *15*, 448, 450, 452, 454.
2. Masetto deGaitani, C., Moraes de Oliveira, A. R., & Bonato, P. S. (2013). Capillary Electromigration Techniques for the Analysis of Drugs and Metabolites in Biological Matrices, a Critical Appraisal, Capillary Electrophor, Microchip Capillary Electrophor, 229–245.
3. Locatelli, M., & Carlucci, G. Advanced Capillary Electrophoresis Techniques in the Analytical Quantification of Drugs, Metabolites and Biomarkers in Biological Samples, Global J. Anal. Chem, *1*, 244–261.
4. Tseng, H. M., Li, Y., & Barrett, D. A. (2010). Bioanalytical Applications of Capillary Electrophoresis with Laser-Induced Native Fluorescence Detection, Bioanalysis, *2*, 1641–1653.
5. Garcia-Campana, A. M., Gamiz-Gracia, L., Lara, F. J., Iruela, M. D. O., & Cruces-Blanco, C. (2009). Applications of Capillary Electrophoresis to the Determination of Antibiotics in Food and Environmental Samples, Anal Bioanal Chem, *395*, 967–986.
6. Pinero, M. Y., Bauza, R., & Arce, L. (2011). Thirty Years of Capillary Electrophoresis in Food Analysis Laboratories: Potential Applications, Electrophoresis, *32*, 1379–1393.
7. Herrero, M., Garcia-Canas, V., Simo, C., & Cifuentes, A. (2010). Recent Advances in the Application of Capillary Electromigration Methods for Food Analysis and Foodomics, Electrophoresis, *31*, 205–228.

8. Bald, E., Kubalczyk, P., Studzinska, S., Dziubakiewicz, E., & Buszewski, B. (2013). Application of Electromigration Techniques in Environmental Analysis Springer Ser, Chem. Phys, *105*, 335–353.
9. Nickerson, B., & Jorgenson, J. W. (1988). High Sensitivity Laser induced Fluorescence Detection in Capillary Zone Electrophoresis, HRC CC, *J. High Resolution*, Chromatogr Chromatogr, Commun, *11*, 878–881.
10. Nickerson, B., & Jorgenson, J. W. (1988). High Speed Capillary Zone Electrophoresis with Laser Induced Fluorescence Detection, HRC CC, *J. High Resolut,* Chromatogr Chromatogr Commun, *11*, 533–534.
11. Brumley, W. C., Grange, A. H., Kelliher, V., Patterson, D. B., Montcalm, A., Glassman, J., & Farley, J. W. (2000). Environmental Screening of Acidic Compounds Based on Capillary Zone Electrophoresis/Laser-Induced Fluorescence Detection with Identification by Gas Chromatography/Mass Spectrometry and Gas Chromatography/High-Resolution Mass Spectrometry, *J. AOAC Int, 83*, 1059–1067.
12. Cai, J., & Henion, J. (1995). Capillary Electrophoresis Mass Spectrometry, *J. Chromatogr. A*, *703*, 667–692.
13. Schmitt-kopplin, P., & Frommberger, M. (2003). Capillary Electrophoresis-Mass Spectrometry, *15* Years of Developments and Applications, Electrophoresis, *24*, 3837–3867.
14. Robledo, V. R., & Smyth, W. F. (2009). The Application of CE-MS in the Trace Analysis of Environmental Pollutants and Food Contaminants, Electrophoresis, *30*, 1647–1660.
15. Hempel, G. (2000). Strategies to improve the Sensitivity in Capillary Electrophoresis for the Analysis of Drugs in Biological Fluids, Electrophoresis, *21*, 691–698.
16. Lucci, P., Pacetti, D., Núñez, O., & Frega, N. G. (2012). Current Trends in Sample Treatment Techniques for Environmental and Food Analysis. In *Chromatography: The Most Versatile Method of Chemical Analysis,* ed. Calderon, L. A., In Tech Publisher, Rijeka (CR), 127–164, ISBN: 978–953–51–0813–9.
17. Qin, W., Wei, H., & Li, S. F. Y. (2002). Determination of Acidic Herbicides in Surface Water by Solid-Phase Extraction followed by Capillary Zone Electrophoresis, *J. Chromatogr. Sci.,* *40*, 387–391.
18. Pedersen-Bjergaard, S., Rasmussen, K. E., & Gronhaug Halvorsen, T. (2000). Liquid-Liquid Extraction procedures for Sample Enrichment in Capillary Zone Electrophoresis, *J. Chromatogr A, 902*, 91–105.
19. Hagberg, J., Dahlen, J., Karlsson, S., & Allard, B. (2000). Application of Capillary Zone Electrophoresis for the Analysis of Low Molecular Weight Organic Acids in Environmental Samples, *Int. J. Environ. Anal Chem*, *78*, 385–396.
20. Fang, L., & Xu, X. (2000). Capillary Electrophoretic Separation and Determination of Chlorophenolic Pollutants in Industrial Waste Waters, *Int. J. Environ. Anal Chem, 77*, 29–38.
21. Perez-Urquiza, M., Ferrer, R., & Beltran, J. L. (2000). Determination of Sulfonated Azo Dyes in River Water Samples by Capillary Zone Electrophoresis, *J. Chromatogr A, 883*, 277–283.
22. Lopez-Avila, V., Van de Goor, T., Gas, B., & Coufal, P. (2003). Separation of Haloacetic Acids in Water by Capillary Zone Electrophoresis with Direct UV Detection and Contactless Conductivity Detection, *J. Chromatogr, A, 993*, 143–152.
23. Kronholm, J., Revilla-Ruiz, P., Porras, S. P., Hartonen, K., Carabias-Martinez, R., & Riekkola, M. L. (2004). Comparison of Gas Chromatography Mass Spectrometry and Capillary Electrophoresis in Analysis of Phenolic Compounds Extracted from Solid Matrices with Pressurized hot Water, *J. Chromatogr. A, 1022*, 9–16.
24. Fukushi, K., Ito, H., Kimura, K., Yokota, K., Saito, K., Chayama, K., Takeda, S., & Wakida, S. I. (2006). Determination of Ammonium in River Water and Sewage Samples by Capillary Zone Electrophoresis with Direct UV Detection, *J. Chromatogr A, 1106*, 61–66.

25. Komarova, N. V., & Kartsova, L. A. (2002). Optimizing Separation conditions for Chloro-phenoxycarboxylic Acid Herbicides in Natural and Potable Water using Capillary zone Elec-trophoresis, *J. Anal* Chem, *57*, 644–650.

26. Wojcik, L., Korczak, K., Szostek, B., & Trojanowicz, M. (2006). Separation and Determina-tion of Perfluorinated Carboxylic Acids using Capillary Zone Electrophoresis with Indirect Photometric Detection, *J. Chromatogr* A, *1128*, 290–297.

27. Vasas, G., Szydlowska, D., Gaspar, A., Welker, M., Trojanowicz, M., & Borbely, G. (2006). Determination of Microcystins in Environmental Samples using Capillary Electrophoresis, *J. Biochem* Biophys Methods, *66*, 87–97.

28. Li, J., Lu, W., Ma, J., & Chen, L. (2011). Determination of Mercury (II) in Water Samples using Dispersive Liquid-Liquid Microextraction and Back Extraction Along with Capillary Zone Electrophoresis, Microchim. Acta, *175*, 301–308.

29. Zhong, S., Tan, S. N., Ge, L., Wang, W., & Chen, J. (2011). Determination of Bisphenol A and Naphthols in River Water Samples by Capillary Zone Electrophoresis after Cloud Point Extraction, Talanta, *85*, 488–492.

30. Amelin, V. G., Bol'shakov, D. S., & Tretiakov, A. V. (2012). Determination of Glyphosate and Aminomethylphosphonic Acid in Surface Water and Vegetable Oil by Capillary Zone Electro-phoresis, *J. Anal* Chem, *67*, 386–391.

31. Paull, B., & King, M. (2003). Quantitative Capillary Zone Electrophoresis of Inorganic An-ions, Electrophoresis, *24*, 1892–1934.

32. Timerbaev, A. R., & Fukushi, K. (2003). Analysis of Seawater and Different Highly Saline Natural Waters by Capillary Zone Electrophoresis, Mar. Chem, *82*, 221–238.

33. Breadmore, M. C., & Haddad, P. R. (2001). Approaches to Enhancing the Sensitivity of Capil-lary Electrophoresis Methods for the determination of Inorganic and Small Organic Anions, Electrophoresis, *22*, 2464–2489.

34. Cugat, M. J., Borrull, F., & Calull, M. (2001). Large-volume Sample Stacking for On Capil-lary Sample Enrichment in the Determination of Naphthalene and Benzenesulfonates in Real Water Samples by Capillary Zone Electrophoresis, Analyst (Cambridge, UK), *126*, 1312–1317.

35. Soto-Chinchilla, J. J., Garcia-Campana, A. M., Gamiz-Gracia, L., & Cruces-Blanco, C. (2006). Application of Capillary Zone Electrophoresis with Large-volume Sample Stacking to the Sensitive Determination of Sulfonamides in Meat and Ground Water, Electrophoresis, *27*, 4060–4068.

36. Quesada-Molina, C., Garcia-Campana, A. M., Olmo-Iruela, L., & Del Olmo, M. (2007). Large Volume Sample Stacking in Capillary Zone Electrophoresis for the Monitoring of the Degra-dation Products of Metribuzin in Environmental Samples, *J. Chromatogr* A, *1164*, 320–328.

37. Quesada-Molina, C., Olmo-Iruela, M., & Garcia-Campana, A. M. (2010). Trace Determina-tion of Sulfonylurea Herbicides in Water and Grape Samples by Capillary Zone Electrophore-sis using Large Volume Sample Stacking, Anal. Bioanal Chem, *397*, 2593–2601.

38. Herrera-Herrera, A. V., Ravelo-Perez, L. M., Hernandez-Borges, J., Afonso, M. M., Palen-zuela, J. A., & Rodriguez-Delgado, M. A. (2011). Oxidized Multiwalled Carbon Nanotubes for the Dispersive Solid-phase Extraction of Quinolone Antibiotics from Water Samples us-ing Capillary Electrophoresis and Large volume sample Stacking with Polarity Switching, *J. Chromatogr.* A, *1218*, 5352–5361.

39. Bernad, J. O., Damascelli, A., Nunez, O., & Galceran, M. T. (2011). Inline Pre-concentration Capillary Zone Electrophoresis for the Analysis of Halo acetic Acids in Water, Electrophore-sis, *32*, 2123–2130.

40. Liu, S., Wang, W., Chen, J., & Sun, J. (2012). Determination of Aniline and Its Derivatives in Environmental Water by Capillary Electrophoresis with On-Line Concentration, *Int. J. Mol. Sci.*, *13*, 6863–6872.

41. Liu, Q., Liu, Y., Guan, Y., & Jia, L. (2009). Comparison of Field Enhanced and Pressure As-sisted Field Enhanced Sample Injection Techniques for the Analysis of Water-Soluble Vita-mins using CZE, *J. Sep. Sci.*, *32*, 1011–1017.

42. Okamoto, H., & Hirokawa, T. (2003). Application of Electrokinetic Supercharging Capillary Zone Electrophoresis to Rare-Earth Ore Samples, *J. Chromatogr., A*, *990*, 335–341.

43. Hirokawa, T., Okamoto, H., & Gas, B. (2003). High-Sensitive Capillary Zone Electrophoresis Analysis by Electrokinetic Injection with Transient Isotachophoretic Pre-concentration: Elec-trokinetic Supercharging, Electrophoresis, *24*, 498–504.

44. Dawod, M., Breadmore, M. C., Guijt, R. M., & Haddad, P. R. (2008). Electrokinetic Super-charging for Online Pre-concentration of Seven Non-Steroidal Anti-inflammatory Drugs in Water Samples, *J Chromatogr* A, *1189*, 278–284.

45. Dawod, M., Breadmore, M. C., Guijt, R. M., & Haddad, P. R. (2009). Counter-Flow Electro-kinetic Supercharging for the Determination of Non-Steroidal Anti-Inflammatory Drugs in Water Samples, *J. Chromatogr.* A, *1216*, 3380–3386.

46. Botello, I., Borrull, F., Aguilar, C., & Calull, M. (2010). Electrokinetic Supercharging Focus-ing in Capillary Zone Electrophoresis of weakly Ionizable Analytes in Environmental and Biological Samples, Electrophoresis, *31*, 2964–2973.

47. Meighan, M. M., Dawod, M., Guijt, R. M., Hayes, M. A., & Breadmore, M. C. (2011). Pres-sure-Assisted Electrokinetic Supercharging for the Enhancement of Non-Steroidal Anti-in-flammatory Drugs, *J. Chromatogr* A, *1218*, 6750–6755.

48. Dawod, M., Breadmore, M. C., Guijt, R. M., & Haddad, P. R. (2010). Electrokinetic Super-charging-Electrospray Ionization Mass Spectrometry for Separation and Online Pre concen-tration of Hypolipidaemic Drugs in Water Samples, Electrophoresis, *31*, 1184–1193.

49. Lu, Y., & Breadmore, M. C. (2010). Analysis of Phenolic Acids by Non-Aqueous Capillary Electrophoresis after Electrokinetic Supercharging, *J. Chromatogr* A, *1217*, 7282–7287.

50. Lu, Y., & Breadmore, M. C. (2010). Fast Analysis of Phenolic Acids by Electrokinetic Super-charging-Nonaqueous Capillary Electrophoresis, *J. Sep. Sci.*, *33*, 2140–2144.

51. Ning, Z., Sui, L., Zhong, H., Li, Y., & Li, R. (2012). Analysis of Achromatic Acids in River Water by Non-aqueous Capillary Electrophoresis with Electro kinetic Supercharging, *Asian J. Chem*, *24*, 805–808.

52. Johnson, R. M., & Pepperman, A. B. (1995). Analysis of Metribuzin and Associated Metabo-lites in Soil and Water Samples by Solid Phase Extraction and Reversed phase Thin Layer Chromatography, *J. Liq.* Chromatogr, *18*, 739–753.

53. Lawrence, J. R., Eldan, M., & Sonzogni, W. C. (1993). Metribuzin and Metabolites in Wiscon-sin (USA) Well Water, Water Res, *27*, 1263–1268.

54. Parker, C. E., Geeson, A. V., Games, D. E., Ramsey, E. D., Abusteit, E. O., Corbin, F. T., & Tomer, K. B. (1988). Thermospray Liquid Chromatographic-Mass Spectrometric Method for the Analysis of Metribuzin and its Metabolites, *J. Chromatogr, 438*, 359–367.

55. Nieto, A., Borrull, F., Marce, R. M., & Pocurull, E. (2008). Pressurized Liquid Extraction of Contaminants from Environmental Samples, Curr Analaysis Chem, *4*, 157–167.

56. Collasiol, A., Pozebon, D., & Maia, S. M. (2004). Ultrasound Assisted Mercury Extraction from Soil and Sediment, Anal. Chim Acta, *518*, 157–164.

57. Rezaee, M., Assadi, Y., Milani Hosseini, M. R., Aghaee, E., Ahmadi, F., & Berijani, S. (2006). Determination of Organic compounds in Water using dispersive Liquid-Liquid Micro extrac-tion, *J. Chromatogr.* A., *1116*, 1–9.

58. Yamini, Y., Rezaee, M., Khanchi, A., Faraji, M., & Saleh, A. (2010). Dispersive Liquid-Liquid Microextraction Based on the Solidification of Floating Organic Drop followed by Inductively Coupled Plasma-Optical Emission Spectrometry as a Fast Technique for the Simultaneous Determination of Heavy Metals. *J. Chromatogr.* A, *1217*, 2358–2364.

59. Herrera-Herrera, A. V., Hernandez-Borges, J., Borges-Miquel, T. M., & Rodriguez-Delgado, M. A. (2010). Dispersive Liquid-Liquid Micro extraction Combined with Nonaqueous Capillary Electrophoresis for the Determination of Fluoroquinolone Antibiotics in Waters, Electrophoresis, *31*, 3457–3465.
60. Yin, X. B. (2007). Dual-Cloud Point Extraction as a Preconcentration and Clean-up Technique for Capillary Electrophoresis Speciation Analysis of Mercury, *J. Chromatogr.* A, *1154*, 437–443.
61. Diaz-Cruz, M. S., Lopez De Alda, M. J., & Barcelo, D. (2003). Environmental behavior and Analysis of Veterinary and Human Drugs in Soils, Sediments and Sludge, TrAC, Trends Anal, Chem, *22*, 340–351.
62. Macia, A., Borrull, F., Calull, M., & Aguilar, C. (2007). Capillary Electrophoresis for the Analysis of Non-Steroidal Anti-Inflammatory Drugs, TrAC, Trends Anal Chem, *26*, 133–153.

CHAPTER 10

FLOW-INJECTION ANALYSIS: PRINCIPLES AND APPLICATIONS

NABIL A. FAKHRE

Deptartment of Chemistry, College of Education,University of Salahalddin, Erbil, Iraq.
E-mail: nabil_fakhri@yahoo.com, havras@yahoo.com

CONTENTS

10.1   INTRODUCTION

Flow-injection analysis (FIA) may be defined as an automated or semiautomated analytical process consisting of a sequential insertion of discrete sample solutions into an unsegmented continuously flowing liquid stream with subsequent detection of the analyte. It is a relatively new analytical process, which shows considerable potential for high-speed precise analysis of discrete samples [1]. This definition, however, was soon considered obsolete and was revised to describe a technique for "information gathering from a concentration gradient formed from an injected, well-defined zone of a fluid, dispersed into a continuous unsegmented stream of a carrier" in order to accommodate new developments in stopped-flow FIA, merging zones, zone sampling and other gradient techniques. This new definition was soon challenged by FI systems which were segmented in one way or another, or which dealt with samples eluted from columns without well-defined boundaries. Furthermore, FIA defined as "A flow analysis technique performed by reproducibly manipulating sample and reagent zones in a flow stream under thermodynamically nonequilibrated conditions." The main important parts of a flow-injection analytical system are a unit for propelling liquids through the manifold system for introducing a sample into a continuously flowing stream of a carrier, an appropriate detector [2], and a transport system linking various elements which make up the FIA system and allowing the sample to attain a suitable degree of dispersion or mixing as it travels through it. When the extent of dispersion is not suitable for the experiment concerned and a reaction or further splitting of the flowing stream is required, the system can be supplemented with accessories such as mixing chambers, reactors and merging points (Fig. 10.1) [3]. In the case of the FIA technique the physical equilibrium (flow homogenization) is never reached at the moment of detection. Moreover, it is not necessary for the chemical equilibrium to be obtained at the moment of detection. The concept of FIA depends on a combination of three factors: reproducible sample injection volumes, controllable sample dispersion, and reproducible timing of the injected sample through the flow system. Except for detector warm-up, the system is ready for instant operation as soon as the sample is introduced. FIA offers several advantages in term of considerable decrease in sample (normally using 10 to 50 µL) and reagent consumption, high sample throughput (50 to 300 samples per hour) reduced residence times (reading time is about 3 to 40 s), shorter reaction times (3 to 60 s), easy switching from one analysis to another (manifolds are easily assembled and/or exchanged), reproducibility (usually less than 2% RSD), reliability, low carry over, high degree of flexibility, and ease of automation. Perhaps the most compelling advantage of the FIA technique is the great reproducibility in the results obtained by this technique that can be set up without excessive difficulties and at very low cost of investment and maintenance. These advantages have led to an extraordinary development of FIA, unprecedented in comparison to any other technique.

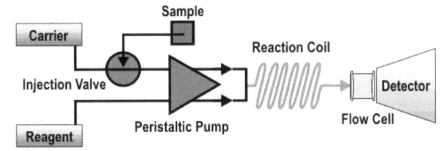

FIGURE 10.1   Schematic diagram of the basic FI system.

10.2   SOME DEFINITIONS

The principle of FIA is deceptively simple, being based on the injection of a definite volume of a liquid sample into a moving, nonsegmented continuous carrier stream of suitable liquid (Fig. 10.2a). The injected sample forms a zone which beings dispersing and reacting with the carrier stream as it is transported towards a detector. A continuous record of the absorbance, electrode potential, or any other physical parameter is made as that parameter continuously changes during passage of the sample material through the flow cell. A typical recorder output has the form of the peak (Fig. 10.2b), the height of which is related to the concentration of the analyte. As the residence time (T) of the sample in the FIA system normally is less than 30 sec, at least two samples can be analyzed per minute. The injected volume may be between 1 mL and 200 mL, which in turn usually requires no more than half a milliliter of reagent per analysis. This makes FIA an automated microchemical technique capable of a sampling rate of at least 100 determinations per hour, with minimum reagent consumption [4].

Peak height, H, which is related to the concentration of the component determined in the injected sample. The peak area, which can be used instead, requires employing an integrator analogous to those used in gas chromatography.

Residence time, T, which is defined as the span elapsed from injection until the maximum signal is attained.

Travel time, ta, which is the period elapsed from injection to the start of the signal (1–2% increase above baseline). The difference between the two parameters, t'=T–ta, is usually very small, by virtue of the characteristics of FIA curves.

Return time, T", which is the period between the appearance of the maximum signal and the return to the baseline.

Baseline-to-baseline, Δt, defined as the interval between the start of the signal and its return to the baseline [5].

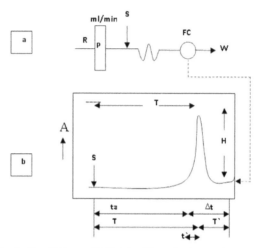

FIGURE 10.2   Single-line FIA manifold (a) with typical recorder output, (b) as obtained with a spectrophotometric flow-through cell. R, carrier stream of a reagent; P, pump; S, sample injection; FC, flow-through cell; W, Waste; H, peak height; and T, residence time.

10.3   HISTORICAL OUTLINE

When a new technique is introduced into the analytical chemistry community, it is usually the case that process is the sum of many individual concepts developed by different investigations. This has been the case with flow-injection analysis [1].

The seeds for development of FIA were presented by the end of 1959. The concept of a sample being injected into continuously flowing stream with continuous recording downstream is inherent in the basic concepts of gas chromatography by James and Martin in 1952. In 1970 Nagy, Feher and Paunger reported on the use of graphic electrodes for the voltametric measurement of samples injected into continuously flowing stream [6]. This is earliest report, which found of what today would be called flow-injected analysis.

The concept of FIA was developed by Ruzicka and Hansen in 1975, they described their concepts of unsegment continuous-flow analysis as an adjunct to the segmented continuously flowing systems [1].

The broad acceptance of FIA is undoubtedly due to its versatility, which allows the method to be used in conjunction with a wide variety of detectors and analytical techniques, and for the assay of a multitude of organic and inorganic substances [7].

10.4   DISPERSION IN THE FIA

The control of dispersion is the most important aspect of FI systems. The dispersion of a fluid zone reproducibly introduced into a nonsegmented flow stream (carrier) during transport of the zone to the detector is the most important physical phenomenon in all FI systems. The specific feature of dispersion processes in FIA is that they are reproducible and controllable through the manipulation of flow parameters and geometrical dimensions of flow conduits. The driving forces active in dispersion of the injected zone into the carrier stream are molecular diffusion and convection, but the effects of convection dominate, and the effects of molecular diffusion may be neglected in most cases. Convection occurs both as result of linear flow-rate differences of fluid elements located at different points along the radial axis of the conduit and as a result of secondary flows created by centrifugal forces perpendicular to the flow direction in nonstraight conduits. A convex parabolic front of the injected zone and a concave parabolic tailing edge are developed with penetration into the carrier stream, the extent increasing with the distance traveled. Thus, under the specific conditions applied in FIA and with a fixed conduit, the acting forces are well under control, so that no random turbulence occurs. The result is that perfectly reproducible concentration time relationships may be obtained which, when recorded and superimposed, precisely overlap each other to form a single curve. This provides the basis for extracting reproducible readout under both physically and chemically nonequilibrium conditions. The dispersion process typical of FIA system is shown in Fig. 10.3.

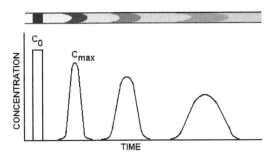

FIGURE 10.3   The dispersion process typical of FIA system.

The injected fluid zones in a nonsegmented flow stream can be manipulated reproducibly to produce various degrees of dispersion. In order to provide a quantitative criterion evaluating the extent of dispersion, the term dispersion coefficient (D) was introduced that being defined as the ratio of the concentration of the constituent of interest in a fluid element of the injected zone before and after dispersion, expressed by:

$$D = C_0/C \qquad (11)$$

where, C_0 is the original concentration of the constituent in the solution before dispersion, and C is the concentration of that fluid element of the dispersed fluid zone from which analytical readout is extracted. When the fluid element with the highest concentration is used for readout, Eq. (2) is expressed as:

$$D = C_0/C_{max} \qquad\qquad (2)$$

where, C_{max} is the concentration of the constituent at peak maximum. D is a dimensionless value, which is equivalent to the dilution factor of the fluid element under consideration. For example, if the sample is diluted 1:1 by carrier, thus the dispersion coefficient is 2. FI systems are categorized into high, medium, and low dispersion systems depending on the degree of dispersion of the injected zone at the read out point. Systems with D above 10 are classified as high, those between 2 and 10 as medium, and those below 2 as low dispersion system. The main experimental parameters influencing the dispersion of an injected fluid zone include sample volume, flow rate of carrier and merging fluid streams, geometrical dimensions and configuration of transport conduits and on line reactor, and pattern of flow segmentation in system with two immiscible phases. The volume of the injected fluid zone, which most cases is the sample, is important factor influencing its dispersion. The dispersion decreases with an increase in sample volume. Ruzicka and Hansen stated that dispersion diminishes with a decrease in flow rate. This happens because decreasing flow rates increase the retention time of the sample awaiting transport to the detector. In this phase the reaction between sample and reagent almost reaches the equilibrium. Hence, the peak signal will be higher in a slower flow rate. Fang said that those conditions are only valid at extremely low flow rate where the rate of molecular diffusion approaches that of convection. This has been experimentally demonstrated by Karlberg and Pacey: with a fixed manifold, dispersion is minimally influenced by flow rate variations within a wide range of 1.6–4.0 mL/min. The geometrical dimensions and configurations of components constituting a FI manifold are important factors influencing the dispersion. The dispersion of the sample zone increases with the square root of the distance traveled through an open narrow tube. This rule is valid only for straight conduits. When the conduits are coiled for the sake of tidiness or knotted to improve radial mixing, the intensity of dispersion is decreased to different degrees, depending on the radius of the coil or knots. This is due to the generation of secondary flows, which limit the axial dispersion while promoting radial dispersion. The dispersion of injected zones is enhanced with increases in the inner diameter of the conduit. No generally applicable quantified relationships are available, however, owing to the complexity of the influences from other parameters.

10.5   DETECTION IN THE FIA

The usual location of a flow injection manifold is on-line and placed before a detector system in order that proper monitoring of the species is performed after it has been subjected to the step developed in a continuous fashion. While the interface between an flow injection (FI) system and a conventional detector is a commercial or laboratory-made flow cell (or simple aspiration to the flame in the case of atomic techniques assisted by this source), the interface to the high resolution detector has a decisive influence on the performance of the hyphenated system, as analytical quality parameters such as reproducibility, accuracy, sensitivity and selectivity are highly dependent on how the coupling is accomplished.

The complexity of the interface is very different depending on whether the measurement is performed in solution, plasma, or vacuum. The pre or postcolumn coupling of FI to LC (usually HPLC) depends on the pursued objective: a precolumn position is mainly used for implementing a continuous separation step prior to chromatographic individual separation and, in a smaller extension, for developing precolumn derivatizing reactions, meanwhile a postcolumn position is most often used for derivatizing purposes. Precolumn (FI-HPLC) assemblies in which the FI manifold includes a separation unit have been mainly devoted to the use of microcolumn with the following objectives:

1. Trace preconcentration, using different materials such as ion exchangers, ion-pairing reagents, ligand-exchangers or size exclusion gels, which provide preconcentration factors up to 10,000;

2. Sample cleanup, by taking advantage of the differences in the interaction between the components of a given sample and the sorbent (e.g., weakly retained phenol compounds or acid substances can be readily separated from phthalate esters strongly adsorbed on a non polar sorbent);

3. Sample storage, by taking advantage of the relatively inert character of many sorbent materials. This is of special interest when samples have to be collected in remote places;

4. Protection of the analytical system, as the solid-phase microcolumn acts as a protective filter, lengthening the usable lifetime of the separation unit;

5. Pre-column derivatization, by using sorbent impregnated with the reagent, a solid redox agent, or a support for retention of a (bio) catalyst. Also other separation steps such as liquid-liquid extraction, membrane extraction, and dialysis have been coupled to HPLC through a FI manifold. The solution containing the analytes once separated from the matrix is injected into the chromatograph. Inexpensive flow-cells (either conventional or demountable microflow cells) with KBr windows and different thickness spacers (0.015–0.22 mm, 0.159 µL volume) have been used in FI-FTIR coupling when organic solvent carriers are involved.

Special attention has been paid to interfacing FI systems and Inductively Coupled Plasma (ICP). A general interface to introduce a liquid into plasma consists of a nebulizer, a spray chamber and a separator. The main shortcomings of using conventional interfaces in FI-ICP couplings related to continuous sample aspiration are the large dead volume and the sample loss involved, as well as the band broadening.

In dealing with conventional spectrometry detectors, one of the most clear integrating effects is achieved when chemiluminescence reactions are involved, as the sample-reagent mixture reaches the luminescence detector in a time as short as necessary for proper detection of the transient emission from the deactivation of the product. Spectrometer array detectors enable more information to be collected faster from a sample injected into the FI manifold, which can be used for improving precision, for enlarging the linear range of the calibration curve, and for multi determinations, among the most important applications. The faster response of charge-coupled detectors also allows the acquisition of chemiluminescence spectra profiles. Lasers provide an excellent excitation source for the typical small volumes in FI.

Atomic spectrometry, either those providing single or multi information, have been widely coupled to FI. Moreover, the multiple aspects of this integration have been exhaustively explained elsewhere. Some examples of: (a) the use of FI coupled to laser-enhanced ionization detection, either for reducing electrical interference or providing optimum dilution levels in matrix interfered determinations and for on-line separation concentration prior to laser assistance; and (b) flow injection for arsenic and antimony hydride generation prior to glow-discharge atomic emission spectrometry detection and for the elimination of the matrix effect in a preconcentration step before flame atomic absorption spectrometry.

Electroanalytical techniques are even more prolific in innovations due to the wider possibilities for the design of probe-sensors that can easily be converted into flow-through sensors through insertion in the appropriate flow cell. Potentiometric electrode arrays have been coupled to FI manifolds with the aim either of improving the determination of a single analyte or simultaneous measurements. Both voltammetric and amperometric measurements have been implemented in FI from the very beginning of the technique. Both conventional and new excitation modes have been applied to both commercial unmodified and chemically modified electrodes. The conventional FI methods have been proposed for the determination of a number of analytes.

10.6   SOLVENT EXTRACTION WITH FIA

Manual liquid-liquid extraction procedures are usually very tedious, involving a large consumption of solvents and chemicals, and are subject to potential contaminants from the atmosphere and chemical glassware. In, addition, the conventional liquid-liquid extraction process requires manipulations with significant volumes of hazardous and / or toxic organic solvents [8]. Therefore, liquid-liquid extraction has been semiautomated or automated by several workers, principally using air-segmented flow systems. The earlier studies of the utility of liquid-liquid extraction in FIA were simultaneously carried out by Karlberg and Bergamin and their co-workers [9, 10] in 1978.

The rapid development of automated liquid-liquid extraction is perhaps due to the broad use and importance of the liquid-liquid extraction process. One of most effective way to shorten the duration of this process has been the construction of dynamic on-line liquid-liquid extraction systems applying the principles of continuous flow analysis [8].

Since its introduction in 1978, solvent extraction-flow-injection has been applied to a wide variety of analytical applications, and many papers have been published describing the use of FIA systems for extraction.

10.6.1 PRINCIPLES OF LIQUID-LIQUID EXTRACTION FIA

Regardless of the way in which the liquid-liquid extraction step is performed via a manual batch procedure or by use of some kind of mechanized or automated system. Especially, solvent extraction FIA involves four operations: (i) the injection of the sample into the aqueous phase; (ii) the mixing of immiscible organic and aqueous phases to form a segmented stream; (iii) the distribution of sample components between the two phases; and (iv) the separation of one or both phases for determination [11].

The aqueous stream of an extractable component is segmented with an organic immiscible solvent stream at the segmenter mixing point, where more or less reproducible droplets of one phase in the other are formed.

The droplets move into the outflow channel after having been formed and tend to minimize their interfacial area with the other phase and to maximize the contact surface area with the wall material of the outflow tubing, thereby wetting it. This process results in the formation of independent, more or less regular segments of both phases in a single moving stream, which then enters the extraction coil.

The extraction process occurs principally in the extraction coil, and to a lesser extent in the segmenter and the phase separator.

The segments of the aqueous and organic phases are subsequently separated in a phase separator into individual streams. The extractable analyte in the receiving phase is determined using a flow-through detection system. The analytical signals are treated in a conventional manner, with the analyte concentration calculated from the peak height, the peak width, or the peak area [8].

10.6.2 BASIC COMPONENTS OF LIQUID-LIQUID EXTRACTION FIA SYSTEMS

The principal operations of liquid-liquid extraction FIA also characterize the three basic components of liquid-liquid extraction FIA systems:

10.6.2.1 SEGMENTERS

Phase segmentation involves dividing the continuous flow of the organic and the aqueous phase into one uniform stream with alternating segments. The immiscible phases are brought together in a narrow tube in a controlled manner so that defined segments of each phase are formed.

Several segmenter types of varying efficiency have been desired in the literature. The most common segmenter types are T-piece segmenters mode of glass [12, 13], stainless steel [14] or glass-lined T-pieces of stainless steel, and combinations of hydrophobic and hydrophilic materials [15, 16].

10.6.2.2 EXTRACTION COILS

There are two principal considerations to be made when choosing an extraction coil: the material the coil is made and the coil dimensions. With respect to the former, the question is whether to use hydrophilic (glass or metal capillary) or lipophilic (fluoroplastics) material. The choice of which coil material to use depends on whether the sample will be extracted from the aqueous into the organic phase or vice versa. The second consideration, concerning the coil dimensions, can affect sample dispersion and extraction efficiency, as related to kinetic efficiency, total extraction yield, and peak broadening. The liquid-liquid extraction process requires that the segmented phases remain in contact while the analyte approaches a state of thermodynamic equilibrium in partitioning between the two phases [8].

10.6.2.3 PHASE SEPARATORS

The phase separation process involves a partitioning of the segment phases after the extraction has been completed in the extraction coil, in such a manner that the unwanted phase is directed to waste while the phase (s) is resampled or pumped through the detection system. In most practical separators, the two-phase system cannot be desegmented totally into two pure individual phases. Typically, phase separation efficiency is 80 to 100% [8].

The current phase separators fall into one of three classes: (i) those based on difference in the density of the two immiscible phases; (ii) those in which solvent affinities with either hydrophobic or hydrophilic inserts enhance the efficiency of the density phase separators; and (iii) membrane phase separators in which a hydrophobic membrane excludes water and a hydrophilic membrane excludes the organic phase from the stream fed to the detector [11].

10.7   SEPARATION AND PRECONCENTRATION IN FIA

Developments in analytical instrumentation allow trace and ultra-trace analysis in diverse kinds of samples. Despite these advances it is still often necessary to use separation and preconcentration procedures prior to detection. Preconcentration steps aim at reducing the limits of detection of the existing analytical techniques by removing interferences and/or increasing the concentration of the species of interest. Conventional preconcentration techniques, such as ion exchange, adsorption, extraction, coprecipitation, a.o., when operated in the batch mode, are time-consuming, labor-intensive, require large sample and reagent volumes and suffer great risks of contamination and analyte loss. With on-line operation using FI techniques the drawbacks of batch-wise operation can be overcome to a great extent and currently on-line preconcentration may be achieved almost as efficiently as a simple AS determination, both in terms of sample throughput and reagent consumption. In fact, up to now the most dramatic improvements achieved in FI-AS have been in the field of on-line preconcentration A number of criteria can be used to better compare the efficiency of the different techniques and procedures. The most frequently used are: (i) the sample throughput, (ii) the sample consumption and (iii) the enhancement factor (EF). The latter is defined as the ratio of the concentrations before and after preconcentration. In practice EF is approximated with the ratio of the slopes of the linear sections of the calibration curves before and after preconcentration.

In FI preconcentration systems, samples may be introduced either on a volume basis or on a time basis. With volume-based sample loading, the amount of sample processed is determined by the sample loop and the filled sample is subsequently transferred from the loop by a suitable carrier. This approach is used when injection of a small defined volume of sample is required. With time-based loading, the amount of sample processed is determined by the sampling flow rate and the sampling time. The time-based loading is often used in separation and preconcentration FI manifolds as it is less time-consuming, easy to accomplish and permits handling larger sample volumes. FI system (Fig. 10.4) was used for the determination of histamine. A mini-column filled with amberlite resin (weak cation exchanger) was introduced to the flow system. A 200 μL of the sample was injected into the carrier stream through the injection valve. The merged streams were passed through a quartz flow cell in a spectrophotometer connected to recorder. Under the optimum conditions, the calibration curve was linear in the range 0.02–1.5 μg mL^{-1} of histamine using the peak height as an analytical signal, while the detection limit was 0.01 μg mL^{-1}. The precision and accuracy of the method were studied depending upon the values of the relative standard deviation and relative error percentage. The selectivity of the method was investigated by studying the effect of interference from other species accompanied with histamine in fish meal. Under the optimum conditions, the system was used for on line separation, preconcentration of histamine. The proposed method was applied for the determination of histamine in fish meal. The results were compared with the standard method and a good agreement between the results was obtained [17].

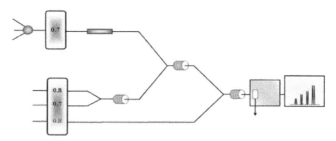

FIGURE 10.4   FI manifold used for separation and preconcentration of histamine, P (pump), S (sample injector), RC (Reaction coils), SV (Selection valve) and W (Waste).

10.8   SEQUENTIAL INJECTION ANALYSIS

Later generations of flow injection analysis technique incorporate many pumps, valves, and tubing to accommodate more complicated chemical reactions that need many reagents. The latest generation, called sequential injection analysis, as shown in Fig. 10.5, has a down scaled system that consumes even smaller volumes of reagents and samples in a few μL levels with the use of a bidirectional syringe pump and multiports selection valve. Reagents and sample can be drawn sequentially and stacked into the mixing coil before mixing while being pushed in reverse direction into the detector. The operational steps from sample introduction, chemical reaction, to detection are fully automated and precisely controlled with computer software. The system can be programmed to stop for a desired period of time; therefore, the study of slow reactions and those that require incubation time such as immunoassay is possible. Accessories such as lab-on-valve (LOV) unit with ports for attaching a fiber optic spectrophotometric detector introduce more areas of applications with real time detection [18–20].

Most research groups have reported that the flow-based systems not only increase sample throughput but also reduce the consumption of sample and reagents. This may be a suitable approach for cases where body fluid/blood samples are limited or need to be divided for various other tests. As compared to most conventional bench top wet chemistry, flow injection requires a lot less sample volume. For example, in titration, sample volume in batch method is in mL whereas in flow-based titration, sample volume injected is in μL [21]. A direct comparison between volumes used can only be made when considering the same analyte and detection methods. Some downscaled batch methods are able to reduce the volume to μL, but in general, FI usually requires relatively less sample volume for a particular analyte or sample being studied. For example, the osmotic fragility test (OFT) of red blood cells normally requires 20 µL of undiluted blood sample in batch spectrometric method whereas only 1.0 µL of undiluted sample is required in the FI system where it is tested in 100-fold dilution [22]. As compared to standard bioassay technique such as ELISA, the volume required by flow-based systems is also usually lower. For example, the assay of hyaluronan in serum using SI required 10 µL of serum sample, as compared to the conventional routine microplate assay that requires 120 µL of serum sample [23].

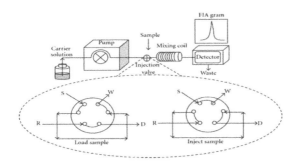

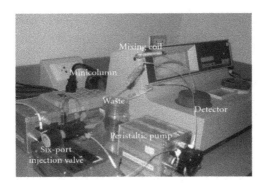

FIGURE 10.5   (a) Diagram of a simple flow injection system (S is sample, W is waste, R is reagent, D is detector) and (b) a picture of a simple flow injection system setup showing a peristaltic pump with pump tubing, a six-port injection valve, a minicolumn chemical reactor, a mixing coil, and a detector.

10.9   MULTISYRINGE FLOW INJECTION ANALYSIS (MSFIA)

Multisyringe flow injection analysis (MSFIA) was introduced by Vector Cerdà and co-workers in 1999 [24] as a robust alternative to its predecessor flow injection techniques, combining the multichannel operation of flow injection analysis [25] with the possibility of flow reversal and selection of the exact volume of sample and reagent required for analysis as presented in sequential injection analysis [26].

Generally, flow injection systems are automation tools where, in opposition to batch conventional assays, physicochemical equilibrium is not attained prior to determination. Hence, flow injection analysis is based in three principles:
1. Reproducible sample injection or insertion in a flowing carrier stream;
2. Controlled dispersion of the sample zone; and
3. Reproducible timing of its movement from the injector point to the detection system.

Since its inception, MSFIA has been the basis for automation of more than 120 different assays, reviewed in several publications [27–30]. This type of automatic flow injection systems is based on the utilization of a multisyringe burette, depicted schematically in Fig. 10.6A and 10.6B. It is a multiple channel piston pump, containing up to four syringes, driven by a single motor of a usual automatic burette and controlled by computer software through a serial port. A two-way commutation valve is connected to the head of each syringe, allowing optional coupling to the manifold lines or to the solution reservoir. Because the four syringes are driven by the same motor, all pistons move at once in the same direction either delivering (dispense operation) or loading the syringes (pickup operation) with liquids. This feature enables that only the necessary amount of reagent solution is introduced into the flow system. Furthermore, when the pistons are moving downwards, it is possible to refill the syringes with solutions present in the respective vessel or to aspirate solutions from the system in order to perform the sampling operation. Syringes with different volumes, ranging from 0.5 to 25 mL are available, enabling the application of a wide range of flow rates. For example, for a 5 mL syringe, flow rates ranging from 0.28 to 15 mL min^{-1} may be attained [31]. Nevertheless, once the flow rate (and volume) is fixed for one syringe, it is also defined for the other channels, and it will depend on the ratio between syringe capacities as different syringes can be placed in any of the four positions. Finally, MSFIA manifolds are not restricted to the syringes and the respective commutation valves. The presence of four digital outputs, each capable of providing 12 V/0.5 A, allows the utilization of up to 12 additional commutation valves, also controlled through the multisyringe apparatus. These extra commutation valves are often necessary to assemble a flow network, where analyte determination and sample treatment can be implemented by including confluences for reagent addition, suitable detectors (spectrophotometers, fluorimeters, flame or atomic emission spectrometers) and devices for mass transfer (gas diffusion or dialysis units), for instance.

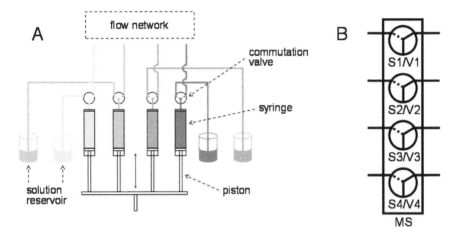

FIGURE 10.6   Schematic representation of multisyringe apparatus, with indication of the different components (A) or simplified (B). Flow management possibilities for one syringe during operation of multisyringe apparatus are also given.

10.10   APPLICATIONS

The concept of flow injection analysis (FIA) was introduced in the mid-seventies. It was preceded by the success of segmented flow analysis, mainly in clinical and environmental analysis. This advance, as well as the development of continuous monitors for process control and environmental monitors, ensured the success of the FIA methodology. As an exceptionally effective means of mechanization for various procedures of wet chemical analysis, the FIA methodology, in use with a whole arsenal of detection methods of modern analytical chemistry, proved to be of great interest to many.

The advantages of flow injection methodology have led to its widespread application in analytical procedures involving less common spectroscopic detection methods. These adaptations usually result in the favorable mechanization of determinations, better precision and a higher sampling rate.

10.10.1   *INORGANIC AND ORGANIC SPECIES*

10.10.1.1   *FORMALDEHYDE IN WATER*

Formaldehyde can be determined in aqueous solution at a rate of 45 samples/h with a small sample requirement (100 μL). The fluorescence of 3,5-diacetyl-1,4-dihydrolutidine formed upon reaction of formaldehyde with ammonium acetate and 2,4-pentanedione (25 s, 95°C) is monitored with a filter fluorometer. The detection limit is 0.1mu. M (3 µg/L) or 10 pmol of HCHO. The response is linear up to 3.3 µM (100 µg/L), the departure from linearity at 0.33 mM is 21%, but high levels are satisfactorily determined with a second-order calibration equation. Interference from S (IV) has been investigated in detail and completely eliminated by addition of H_2O_2 before rendering the sample alkaline. There are no effects from commonly occurring metal ions and anions; the method is very selective to formaldehyde compared to other carbonyl compounds. A S (IV)-containing preservative has been formulated for the stabilization of low concentrations of HCHO. Results are presented for fog water samples [32].

10.10.1.2 DITHIOCARBAMATE PESTICIDES IN SOLID SAMPLES

A flow injection Fourier transform infrared spectrometric procedure has been developed for the determination the dithiocarbamate pesticides Ziram and Thiram in solid samples. All the operations involved, such as extraction, filtration and measurement, were integrated in the experimental set-up in order to avoid excessive manipulation of samples and standards. Ultrasonic assisted and mechanical extraction were evaluated for the solubilization of the analytes and, additionally, the effect of carrier flow rate, sample loop volume and the ratio between sample mass and volume of solvent employed were studied. Quantitative extractions with chloroform were obtained for both Ziram and Thiram, after 5 and 2 min, respectively, of mechanical shaking of sample slurries. Absorbance measurement, in the wave number range of 1600–1460 cm^{-1} for Ziram and 1400–1315 cm^{-1} for Thiram, was carried out, and the area values of the peaks obtained, as a function of time, were interpolated in external calibration lines prepared from standard solutions of Ziram and Thiram in chloroform. Analyzes of commercial formulations and spiked soil samples incubated two weeks were in a good agreement with values found by other methodologies. Absolute detection limits of 400 micrograms for Ziram and 785 micrograms for Thiram and variation coefficients of 6.4% and 2.5% were obtained by use of the aforementioned methodology [33].

10.10.1.3 SPECIATION AND PRECONCENTRATION OF INORGANIC ANTIMONY AND MANGANESE IN WATERS USING MICROCOLUMN-FLOW INJECTION SYSTEM AND DETERMINATION BY ATOMIC ABSORPTION SPECTROMETRY

The purpose of this study is to develop a microcolumn-flow injection system for the determination of Sb(III) and Sb(V) in waters including drinking water. For this purpose, chelating resins and natural and synthetic zeolites were tried as sorbents and their selectivity towards Sb(III) and Sb(V) were examined. The efficiency of the sorbent was studied as a preconcentration agent and then was examined by recovery studies. The samples prepared using the mentioned methodology was analyzed by hydride generation atomic absorption spectrometry (HGAAS) since this technique is one of the most sensitive techniques in Sb determinations. Availability of atomic absorption spectrometry in our laboratory was another factor in choosing this method. Several important parameters in HGAAS technique such as the concentration of the acid used, concentration of NaBH4 reagent, concentration of the prereducing agent, etc. that affect the sensitivity for the determination of antimony were examined. A similar system was investigated for the determination of manganese (Mn) and initial results were obtained. Manganese concentrations were determined by flame AAS [34].

10.10.1.4   DETERMINATION OF MOBILE THALLIUM IN SOIL BY FLOW INJECTION DIFFERENTIAL PULSE ANODIC STRIPPING VOLTAMMETRY

A procedure for the determination of mobile thallium in soil has been developed. Free thallium (I) was extracted with deionized water, while the sorbed and exchangeable metal was extracted with 1 M ammonium nitrate. Thallium in extracts was determined by differential-pulse anodic stripping voltammetry in a flow-injection system with medium circulation. 0.05 M EDTA was the base electrolyte. The procedure was applied for the determination of mobile thallium in a sample of ground from a waste dump, which was a potential source of thallium pollution. The total concentration of thallium in the sample was 17.9 $\mu g\ g^{-1}$. A 3.97±0.25% of total thallium was found to be water soluble and 12.4±1.1% was sorbed and exchangeable. An attempt to determine the thallium bound to the carbonate fraction failed because of the high accumulation of interfering lead in this fraction. A single extraction cycle recommended by the literature was not sufficient to extract the whole thallium content of the fraction. Minor rivers of the area of the waste dump have a thallium concentration higher by two orders of magnitude than in reference rivers [35].

10.10.1.5   FLOW INJECTION CATALYTIC SPECTROPHOTOMETRIC SIMULTANEOUS DETERMINATION OF NITRITE AND NITRATE IN NATURAL WATERS

A new flow injection catalytic spectrophotometric method is proposed for the simultaneous determination of nitrite and nitrate based on the catalytic effect of nitrite on the redox reaction between crystal violet and potassium bromate in phosphoric acid medium and nitrate being on-line reduced to nitrite with a cadmium-coated zinc reduction column. The redox reaction is monitored spectrophotometrically by measuring the decrease in the absorbance of crystal violet at the maximum absorption wavelength of 610 nm. A technique of inserting a reduction column into sampling loop is adopted and the flow injection system produces a signal with a shoulder. The height of shoulder in the ascending part of the peak corresponds to the nitrite concentration and the maximum of the peak corresponds to nitrate plus nitrite. The detection limits are 0.3 ng ml^{-1} for nitrite and 1.0ng ml^{-1} for the nitrate. Up to 32 samples can be analyzed per hour with a relative standard deviation of less than 2%. The method has been successfully applied for the simultaneous determination of nitrite and nitrate in natural waters [36].

10.10.1.6   STABLE AND RADIOACTIVE STRONTIUM IN WATER, MILK AND SOIL

A multisyringe flow injection (MSFIA) method for the determination of stable and radioactive strontium, using a solid phase resin (Sr-Resin), has been developed. Strontium concentrations are determined by atomic emission spectroscopy and by a low background proportional counter. The method has been applied to different samples (water, milk and soil) of environmental interest. The LLD of the stable and radioactive Sr were 10 microg/L and 0.01 Bq, respectively. The standard deviation of the separation procedure is 2% (n=10) [37].

10.10.1.7   SIMULTANEOUS DETERMINATION OF NITRITE AND NITRATE IN WATER

A novel spectrophotometric reaction system was developed for the determination of nitrite as well as nitrate in water samples, and was applied to a flow-injection analysis (FIA). The spectrophotometric flow-injection system coupled with a copperised cadmium reductor column was proposed. The detection was based on the nitrosation reaction between nitrite ion and phloroglucinol (1,3,5-trihydroxybenzene), a commercially available phenolic compound. Sample injected into a carrier stream was split into two streams at the Y-shaped connector. One of the streams merged directly and reacted with the reagent stream: nitrite ion in the samples was detected. The other stream was passed through the copperised cadmium reductor column, where the reduction of nitrate to nitrite occurred, and the sample zone was then mixed with the reagent stream and passed through the detector: the sum of nitrate and nitrite was detected. The optimized conditions allow a linear calibration range of 0.03–0.30 µg NO_2^-NmL^{-1} and 0.10–1.00 µg NO_3^-NmL^{-1}. The detection limits for nitrite and nitrate, defined as three times the standard deviation of measured blanks are 2.9 ng NO_2^-NmL^{-1}and 2.3 ng NO_3^-NmL^{-1}, respectively. Up to 20 samples can be analyzed per hour with a relative standard deviation of less than 1.5%. The proposed method could be applied successfully to the simultaneous determination of nitrite and nitrate in water samples [38].

10.10.1.8   FLOW-INJECTION CHEMILUMINESCENCE STUDY OF LUMINOL-HYDROGEN PEROXIDE-CARBENDAZIM SYSTEM IN TAP WATER SAMPLES

A new flow-injection chemiluminescence (CL) method is described for the determination of carbendazim. The method is based on the CL reaction of 300 uminal and hydrogen peroxide (H_2O_2). Carbendazim can greatly enhance the chemiluminescence intensity in sodium hydroxide-sodium dihydrogen phosphate (NaOH–NaH$_2$PO$_4$) medium (pH=12.6). Under the optimum conditions, the linear range for the determination of carbendazim is 2.00×10 to 2.00×10 g mL with a detection limit (S/N=3) of 7.24×10 g mL. The relative standard deviation is 1.8% for 1.0×10 g mL carbendazim (n=8). The proposed method has been applied to the determination of carbendazim in tap water samples. Furthermore, the possible enhanced CL mechanism is discussed by examining the CL spectra and fluorescence spectra [39].

10.10.1.9   SULFATE, NITRITE AND NITRATE MONITORING IN DRINKING WATER AND WASTEWATER

An automated monitoring system for sulfate, nitrite and nitrate based on sequential injection analysis (SIA) was developed. For nitrite determination the modified Griess-Ilosvay method was used, whereas nitrate was previously reduced to nitrite using a cadmium column followed by nitrite determination. A turbidimetric method was carried out in order to determine sulfate. The results showed that the proposed SIA monitoring system constitutes an effective approach for nitrite, nitrate and sulfate determination since it is able to determine levels required by international agencies that regulate these parameters in water. Detection limits of 0.0207 mgNL^{-1}, 0.0022 mgNL^{-1} and 3.0 mgSO$_4^{2-}$L^{-1} were obtained for nitrate, nitrite and sulfate, respectively. The developed method offers also typical characteristics of the multicommutated systems, as portability, low reagents consumption and the subsequently minimization of waste generation. The proposed system was successfully applied to drinking water and wastewater samples and validated with a certified river water sample [40].

10.10.1.10   HEXAVALENT CHROMIUM WITH ON-LINE PRECONCENTRATION ON AN ANION IMPRINTED POLYMER IN DIFFERENT ENVIRONMENTAL SAMPLES SUCH AS SEA AND RIVER WATERS, SOILS AND SEDIMENTS

A flow injection preconcentration system for the flame atomic absorption spectrometric determination of hexavalent chromium has been developed. The method employs on-line preconcentration of Cr (VI) on a minicolumn packed with Cr(VI)-

imprinted poly(4-vinyl pyridineco2-hydroxyethyl methacrylate) placed into a flow injection system. Hexava-lent chromium was eluted with a small volume of diluted hydrochloric acid into the nebulizer-burner system of a flame atomic absorption spectrometer. An enrichment factor of 550 and a 3σ detection limit of 0.04 $\mu g \cdot L^{-1}$ along a sampling frequency of 4 h-1 at a sample flow rate of 3.5 $mL \cdot min^{-1}$. The relative standard deviation is 2.9% for 1 $\mu g \cdot L^{-1}$ Cr (VI) (n = 11). The flow injection system proposed has the advantage of being simpler because the use of expensive and sophisticated instruments is avoided. Ease of use, continuous process and selectivity make this method suitable for Cr (VI) determination in different environmental samples such as sea and river waters, soils and sediments [41].

10.10.2   SELECTED PROCEDURES FOR INORGANIC AND ORGANIC SPECIES

10.10.2.1   DETERMINATION OF ACETOCHLOR IN FOOD SAMPLES

Acetochlor is an herbicide widely used for controlling grass weeds in various crops. Acetochlor was first used in 1994 in the U.S.A. and in Europe in 2000. As a member of the chloroacetanilide class of broad leaf herbicides, it is applied to the soil as a pre and postemergence treatment. Acetochlor is mainly absorbed by the roots and leaves, inhibiting photosynthetic electron transport of the host plant. The pollution problems associated with acetochlor have increasingly attracted attention. The toxicity of acetochlor, such as inducing sister chromatid exchanges in cultured human lymphocytes, mutagenizing germ cells of male rats, and altering thyroid hormone-dependent gene expression in Xenopus laevis, has been reported. The typical half-life of chloroacetanilide herbicides under natural conditions ranges from 15 to 30 days. The presence of acetochlor and its metabolites has become a significant pollution problem, and effective methods for their removal or treatment need to be pursued. Various techniques such as HPLC, LC-MS, and GC are used for the determination of acetochlor and its metabolites in soil and water. The aim of the present study was to develop a simple and quick method for the determination of acetochlor in formulations and food samples.

10.10.2.1.1   MATERIALS AND METHODS

Instruments. The flow injection (FI) system comprised a peristaltic pump (Becton Dickinson, Franklin Lakes, NJ) with PTFE (1.19 mm i.d.) and silicon (1.71 mm i.d.) flow tubes, V-450 six-port injection valve (Upchurch Scientific Inc., Oak Harbor, WA), and UNICO UV-2100 UV-VIS spectrophotometer(United Products and Instruments Inc., Dayton, NJ) as the detector. A schematic of the FI system is given in Fig. 10.7.

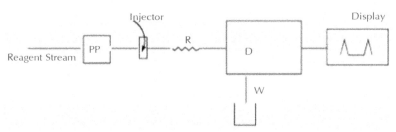

FIGURE 10.7 Single-channel FI for the spectrophotometric determination of acetochlor. R = coupling coil length, D = detector, W = waste, pp = peristaltic pump.

10.10.2.1.2 REAGENTS AND CHEMICALS

All chemicals used were of analytical reagent-grade purity. Sodium nitrite, aniline (Merck, Darmstadt, Germany), concentrated hydrochloric acid (Merck), and ethanol (Merck) were used for this work. Standard reference material was purchased from Dr. Ehrenstorfer GmbH (Augsberg, Germany). A commercial sample containing acetachlor was purchased commercially from a local market.

10.10.2.1.3 SOLUTION PREPARATIONS

Aniline solution (1%): 1% aniline solution was prepared by adding 1 mL of aniline in 50 mL of ethanol and diluted with distilled water up to 100 mL.

Nitrite solution (0.15%): Nitrite solution was prepared by dissolving 0.225 g sodium nitrite in distilled water and diluted up to 100 mL.

Diazotized aniline solution: Diazotized aniline solution was prepared by mixing nitrite solution (0.15%) and aniline solution (1%) in a 3:1 ratio.

Standard acetochlor solution: Acetochlor solution was prepared and hydrolyzed by taking a known volume of standard acetochlor in a beaker followed by the addition of 0.2 M of HCl and a few milliliters of ethanol. This was heated in a boiling water bath for 10 min, cooled, and diluted up to 10 mL with ethanol. Working standard solutions were prepared in the range of 2–0.009 ppm by suitable dilution of the stock standard solution.

Batch procedure: To different concentrations of hydrolyzed acetochlor working standard solution was added 6 mL of diazotizing reagent at room temperature and kept for 10 min, and azo dye was formed. The dilution was made with distilled water up to 10 mL, and absorbance was measured at 400 nm against reagent blank.

FI procedure: The diazotized reagent was continuously pumped as a carrier stream and used as blank. One milliliter of hydrolyzed acetochlor standard solution in the concentration range of 2–0.009 ppm was injected into the carrier stream, and absorbance of the azo dye was measured continuously at 400 nm.

10.10.2.1.4 PROCEDURE FOR ACETOCHLOR IN COMMERCIAL FORMULATIONS

For the determination of acetochlor in commercial formulations, 0.1 mL of commercial sample was hydrolyzed in the same way as standard. Working standards were prepared by dilution with distilled water. The color was developed by the addition of 6 mL diazotizing reagent and diluted with distilled water up to 10 mL in a volumetric flask, and absorbance was measured at 400 nm. The amount of acetochlor present in each preparation was determined from the calibration plot using a standard curve.

Fifteen grams of homogenized food sample and 20 mL of sugarcane juice were taken in a beaker, and 50 mL of solvent (petroleum ether and acetone, 1:1) was added to it for extraction of herbicides. The mixture was shaken for 2 hr. After equilibration, the sample was filtered and passed through 10 g of sodium sulfate. The filtrate was evaporated on a rotary evaporator up to a volume of 5 mL and hydrolyzed with 0.2 M hydrochloric acid by heating in a boiling water bath for 10 min. Two and one milliliter from the hydrolyzed extract were used for batch and flow injection analysis, respectively. Each sample was analyzed in triplicate.

A recovery test was performed on control samples and fortified with a known concentration of acetochlor solution. The solvent was evaporated at room temperature for 1 hr. Six replicates of the fortified samples were analyzed [42].

10.10.2.2 HISTAMINE IN FISH MEAL USING CATION-EXCHANGER RESIN

The FI manifold used in this work is shown in Fig. 10.4. Two peristaltic pumps were used in the system. The first one propels 0.1% p-nitroaniline (prepared in 0.25M H_2SO_4) solution, 0.5% sodium nitrite solution and 1.5M potassium hydroxide solution, with flow rates 0.8, 0.7 and 0.8 mL/min, respectively.

The second pump propels one of the three following solutions: buffer solution, sample solution or hydrochloric acid solution. Three reaction coils were used in the system with lengths 40cm (RC1), 20cm (RC2) and 30cm (RC3). A mini–column (preconcentration column) 3.5 cm in length and 2.5 mm I.d. containing Amberlite resin (weak cation exchanger) was used for histamine preconcentration. The loaded time was 5 min. A selection valve was used for changing the pumped solution between hydrochloric acid, buffer solution and the sample solution.

The mini-column was conditioned through 30s washing with the buffer solution, before starting any sample loading. After the sample loading the selection valve turned and the mini-column was washed with the buffer solution until the signal returns to the base line. When the analytical signal returns to the base line the selection valve turned and hydrochloric acid solution elute all retained histamine in the mini-column. The detection takes place with spectrophotometer containing flow cell connected with recorder.

10.10.2.2.1   CALIBRATION CURVE

Under the conditions established, the calibration curve of the determination of histamine was obtained. The calibration curve was linear over the range 0.02–1.5 µg/mL $(1.8 \times 10^{-7} – 1.35 \times 10^{-5}$ M) with detection limit 0.02 µg/mL.

10.10.2.2.2   PRECISION AND ACCURACY

The precision and accuracy of the proposed method were checked by three replicate analyzes of three different concentrations. The results are shown in Table 10.1. It seems that the reproducibility of the lower concentration is less than higher concentrations.

TABLE 10.1   Accuracy and Precision Data of On-line Separation and Preconcentration FI Method

Conc. (µg/mL)	RSD%	Erel%
0.02	5.23	+2.18
0.7	2.10	+3.30
1.5	0.69	−2.08

10.10.2.2.3   INTERFERENCE

The selectivity of the method was investigated by studying the effect of interference from other species accompanied with histamine in fish meal. This effect was studied by adding a known amount of the interference species to a solution containing histamine (0.75 µg/mL). Table 10.2 shows the tolerance levels of interfering species for determination of histamine. Strong interference showed by basic amino acids (lysine, arginine): this was probably due to the similarity in their basicity with histamine.

TABLE 10.2   Effect of Interfering Species for Histamine Determination

Foreign species	Conc.(µg/mL)	Erel %
Albumine	50.0	+1.78
Alanine	400.0	+2.44
Arginine	10.0	+2.90
Aspargine	25.0	+4.10
Aspartic acid	25.0	−2.78

Cysteine	100.0	+4.54
Histidine	25.0	+3.93
Lysine	10.0	+3.56
Methionine	100.0	+4.02
Proline	50.0	+2.37
Phenyalanine	100.0	+4.55
Tryptophane	25.0	−2.96
Tyrosine	100.0	+3.50
Ca2+	40	+4.11

10.10.2.2.4   APPLICATION OF THE METHOD

The proposed method was applied for the determination of histamine in fish meal samples. In order to validate the results obtained by the proposed flow method the samples were analyzed also using a standard method. The results of proposed methods were compared with that obtained by standard method employing t-test and F-test. The comparison indicated that there was no significant difference between the accuracy and precision of two methods at 99% level of confidence [17].

10.10.2.3   PHENOL AND O-CRESOL IN SOIL EXTRACTS

In consequence of the large use of phenolic compounds, allied to their high toxicity and the mobility in soil, the analytical methods for phenolic compound determination in water and soil are of prime importance. Phenol and o-cresol determination by flow injection analysis (FIA) with spectrophotometric detection, employing the 4-aminoantipyrine reaction, is proposed in this work in order to quantify these species in soil extracts. The method was improved by a factorial planning, being verified a higher sampling rate in comparison with the conventional method, and recovery values, limits of detection and quantification similar for both methods. These aspects suggest this method as a feasible alternative for phenolic compound determinations in soil extracts.

This method is based on the 4-aminoantipyrine reaction with phenolic species in the presence of potassium ferricyanide (at pH 7.90), providing an intense red color development of the pyrazolones group, with maximum absorption in the wavelength of 510 nm. As the 4-AAP reaction is an official and well-established method for phenolic compound determinations in water samples, it has been used for comparison purposes in the development of new methods. Many times, the method is tedious and time-consuming since a distillation step is necessary to separate the phenolic species from the aqueous matrices due to spectral interferences, especially by organic matter.

A valuable approach to analytical method automation is the system based on flow analysis, especially flow injection analysis (FIA) and sequential injection analysis (SIA), coupled to various detection systems. The flow systems minimize the analyst intervention, increase the sampling rate and improve the precision of the measurements. Moreover, they show great potential to the development of cleaner analytical methods owing to the lower waste generation. The methods employing flow analysis have been successfully applied to the phenolic compound determination in water samples, coupled to electrochemical detectors and mainly to spectrophotometric detectors.

A previous evaluation of some parameters was made in order to obtain the better absorbance signal. The flow rate, concentrations of the 4-AAP and $K_3[Fe(CN)_6]$ solutions, pH of the $K_3[Fe(CN)_6]$ solution, volume of the sample loop and of the reaction coil were investigated. Afterwards, these three last factors were studied by a 23 factorial planning in two different levels: (i) pH of the $K_3[Fe(CN)_6]$ solution (6.2 and 11.0), and (ii) volumes of the sample loop (200 and 400 mL) and of the reaction coil (150 and 300 mL). A simple FIA system was used for this study, as showed in Fig. 10.8. Six standard phenol or o-cresol solutions with concentrations between 0.50 and 16.00 mg L^{-1} were used for construction of the analytical curves, being each standard or sample injected three times.

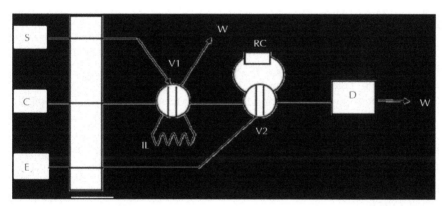

FIGURE 10.8 FI manifold proposed for the continuous determination of hexazinone in water: sample solution (S), carrier solution (C), eluent (E), peristaltic pump (PP), injection valves 1 and 2 (V), injection loop (IL), retention column (RC), detector (D), and wastes (W).

Determination of phenol and o-cresol in soil extracts a mass of 10 g (± 0.1 mg) of the soil sample was maintained in contact with 300 mL of 0.01 mol L^{-1} $CaCl_2$ solution in stoppered amber glass under gentle orbital shaking for 96 h. Then, the soil suspension was centrifuged at 2500 rpm for 10 min, being the supernatant phase reserved for the experiments. An appropriate volume of 25.0 mL was spiked with phenol in order to provide concentrations of 1.00, 10.00 and 50.00 mg L^{-1}, and the same procedure was carried out for *o*-cresol. These solutions were kept under shaking for 15 min, and after that, the analytes were quantified by the conventional and FIA methods, with suitable dilution when necessary. A blank experiment was done in parallel. All these procedures were performed in triplicate [43].

10.10.2.4   HEXAZINONE IN WATER BY SOLID PHASE EXTRACTION

In this work, the continuous determination of hexazinone in water was carried out by developing an online solid phase extraction by flow injection coupled with spectrophotometric detection. Variables related to hydrodynamic conditions were optimized by a full factorial design 3^2, of which the results were analyzed through an analysis of variance. Under the proposed optimal conditions, the principal figures of merit were a working range between 0.50 to 7.00 µg mL^{-1} of hexazinone; a precision of 4.6% expressed as variance coefficient; a limit of detection of 0.05 µg mL^{-1}, and a limit of determination of 0.16 µg mL^{-1}. Univariant calibrations based on the height or area of transitorial signals were compared to identify the best conditions for quantification. Samples of well and sea water were analyzed, obtaining satisfactory results in all cases in terms of precision and accuracy.

With the two valves in load position, the 2000 µL loop of the injection valve (V1) was filled with the sample, meanwhile the carrier solution passed through the adsorption minicolumn positioned in the loop of the second valve in order to prepare the stationary phase for the retention of HEXA; also, an aqueous MeOH 70% v/v solution (eluent) reached the flow cell, allowing for baseline recording. Then valve 1 was changed to injection position for the preconcentration of the analyte on the C18 support. Later, the change of valve 2 to injection position facilitated that the eluent passed through the column, desorbing the retained species and sending them to the detector. Finally, valves 1 and 2 were returned to the load position to begin the analysis of a new sample. All experiments were done in triplicate [44].

KEYWORDS

- **Applications**
- **Definitions**
- **Detection**
- **Multisyringe FIA**
- **Sequential Injection**
- **Solvent Extraction FIA**

REFERENCES

1. Stewart, K. K. (1981). Flow Injection Analysis, A Review of Its Early History, Talanta, *28*, 789–797.
2. Novic, M., Berregi, I., Rios, A., & Valcarcel, M. (1999). A New Sample-Injection/Sample-Dilution System for the Flow-Injection Analytical Technique, Anal. Chim. Acta, *381*, 287–295.
3. Valcarcel, M., & Laque de Castro, M. D. (1988). "Automatic Methods of Analysis", Elsevier Science Publishing, 166.
4. Fakhre, N. A. (2000). "Extraction Flow Injection Spectrophotometric Determination of Permanganate and Chromate Using Crown Ethers", Ph D. Thesis, University of Baghdad, 5.
5. Valcarcel, M., & Laque de Castro, M. D. (1987). "Flow Injection Analysis, Principles and Applications", Ellis Horwood Ltd, *43*, 378.
6. Nagy, G., Fehery, Z. S., & Pungor, E. (1970). Application of Silicone Rubber-Based Graphite Electrodes for Continuous Flow Measurements: Part II. Voltammetric Study of Active Substances Injected Into Electrolyte streams, Anal. Chim. Acta, *52*, 47–54.
7. Ruzicka, J., & Hansan, E. H. (1986). The First Decade of Flow Injection Analysis: from Serial Assay to Diagnostic Tool, Anal. Chim. Acta, *79*, 1–58.
8. Kuban, V. (1991). Liquid-Liquid Extraction Flow Injection Analysis, Critical Reviews in Analytical Chemistry, *22*, 477–557.
9. Karlberg, B., & Thelander, S. (1978). Extraction Based on the Flow-Injection Principle: Part I. Description of the Extraction System Anal Chim. Acta, *98,* 1–7.
10. Bergamin, F. H., Madeiros, J. X., Reis, B. F., & Zagatto, E. A. (1978). Solvent Extraction in Continuous Flow Injection Analysis: Determination of Molybdenum in Plant Material, Anal. Chim. Acta, *101*, 9–16.
11. Lucy, C. A., & Yeung, K. K. C. (1994). Solvent Extraction-Flow Injection without Phase Separation Through the use of Differential Flow Velocities Within the Segmented Flow, Anal. Chem, *66*, 2220–2225.
12. Kawase, J., Nakae, A., & Yamanaka, M. (1979). Determination of Anionic Surfactants by Flow Injection Analysis based on Ion-Pair Extraction, Anal. Chem., *51*, 1640–1643.
13. Kawase, J. (1980). Automated Determination of Cationic Surfactants by Flow Injection Analysis Based on Ion-Pair Extraction, Anal. Chem., *52*, 2124–2127.
14. Fujiwara, K. (1988). Application of a Wave-Guide Capillary Cell in the Determination of Copper by Flow Injection Analysis, Anal. Chim. Acta, *212*, 245–251.
15. Gallego, M., & Valcarcel, M. (1985). Indirect Atomic Absorption Spectrometric Determination of Perchlorate by Liquid-Liquid Extraction in a Flow-Injection System, Anal. Chim. Acta, *169*, 161–169.
16. Gallego, M., Silva, M., & Valcarcel, M. (1986). Determination of Nitrate and Nitrite by Continuous Liquid-Liquid Extraction with a Flow-Injection Atomic-Absorption Detection System, Fresenius, Z. Anal. Chem, *323*, 50–53.
17. Fakhre, N. A., & Abdullah, M. S. (2011). On-Line Separation and Preconcentration for Histamine Determination in Fish Meal Using Cation-Exchanger Resin, *J. Food Sci.* Engin, *1*, 282–288.
18. Luque de Castro, M. D., Ruiz-Jim´enez, J., & P´erez–Serradilla, J. A. (2008). "Lab-On-Valve: a Useful Tool in Biochemical Analysis," Trends Anal. Chem, *27(2)*, 118–126.
19. Vidigal, S. S. M. P., Oth, I. V. T., & Rangel, A. O. S. S. (2010). Sequential Injection Lab-on-Valve System for the Determination of the Activity of Peroxidase in Vegetables, *Journal of Agricultural and Food Chemistry, 58(4)*, 2071–2075.
20. Lee, P. L., Sun, Y. C., & Ling, Y. C. (2009). Magnetic Nano-Adsorbent Integrated with Lab-On-Valve System for Trace Analysis of Multiple Heavy Metals, *J. Anal. Atom.* Spectrosc, *24(3)*, 320–327.

21. Jakmunee, J., Pathimapornlert, L., Kradtap Hartwell, S., & Grudpan, K. (2005). Novel Approach for Mono-Segmented Flow Micro-Titration with Sequential Injection using a Lab-On-Valve System, a Model Study for the Assay of Acidity in Fruit Juices, Analyst, *130(3)*, 299–303.

22. Khonyoung, S., Kradtap Hartwell, S., Jakmunee, J., Lapanantnoppakhun, S., Sanguansermsri, T., & Grudpan, K. A. (2009). Stopped Flow System with Hydrodynamic Injection for Red Blood Cells Osmotic Fragility Test, Possibility for Automatic Screening of Beta-Thalassemia Trait, Anal. Sci., *25(6)*, 819–824.

23. Kradtap Hartwell, S., Boonmalai, A., Kongtawelert, P., & Grudpan, K. (2010). Sequential Injection-Immunoassay System with a Plain Glass Capillary Reactor for the Assay of Hyaluronan, Anal. Sci., *26(1)*, 69–74.

24. Cerdà, V., Estela, J. M., Forteza, R., Cladera, A., Becerra, E., Altimira, P., Sitjar, P. Altimira, P., & Sitjar, P. (1999). Flow Techniques in Water Analysis, Talanta, *50*, 695.

25. Ruzicka, J., & Hansen, E. H. (1975). Flow Injection Analyses, *1*, New Concept of Fast Continuous-Flow Analysis, Anal. Chim. Acta, *78(1)*, 145–157.

26. Ruzicka, J., & Marshall, G. D. (1990). Sequential Injection A New Concept for Chemical Sensors, Process Analysis and Laboratory Assays, Anal. Chim. Acta, *237(2)*, 329–343.

27. Almeida, M. I. G. S., Estela, J. M., & Cerdà, V. (2011). Multisyringe Flow Injection Potentialities for Hyphenation with different Types of Separation Techniques, Anal. Lett., *44(1–3)*, 360–373.

28. Magalhmes, L. M., Ribeiro, J. P. N., Segundo, M. A., Reis, S., & Lima, J. L. F. C. (2009). Multi- Syringe Flow-Injection Systems Improve Antioxidant Assessment, Trac-Trends Anal. Chem., *28(8)*, 952–960.

29. Maya, F., Estela, J. M., & Cerdà, V. (2010). Interfacing Online Solid Phase Extraction with Monolithic Column Multisyringe Chromatography and Chemiluminescence Detection: an Effective Tool for Fast, Sensitive and Selective Determination of Thiazide Diuretics, Talanta, *80(3)*, 1333–1340.

30. Segundo, M. A., & Magalhes, L. M. (2006). Multisyringe Flow Injection Analysis: State-of the-Art and Perspectives, Anal. Sci., *22(1)*, 3–8.

31. Miro, M., Cerdà, V., & Estela, J. M. (2002). Multisyringe Flow Injection Analysis, Characterization and Applications, Trac-Trends Anal. Chem., *21(3)*, 199–210.

32. Dong, S., & Dasgupta, P. K. (1987). Fast Fluorometric Flow Injection Analysis of Formaldehyde in Atmospheric Water, Environ., Sci. Technol., *21(6)*, 581–588.

33. Cassella, A. R., Cassella, R. J., Garrigues, S., Santelli, R. E., de Campos, R. C., & De la Guardia, M. (2000). Flow Injection-FTIR Determination of Dithiocarbamate Pesticides, Analyst, *125(10)*, 1829–1833.

34. Erdem, A. (2003). Speciation and Preconcentration of Inorganic Antimony and Manganese in Waters Using Microcolumn-Flow Injection System and Determination by Atomic Absorption Spectrometry, M Sc Thesis, Chemistry İzmir Institute of Technology, İzmir, Turkey September.

35. Lukaszewski, Z., Karbowska, B., & Zembrzuski, W. (2003). Determination of Mobile Thallium in Soil by Flow Injection Differential Pulse Anodic Stripping Voltammetry, Electroanalysis, *15(5–6)*, 480–483.

36. Yue, X. F., Zhang, Z. Q., & Yan, H. T. (2004). Flow Injection Catalytic Spectrophotometric Simultaneous Determination of Nitrite and Nitrate in Natural Waters, Talanta, *62(1)*, 97–101.

37. Fajardo, Y., Gómez, E., Mas, F., Garcias, F., Cerdà, V., & Casas, M. (2004). Multisyringe Flow Injection Analysis of Stable and Radioactive Strontium in Water, Milk and Soil Samples of Environmental Interest, Appl. Radiat. Isot, *61(2–3)*, 273–277.

38. Burakham, R., Oshima, M., Grudpan, K., & Motomizu, S. (2005). Simple Flow-Injection System for the Simultaneous Determination of Nitrite and Nitrate in Water Samples, Talanta, *64(5),* 1259–65.

39. Liao, S., & Xie, Z. (2006). Flow-Injection Chemiluminescence Study of Luminol-Hydrogen Peroxide-Carbendazim System in Tap Water Samples, Spectrosc. Lett., *39(5),* 473–485.

40. Ayala, A., Leal, L. O., Ferrer, L., & Cerdà, V. (2012). Multiparametric Automated System for Sulfate, Nitrite and Nitrate Monitoring in Drinking Water and Waste water Based on Sequential Injection Analysis, *Microchem. J.*, 100, 55–60.

41. Carmen, M., Biurrun, Y., Castro-Romero, J. M., & Carro-Mariño, N. (2012). Flow-Injection Flame Atomic Absorption Determination of Hexavalent Chromium with On-Line Preconcentration on an Anion Imprinted Polymer in Different Environmental Samples such as Sea and River Waters, Soils and Sediments, *Am. J. Anal.* Chem, 3 *(11)*, 755–760.

42. Shah, J., Jan, M. R., & Bashir, N. (2008). Flow Injection Spectrophotometric Determination of Acetochlor in Food Samples, American Lab., March 24.

43. Dolatto, R. G., Messerschmidt, I., Pereira, B. F., Silveirac, C. A. P., & Abate, G. (2012). Determination of phenol and O-Cresol in Soil Extracts by Flow Injection Analysis with Spectrophotometric Detection, *J. Braz. Chem.* Soc, *23(5)*, 970–976.

44. Amador-Hernándeza, J., Velázquez-Manzanaresa, M., Gutiérrez-Ortizb, M., & Márquez-Reyesb, J. M. (2012). On-Line Determination of Hexazinone in Water by Solid Phase Extraction UV/Vis Spectrophotometry, *Eurasian J Anal Chem.*, *7(2)*, 96–103.

CHAPTER 11

ENVIRONMENTAL APPLICATIONS OF SPECTROCHEMICAL ANALYSIS

MAHMOOD M. BARBOOTI

Department of Applied Chemistry, School of Applied Sciences, University of Technology, P.O. Box 35045, Baghdad, Iraq; E-mail: brbt2m@gmail.com

CONTENTS

11.1   INTRODUCTION

Spectrophotometric methods of analysis are based on the interaction of light with matter. Compounds respond differently to the incident radiation depending on the wavelength range and hence the energy of the radiation, in addition to the electronic structure and types of bonding. The UV/Vis radiation (200–750 nm) is considered an energetic radiation that can interact with valence electrons of molecules and ions to excite them to a higher energy level. Thus, the absorption in this region gives important information about the electronic nature of the compounds. In the meantime, the infrared radiation in the region (1–1000 μm) is less energetic than UV/Vis and best interact with vibration of atoms and can reflect the nature of bonding and functional groups of a compound.

11.2   APPLICATION OF UV/VIS METHODS

11.2.1   GENERAL

Many important UV/Vis methodologies have been developed for the determination of organic, inorganic and complex ionic structure in various matrices. Heavy metal determination can be done successfully following complex formation with distinct color development by association with selected ligands [1]. Specific procedures have been developed for the detection and determination of some inorganic anions like cyanide in water, soil, and living tissues. However, more specific spectro-chemical techniques like atomic absorption and induced coupled plasma emission were developed for the determination of heavy metals. This chapter will not include heavy metal determination since the matter was discussed in details in Chapter 13 of this book. As early as 1972, the UV absorption method was employed for the identification of petroleum products including distillate and residual fuel oils and lubricating oils in marine environment [2]. The results suggested a basis for an analytical procedure, which may be useful for the identification of oil spills.

  Although the main task in spectrophotometric analysis is the determination of various chemical species, Spence [3] proposed a rapid spectrophotometric detection for analysis of bacterial contamination in water. Bacterial contamination in water is a health hazard worldwide. Traditional water testing techniques detect living bacteria in approximately 48 h, while they demonstrated that optical techniques can detect bacteria in as little as six hours. The Beer-Lambert Law, as applied to spectrophotometric turbidity studies, correlates the concentration of organism growth in a solution to the absorption of visible light. The change in transmittance over time correlates inversely to the bacterial growth curve, a sharp drop in transmittance signifying the exponential growth phase. They observed this change within six to 12 h following the inoculation of Escherichia coli into samples, using both a standard monochromator and a device engineered specifically for this study, the Optical

Bacteria Detector (OBD). They have employed cell counting algorithms to establish a baseline for their data, and bioreactors to control the initial rate of growth of our cultures. OBD is designed to be an effective and inexpensive field device, with minimum use of consumables and waste generation. It uses a phototransistor as a sensor and a *light emitting diodes*, LED with peak output wavelength of 523 nm. OBD can be tuned to test for other bacteria, such as Salmonella sps., by changing the wavelength of the LED light source.

11.2.2 WATER ANALYSIS

11.2.2.1 OVERVIEW

The examination of water serves many purposes. The target use of the water determines the type of tests required. The analysis of drinking water implies the bacteriological examinations to ensure the absence of disease causing microorganism before any other chemical test may be done. Many chemical species have detrimental health effects that necessitate precise and comprehensive analyzes of the drinking water to ensure safe water supplies to the community [4]. Industrial and boiler water requirements lie mostly with the hardness substances and dissolved oxygen to ensure limited rates of corrosion and minimum scale formation. Thus, the design of analytical procedures depends on the purpose for which the water to be used. Residual amounts of pesticides and fertilizers will be passed to the ground and surface waters by infiltration and run off. The industrial wastewater analysis results must fall within the environmental and health requirements. Thus, most of the atmospheric and land pollution finds its way to the surface and ground water sources. In the following paragraphs, the application of spectrophotometric methods in the determination of pollutants in water and wastewater will be discussed. However, the analysis of water samples for some nutrients like phosphate will not be covered by this discussion.

Ascorbic acid is unstable in solution specially in the presence of some metal ions, like Fe, Mg, Cu, or Ni, that accelerate the degradation rate of the acid [5, 6]. Thus, the rapidity of degradation due to specified factors depend on water quality. A screening test was developed by Jezierska, et al. [7] for the estimation of water quality using L-ascorbic acid as an indicator chemical, due to its sensitivity to pollutants. They investigated the absorption spectra of L-ascorbic acid dissolved at different concentrations in water from different sources. Water quality index (WQI) was defined as the change in maximum L-ascorbic acid absorbance at 265 nm over two arbitrarily chosen time periods,, that is, between the 1st and 10th minutes and 1st and 20th minutes. They found that a high WQI value was significantly associated with low water quality, and vice versa. The proposed technique is a quick, simple and inexpensive method for obtaining a preliminary estimate of water quality.

The advances in fluorescence spectrophotometry enable the analysis of organic matter dissolved in river water. Baker, et al. [8] investigated the potential of detecting sewage pollution in a small, urbanized catchment. Downstream sampling highlighted a summer maximum in tryptophan fluorescence intensity during low flow. No correlation was observed between ammonia and tryptophan fluorescence intensity. In contrast, two sewage related point-pollution events had both high tryptophan fluorescence intensity and ammonia, suggesting that the summer tryptophan increase does not original from foul sewage. Sewage inputs to the river were therefore monitored at summer base flow. This demonstrated that >10% of the rivers' discharge is provided by sewerage inputs and that these inputs could be grouped by their fluorescence and ammonia properties: (i) clean' storm waters with low ammonia and tryptophan intensity; (ii) gray' waters with high tryptophan intensity and low ammonia concentration; and (iii) foul waters with high tryptophan intensity and ammonia concentration. All three types of sewerage input occurred irrespective of flow conditions, suggesting that sewerage cross connections are occurring.

11.2.2.2   INORGANIC ANIONS

11.2.2.2.1   FLUORIDE

Fluoride is present in surface waters at various levels. According to WHO [9], low concentrations of fluoride in drinking water have been considered beneficial to prevent dental carries, but excessive exposure to fluoride in drinking water can give rise to a number of adverse effects. WHO [10) has set a limit value of 1.5 mg·ℓ^{-1} for fluoride in drinking water. There is a narrow margin between the desired and harmful doses of fluoride in drinking water [11]. The establishment of a rapid, sensitive and simple determination method is important. Many spectrophotometric methods have been proposed and tested for F determination. Alizarin red method was accepted for fluoride determination in water samples for a long time. Samples are treated with alizarin red reagent and the complex is extracted with pentanol and the absorbance is measured at 430 nm [12]. The method was employed for the estimation of the fluoride levels of water, soil and vegetables from four irrigation farms at the bank of Basawa River, Zaria, Nigeria [13]. The soil samples from the four farms had a mean soil leachable fluoride in the range of 0.091 mg Kg^{-1} to 0.135 mg Kg^{-1}, while the river water had mean fluoride levels in the range 0.081 to 0.191 mg Kg^{-1}.

Barghouthi and Amereih [14] developed a sensitive spectrophotometric method for the determination of fluoride in drinking water using aluminum complexes of triphenylmethane dyes (chrome azurol B) as spectrophotometric reagents. Fluoride reacts with the dark pink aluminum chrome azurol B complex to produce a colorless aluminum fluoride complex by replacement of the chrome azurol B by fluoride and liberation of the free ligand. This leads to a change in color from that of the complex, dark pink, to that of the free ligand, dark orange. The method allowed a reliable determination of fluoride in the range of 0.5–4.0 mg·l^{-1}. The molar absorptivity at 582 nm is 1.44×10^4. The sensitivity, detection limit, quantitation limit, and percentage recovery for 1.5 mg·L^{-1} fluoride was found to be 0.125 ± 0.003 µg·mL^{-1}, 0.2 mg·ℓ$^{-1}$, 0.5 mg·L^{-1}, and 97.1 ± 4.2, respectively.

Recently, Marques and Coelho [15] described a flow system for the determination of fluoride in natural waters, based on its reaction with zirconium ions and 2-(para-sulfophenylazo)-1,8-dihydroxy-3,6-naphthalene-disulfonate   (SPADNS). Under optimized conditions, a linear response was observed within the range of 0.1–2.2 mg.L^{-1}, with the detection limit, coefficient of variation and sampling rate estimated as 0.02 mg.L^{-1}), 4.1% and 60 determinations per hour, respectively. In order to analyze samples containing high fluoride content a wider linear range (0.3–6.6 mg.L^{-1}) can be obtained by using a low sample flow rate and low sample volume. This method is fast, amenable to automation, environmentally friendly and of low-cost. In addition, it could be successfully applied to the determination of fluoride in water samples; the results obtained being in agreement with those of the ion selective, ISE, method.

A novel solid phase extraction, SPE, method was proposed and used by Faraj-Zadeh and Kalhor [16] for the preconcentration and determination of fluoride in water. They used octyl chemically bonded SPE cartridge saturated with aluminum-oxinate chelate, where the water samples was passed through. The excess of oxine on the sorbent is washed with acetate buffer and finally the residual of aluminum-oxinate chelate on the sorbent is eluted by ethanol and its absorbance is read at 375 nm. The decrease of adsorbed aluminum-oxinate chelate on sorbent is proportional to the fluoride ion concentration in the water samples. The detection limit of this method is 80 ng/mL and the linear dynamic range is between 0.1–2 µg/mL. Sensitivity of the method is excellent and absorbance variation for each µg/ml fluoride ion is 0.6 (approximately equal to Eriochrome cyanine R-zirconium standard method and ten times better than the Alizarin complexon standard method). Results of this method have good agreement with the Eriochrome cyanine R-zirconium standard method. Also this method has less interference as compared to SPADNS standard method.

11.2.2.2.2   CYANIDE

Cyanide is a highly toxic pollutant and its determination in the environment receives increasing interest. Cyanide has been listed as one of the toxic pollutants that need to be monitored in the environment. Some of the proposed methods necessitate the separation of cyanide from the medium by distillation prior to the reaction with the coloring reagent [17]. Other methods involve a series of steps to form an intermediate product, which undergo a specific reaction, and production of a colored product. The measurement therefore does not involve the cyanide ion itself but a product with which cyanide has certain stoichiometry [18]. Table 11.1 shows a list of spectrophotometric methods for cyanide determination.

TABLE 11.1   Spectrophotometric Determination Methods of Cyanide

Medium	Wave-length, nm	Reagents and Methods	Linear range	Detection Limit or Sensitivity	Ref.
Wastewater	437	Oxidation by Cl_2 Indirect by Residual Cl_2 determination by color reaction with o-toluidine (3,3'-dimethylbenzidine).	0–0.01 mg..mL^{-1}	DL: 9.4 ppb	17
Air, industrial effluent, biological samples, pesticide and acrylonitrile.	445	pyridine to form glutaconic aldehyde. Aldehyde coupling with p-aminoacetophenone.	0.01–0.16 mg.L^{-1}	Sandell's sensitivity: 0.0001 g.cm^{-2}	18
Wastewater	578	Distillation, followed by reaction with pyridine-barbituric acid			19
Environmental waters		Phenolphthalein-EDTA	0.01–3.0 mg.L^{-1}	DL: 5 ppb	20
Water and soil extracts	575	Add phosphate buffer to CN solutions, then chloramines-T reagent. Add pyridine and barbituric acid solutions, stand for 1 hr.	0–100 mg.kg^{-1} soil	0.05 μg	21

Environmental protection Agency, EPA, offers *EPA Method 335.2:* for the determination of total cyanide in Water. The official Name of the method is Cyanide, Total (Titrimetric; Spectrophotometric). The method involves the release of cyanide as hydrocyanic acid (HCN) from cyanide complexes by means of a reflux-distillation operation and absorbed in a scrubber containing sodium hydroxide solution (Fig. 11.1). The cyanide ion in the absorbing solution is then determined by volumetric titration using silver nitrate or colorimetrically using pyridine-barbituric acid procedure. Spectrophotometer is suitable for measurements at 578 nm or 620 nm with a 1.0 cm cell or larger. Sulfide may be removed by reflux distillation.

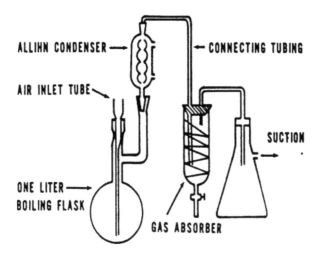

FIGURE 11.1   Cyanide Distillation Apparatus.

Recently, cobalt (II) phthalocyanine tetracarboxylate [Co (II)Pc-COOH] has been prepared and used in aqueous solutions as a novel chromogenic reagent for the spectrophotometric determination of cyanide ion [22]. The method is based on measuring the increase in the intensity of the monomer peak in the reagent absorbance at 682 nm due to the formation of a 1:2 [Co (II)Pc-COOH]: [CN] complex with a relatively high formation constant.. Interference by most common ions is negligible, except that by sulfite. The proposed method is used for determining cyanide concentration in gold, silver and chromium electroplating wastewater bath solutions after a prior distillation with 1:1 H_2SO_4 and collection of the volatile cyanide in 1 M NaOH solution containing lead carbonate as recommended by ASTM, USEPA, ISO and APAHE separation procedures. The results fairly well agree with potentiometric data obtained using the solid state cyanide ion selective electrode.

11.2.2.2.3   NITRITE

Nitrogen is present in the air and water environments in various chemical forms. Nitrite and nitrate anions are common in the environment. Bacterial as well as chemical factors are involved in the conversion of nitrogen from one chemical species to another. In aquatic systems, nitrite is formed by biological oxidation of ammonium, and converted subsequently to nitrate. High concentrations of nitrite are present in industrial water, sewage and in biologically purified effluents and in polluted streams. Traces of nitrite and nitrate in drinking water may lead to methemoglobinemia in infants [23]. Carcinogenic nitrosamines may form from the interaction of nitrite with secondary amines of the body. Thus, it is a health and environmental requirement to establish a reliable, simple and fast method for the estimation of nitrite in water. General procedures for the determination of nitrites are usually based up on some form of diazotization reaction. Cherian and Narayana [24] criticized the AOAC official method of analysis for nitrite and nitrate determination [25] for involving the use or production of carcinogenic intermediate compounds and need for careful control of acidity in each step of the process. They summarized the spectrophotometric methods for the determination of nitrite.

The authors of Ref. [24] used diazotization procedure to develop a selective and rapid spectrophotometric determination of nitrite in various water and soil samples. The method is based on the reaction of nitrite with p-nitroaniline in acid medium to form diazonium ion, which is coupled with ethoxyethylenemaleic ester or ethylcyanoacetate in basic medium to form azo dyes, showing absorption maxima at 439 and 465 nm, respectively. The method obeys Beer's law in the concentration range of 0.5–16 μg mL^{-1} of nitrite with ethoxy-ethylenemaleic ester and 0.2–18 μg mL^{-1} of nitrite with ethylcyano-acetate. The molar absorptivity and Sandell's sensitivity of p-nitro-aniline-ethoxyethylenemaleic ester and p-nitroaniline-ethylcyano-acetate azo dyes are 5.04×10^4 L mol^{-1}cm^{-1}, 0.98×10^{-2} μg.cm^{-2} and 1.21×10^4 L mol^{-1}. cm^{-1}, 0.98×10^{-2} μg.cm^{-2}, respectively. Heavy metals do not interfere and concentration levels of 20–100 folds were found tolerable with the method. In the meantime, there was no appreciable interference from anions at concentration levels of 200 to 400 folds of the analyte.

Nagaraja, et al. [26] introduced a new diazotizing reagent for the spectrophotometric determination of nitrite. The method is based on diazotization-coupling reaction between dapsone, an antibiotic compound, and phloroglucinol in hydrochloric acid medium. The reactions were conducted at room temperature, the molar absorptivity at 425 nm is 4.28×10^4 l.Mol.cm^{-1} and was stable for 50 h. Beer's law was obeyed in the nitrite range of 0.008–1.0 μg.ml^{-1}. Tolerance limits were tested for 33 species. The method has been found to be applicable for the determination of nitrite in natural and wastewater.

Revanasiddappa, et al. [27] used the diazotization with p-nitroaniline and coupling with acetyl acetone as the basis of the procedure of nitrite determination. The dye showed an absorption maximum at 490 and obeyed Beer's law over the range 0.05–0.14 μg.mL^{-1}. The molar absorptivity was 3.2×10^4 L.mol^{-1}.cm^{-1}. The method has been suggested for the determination of nitrite in water and soil samples.

A kinetic method for the spectrophotometric determination of nitrite in natural water was developed by Okutani, et al. [28]. They used the oxidation of 2,2'-azinobis(3-ethylbenzothiazoline-6-sulfonic acid by nitrite in an acid medium beyond the amounts of stoichiometric completion. A concentration of 10 ppb of nitrite could be determined.

11.2.2.3   ORGANIC POLLUTANTS

11.2.2.3.1   PHENOLS

Phenol is used in many industrial purposes like plastic and plywood manufacturing. Phenol has some toxic properties and any residual amount in wastewater may be hazardous. Determination of phenolic compounds in water is an important part of water quality measurement. The levels of phenol and phenolic compounds give an indication of the presence of pollution from industrial sources such as petroleum products, insecticide, herbicide, fungicide and pesticide residues. The presence, even in concentration of 1 ppb, of some phenols in drinking water supplies may lead, on chlorination, to the formation of objectionably tasting and odoriferous chlorophenols.

A colorimetric method was used for the monitoring of phenol removal from water, by sorption on rice husk, was developed by Munaf, et al. [29]. The method was based on oxidizing coupling of 4-aminoantipyrine (4-APP) with phenol. Phenol concentration was determined by measuring the absorption of the colored product at 506 nm. The method aided the evaluation of the role of several parameters on the phenol uptake such as particle size, pH and concentration of reagents. At the optimal conditions, phenol substance removal from aqueous solution is 92%. The method was applied to removal phenol substance present in plywood industry and hospital waste waters. Katsaounos, et al. [30] reevaluated the method in combination with micellar assisted preconcentration (cloud point extraction). The method employs the conventional reaction pathway while extraction was facilitated by surfactant based precipitation, during which the nonpolar derivative of 4-AAP-phenol was entrapped in the micelles and concentrated into a surfactant-rich phase. The latter is the resolubilized and the complex was quantified spectrophotometrically in the presence of a surfactant. Compared to the traditional method, the modification offered certain analytical advantages like massive analysis of many samples, lower detection limits and shorter time of analysis. The method was applied in various samples of different origin with satisfactory results.

The spectrophotometric properties of porphyrins are altered upon interaction with chlorophenols and other organochlorine pollutants. The interaction of many porphyrin compounds and their metal complexes with pentachlorophenol (PCP) have been shown by Awawdeh and James Harmon [31] to induce a red shift in the Soret spectrum with absorbance losses at various wavelengths, and the appearance of new peaks at different wavelengths. The intensity of the Soret spectral change is proportional to the pentachlorophenol concentration with a detection limit of 0.5–1.16 ppb. However, the dependence for concentrations less than 4 ppb the dependence was log-linear and linear for concentrations greater than 4 ppb. Monosulfonate tetraphenylporphyrin immobilized as a monolayer on a Kimwipe tissue exhibits an absorbance peak in the Soret region at 422 nm. The interaction of the porphyrin with PCP induces a red shift in the Soret spectrum with absorbance loss at 419 nm and the appearance of new peaks at 446 nm. The intensity of the Soret spectral change is proportional to the log of PCP concentration. The detection limit with immobilized TPPS$_1$ for PCP is 0.5 ppb. These results suggest the potential for development of spectrophotometric chemosensor for PCP residues in water with detection limits less than US EPA maximum contaminate level (MCL) of 1 ppb. The immobilized TPPS$_1$ on the Kim wipe will make it possible to develop wiping sensors to monitor the PCP or other pesticides residues on the vegetables or wood products.

11.2.2.3.2 DETERGENTS

Detergents are important category of domestic and industrial chemicals. They are used in huge quantities. Some of them are biodegradable under aerobic conditions and considered safe for the environment. The nonbiodegradable surfactants present an environmental problem. Alkyl benzene sulfonates are among the most used surfactants and they are released in high quantities to the surface water systems. Many authorities issued certain limits for these surfactants. They need to be quantified precisely and reliable methods for the determination of detergents in water must be available. Ghiasvand, et al. [32] described a simple, sensitive and selective spectrophotometric method for the determination of sodium dodecyl sulfate (SDS) after its separation by liquid- liquid extraction (LLE). Trace amounts of SDS were extracted into dichloromethane, based on ion-paired complex formation between SDS and Toluidin blue in pH=2.5, in a single step extraction. Under the optimal experimental conditions, absorbance of the organic extract obeyed Beer's law over the range of 0.05–4.00 µg mL^{-1} of SDS and the limit of detection, LOD, was 33.0 ng.mL^{-1}. The proposed methods were applied successfully for the determination of SDS in wastewater samples.

Adak et al. [33] used the ability of sodium dodecyl sulfate (SDS) to form yellow colored complex with acridine orange to establish a rapid procedure for its spectrophotometric determination in wastewater. The complex was extracted into toluene and the absorbance was measured at 467 nm. The linear range of the method was 0–6.0 ppm.

11.2.2.3.3 PESTICIDES RESIDUES

A sensitive and selective method for the preconcentration and determination of carbaryl, chlorpyrifos, linuron, and thiram was developed by Tunçeli, et al. [34]. The column sorption method was used for the preconcentration studies. Several parameters, such as amount of sorbent, pH, flow rate, volume of elution solution, and interferences that can influence the retention of pesticides on Saccharomyces cerevisiae immobilized on sepiolite were investigated. Results showed that it was possible to achieve quantitative analysis when the sample pH was in the range 4–6 for carbaryl and thiram, 4–8 for linuron and 6 for chlorpyrifos using 100 mL of sample solution containing 20 micrograms of pesticide and 5 mL of eluent. Recoveries of carbaryl, chlorpyrifos, linuron, and thiram were 93.2 ∓ 0.4%, 97.1 ∓ 0.3%, 98.5 ∓ 0.4%, and 96.1 ∓ 0.2%, respectively, at 95% confidence level under optimum conditions. The sorption capacity was 41, 28, 35, and 46 mg.g^{-1} for carbaryl, chlorpyrifos, linuron, and thiram, respectively. Saccharomyces cerevisiae immobilized on sepiolite is suitable for repeated use without loss of capacity up to 20 five cycles. The pesticides studied have been determined in river water with high precision and accuracy.

Multichannel spectrophotometry was performed an assay for paraoxon in spiked beverages. A 96-well microplate was used for this purpose. The measuring protocol was based on inhibition of the enzyme acetyl cholinesterase by paraoxon that resulted in decreased or no reaction of the enzyme product thiocholine with Ellman's reagent (5,5-dithiobis [2-nitrobenzoic acid]). The above assay was practically tested using spiked drinking water, mineral water, and coffee. Analytical parameters such as the limit of detection, time, and sample size consumption were adequate. The limit of detection for beverages ranged from 32 to 48 ppb corresponding to 0.32–0.48 ng of paraoxon in absolute values. The described assay seems to be convenient in terms of practical use [35].

Assaker and Rima [36] developed a simple spectrophotometric method for the determination of atrazine. This method is based on the complexation of atrazine derivatization (dechlorinated atrazine [DA]) with a mixture of formaldehyde and ketone compound, as described by Mannich reaction. The complex was determined by UV-Vis absorption measurement and the ketone compound used was the uranine due to its high coefficient absorption. The UV spectrum of the complex shows maxima of absorption at 207 nm and at 227 nm. An internal standard was used to quantify the atrazine. There is a good linearity between the absorbance and the concentration in the range of 0.1–10 μg.mL^{-1} of atrazine. The recovery value was 97% and the limit of detection was 0.01 μg.mL^{-1}. Real samples collected from irrigation local area were analyzed using this method and the estimated concentration of atrazine found in the mentioned river is 0.29 ± 0.011 μg.mL^{-1}.

Grahovac, et al. [37] developed a sensitive and simple kinetic-spectrophotometric method for the determination of the insecticide dimethoate [O,O-dimethyl-S-(N-methyl-carbomoylmethyl)-phosphoro-dithioate] is developed. The method was based on the inhibited effect of dimethoate on the oxidation of sulfanilic acid (SA) by potassium periodate in acetate buffer in the presence of Fe(III) ion and 1,10-phenantroline. Dimethoate was determined with linear calibration graph in the interval from 28.10 to 196.70 ng/mL. The reaction was followed spectrophotometrically at 368 nm. The developed procedure was successfully applied to the rapid determination of dimethoate in spiked milk and water samples. Liquid-liquid extraction and solid-phase extraction (SPE) was used for extraction of dimethoate from milk and water samples with Chroma bond (Macherey- Nagel) C18 cartridges. The HPLC method was used for comparison to verify the results. The results obtained by two different methods showed good agreement.

11.2.3 ATMOSPHERIC ANALYSIS

One of the early reports on the determination of atmospheric pollutants was the work of Philip W. West [38]. Among the various methods discussed were those of ozone and nitrogen oxides determination. For ozone determination, the concentration of pyridine-4- aldehyde produced by ozonolysis can be measured spectrophorometrically. Nitrogen oxides can be determined by utilization of diazotization with sulfanilic acid and N naphthylamine ethylene diamine for coupling. Analytical methods usually use chemical and/or physical properties of the analyte to establish a reliable method for the determination. Sulfur dioxide determination methods as an example make use of the reducing characteristic, acid behavior, and the specific reaction with complexing ligands [34]. For hydrogen sulfide determination, analysts can use either the reduction methods, or its capacity to form colored compounds with variety of heavy metals [39].

The need for gas compound measurement concerns overall three domains: environmental monitoring, emission measurement and risk assessment. These fields are different because of concentration range (from 10^{-3} to thousands mg.m^{-3}). A fast technique has been developed by Dupuit, et al. [40] based on UV spectrophotometry. Simple robust optics and absence of interference from water vapor and carbon dioxide are two of the main benefits of this method. All measurements are performed with a quartz flow cell of 10 cm path length. Under such conditions, the detection limits of various compounds (ammonia, hydrogen sulfide, sulfur dioxide and BTEX (benzene, toluene, ethylbenzene and p-xylene)) vary between 30 and 100 mg.m^{-3}. This UV spectrometry system has been tested with success in two applications. The first one is during gaseous ammonia dispersion, simulating a chemical accident. The second one is BTEX monitoring measurement in a process control of soil remediation. In this case, UV is associated with spectral data treatment software. All results are compared with reference methods (Nessler reagent for ammonia, gas chromatography for BTEX). An acceptable agreement was found.

Tyras [41] developed a spectrophotometric method for determination of methyl alcohol in the atmosphere. The method is based on adsorption of methyl alcohol on activated charcoal, desorption with carbon tetrachloride, oxidation of methyl alcohol to formaldehyde and spectrophotometric determination with chromotropic acid.

A simple and sensitive spectrophotometric method was developed for the determination of trace amounts of sulfur dioxide by Gayathri and Balasubramanian [42]. The method is based on the reaction of SO_2 with a known excess of ICl as the oxidant. The unreacted ICl iodinates thymol blue under acidic conditions. The λ_{max} of thymol blue is at 545 nm under acidic conditions, and on iodination λ_{max} shifts to 430 nm. This shift results in a decrease in the absorbance at 545 nm. The amount of uniodinated thymol blue present depends on the concentration of unreacted ICl, which in turn depends on the SO_2 concentration. The system obeys Beer's law in the range 0–30 mg SO_2 in a final volume of 25 mL, having a molar absorptivity of 3.2×10^4 L.mol^{-1}.cm^{-1} with a relative standard deviation (RSD) of 2% at 24 mg SO_2 (n = 10). The uniodinated dye can be extracted into 5 mL isoamyl alcohol under acidic conditions for measurement of the absorbance. The extraction method obeys Beer's law in the range 0–5 mg SO_2, having a molar absorpitivity of 4.16×10^4 L/mol·cm with an RSD of 1.9% at 4 mg SO_2 (n = 10). The method has been successfully applied to the determination of atmospheric SO_2.

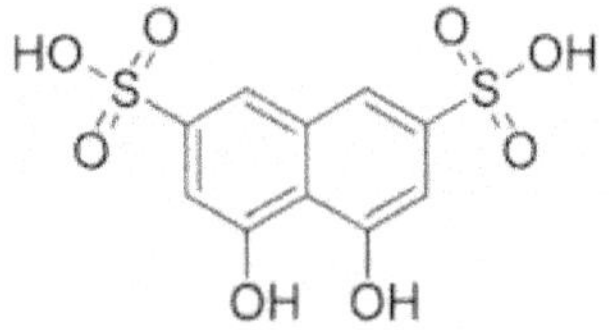

SCHEME 11.1

Good precision has been obtained. Coefficient of variation was found to be $\mp$ 4.96% for n = 24. The minimum determinable amount of methyl alcohol in the atmosphere was found to be 0.005 mg.m^{-3} by passing 720 L of air.

The use of a *p*-aminophenylazoic dye, *viz.* 4-(4-aminophenylazo)-1-naphthylamine, is proposed by Baiulescu, et. al. [43] as a spectrophotometric reagent for the determination of atmospheric sulfur dioxide absorbed in sodium tetrachloromercurate solutions. The determinations were made in ethanol-dye-formaldehyde systems displaying a red color at a pH value of 1.3. In the presence of sulfur dioxide solutions the red color turns to a blue one, which has a maximum absorption between 600 and 640 nm. The color development is instantaneous and is sufficiently stable to permit absorbance measurements. The reaction is subjected to interferences of nitrogen dioxide but this can be avoided by the use of an appropriate masking agent. The method allows between 0.07 and 2A mcg/mL of sulfur dioxide to be determined. The ratio of sulfur dioxide to ligand in the compound was found to be 1:1. The average of the instability constant was calculated to be 2.5×10^{-5}.

A spectrophotometric method was developed for the determination of dimethyl disulfide (DMDS) in complex gas mixtures emitted by the rubber industry [44]. The method allows organic disulfides to be quantitated in amounts exceeding 1.5 mg/m^3 gas (expressed as DMDS) and can be applied to the determination of disulfides in air at workplace as well as in gaseous products of thermal and biological decomposition of solid substances. The method enables disulfides and mercaptans to be determined simultaneously.

Bergshoef, et al. [45] has evaluated a simple, inexpensive and reliable manual spectrophotometric method for the determination of ozone in ambient air. Air stream is passed through indigo disulfonate (IDS) solution in a special double sintered disc absorber. The reduction in the blue color of the IDS solution can be measured at 610nm as the analytical signal of ozone. The authors reported that the stoichiometry of the reaction is independent of sample flow-rate and the amount of ozone absorbed. The decrease of absorbance caused by ozone is 21,960 L mol^{-1} cm^{-1}. Thus, the method can be used for simple determinations without calibration when an inaccuracy up to about 20% is admissible. Otherwise, a calibration against a reference method is necessary. In comparison with gas phase titration, the proposed method gave an apparent stoichiometry factor of 1.13. Nitrogen dioxide gives a slightly positive interference (6% of its mass) but the other common air pollutants, in ambient concentration, do not interfere. The relative standard deviation for determinations of ozone concentrations between 50 and 1000 µg m^{-3} is less than 5%. The limit of detection is about 8 µg.m^{-3}.

11.2.4   ATMOSPHERIC ANALYSIS

The amount of pesticides in agricultural products has been measured by different methods. Pesticide residues in soil, milk, and water can be analyzed successfully by spectrophotometric methods. Glyphosate is a broad spectrum herbicide and its behavior in soil is an important consideration for the evaluation of its environmental toxicity. The herbicide was separated by adsorption soils. The amino group of glyphosate was first transformed into the corresponding dithiocarbamate derivative in aqueous acetonitrile and allowed to react with copper (I) perchlorate to form yellowish green colored complex, showing λ_{max} at 392 nm. The method is quite sensitive and the molar absorptivity (ε) and Sandell's sensitivity were found to be 1.85×10^3 L.mol^{-1}.cm^{-1} and 0.091 µg cm^{-2}, respectively. The leaching potential of this herbicide vis-à-vis associated environmental pollution risk was evaluated by Groundwater Ubiquity Score (GUS) model. Its value has been observed < 1.8, which classifies glyphosate as nonleacher pesticide in terms of leaching behavior [46].

A new and highly sensitive spectrophotometric method was developed by Janghel, et. al. [47] for the determination of widely used organophosphorus pesticide monocrotophos at ppm levels. The pesticide was hydrolyzed in alkaline medium to N-methylaceto-acetamide followed by coupling with diazotized p-amino acetophenone. The absorption maxima of the reddish-violet colored compound formed is measured at 560 nm. Beer's law was obeyed over the concentration range of 1.2 to 6.8 µg in a final solution volume of 25 mL. The molar absorptivity and Sandell's sensitivity were found to be 7.1×10^5 (± 100) L mole^{-1} cm^{-1} and 0.008 µg.cm^{-2}, respectively. The standard deviation and relative standard deviation were found to be $\pm$ 0.005 and 2.05%, respectively. The method is simple, sensitive and free from interferences of other pesticides and diverse ions. The method has been satisfactorily applied to the determination of monocrotophos in environmental, agricultural and biological samples.

Alvarez-Rodríguez [48] used the spectrophotometric method for the determination of carbamate pesticides carbaryl, bendiocarb, carbofuran, methiocarb, promecarb and propoxur. The pesticides are hydrolyzed in alkaline medium to 1-naphthol or phenolates, which were coupled with diazotized trimethylaniline (TMA) in a sodium dodecyl sulfate micellar medium to form the azo dyes. The micellar medium increased the solubility of TMA and enhanced the sensitivity. The performance of TMA was compared with that of sulfanilic acid, which is the conventional reagent used in these determinations. Although the absorbance of the azo dyes of sulfanilic acid was also enhanced in the micellar solution, the sensitivity was still greater with TMA. Further, at the pH required for the coupling reaction, 9.5, the absorbance of the blank was negligible with TMA but large with sulfanilic acid. Using TMA, the limits of detection were in the range of 0.2–2 $\mu g.cm^{-3}$. The procedure was applied to the determination of carbaryl in spiked tap, river and pond water samples and to the evaluation of various carbamates in commercial pesticide formulations.

Polyacrylamide (PAM) is used to reduce soil erosion in irrigated land. The method finds increased attention recently and so was the concern of release of excess PAM to the water system. This implies the need for the establishment of a simple and reliable method to measure the residual PAM concentration in waters to assess the fate and efficiency of PAM application. Lu and Wu [49] developed and tested an analytical method to determine the PAM concentration of waters with correction for dissolved organic matter (DOM) interference. The method is based on a combination of determining the total concentration of amide groups by the N-bromination method (NBM) and determining the DOM content spectrophotometrically. The total concentration of amide groups of both PAM and DOM was determined by NBM at 570 nm. The DOM moiety, which is proportional to DOM concentration, was determined by spectrophotometry using a UV 254-nm wavelength. The actual PAM concentration of a water sample (soil extract containing PAM in this study) was obtained from NBM readings subtracted by the interferential DOM contribution using a correction curve. Analysis of PAM in two soil-water samples showed that the recoveries ranged from 94 to 100.3% for the 2 mg/L PAM sample and from 98.4 to 101.4% for the 10 mg/L PAM sample with various DOM concentrations. The coefficients of variation were <6% in all cases.

The color reaction of phenol with 1,2-naphthaquinone-4-sulfonic acid (NS) in alkaline medium (pH 8.7–9.3) was used by Chakravarty, et al. [50], for the determination of phenol in industrial wastewater. The reaction resulted in the formation of a bluish green indophenols dye absorbing specifically at 650 nm. The method obeyed Beer's law up to a concentration of 6.0 $\mu g.mL^{-1}$. The detection limit was 0.05 $\mu g.mL^{-1}$. The effects of the presence of other phenol and aromatic compounds are proved tolerable at levels of 3–8 folds of the analyte. The least effective compounds were the 3 and 4-aminophenols.

Higher-order derivative spectrophotometry (HODS) is a very good tool for the fine-resolution of spectra and other electric signals. This method allows one to separate superimposed curves for quantitative measuring. Talsky [51] demonstrated the advantages of the HODS for quantitative measurements of pollutants in water, air and soils. The author discussed the simultaneous estimation of aniline and phenol in waste water, the quantitative determination of PCP in polluted drinking water, phenol in turbid samples, the identification of aromatic amides and phenols in air after absorption in solvents.

A new spectrophotometric method has been reported by Murillo [52] for resolving binary mixtures of cephradine (CED) and clavulanic acid (CA). The method is based on the use of the first derivative of the ratio spectra obtained by dividing the amplitudes, at appropriate wavelengths, of the absorption spectra of the mixtures by those of a standard solution of clavulanic acid for determining cephradine and a standard solution of cephradine for determining clavulanic acid. Beer's law is obeyed for cephradine concentrations up to 56.0 mg.L^{-1} and for clavulanic acid concentrations up to 32.0 mg.L^{-1}. They claimed recovery values of cephradine and clavulanic acid in their 6:1 and 1:4 (m/m) mixtures.

11.3 APPLICATION OF FTIR METHODS

11.3.1 GENERAL

The Fourier Transform infrared, FTIR, spectroscopy, which is characterized by limited or no sample preparation, proved unique in the environmental applications for soil and atmospheric samples. The FTIR spectrophotometry is advantageous over the time consuming chemical methods for monitoring of routine composting processes. The method can be used to determine the efficiency of organic compounds decomposition in waste materials [53] as well as for process and quality control, for the assessment of abandoned landfills and for monitoring and checking of the successful landfill remediation [54].

11.3.2 SOIL ANALYSIS

As mentioned in Chapter 2, the IR absorption is correlated with vibrational changes of molecules. Inorganic as well as organic groups exhibit such vibrations and, thus, can be detected from their specific absorption. The intensity of the absorption is directly correlated with the concentration of the species. Cox et al. [55], used a novel FTIR technique to study the soil organic matter, SOM. The technique involved recording the spectra of soils and subtracted the spectra of the pyrolyzed soils. The resultant IR spectrum represents the organic portion of the sample. The use of organic components increases the discrimination in soils. The soil subtraction spectrum exhibited absorption bands at 3400, 2925, 2858, and 1730 cm^{-1} which are characteristic to soil humic compounds. The sharp band at 1050 cm^{-1} can be assigned to the Si–O stretch, which has not been completely eliminated in the subtraction process.

Jahn, et al. [56], developed a Mid-infrared (mid-IR) spectroscopy method for the analysis of nitrate in various soils. They used the FTIR attenuated total reflectance (ATR) technique in the range of 1–140 ppm. Three-dimensional plots were created by graphing the wavelet deconvoluted values at 32 scales for each sample. From each plot, the volume of the nitrate peak was determined and correlated to nitrate concentrations. Results of the laboratory experiments indicated values for the coefficient of determination R^2 as high as 0.98 and standard errors as low as 24 ppm.

Janik et al. [57], applied mid-infrared (MIR) spectroscopy and partial least-squares (PLS) analysis to predict the concentration of organic carbon fractions present in soil. The PLS calibrations were derived from a standard set of soils that had been analyzed for total organic carbon (TOC), particulate organic carbon (POC), and charcoal carbon (char-C) using physical and chemical means. PLS calibration models from this standard set of soils allowed the prediction of TOC, POC, and char-C fractions with a coefficient of determination (R^2) of measured versus predicted data ranging between 0.97 and 0.73. For the POC fraction, the coefficient of determination could be improved ($R^2=0.94$) through the use of local calibration sets. The capacity to estimate soil fractions such as char-C rapidly and inexpensively makes this approach highly attractive for studies where large numbers of analyzes are required. Inclusion of a set of soils from Kenya demonstrated the robustness of the method for total organic carbon and charcoal carbon prediction.

The FTIR spectroscopy can be used to describe soil characteristics in the form of complex multivariate datasets. The FTIR was successfully used to investigate soils at different stages of recovery from degradation following opencast mining and from undisturbed land [58]. Viscarra Rossel, et al. [59], reported the advantage of qualitative analysis in the mid-IR, MIR, of soil properties in comparison with other spectroscopic analyzes and with the standard methods. They examined soil pH, organic carbon, OC, cation exchange capacity, phosphorus, and electrical conductivity and found that accurate predictions can be obtained with MIR analysis.

When a FTIR spectrometer was used to determine gases released from soils and rock formations, no other gases than CO_2 have been detected except CO in the open-path compartment dedicated to atmosphere analysis [60]. The equipment consisted of packers isolating a chamber for gas collection. A pump allows gas transfer towards FT-IR sensors located at the surface. Such analytical approach shows several advantages for gas monitoring in boreholes: it allows variation detection with time of the partial pressure of gases; detection and evolution with time of the concentration of annex gases or markers; application to the injection and postinjection periods to determine possible deviations from a previously recorded baseline; and possible use of several boreholes in networks.

Bioremediation processes of polluted soils by microbial strains can be monitored by FTIR spectroscopy. Parikh and Chorover [61], studied the biomineralization by FTIR, which provided simultaneously molecular-scale information on both organic and inorganic constituents of a sample. The processes can be analyzed for cell adhesion, biofilm growth, and biological Mn-oxidation by *Pseudomonas putida* strain GB-1. Bhat et al. [62], employed FTIR to follow the microbial degradation of hydrocarbon contaminated Soil. The FTIR spectra of treated samples after several time intervals were compared with the initial spectra of the contaminated soils. The spectra of the treated samples revealed the presence of new bands pertaining to aliphatic and polycyclic aromatic hydrocarbons including various alcohols, aldehydes and ketones.

Soil can act as sinks as well as sources of carbon. A major fraction of carbon in soils is contained in the soil organic matter (SOM). It contributes to plant growth through its effect on the physical, chemical, and biological properties of the soil. DRIFTS spectroscopy can be used for the characterization of SOM for the determination of the overall quality of soils. It takes only a few Minutes and is much faster than fractionating the soil by wet chemical and/or physical methods [63].

Nault et al. [64], used DRIFT spectroscopy to compare changes in organic chemistry of 10 species of foliar litter undergoing *in situ* decomposition for 1 to 12 years. This study demonstrated that DRIFT spectroscopy is a fast and simple method for the analysis of large numbers of samples to give good estimates of litter chemistry. Thus, DRIFTS spectroscopy is considered a faster technique to analyze the composition and the dynamics of organic matter in soils compared to FTIR spectroscopy [65].

11.3.3 ATMOSPHERIC ANALYSIS

The basic principle and methods of FTIR spectroscopy for the study of atmosphere were presented by Bacsik, et al. [66] in a review article. The authors summarized the basic literature in the field of special environmental applications of FTIR spectroscopy, such as power plants, petrochemical and natural gas plants, waste disposals, agricultural, and industrial sites, and the detection of gases produced in flames, in biomass burning, and in flares [67].

FTIR showed advantages over traditional point-measurement methods by providing detection over large sampling areas [68]. This technique has increasingly been accepted by different environmental agencies as a tool in the measurement and the monitoring of the atmospheric gases [69]. FT-IR spectrometers can detect over a hundred volatile organic compounds (VOC) emitted from industrial and biogenic sources. Gas concentrations in stratosphere and troposphere were determined using FT-IR spectrometers [70].

Incorrect application of manure may introduce some air pollutants like ammonia, and nitrogen oxides, NOx to the atmosphere. In this respect, Galle, et al. [68], made some area-integrated measurements of ammonia emissions after spreading of pig slurry on a wheat field, based on gradient measurements using FTIR spectroscopy. They concluded that the gradient method is valuable for measurement of ammonia emissions from wide area, although the detection limits of the system limits its use to the relatively high emissions.

Volcanoes are considered important natural sources of air pollution. Mori and co-workers [71], used an FT-IR for the determination of HCl and SO_2 in volcanic gas. The most abundant gas typically released into the atmosphere by volcanoes is water vapor (H2O), followed by carbon dioxide (CO_2) and sulfur dioxide (SO_2). The observations over 15 years suggest that HCl/SO_2 and HF/HCl ratios are the most promising parameters reflecting volcanic activity among various parameters observable in remote FTIR measurements [72].

The FTIR can serve passive remote sensing and allows the detection and identification of pollutant clouds in the atmosphere. Beil, et al. [73] described the measurement technique and a data analysis method that does not require a previously measured background spectrum. They measured the spectral radiance of the environment after a radiometric calibration of the FTIR spectrometer with IR reference sources. With the inverse function of Planck's radiation law, the (spectrally resolved) brightness temperature is computed. The temperature spectrum has a constant baseline for many natural materials that serve as the background in field measurements (forest, etc.) because their emission is high and almost constant in the spectral range 800–1200 cm^{-1}. Enhancement of the signal to noise ratio have been shown to be achieved by the alignment of the spectrometer to backgrounds with a high temperature difference to the environment.

11.3.3.1   ANTHROPOGENIC AIR POLLUTION

11.3.3.1.1   AIR FLIGHT

Unburnt hydrocarbons, carbon monoxide, and nitrogen oxides are the main emissions from aircrafts. The improvement of turbine engine reduced the level of these pollutants in air. However, the concentration of such gases in air is relatively high near airports. For this reason a nonintrusive FTIR method to detect hydrocarbons in emissions from gas turbine engines [74], where no sampling system is required and there is no physical interference with the exhaust plume. Several species can be simultaneously monitored. The equipment is portable and can be simply set up and used outside the laboratory in engine test facilities, airfields.

The following classes of compounds: singly and doubly nitrogen substituted aromatic, terpenes, hemi-terpenes, retenes and other pyrolysis biomarker compounds, carboxylic acids and dicarboxylic acids were identified in gases from biomass burning [75]. Burling et al. [76], used OP-FTIR to detect and quantify 19 gas-phase species in biomass fires: CO_2, CO, CH_4, C_2H_2, C_2H_4, C_3H_6, HCHO, HCOOH, CH_3OH, CH_3COOH, furan, H_2O, NO, NO_2, HONO, NH_3, HCN, HCl, and SO_2.

11.3.3.1.2   BIOAEROSOLS

The significance of bioaerosols has been discussed in environmental and occupational hygiene. Identification of microorganisms using cultivation and microscopic examination is time consuming and alone does not provide sufficient information with respect to the evaluation of health hazards in connection with bioaerosol exposure. FTIR spectroscopy has widely been used for the characterization and identification of bacteria and yeasts, due to the fact that they are hydrophilic microorganisms and can easily be suspended in water for sample preparation [77]; Duygu et al. [78]. The identification of airborne fungi using FTIR spectroscopy was described by Fischer et al. [79]. They found that the method was suited to reproducibly differentiate *Aspergillus* and *Penicillium* species. The results obtained can serve as a basis for the development of a database for species identification and strain characterization of microfungi [79].

Earth's atmosphere contains aerosols of various types and concentrations: anthropogenic products, natural organic and inorganic products. The aerosol components interact with Earth's radiation budget and climate. As a result, of sunlight will be scattered and reflected back to the space. In lower atmosphere aerosols may modify the size of cloud particles, and hence affects the reflection and absorption of light by clouds. Aerosols are also chemically active and may result in change in the concentration of important species like ozone in the stratosphere. The inorganic components of aerosols consist of inorganic salts (e.g., sulfate, nitrate, and ammonium). Ion chromatography is used for the estimation of such species [80]. However, the method is destructive and time consuming in sample preparation and analysis. The FTIR is advantageous over the ion chromatography in this respect. Tsai and Kuo [81], employed the method to determine on-site chemical composition of aerosol samples and to investigate the relationship between particle compositions and diameters. They used DRIFTS for quantitative determination of SO_4^{2-}, NO_3^- and NH_4^+ in atmospheric aerosols.

Nitrogen dioxide determination in the atmosphere can be done by FTIR favorably over other methods like chemiluminscence and fluorescence. These methods involve multireagent procedures with the increased possibility of the experimental errors, in comparison with DRIFTS, which uses NaOH–sodium arsenite solution as an absorbing reagent. Consequently, nitrite can be determined by DRIFTS at submicrogram level [82].

The open-path FT-IR Spectroscopy is conventionally used for monitoring gaseous air pollutants. Air is allowed to enter the absorption cell and enters the optical path where species of interest will absorb the IR radiation [83]. Absorption will be recorded at predetermined frequencies that are specific for certain gases. Thus, no sample collection, handling or preparation is necessary. Real time data can be obtained for daily, weekly and seasonal reporting. However, OP-FTIR requires relatively high concentrations of gases such as stack measurement and landfill measurement [84]. The method is used *for in situ* application and the data can be stored and reanalyzed for a diverse range of volatile or nonvolatile compounds; cost effectiveness [85]. Perry et al. [86] and Tso and Chang [87], applied OP-FTIR to determine the VOC and ammonia concentrations in industrial areas, the concentration of pollutants at a level of 0.1 ppm. The applications of OP-FTIR in the monitoring of gaseous pollutants are shown in Table 11.2.

TABLE 11.2   Application of OP-FTIR

Media	Gases monitored	Reference
Air	Aerosols	[88]
Landfill gases	CH_4 and CO_2	[89]
Concentrated swine production facility	CO, CO_2, CH_4, NH_3 and N_2O	[90]

11.3.4   WATER POLLUTION

Lu, et al. [91], employed highly sensitive Mid-Infrared Sensor Technology in the determination of chlorinated aliphatic hydrocarbons and chlorinated aromatic hydrocarbons (CHCs) in water. These hydrocarbons are considered among toxic and carcinogenic contaminants commonly found in environmental samples. They used FTIR-attenuated total reflectance (FTIR-ATR) sensor for in-situ and simultaneous detection of multiple CHCs, including monochlorobenzene, 1,2-dichlorobenzene, 1,3-dichlorobenzene, trichloroethylene, perchloroethylene, and chloroform. For the detection the polycrystalline silver halide sensor fiber coated with an ethylene/propylene copolymer membrane acted as a solid phase extractor. This system exhibited a high detection sensitivity towards the CHCs mixture at a wide concentration range of 5~700 ppb and has a high potential to be used as a trace-sensitive on-line device for water contamination monitoring.

Grube et al. [92], used FTIR to analyze relevant amount of compositional and structural information concerning environmental samples. The analysis can be used to determine the nature of pollutants, and the bonding mechanism by which pollutants are removed by sorption processes. FTIR spectroscopy is used to characterize the sorbents or to establish the mechanism involved in sorption processes of heavy metals and organic compounds from wastewater [93–95].

Biosorption is considered as an alternative process for the removal of heavy metals, metalloid species, compounds and particles from aqueous solution onto biological materials. They have many advantages such as reusability, low operating cost, improved selectivity for specific metals of interests, removal of heavy metals found in low concentrations in wastewaters, short operation time, and no production of secondary compounds which can be toxic [96]. FTIR spectroscopy can be used for the characterization of biomaterials used in de-polluting processes, but also to characterize materials obtained after chemical modification of them. Thus, they used FTIR spectroscopy to characterize the material obtained after chemical modification of chitosan with glutardialdehyde in order to obtain a product with good sorption properties [97], but also to characterize the materials obtained after alkaline treatment of bentonite to increase its capacity to retain ammonium ions from synthetic solutions [98].

FT-IR spectroscopy has been used to identify the nature of interaction of the sorbent (biosorbent) with the pollutants (heavy metals, inorganic compounds, organic compounds). For copper removal by fungal biomass, FTIR spectra indicated that the stretching vibration of OH group was shifted from 3393 cm^{-1} to higher values in the presence of copper ions. Thus, the chemical interactions between the copper ions and the hydroxyl groups occurred on the biomass surface. The carboxyl peak observed for unloaded biomass at 1638 cm^{-1} is shifted to a lower value that indicated copper carboxylate formation [99, 100].

Brechbühl, et al. [101], employed ATR-FTIR to study the change in the sorption of hazardous arsenic species by competition with carbonate ions on hematite. The experimental conditions covered an acidic to neutral solution and fixing the CO_2 partial pressure to maintain various dissolved carbonate. Sorption data were modeled with a one-site three plane model considering carbonate and arsenate surface complexes derived from ATR-FTIR spectroscopy analyzes. The competitive effect of carbonate increased with increasing CO_2 partial pressure and decreasing arsenic concentrations. The results imply that in natural arsenic-contaminated systems where iron oxide minerals are important sorbents, dissolved carbonate may increase aqueous arsenite concentrations, but will affect dissolved arsenate concentrations only at neutral to alkaline pH and at very high CO_2 partial pressures.

Using ATR-FTIR spectroscopy, Christl, et al. [102], proved that silicate polymerizes on hematite surface and resulted in appreciable changes in the arsenic sorption. The competitive effect of silicate on arsenate and arsenite sorption increased with increasing silicate preequilibration time. They concluded that the long-term exposure to dissolved silicate can decrease the potential of natural iron (oxyhydroxides for adsorbing inorganic arsenic.

The FTIR and Attenuated Total Reflectance (ATR) spectroscopy in the mid infrared (MIR) wavelength range (2500–16,000 nm) have been also developed for contaminant detection in water. The authors tested the near infrared spectroscopy (NIRS) for the detection and quantification of pesticides including Alachlor and Atrazine in aqueous solution. Calibration models were built to predict pesticide concentration using PLS regression (PLSR). The proposed method shows potential for direct measurement of low concentrations of pesticides in aqueous solution. The research was performed in the laboratory conditions, and it is well known that the NIR spectrum of aqueous samples is susceptible to changes in the environment (e.g., temperature, humidity) and sample (e.g., pH, turbidity). Thus further experiments are necessary to test the effect of such perturbations on predictive ability [103].

For all real samples analyzed two-point baseline corrections were performed to obtain the quantitative absorption peak for sulfate at around 617 cm^{-1} [104]. Ma, et al. [105], studied the chemical composition, reactivity to SO_2 and hygroscopic properties reported a case study of Asian dust storm particles, using various techniques including FTIR. One of the most important applications of DRIFTS spectroscopy is to investigate sorption of some gases on various materials as summarized in Table 11.3.

TABLE 11.3 Application of DRIFTS in Sorption Studies

Gas	Sorbent	Reference
N_2 and O_2	Synthetic and natural mordenites, and on MS 4A, 5A and 13X	[106]
N_2 and mixtures with O_2	Zeolites NaLSX and NaZSM-5	[107]
CO_2	Silicate	[108]
SO_2	Iron oxides	[109]

Dust particles can react with gaseous components or pollutants from the atmosphere such as sulfur dioxide. The major components of Asian dust storm particles were aluminosilicate, SiO_2 and $CaCO_3$ mixed with some organic and nitrate compounds. The particles analyzed by Ma and co-workers [105] are coming from anthropogenic sources and local sources after long transportation. Between SO_2 uptake coefficient and mass was established a linear dependence.

From this study it can be concluded that adsorbed SO_2 could be oxidized on the surface of most iron oxides to form a surface sulfate species at ambient temperature, and the surface hydroxyl species on the iron oxides was the key reactant for the heterogeneous oxidation [109].

Heterogeneous reaction of NO_2 with carbonaceous materials (commercial carbon black, spark generator soot, diesel soot from passenger car and high-purity graphite) at elevated temperature (400°C) was studied using DRIFT spectroscopy. Different infrared signals appear when NO_2 is adsorbed either on aliphatic or graphitic domains of soot [110].

Gas sensors are playing an important role in the detection of toxic pollutants such as CO, H_2S, NO_x, SO_2, and inflammable gases such as hydrocarbons, H_2, CH_4. Undopped and Pd-doped SnO_2 sensor surfaces were characterized by recording DRIFTS spectra in parallel with electrical resistance measurements. It appeared that several reactions take place in the presence of CO depending both on temperature and humidity. It was found that all surface species are involved in the reactions and it is supposed that parallel and consecutive CO reactions take place on the surface [111]. All parts of the sensor (sensing layer, electrodes, and substrate) have influence to the gas detection and their role has to be taken into consideration when one attempts to understand how a sensor works [112].

Techniques for measuring gas pollutants such as continuous air pollutants analyzer (SO_2, NO_2, O_3 and NH_3), on-line gas chromatography (GC) used simple real-time instruments to quantify gas pollutants. They need to use several sensors in order to analyze multiple gas pollutants simultaneously.

11.4   APPLICATIONS OF TURBIDITY MEASUREMENTS

Turbidity is the amount of cloudiness in the water. The turbidity of domestic water supplies is an important parameter and must be measured as one the quality factors. Disposal of treated wastewater is often depends on the turbidity values. Turbidity of drinking water even at low levels will prevent the germs killing ability of chlorination. In chemical analysis, and as described in Chapter 2, the turbidity measurement cane used for the determination of chemical species like sulfate in water after precipitation as barium sulfate and induce the suspension by stirring in the presence of acidic sodium chloride solution. This necessitates the stability of the suspension through which radiation is passed for a definite time interval to complete the measurement. In some cases, stabilizers and thickeners are added to the suspensions to ensure reliable measurements.

The water of Senegal River showed the existence of the pollution from the turbidity during August; corresponding at rainy season. N'diaye, et al. [113] carried out a study to assess the water quality of the river by turbidity and Chemical parameters during the rainy season. They studied many samples for ammonium, (NH_4^+, Ortho-phosphates (PO_4^{3-}), Silica (SiO_2), oxidizable matters (OM), aluminum (Al), iron (Fe), manganese (Mn), zinc (Zn) and lead (Pb). Analysis observation revealed variation in the values of NH_4^+ from 0.01–1.12 mg.L^{-1}, PO_4^{3-} from 0.49–1.70 mg.L^{-1}, SiO_2 from 0.04–1.02 mg.L^{-1}, OM from 1.28–3.84 mg.L^{-1}, Al from 20–500 µg.L^{-1}, Fe from 170–320 µg/L^{-1}, Mn from 1–6 µg.L^{-1}, Zn from 20–100 µg.L^{-1} and Pb from 0.5–10.2 µg.L^{-1}. The use of turbidimetry for assessing water pollution is positive and motivating by its permanence, instantaneous but not sufficient.

Néstor Zárate, et al. [114], proposed an automatic procedure for turbidimetric determination of sulfate in rainwater, based on the multicommuted flow analysis approach. They developed the photometer set-up to include: a light-emitting diode, LED ($\lambda = 420$ nm), a long pathlength flow cell (100 mm), and a photodiode. A pre-concentration step is included in order to improve sensitivity. After establishing the adequate operational conditions, the proposed procedure presented the following useful features: a linear response ranging from 0.1 mg L^{-1} up to 2.0 mg L^{-1} sulfate, a detection limit (3σ criterion) of 0.04 mg L^{-1} sulfate, a relative standard deviation of 1.5% for 0.5 mg L^{-1} sulfate solution (n = 7), barium chloride consumptions of 17.0 mg per determination, and waste generation of 7.3 mL *per* determination. Accuracy assessment was achieved by applying the paired t-test between results obtained using the proposed procedure and ICP-OES and showed that there is no significant difference at 95% confidence level.

Lead, which can be found in old paint, soil, and dust, has been clearly shown to have adverse health effects on the neurological systems of both children and adults. Studabaker, et al. [115], described a new method for measuring lead extracted from paint with 25% (v/v) of nitric acid that is based on turbidimetry. An aliquot of the filtered extract is mixed with an aliquot of solid potassium molybdate in 1 M ammonium acetate to form a turbid suspension of lead molybdate. The lead concentration is determined using a portable turbidity meter. The method has a response of approximately 0.9 Nephelometric Turbidity Units (NTUs) per µg lead per mL extract, with a range of 1–1000 Nephelometric Turbidity Units (NTUs). Precision at a concentration corresponding to the EPA-mandated decision point of 1 mg of lead per cm^2 is <2%. This method suffer from interferences from Ba^{2+}, Ca^{2+}, Mg^{2+}, Fe^{3+}, Co^{2+}, Cu^{2+}, and Cd^{2+}, that might be found in the paint at concentrations of 10 mg.mL^{-1} or to Zn^{2+}, at 50 mg.mL^{-1}. The results from the analysis of 14 samples and reference materials showed a correlation coefficient of 0.97, with those obtained by inductively coupled plasma-atomic emission spectroscopy (ICP-AES). The average relative percent difference between the turbidimetric method and the ICP-AES method for the 24 sets measured as milligrams of lead per cm^{-2} is –0.63 +/– 32.5%; the mean difference is –2.1 +/– 7.0 mg lead per cm^2.

A rapid method for the determination of petroleum hydrocarbons in soil samples was described by Barbooti [116]. The method is based on the extraction of hydrocarbons by a solvent and the treatment of the solution with an aqueous solution of a surfactant to release the hydrocarbons to the water phase in the form of a stable emulsion. The emulsion is then used to measure the hydrocarbon content by turbidimetry. The effects of various operating parameters including the surfactant solution composition and time of extraction and time of mixing with the releasing solution are investigated. The normal method of stabilizing the emulsion was performed in acid solution of electrolyte. The turbidity values (T) were related with hydrocarbon concentration in the extract (C) by the following equation.

$$\text{Turbidity} = 2.75\ C + 205.7 \qquad (1)$$

With R^2 = 0.9929. The soil hydrocarbon content (SHC) measured in µg/g can then be calculated using the formula:

$$\text{SHC} = \frac{[Extract\,Vol.(mL) \times C]}{Sample\,weight,\ \text{g}} \qquad (2)$$

The results correlated well with the results of total hydrocarbons in soils determined by standard methods. The method was applied for the estimation of hydrocarbons in Passaic river sediments taken from various locations and depths. For field work the method was used to supply data on the hydrocarbon contamination of soil samples taken within an oil refinery and a monitoring well drilled within heavy hydrocarbon waste dumping location.

11.5   APPLICATIONS OF X-RAY FLUORESCENCE

XRF spectrometry ideally suits very fast qualitative and quantitative elemental analysis. Typically all elements in the periodic table from sodium (Na) through to uranium (U) can be detected simultaneously, with good quality spectra obtained in seconds/minutes. While X-ray fluorescence, XRF, is less sensitive than atomic spectrometry methods such as ICP-AES and ICP-MS, it offers a number of significant advantages including minimal sample preparation, rapid analysis times, multielement detection, and true field use using hand-held analyzers. Analysis using XRF technique can make an essential contribution to a wide range of applications including:

11.5.1   AIR POLLUTION AND CONTAMINATION OF WORK PLACE

Ambient aerosols consist of a wide variety of elements and compounds that can either condense to form clusters and particles up to sizes of ten micrometers or that were already abraded from a surface by some mechanical process. These particles remain suspended within the atmosphere for a time that depends on their particle size, weight, solubility and other physical factors. The connection between particle size and human health has been demonstrated in many studies. Smaller particles penetrate more deeply into the lung and are more difficult to be removed from the respiratory tract. But the effect of chemical composition on health is less well known and still remains a subject of study.

The main sources of air pollution are:
- Fuel combustion and emissions from motor vehicles, since loaded gasoline is in use.
- High loading from elements expected to be related to soil dust, such as calcium and Iron.

Various elements have adverse effects caused by their toxicity. Such elements are found in aerosol in small concentrations, and hence, they are not easy to measure. Therefore, the analysis of the trace elements with understanding their distribution is required to identify their emission sources and to develop a strategy for their reduction.

- Many air monitoring programs that look for respirable particulates in the air use suitable filters or collection membranes. Loaded filters can be presented directly to an XRF instrument, constitutes a major advantage of XRF over nearly all other trace analysis techniques like AAS, ICP, AES, MS, etc. [117]. Therefore the use of XRF should be encouraged in this field, even if other analysis method is available. Aerosol particles can be collected from atmospheric air either by filtration or by impaction [118].

11.5.2 SOILS

The examination of elemental distribution in soil is very important because under environmental conditions it can accumulate or release them. Assessment of metal contents in soils and the risks due to exposures are important in environmental management and overall protection of human health [119]. Elemental concentrations in soils arise from both natural processes and anthropogenic pollution sources. Trace elements such as zinc, chromium, copper, cobalt, and iron are beneficial for both plants and humans, but toxic effects may manifest at concentrations higher than certain threshold for each element. In particular, childhood lead poisoning remains a serious environmental health problem, affecting the central nervous system and acting as cofactor in many other illnesses [120]. X-ray fluorescence spectrometric methods may be more suitable for rapid soil environmental applications because of their element specificity, high measurement precision, minimal sample preparation, and the high sample throughput, and in some cases, their portability to make onsite multielemental analyzes [121].

KEYWORDS

- **Atmospheric Analysis**
- **FTIR Applications**
- **Soil Analysis**
- **Turbidity Measurements**
- **UV/Vis Applications**
- **Water Analysis**
- **X-Ray Fluorescence**

REFERENCES

1. Marczenko, Z. (1986). Separation and Spectrophotometric Determination of Elements, Horwood, E., Series.
2. Levy, E. M. (1972). The Identification of Petroleum Products in the Marine Environment by Absorption Spectrophotometry, Water Res, *6(1)*, 57–69.
3. Spence, S. L. (2011). Rapid Spectrophotometric Detection for Analysis of Bacterial contamination in Water, M.Sc. Thesis, Physics Dept., Colorado State University.

4. Barbooti, M. M., Bolzoni, G., Mirza, I. A., Pelosi, M., Barilli, L., Kadhum, R., & Peterlongo, G. (2010). Evaluation of Quality of Drinking Water from Baghdad, *Science World Journal*, *5(2)*, 35–46.

5. Davey, M. W., Montagu, M. V., Inze, D., Sanmartin, M., Kanellis, A., Smirnoff, N., Benzie, I. J.J., Strain, J. J., Favell, D., & Fletcher, J. (2000). Review plant L-Ascorbic Acid: Chemistry, Function, Metabolism, Bioavailability and Effects of Processing. *J. Sci. Food Agric*, *80*, 825–860.

6. Deutsch, J. C. (1998). Spontaneous Hydrolysis and Dehydration of Dehydroascorbic Acid in Aqueous Solution, *Anal Biochem*, 260, 223–229.

7. Jezierska, K., Gonet Podraza, W., & Domek, H. (2011). A New Method for the Determination of Water Quality, Water SA, Pretoria, *37(1), Online version* ISSN 1816–7950.

8. Baker, A., Inverarity, R., Charlton, M., & Richmond, S. (2003). Detecting River Pollution using Fluorescence Spectrophotometry, Case Studies from the Ouseburn, NE England, Environmental Pollution, *124(1)*, 57–70.

9. WHO, Fluoride in Drinking-Water. IWA Publishing, London, (2006).

10. WHO, Guidelines for Drinking-Water Quality (2nd edn.) *1*. World Health Organization, Geneva, 375–377, (2004).

11. Czarnowski, W. R., Zesniowska, K., & Krechniak, J. (1996). Fluoride in Drinking Water and Human Urine in Northern and Central Poland, Sci. Total Environ, *191*, 177–184.

12. Meyling, A. H., & Meyling, J. (1963). A Modified Zirconium-Alizarin Method for determining Fluoride in Natural Waters, the Analyst, *88*, 84.

13. Paul, E. D., Gimba, C. E., Kagbu, J. A., Ndukwe, G. I., & Okibe, F. G. (2011). Spectrometric Determination of Fluoride in Water, Soil and Vegetables from the Precinct of River Basawa, Zaria, Nigeria, *J. Basic* Appl. Chem., *1(6)*, 33–38.

14. Barghouthi, Z., & Amereih, S. (2012). Spectrophotometric Determination of Fluoride in Drinking Water Using Aluminium Complexes of Triphenylmethane Dyes, Water SA, *38(4)*, 543–548.

15. Marques, T. L., & Coelho, N. M. (2013). Proposed Flow system for Spectrophotometric Determination of Fluoride in Natural Waters, Talanta, *105*, 69–74 doi: 10.1016/j.talanta.2012.11.071.

16. Faraj-Zadeh, M. A., & Kalhor, E. G. (2001). Extraction Spectrophotometric Method for Determination of Fluoride in the Range of Microgram per Liter in Natural Waters, Microchim Acta, *137(3–4)*, 169–171.

17. Gumus, G., Demirata, B., & Apak, R. (2000). Simultaneous Spectrophotometric Determination of Cyanide and Thiocyanate after Separation on Melamine-Formaldehyde Resin, Talanta, *53*, 305–315.

18. Agrawal, O., Sunita, G., & Gupta, V. K. (2005). A Sensitive Colorimetric Reagent for the Determination of Cyanide and Hydrogen Cyanide in Various Environmental Samples, *J. Chinese Chem. Soc*, *52*, 51–57.

19. Osobamiro, M. T. (2012). Determination of the Concentration of Total Cyanide in Waste Water of a Tobacco Company in Southwestern Nigeria, *J. Appl. Sci. Environ. Manag*, *16(1)*, 61–63.

20. Cacace, D; Ashbaugh, H; Kouri, N; Bledsoe, S., Lancaster, S., & Chalk, S. (2007). Spectrophotometric Determination of Aqueous Cyanide using a Revised Phenolphthalin Method, *Anal. Chim Acta*, *589*, 137–141.

21. Environment Agency, The Determination of Cyanide and Thiocyanate in Soils and Similar Matrices, in "Methods for the Examination of Waters and Associated Materials", (2011), 57 pages.

22. El-Nemma, E. M., Abd-Rabboh H. S. M., & Hassan, S. S. M. (2010). A Novel Spectrophotometric Method for Determination of Cyanide using Cobalt (II) Phthalocyanine Tetracarboxylate as a Chromogen, *Internat. J. Environ. Anal Chem, 90(2)*, 148–158.

23. World Health Organization, WHO, Nitrate and Nitrite in Drinking-Water, Background Document for Development of WHO Guidelines for Drinking-Water Quality, Geneva, (2011).

24. Cherian, T., & Narayana, B. (2006). A New System for the Spectrophotometric Determination of Trace Amounts of Nitrite in Environmental Samples *J. Braz. Chem. Soc, 17(3)*, 577–581.

25. AOAC Method. Official Methods of Analysis, 16th ed., AOAC, (1995), 8–9.

26. Nagaraja, P., Kumar, M. S. H., & Mallikarjuna, N. N. (2002). Dapsone *a New Diazotizing Reagent for the Spectrophotometric Determination of Nitrite in Waste and Natural Water Samples*, In: Annali di Chimica, *92 (1–2)*, 127–134.

27. Revanasiddappa, H. D., Kumar, K., & Bilwa, M. (2001). A Facile Spectrophotometric determination of Nitrite using Diazotization with P-Nitroanaline and coupling with Acetyl Acetone, *Microchimica Acta, 137(3–4)*, 249–253.

28. Okutani, T., Sakuragawa, A., Kamikura, S., & Shimura, M. (1991). Spectrophotometric Determination of Nitrite Based on Its Catalytic Effects on the Oxidation of 2, 2'-Azinobis (3-Ethylbenzothiazoline-6-Sulfonic Acid. *Anal Sci., 7*, 793–797.

29. Munaf, E., Zein, R., Kurniadi; R., & Kurniadi, I. (1997). The Use of Rice Husk for Removal of Phenol from Waste Water as Studied Using 4-Aminoantipyrine, Spectrophotometric Method, *Environ. Technology, 18(3)*, 355–358.

30. Katsaounos, C. Z., Paleologos, E. K., Giokas, D. L., & Karayannis, M. I. (2003). The 4-Aminoantipyrine Method Revisited: Determination of Trace Phenols by Micellar Assisted Preconcentration, Internat. *J. Environ. Anal Chem., 83(6)*, 507–514.

31. Awawdeh, A. M., & James Harmon, H. (2005). Spectrophotometric Detection of Pentachlorophenol (PCP) in Water using Immobilized and Water soluble Porphyrins, *Biosensors and Bioelectronics, 20(8)*, 1595–1601.

32. Ghiasvand, A. R., Taherimaslak, Z., & Allahyari, M. (2009). Sensitive and Selective Spectrophotometricand a New Adsorptive Stripping Voltammetric Determination of Sodium Dodecyl Sulfate in Nonaqueous Solution after Its Extraction Using Toluidine Blue, *Int. J. Electrochem. Sci, 4*, 320–335.

33. Adak, A., Pal, A., & Bandyopadhyay, M. (2005). Spectrphotometric Determination of Anionic Surfactants in Wastewater using Acridine Orange, *Indian J. Chem. Technol, 1*, 145–148.

34. Tunçeli, A., Bag, H., & Türker, A. R. (2001). Spectrophotometric Determination of Some Pesticides in Water Samples after Preconcentration with Saccharomyces Cerevisiae Immobilized on Sepiolite, *Fresenius J Anal Chem, 71(8)*, 1134–1138.

35. Pohanka, M., Zdarova Karasova, J., Kuca, K., & Pikula, J. (2010). Multichannel Spectrophotometry for Analysis of Organophosphate Paraoxon in Beverages, *Turk J Chem, 34*, 91–98.

36. Assaker, K., & Rima, J. (2012). Improvement of Spectrophotometric Method for the Determination of Atrazine in contaminated Water by Inducing of Mannich Reaction, *J Food Res, 1(4)*, 17–26 ISSN 1927–0895.

37. Grahovac, Z. M., Miti, S. A. S., Pecev, E. T., & Pavlovi, A. N. (2010). Development of New Kinetic-Spectrophotometric Method for Determination Insecticide Dimethoate in Milk and Water, *J. Chinese Chem. Soc, 57*, 1027–1034.

38. West, P. W., http://pac.iupac.org/publications/pac/pdf/1970/pdf/21.

39. Nair, E. M. S., & Kuriakose, V. (1999). Detection Sampling and Analysis of Air Pollutants, Proc. 1st International Seminar, SAFE '99, Safety & Fire Engineering, Cochin, India, Nov, 24–26.

40. Dupuit, E., Dandrieux, A., Kvapil, P., Ollivier, J., Dusserre, G., & Thomas, O. (2000). UV Spectrophotometry for Monitoring Toxic Gases, *Analusis*, *28*, 966–972.
41. Tyras, H. (1989). Spectrophotometric Determination of Methyl-Alcohol in the atmosphere, Gesamte, Z., Hyg, *35(2)*, 96–97.
42. Gayathri, N., & Balasubramanian, N. (2001). Spectrophotometric Determination of Sulfur Dioxide in Air, using Thymol Blue, *J. AOAC International*, *84(4)*, 1065–1069.
43. Baiulescu, G. E., Marcuta, P. C., & Marinescu, D. M. (1973). Spectrophotometric Determination of Atmospheric Sulfur Dioxide with 4(4-aminophenylazo)-1-Naphthylamine, *Internat. J. Environ Anal. Chem*, *2(3)*, 203–211.
44. Turek, A., Skrzydlinska, M., & Ptaszynski, B. (1996). Spectrophotometric Determination of Dimethyl Disulfide in Air Pollutants from the Rubber Industry, *Collect. Czech Chem. Communication*, *61*, 1738–1744.
45. Bergshoef, G., Roelof, W., Lanting, R. W., Ham, J. V., Prop, J. M. G., & Reijnders, H. F. R. (1984). Spectrophotometric Determination of Ozone in Air with Indigo Disulphonate, *Analyst*, *109*, 1165–1169.
46. Sharma, D. K., Gupta, A., Kashyap, R., & Kumar, N. (2012). Spectrophotometric Method for the Determination of Glyphosate in Relation to Its Environmental and Toxicological Analysis, *Arch. Environ. Sci*, *6*, 42–49.
47. Janghel, J. K., Rai, M. K., & Rai, V. K. (2006). A New and Highly Sensitive Spectrophotometric Determination of Monocrotophos in Environmental, Agricultural and Biological Samples, *J. Chinese Chem. Soc*, *53(2)*, 343–347.
48. Alvarez-Rodríguez, L., Monferrer-Pons, L. l., Esteve Romero, J. S., García-Alvarez-Coque, M. C., & Ramis-Ramos, G. (1997). Spectrophotometric Determination of Carbamate Pesticides with Diazotized Trimethylaniline in a Micellar Medium of Sodium Dodecylsulfate, *Analyst*, *122*, 459–463.
49. Lu, J. H., & Wu, L. (2001). Spectrophotometric Determination of Polyacrylamide in Waters Containing Dissolved Organic Matter, *J Agric Food Chem*, *49(9)*, 4177–4182.
50. Chakravarty, S., Deb, M. K., & Mishra, R. K. (1994). Simple Spectrophotometric Determination of Phenol in Industrial Wastewater, *Asian J. Chem*, *6(4)*, 766–770.
51. Talsky, G. (1983). Higher-Order Derivative Spectrophotometry in Environmental Analytical Chemistry, Internat, *J. Environ. Anal Chem.*, *14(2)*, 81–89.
52. Murillo, J. A., Lemus, J. M., & Garcia, L. F. (1993). Application of the Ratio Spectra Derivative Spectrophotometry to the Analysis of Cephradine and Clavulanic Acid in Binary Mixtures, *Fresenius' J.Anal Chem.*, *347 (3–4)*, 114–118.
53. Smidt, E., & Meissl, K. (2007). The Applicability of Fourier Transform Infrared (FT-IR) Spectroscopy in Waste Management, Waste Management, *27*, 268–276.
54. Grube, M., Muter, O., Strikauska, S., Gavare, M., & Limane, B. (2008). Application of FT-IR Spectroscopy for Control of the Medium Composition during the Biodegradation of Nitro Aromatic Compounds, *J. Ind. Microbiol Biotechnol*, *35*, 1545–1549.
55. Cox; R. J., Peterson, H. L., Young, J., Cusik, C., & Espinoza, E. O. (2000). The Forensic Analysis of Soil Organic by FTIR, *Forensic Sci. Internat*, *108*, 107–116.
56. Jahn, B. R., Linker, R., Upadhyaya, S. K., Shaviv, A., Slaughter, D. C., & Shmulevich, I. (2006). Mid-infrared Spectroscopic Determination of Soil Nitrate Content, *Biosystems Eng.*, *94(4)*, 505–515.
57. Janik, L. J., Skjemstad, J. O., Shepherd, K. D., & Spouncer, L. R. (2007). The Prediction of Soil Carbon Fractions using Mid-Infrared-Partial Least Square Analysis, *Australian J. Soil Res*, *45*, 73–81.
58. Elliott, G. N., Worgan, H., Broadhurst, D., Draper, J., & Scullion, J. (2007). Soil Differentiation Using Fingerprint FTIR Infrared Spectroscopy, Chemometrics and Genetic Algorithm-based Feature Selection Soil. *Biol. Biochem*, *39*, 2888–2896.

59. Viscarra Rossel, R. A., Walvoort, D. J. J., McBratney, A. B., Janik, L. J., & Skjemstad, J. O. (2006). Visible, Near Infrared, Mid Infrared or Combined Diffuse Reflectance Spectroscopy for Simultaneous Assessment of Various Soil Properties, *Geoderma, 131*, 59–75.

60. Pironon, J., de Donato, P., Barrès, O., Garnier, C. (2009). Online Greenhouse Gas Detection from Soils and Rock Formations, *Energy Procedia, 1*, 2375–2382.

61. Parikh, S. J., & Chorover, J. (2005). FTIR Spectroscopic study of Biogenic Mn-Oxide formation by Pseudomonas Putida GB-1, *Geomicrobiol J, 22*, 207–218.

62. Bhat, M. M., Shankar, S., Shikha, Y. M., & Shukla, R. N. (2011). Remediation of Hydrocarbon Contaminated Soil through Microbial Degradation FTIR Based Prediction, *Adv. Appl. Sci. Res, 2(2)*, 321–326.

63. Zimmermann, M., Leifeld, J., & Fuhrer, J. (2007). Quantifying Soil Organic Carbon Fractions by Infrared Spectroscopy, *Soil Biol Biochem, 39*, 224–231.

64. Nault, J. R., Preston, C. M., Trofymow, J. A. T., Fyles, J., Kozak, L., Siltanen, M., & Titus, B. (2009). Applicability of Diffuse Reflectance FTIR Spectroscopy to the Chemical Analysis of Decomposing Foliar Litter in Canada, Forests, *Soil Sci., 174(3)*, 130–142.

65. Tremblay, L., & Gagné, J-P. (2002). Fast Quantification of Humic Substances and organic Matter by Direct Analysis of Sediments using DRIFT Spectroscopy, *Anal Chem., 74*, 2985–2993.

66. Bacsik, Z., & Mink, J. (2007). Photolysis-Assisted, Long-Path FT-IR Detection of Air Pollutants in the Presence of Water and Carbon Dioxide, *Talanta, 71*, 149–154.

67. Bacsik, Z., Mink, J., & Keresztury, G. (2005). FTIR Spectroscopy of the Atmosphere Part 12 Applications, Appl. Spectrosc. Rev. (2004) *39*, 295–363 and Part 2, *40*, 327–390.

68. Galle, B., Klemedtsson, L., Bergqvist, B., Ferm, M., Törnqvist, K., Griffith, D. W. T., Jensen, N. O., & Hansen, F. (2000). Measurements of Ammonia Emissions from Spreading of Manure using Gradient FTIR Techniques, Atmos. Environ, *34*, 4907–4915.

69. Russwurm, G. M. (1999). Compendium of Methods for the Determination of Toxic Organic Compounds in Ambient Air, 2nd Ed. Long-Path Open-Path Fourier Transform Infrared Monitoring of Atmospheric Gases–Method, *16*, 161–164.

70. Puckrin, E., Evans, W. F. J., & Adamson, T. A. B. (1996). Measurement of Tropospheric Ozone by Thermal Emission Spectroscopy, Atmospheric Environ, *30(4)*, 563–568.

71. Mori, T., Notsu, K., Tohjima, Y., & Wakita, H. (1993). Remote Detection of HCl and SO_2 in Volcanic Gas from Unzen Volcano, Japan Geophys Res. Lett., *20*, 1355–1358.

72. Notsu, K., & Mori, T. (2010). Chemical Monitoring of Volcanic Gas Using Remote FT-IR Spectroscopy at several Active Volcanoes in Japan Appl Geochem., *25*, 505–512.

73. Beil, A., Daum, R., Matz, G., & Harig, R. (1998). Remote Sensing of Atmospheric Pollution by Passive FTIR Spectrometry" in Spectroscopic Atmospheric Environmental Monitoring Techniques, Klaus Schäfer, Herausgeber, Proc. SPIE, *3493*, 32–43.

74. Arrigone, G. M., & Hilton, M. (2005). Theory and Practice in Using Fourier Transform Infrared Spectroscopy to Detect Hydrocarbons in Emissions from Gas Turbine Engines. Fuel, *84*, 1052–1058.

75. Johnson, T. J., Profeta, L. T. M., Sams, R. L., Griffith, D. W. T., & Yokelson, R. L. (2010). An Infrared Spectral Database for Detection of Gases emitted by Biomass Burning, Vibrational Spectrosc, *53*, 97–102.

76. Burling, I. R., Yokelson, R. J., Griffith, D. W., Johnson, T. J., Veres, P., Roberts, J., Warneke, C., Urbanski, S. P., Reardon, J., Weise, D. R., Hao, W., & De Gouw, J. A. (2010). Laboratory Measurements of Trace Gas Emissions from Biomass Burning of Fuel Types from the Southeastern and Southwestern United States, *Atmospheric Chem. Phys., 10(22)*, 11115–11130.

77. Essendoubi, M., Toubas, D., Bouzaggou, M., Pinon, J. M; Manfait, M., & Sockalingum, G. D. (2005). Rapid identification of Candida Species by FT-IR Microspectroscopy, *Biochim Biophys Acta, 1724(3)*, 239–247.

78. Duygu, D. (Yalcin); Baykal, T., Açikgöz, I., & Yildiz, K. (2009). Fourier Transform Infrared (FT-IR) Spectroscopy for Biological studies, *G.U. J. Sci.*, *22(3)*, 117–121.

79. Fischer, G., Braun, S., Thissen, R., Dott, W. (2006). FT-IR Spectroscopy as a Tool for Rapid Identification and Intra-species Characterization of Airborne Filamentous Fungi, *J. Microbiol. Methods*, *64*, 63–77.

80. Chen, Y., & Wang, J. (2012). Removal of Radionuclide Sr²⁺ Ions from Aqueous Solution using Synthesized Magnetic Chitosan Beads, *Nucl Eng. Design*, *242*, 445–451.

81. Tsai, Y. I., & Kuo, S. C. (2006). Development of Diffuse Reflectance Infrared Fourier Transform Spectroscopy for the Rapid Characterization of Aerosols, *Atmospheric Environment*, *40*, 1781–1793.

82. Verma, S. K., Deb, M. K., & Verma, D. (2008). Determination of Nitrogen Dioxide in Ambient Air Employing Diffuse Reflectance Fourier Transform Infrared Spectroscopy, *Atmos. Res*, *90*, 33–40.

83. Minnich, T. R., & Scotto, R. L. (1999). Use of Open-Path FTIR Spectroscopy to Address Air Monitoring needs during Site Remediation, *Remediation J, 9(3)*, 79–92.

84. Hong, D. W., Heo, G. S., Han, J. S., & Chao, S. Y. (2004). Application of the Open Path FTIR with COLISB to Measurement of Ozone and COCs in the Urban Area, Atmos. *Environ*, *38*, 5567–5576.

85. Marshall, T. L., Chaffin, C. T., Hammaker, R. M., & Fateley, W. G. (1994). An Introduction to Open-Path FT-IR, Atmospheric Monitoring. *Environ Sci. Technol*, *28(5)*, 224A-232A.

86. Perry, S. H., McKane, P. L., Pescatore, D. E., DuBois, A. E., & Kricks, R. J. (1995). Maximizing the Use of Open-Path FTIR for 24-h Monitoring around the process Area of an Industrial Chemical Facility, AWMA Conference on Optical Remote Sensing for Environmental and process Monitoring SPIE, *2883*, 333–344.

87. Tso, T. L., & Chang, S. Y. (1996). Unambiguous Identification of Fugitive Pollutants and the Determining of Annual Emission Flux as a Diurnal Monitoring Mode using Open-Path Fourier Transform Infrared Spectroscopy, *Analytical Sciences*, *12*, 311–319.

88. Wu, C. F., Chen, Y. L., Chen, C. C., Yang, T. T., & Chang, P. E. (2007). Applying Open-Path Fourier transform Infrared Spectroscopy for Measuring aerosols, *J. Environ Sci. Health*, Part A, *42(8)*, 1131–1140.

89. Hegde, U., Chang, T. C., & Yang, S. S. (2003). Methane and Carbon Dioxide emissions from Shanchu-ku Landfill site in Northern Taiwan, *Chemosphere*, *52*, 1275–1285.

90. Childers, J. W., Thompson, Jr., E. L., Harris, D. B., Kirchgessner, D. A., Clayton, M., Natschke, D. F., & Phillips, W. J. (2001). Multi Pollutant Concentration Measurements around a Concentrated Swine Production Facility using Open-Path FTIR Spectrometry, *Atmos. Environ*, *35(11)*, 1923–1936.

91. Lu, R., Mizaikoff, B., Li, W. W., Qian, C., Katzir, A., Raichlin, Y., Sheng, G., & Yu, H. Q. (2013). Determination of Chlorinated Hydrocarbons in Water using Highly Sensitive Mid Infrared Sensor Technology, *Sci. Rep.*, *3*, 2525.

92. Grube, M., Muter, O., Strikauska, S., Gavare, M., & Limane, B. (2008). Application of FT-IR Spectroscopy for Control of the Medium Composition during the Biodegradation of Nitro Aromatic Compounds, *J. Ind. Microbiol Biotechnol*, *35*, 1545–1549.

93. Chen, Y., & Wang, J. (2012). Removal of Radionuclide Sr2+ Ions from Aqueous Solution using Synthesized Magnetic Chitosan Beads Nucl Eng. *Design*, *242*, 445–451.

94. Jordan, N., Foerstendorf, H., Weiß, H. K., Schild, D., & Brendler, V. (2011). Sorption of Selenium (VI) Onto Anatase: Macroscopic and Microscopic Characterization. Geochim Cosmochim. *Acta*, *75(6)*, 1519–1530.

95. Parolo, M. E., Savini, M. C., Vallés; J. M., Baschini, M. T., & Avena, M. J. (2008). Tetracycline Adsorption on Montmorillonite: pH and Ionic Strength Effects. *Appl. Clay Sci.*, *40*, 179–186.

96. Mungasavalli, D. P., Viraraghavan, T., & Chunglin, Y. (2007). Biosorption of Chromium from Aqueous Solutions by Pretreated Aspergillus Niger: Batch and Column Studies, Colloids. Surf A Physicochem. Eng. *Aspects*, *301*, 214–223.

97. Deleanu, C., Simonescu, C. M., & Căpăţînă, C. (2008). Comparative Study on the Adsorption of Cu(II) Ions Onto Chistosan and Chemical Modified Chitosan, Proc. 12th Conf. Environ Mineral Proc., Part III–5. 7.6. Ostrava, Czech Republic, 201–207.

98. Simonescu; C. M., Deleanu, C., Bobirică, L., Melinescu, A., & Giurginca, M. (2005). Bentonite and Nabentonite Used in Ammonium Removal from Wastewaters, Proc. 14th Romanian Internat Conf Chem Chem.Eng., Ed. Printech, Bucharest 22–24 Sep, *6*, S, 206-262.

99. Yee, N., Benning, L. G., Phoenix, V. R., & Ferris, F. G. (2004). Characterization of Metal-Cyanobacteria Sorption Reactions, A Combined Macroscopic and Infrared Spectroscopic Investigation, *Environ. Sci. Technol.*, *38*, 775–782.

100. Burnett, P. G. G., Daughney, J. C., & Peak, D. (2006). Cd adsorption onto Anoxybacillus Flavithermus: Surface Complexation Modeling and Spectroscopic Investigations, Geochim. *Cosmochim Acta, 70*, 5253–5269.

101. Brechbühl, Y., Christl, I., Elzinga, E. J., & Kretzschmar R. (2012). Competitive Sorption of Carbonate and Arsenic to Hematite: Combined ATR-FTIR and Batch Experiments, *J Colloid Interface Sci, 1, 377(1)*, 313–21.

102. Christl, I., Brechbühl, Y., Graf, M., & Kretzschmar, R. (2012). Polymerization of Silicate on Hematite Surfaces and its Influence on Arsenic Sorption, *Environ. Sci. Technol, 46*, 13235–13243.

103. Gowen, A., Tsuchisaka, Y., O'Donnell, C., & Tsenkova, R. (2011). Investigation of the Potential of near Infrared Spectroscopy for the Detection and Quantification of Pesticides in Aqueous Solution, *Am. J. Anal Chem, 2*, 53–62.

104. Verma, S. K., & Deb, M. K. (2007). Direct and Rapid Determination of Sulfate in Environmental Samples with Diffuse Reflectance Fourier Transform Infrared Spectroscopy using KBr Substrate, *Talanta, 71*, 1546–1552.

105. Ma, Q., Liu, Y., Liu, C., Ma, J., & He, H. (2012). A Case Study of Asian Dust Storm Particles: Chemical Composition, Reactivity to SO2 and Hygroscopic Properties. *J. Environ. Sci., 24(1)*, 62–71.

106. Valyon, J., Lónyi, F., Onyestyák, G., & Papp, J. (2003). DRIFT and FR Spectroscopic Investigation of N_2 and O_2 adsorption on Zeolites, Microporous and Mesoporous Materials, *61(1–3)*, 147–158, ZEOLITE '02 (Proc. 6th International Conference on the occurrence), Properties and Utilization of Natural Zeolites.

107. Kazansky, V. B., Sokolova, N. A., & Bülov, M. (2004). DRIFT Spectroscopy Study of Nitrogen Sorption and Nitrogen-Oxygen Transport Co-Diffusion and Counter-Diffusion in NaLSX and NaZSM-5 Zeolites. *Microporous Mesoporous Mater., 67*, 283–289.

108. Llewellyn, P. L., & Theocharis, C. R. (1991). A Diffuse reflectance Fourier Transform Infrared Study of Carbondioxide Adsorption on Silicalite-I, *J. Chem. Technology Biotechnology, 52*, 473–480.

109. Fu, H., Wang, X., Wu, H., Yin, Y., & Chen, J. (2007). Heterogeneous Uptake and Oxidation of SO₂ on Iron Oxides, *J. Phys. Chem. C, 111*, 6077–6085.

110. Muckenhuber, H., & Grothe, H. (2007). Drifts Study of the Heterogeneous Reaction of NO_2 with Carbonaceous Materials at Elevated Temperature, *Carbon, 45(2)*, 321–329.

111. Harbeck, S. (2005). Characterization and Functionality of SnO_2 Gas Sensors using Vibrational Spectroscopy, Thesis, Tubingen.

112. Bârsan, N., & Weimar, U. (2003). Understanding the Fundamental Principles of Metal Oxide Based Gas Sensors; the Example of CO Sensing with SnO_2 Sensors in the Presence of Humidity, *J. Phys.: Condens Matter, 15* R813 doi:10.1088/0953–8984/15/20/201.

113. N'diaye, A. D., Dhaouadi, H., El-Kory, M. B., & Ould Kankou, M. O. S. A. (2013). Water Quality Assessment of Senegal River in Mauritania by Turbidity and Chemical Parameters Analysis during Rainy Season, Issues *in Biol. Sci. Pharmac Res*, *1(2)*, 016–021.
114. Néstor Zárate, N., Pérez-Olmos, R., & Freire Dos Reis, R. (2011). Turbidimetric Determination of Sulfate in Rainwater Employing a LED Based Photometer and Multicommuted Flow Analysis System with In-Line Preconcentration, *J. Braz. Chem. Soc*, *22(6)*, http://dx.doi.org/10.1590/S0103–50532011000600002.
115. Studabaker, W. B., McCombs, M., Sorrell, K., Salmons, C., Brown, G. G., Binstock, D., Gutknecht, W. F., & Harper, S. L. (2010). Field Turbidity Method for the Determination of Lead in Acid Extracts of Dried Paint, *J. Environ Monit. 12(7)*, 1393–403.
116. Barbooti, M. M. (2011). Turbidimetric Determination of Hydrocarbon Contamination in Passaic River Sediments and Refinery Polluted Soils, *J. Environ. Protection, 2*, 915–922.
117. Steinhoff, G., Haupt, O., & Dannecker, W. (2000). Fast Determination of Trace Elements on Aerosol Loaded Filters by XRF Analysis *Considering the Inhomogeneous Elemental Distribution. J. Anal. Chem, 366*, 174–177.
118. Clark, S., Menrath, W., Chen, M., Roda, S., & Succop, P. (1999). Use of a Field Portable X-Ray Analyzerto Determine the Concentration of Lead and Other Metals in Soil Samples, Annals of Agricultural and Environmental Medicine, *6*, 27–32.
119. Biasioli, M., Gronan, M., Kralj, H., Madrid, T., Diaz Barrientos, E., & Ajmone-Marsan, E. (2007). Potentially Toxic Elements Contamination in Urban Soils, J. Environ Qual, 36, 70–79.
120. Brewster, U. C., & Perazella, M. A. (2004). A Review of Chronic Lead Intoxication, Am. J. Med. Sci, 327(6), 341–347.
121. Dao, T. H., & Miao, Y., & Zhang, F. (2011). X-Ray Fluorescence based Approach to Precision Measurements of Bioavailable P in Soil Environments, J. Soil Sediments, 11, 577–588.

CHAPTER 12

APPLICATIONS OF ELECTROANALYTICAL METHODS

FARNOUSH FARIDBOD, PARVIZ NOROUZI, and
MOHAMMAD REZA GANJALI*

Center of Excellence in Electrochemistry, Faculty of Chemistry, University of Tehran,
Tehran, Iran; *E-mail: ganjali@khayam.ut.ac.ir

CONTENTS

12.1   INTRODUCTION

Potentiometric and potentiostatic methods have been applied for determination of various inorganic and organic pollutants in atmosphere, water and soils samples. Electroanalytical methods offer advantages of selectivity, sensitivity, simplicity, and inexpensive. Also, they have a wide linear range of concentration, which can be extended from Pico molar to micromolar by select of appropriate analytical parameters. Moreover, they have a minimum sample treatment. The analytical instrumentation used for electroanalysis could be easily portable, automated and used for on-line measurements.

Potentiometry using ionophore-based indicator electrodes are well-established method routinely used for the selective, online and direct measurements of a wide variety species (inorganic/organic cations or anions) in the environmental samples.[1,2]

There are some general experimental points, which should be considered before a technique is going to be applied.

Almost every species, which is able to form ionic species, can be determined by direct potentiometric method.

In potentiostatic techniques, the species should be dissolved in an appropriate liquid solvent and capable of being reduced or oxidized within the potential range of the technique and electrode material. If the species are not electroactive but they can react with electroactive materials, it is possible to determine them indirectly.

The amounts of sample, which are needed to measure the concentrations, vary greatly with the applied technique. Volumes of the samples can vary from about 20 mL to less than one microliter. In a normal experiment 5–10 mL of analyte solution is required. The required volume of the sample depends on the shape of the electrodes. Nowadays, by advanced developments in the field of electrochemistry as well as electronics and biology, sensitive electroanalytical devices (such as sensors or biosensors) are made which are able to do the analysis using small amount of sample as much as a drop of blood (about 5 μL).

Without considering the time spent on the sample preparation, the time required to obtain a signal varies from a few seconds (in single-sweep square-wave voltammetry), to a couple of minutes (in a cyclic voltammetry), to 30 min (or more) (in a very-low-concentration Adsorptive Striping Voltammetry).

Accuracy of the electrochemical methods varies with the used technique from 1 to 10%.

12.2   CLASSIFICATION OF SAMPLES AND ANALYTES IN ENVIRONMENTAL WORKS

The pollutants in the environments can be found mostly in the water and soil. However, sometimes some pollutants in biological samples like leaf, or vegetables, … may be analyzed too. A variety of inorganic and organic pollutants are spread in the air, water and ground. Depends on the type of sample and kind of analyte, sampling process and its storage are different.[3] In Table 12.1, a nice classification of the type of sample and analytes in environmental works are shown.

TABLE 12.1   Classification of Samples and the Analytes in Environmental Works

Phase of sample	Type of sample	Major analytes

Gas sample	• Atmospheric gases
	• Indoor air
	• Exhaust gases from ve-
	  hicles
	• Industrial gases
	• Pollutant gases
		• In form of gas or vapor
		  – Inorganic gases
		  – Organic gases
		• In form of dust, aerosol cations,
		  anions, heavy metal, some or-
		  ganic compound

Liquid sample	• Drinking water
	• Tab water
	• Mineral water
	• River water
	• Lake water
	• Sea water
	• Rain water
	• Ground water
	• Waste water
		• Dissolved inorganic compounds
		  – Inorganic gases
		  – Organic gases
		  – Heavy metals
		  – Cations
		  – Anions
		• Dissolved organic compounds
		  – Surfactants
		  – Pesticides
		  – Insecticides
		  – Fertilizer

Solid sample	• Soil
	• Sediment
	• Dusts
	• Dusts from electronic
	  devices
	• Plant materials
	• Fly ashes
	• Ashes
	• Wastes
	• Tissue or organs of liv-
	  ing organism
		• Inorganic compounds
		  – Anions
		  – Cations
		  – Heavy metals
		• Organic compounds
		  – Pesticides
		  – Insecticides
		  – Fertilizers
		  – Dioxin

Most of the analytes listed in above table including cations, anions, heavy metals, some pesticides, insecticides or fertilizers can be determined by electroanalytical methods.

12.3   GENERAL SAMPLE PREPARATIONS FOR ELECTROANALYTICAL MEASUREMENTS

Sample preparation is the step that plays an important role in the quality of analytical results especially when they are used for decision making in the area of environmental protection and management.

Some sample preparation is required in electroanalytical techniques depends on the sample and the applied technique. For example, in determination of Pb^{2+} and Cd^{2+} ion in seawater with a microelectrode and square-wave anodic stripping voltammetry (ASV), no preparation is required. In fact, the simplest preparation is in case of natural water. In contrast, determination of pesticides in blood plasma at a glassy carbon electrode with differential pulse voltammetry (DPV) requires that the sample first be pretreated with several reagents, buffered, and separated.

In general, all electroanalytical methods used in environmental monitoring can be divided into two main groups.

12.3.1   DIRECT METHOD

When analyte can be determined directly in a sample: This approach is applicable only when the matrix is relatively simple (no interfering species or it is not require to remove or mask the interference). Also the analyte concentration should be the higher than the detection limit of the method or instrument. Because of this, direct methods find only limited applications in analysis of environmental samples. Potentiometric techniques (ion selective electrodes) used in the analysis of drinking water or river water is a good example of this approach.

12.3.2   INDIRECT METHOD

When trace or ultra-trace levels of an analyte are analyzed: In this case the analyte is analyzed in a matrix obtained from sample preparation or analyte isolation or pre-concentration. The sample preparations and treatments are mostly performed in case of increasing the analyte concentration to a level higher than the detection limit of the used technique or for removing the interferences, or in order to store the sample for prolonged periods of time.

Since electrochemical methods have been mostly used for soil and water samples, here, these samples are more discussed.

12.3.3   WATER SAMPLES

An advantage of electrochemical analysis for water samples is that many of such samples like mineral, tab, drinking water samples require no pretreatment and the only measuring is required. However, before starting the analysis, would be better the samples pass through 0.45 μm filters to remove particulates. The filtrate is then acidified to pH 2 by the addition of ultra-pure hydrochloric acid. This prevents adsorption of the analyte ions on the walls of the container, and causes dissociation of metal ions from some complexes, thereby making these ions available for the analysis. The acidified samples can be stored at $-20°C$. In some water samples, there is a substantial concentration of organic material, which can form stable complexes with the metal ions and can also adsorb onto the electrode surface. These interfering effects can be eliminated by destroying the organic material by either UV-irradiation or acid digestion. Acid digestion or low-temperature ashing can also be used for pretreatment of the suspended and particulate matter removed in the filtration. These pretreatments do not necessarily have to be used for all environmental samples. For example, tap, and rain waters do not have particulate matter or significant organic content, so the filtration and UV-irradiation steps are not necessary. For samples like wastewater, which have a rather high ionic strength, only a dilution step would be required after filtration.

12.3.4   SOLID SAMPLES

Solid environmental samples include soils or sludge. Such samples cannot be introduced to the electroanalysis system and require additional pretreatments, like extraction, digestion, or alkali fusion etc.

Since most ionic species can readily dissolve in water, water extraction can be used to transfer such species to an aqueous phase and their prior to analysis. To do this a known amount of the sample is added to a known volume of an extracting solvent, and mixed. Mixtures of water and miscible solvents like light alcohols or dilute solutions acids, bases, and salts like potassium chloride or phosphate buffers can be used in the extraction step.

It is clear that simple extraction might be inefficient for rather solid samples due to the low recovery of the analyte. Under such circumstances acid digestion or alkali fusion are used. To perform acid digestion concentrated acids or their mixtures is used.

Another alternative method for the preparation such samples is alkali fusion. In general the sample and a proper alkaline compound (e.g., Na_2CO_3, Na_2O_2, lithium tetraborate, or NaOH) are mixed and heated to until they are molten. Next the melt is cooled before being dissolved in a digestion solution. The resulting solution might require further treatment before being injected to the electroanalysis system.

Another conventional method for preparing the solid samples for electroanalyzes involves complete combustion of the sample in the presence of oxygen. As a result the metallic elements remain in the ash, while the nonmetallic ones are converted to gaseous compounds, which can be absorbed by a scrubbing solution. Combustion methods have been used for the determination of halides and total sulfur, nitrogen, and phosphorus in samples such as plant materials, silicate rocks, coal, and oil shale.

12.4   GENERAL STEP-BY-STEP MEASUREMENT GUIDE

Here a general step by step guideline is presented used in an electroanalytical method. To have a successful analysis, a number of preparative steps need to be made before a simple concentration measurement can be taken. Besides, the experimental procedure must be evaluated. All glassware must be clean and any analytical equipment set up according to the manufacturer's instructions [1, 2].

STEP 1. PREPARATION THE STANDARD SOLUTIONS

Standard solutions are a range of solutions with known concentration, which are prepared from the dilution of the stock solution. The pH and the ionic strength of the standard solution as much as possible should be the same with the unknown solution.

STEP 2. PREPARATION OF THE SAMPLE SOLUTION

Based on the type of the sample, different sample preparation is needed to digest the sample. Sometimes is needed to dilute the sample solution, which it can be placed in the calibration series.

STEP 3. ADJUSTMENT OF THE IONIC STRENGTH, PH OF THE SOLUTIONS AND SELECT A SUITABLE ELECTROLYTE

It is now necessary to adjust the solution ionic strength by addition of an ionic strength adjustment buffer (ISAB). An example would be in potassium determinations. It is necessary to add 2 mL of 2M Ammonium Sulphate ISAB per 100 mL of potassium standard or sample. Also, the pH of the solution should be adjusted in the applicable pH range of the used method. Selection of a suitable electrolyte depends on the type of analyte, the applied method and the pH of the measurements. Many voltammetric measurements used buffer solution as both electrolyte and adjustment of the solution pH.

STEP 4. ELECTROCHEMICAL MEASUREMENTS

A galvanic or electrolytic cell assembly can be applied based on the selected technique.

Numerous types of electrodes can be used as working electrodes. However, some times the modification of the electrode surface is needed to do more sensitive and selective determinations.

In case of potentiometry, place the ion selective electrode and the reference electrode into a known quantity of the least concentrated standard. Allow sample to reach room temperature and pressure. Stir quietly. Measure the electrode potential.

In case of potentiostatic techniques, the major components are a three-electrode cell, voltage generator (potentiostat) and recorder (or computer). Be sure all the connections are connected. The working, reference and auxiliary electrodes (WE, RE and AE, respectively) are immersed in the sample containing the analyte and a supporting electrolyte. The voltage generator controls the potential of the WE during the preconcentration step (deposition) and measuring cycle (potential scan). The potential of the RE remains constant. The AE serves to conduct current from the source to the WE. During the potential scan, this current is measured and recorded as a function of the potential difference between WE and RE. Setup the important parameters of the used technique in the software of instrument (such as, scan rate, potential window, amplitude, ...). In most case, the voltage is changed linearly at a fixed rate (mV/s) during the potential scan. Other waveforms, for example square wave (SW) and differential pulse (DP), have been developed to improve the separation between capacitative and faradic components in the current signal. The faradic current is proportional to the concentration of the analyte in solution, while the capacitative component is the desired result of the electrical double layer, formed at the electrode surface. Other advantages of these waveforms are increased speed and sensitivity of the analysis, improved peak separation between analytes and reduced interferences caused by surface active compounds in solution.

Now, allow sample to reach room temperature and pressure. Then start the measurement. It should be noted that, in case the solved oxygen in the solution interferes in the redox process of the analytes, it is needed before the measurements, nitrogen or any inert gas pass through the solution. Also, some times is needed the solution heated or even stirred gently.

STEP 5. CLEAN THE ELECTRODES

Rinse the electrodes with distilled water and if needed clean the surface by alumina (in case of solid electrodes), repeat step 6 using the next concentration of standard solution, in series, until all the standards have been measured.

STEP 6. PLOT THE CALIBRATION CURVE

This is obtained by measuring the electrochemical cell potentials difference in standard solutions.

STEP 7. SAMPLE MEASUREMENT

Take the same volume of sample as standards and place both electrodes in that sample; stir and allow reaching the room conditions. The electrode potential of the unknown sample is measured and the concentration displayed on the meter. Using this method, known as direct measurement, it is possible to determine the concentration of numerous samples in a very short time.

It should be noted that, procedure for striping voltammetry is a little bit different from the above mentioned. In stripping voltammetry (SV) a potential is applied to the electrode surface, which causes the species, accumulate on the electrode surface, then another potential is applied to the working electrode to strip the species from the electrode surface to the solution. In adsorptive striping voltammetry (AdSV) methods, a specific ligand can be applied to help the collection of the species from the bulk. Depends on the analyte, a suitable ligand is selected to modify the surface of the electrode. Apply a constant potential to preconcentrate analyte on the electrode surface usually under conditions of forced convection (by stirring). For example in an AdCSV, an electrode surface can be modified by a suitable ligand to form a complex with a certain metal cation. The deposition potential is slightly more positive (by ≥ 0.1 V) than the reduction potential of the analyte. During deposition, a fraction of the complex forms a mono-molecular layer on the electrode surface. After a resting period (equilibration time), a potential scan to more negative voltages is carried out in the quiet solution. At a potential specific to the analyte-ligand complex, the adsorbed metal is reduced and stripped from the electrode surface. The recorded reduction current is measured as the height or area under the peak above the baseline, which is proportional to the metal concentration in the cell.

12.5   ANALYSIS OF POLLUTANTS IN WATER SAMPLES

Water is one of the vital environments, which can easily undergo of pollutants. Water pollutions can be the source of many diseases in human, and animals. As well, they can affect the life of microorganisms and plants. Water pollution occurs when pollutants are directly or indirectly discharged into water bodies without adequate treatment to remove harmful compounds. Pollution can enter to drinking, tab, mineral, lake, river, sea, oceans, rain, ground waters and waste water from industrial and agricultural area. The pollutants can be inorganic species or even organic ones. Many instrumental analytical methods have been used for analysis of water pollutants. Besides, some electroanalytical techniques are also used for pollutants monitoring in water samples. Among the techniques stripping voltammetry and potentiometric methods are used more than the others. The first one is used because of the very low detection limits and the second one because of the simplicity, portability and inexpensively.

12.5.1. INORGANIC POLLUTANTS IN WATER

Inorganic water pollutants which can be analyzed by electrochemical methods include dissolved inorganic anion and cations and heavy metals. Normally the concentration of the heavy metals in water samples like, drinking, mineral and sea water is too low. Since some of heavy metals are toxic for living organism, trace analysis of them in environmental samples is of great importance. Toxic cations can be measured by even potentiometric or potentiostatic techniques. The detection limit of potentiometric electrodes in the best condition can be 1.0×10^{-8} M. However, for determination of lower amounts of these elements stripping voltammetric techniques can be used.

Several analytical techniques used for trace metal analysis includes Inductively Coupled Plasma Optical Emission Spectroscopy (ICP-AES), Inductively Coupled Plasma-Mass spectrometry (ICP-MS), and Graphite Furnace Atomic Absorption Spectroscopy (GFAAS). However, these methods need preconcentration and matrix removal procedures to remove major ion interferences and to lower the limit of detection. For example, in preparation for sea water analysis using GFAAS, samples may be subjected to the complexation of the analyte with dipyrrolidine dithiocarbamate/ammonium pyrrolidine dithiocarbamate, then extract the metal complex into chloroform, and back extraction to nitric acid. In this way, trace metals may be concentrated several hundred fold.

In comparison to these methods, stripping voltammetry is particularly suited for trace metal analysis, because a preconcentration step is an integral part of the measuring cycle. The use of the adsorptive cathodic stripping voltammetry (AdCSV) and its application for dissolved trace metal determination in different matrixes (i.e., seawater, freshwater, rocks, blood, biological samples) is known since several years [4–6].

Years after years, new methods have been developed for the determination of almost each metal and nowadays, around 20 elements can be determined at trace levels in natural waters using adsorptive cathodic stripping voltammetry (AdCSV) [6]. Although the pretreatment of the sample is simple (the determination is almost direct), it has low detection limits (Pico molar to nanomolar) and needs a low volume of sample (around 10 mL), it takes a long time to analyze one sample in comparison with other techniques like inductively coupled plasmamass spectroscopy (ICP-MS) or flow injection analysis (FIA). The multielement detection capability is one of the main advantages of the AdCSV technique.

A methodology has been reported for a direct simultaneous determination of Co, Cu, Fe, Ni and V in sediment pore waters using a mixture of ligands dimethyl-glyoxime (DMG) and catechol by AdCSV in a single scan [6]. By this method, the time of analysis and volume of sample needed were reduced. Detection limits of the technique were 0.04 nM for Co, 0.09 nM for Cu, 1.29 nM for Fe, 0.46 nM for Ni and 2.52 nM for V which making the method suitable for the direct simultaneous determination of these five metals in pore waters, estuarine waters and probably coastal waters. The instrument used was a Metrohm VA-797 computerize equipped with a hanging mercury dropping electrode (HMDE) as working electrode (drop surface area 0.38 mm^2), an Ag/AgCl as reference electrode and a Pt wire as auxiliary electrode. Samples were deoxygenated withN$_2$ (N-50 grade), presaturated with water vapor by passage through a gas scrubbing tube containing Milli-Q (Millipore) water. Short sediment cores (approx. 10 cm) were sampled in 2005 onboard the R/V Mytilus (IIM-CSIC) in the main channel of the Vigo Ria (NW Iberian Peninsula). Acid-washed methacrylate tubes placed inside a Rouvilloise grab sampler were used to collect the cores. Once at the onshore lab, sediment cores were stored in plastic bags –18°C. In order to preserve the redox conditions, all sample treatment was undertaken inside a glove box filled with N2 (815-PGB, Plas-Labs). Frozen cores were extruded from the tube and the overlying water was placed into acid-cleaned plastic bottles whereas the sediment was sliced at 2–3 cm layers and allows thawing inside acid-cleaned polyethylene centrifuge tubes. In order to extract the pore water, sediments were centrifuged at 3000 rpm for 40 min, and the supernatant was passed through acid-washed 0.45μm acetate cellulose syringe filters and collected in acid-washed 50mL polyethylene bottles and acidified (pH 2) with concentrated HNO$_3$. The overlying waters were filtered and acidified as above. Optimization experiments were carried out using a composite sample of the above mentioned cores. This sample was diluted (×10) and UV-digested for 2 h in the presence of H$_2$O$_2$ using a UV-Digestor equipped with a high-pressure mercury lamp of 200W, in order to remove interfering organics and to breakdown organic complexing ligands. From the tests it was concluded that 2 h of irradiation time with a volume of 20 μL for each 20 mL sample are the optimal conditions. Pore water thus treated is hereafter termed "UVPW." Dissolved metal concentrations in the sample were 0.92 nM (Co), 7.14nM (Cu), 49,38nM (Fe), 7.22nM (Ni) and 11.3 nM (V). Aliquots of 9 mL of UVPW were pipetted into 10 mL PFA tubes (Nalgene). The pH was adjusted near neutral with appropriate amounts of diluted ammonia solution and HEPES buffer and ligands (catechol and DMG) were added. The sample was then transferred to the voltammetric cell and purged for 10 min. The stirrer rod was switched to 2000 rpm, three mercury drops were discarded and the deposition period, at the appropriate potential, was initiated after extrusion of the fourth. At the end of the adsorption period, the sample was allowed to rest for 15 s and the voltammogram was recorded as the potential was scanned in the negative direction to –1.15V in the differential pulse mode. The scanning parameters used were pulse amplitude of 50 mV, pulse duration of 40 ms, a pulse frequency of 5 Hz and scan rate of 20 mV/s.

    Another example is a sensitive, simple and fast adsorptive stripping voltammetric procedure for trace determination of cadmium in natural samples containing high concentrations of surface active substances and humid substances [7]. The method is based on adsorptive accumulation of the Cd(II)-cupferron complex onto a hanging mercury drop electrode, followed by the reduction of the adsorbed species by a voltammetric scan using differential pulse modulation. The detection limit was 3×10^{-10} M with an accumulation time 30 s. The procedure was applied to the determination of Cd(II) in natural water samples without any pretreatment. All voltammetric experiments were carried out with an Autolab PGSTAT 10 analyzer (Utrecht, The Netherlands), with hanging mercury drop electrode (1.4 mm^2), a Pt auxiliary electrode and an Ag/AgCl reference electrode. Acetate buffer (pH 6.1) was prepared from acetic acid and sodium hydroxide. For voltammetric measurement, 5 mL of the mixed solution was pipetted into the electrochemical cell and 100 mL of 1×10^{-1} M cupferron, 0.5 mL of 1 M acetate buffer pH 6.1, 4.4 mL of H$_2$O were added and deoxygenation by nitrogen for 5 min. The accumulation potential -0.35 V for 30 s was applied to a fresh mercury drop in the stirred solution, and adsorption of the Cd(II)-cupferron complex took place. After the equilibration time of about 5 s, a differential pulse voltammogram was recorded, while the potential was scanned from -0.35 V to -0.70 V. The scan rate and pulse height were 20 mV/s and -50 mV, respectively.

In another case, Arsenic(III), selenium(IV), copper(II), lead(II), cadmium(II), zinc(II) and manganese(II) have been determined in environmental samples by differential pulse cathodic and anodic stripping voltammetry.[8] The voltammetric measurements were carried out using a stationary mercury electrode, as the working electrode, a platinum electrode as the auxiliary and a Ag/AgCl/KCl(sat) electrode as reference electrode. Before the voltammetric determinations, sea water samples were buffered at pH 9.0 by addition of an appropriate amount of hydrochloric acid and ammonia solution. The buffer was also employed as the supporting electrolyte. For preparation of sediment samples, approximately 0.5–0.8 g of sediment, accurately weighed in a Pyrex digestion tube, were dissolved in 7 mL of 37% (m/m) hydrochloric acid and 5 mL of 69% (m/m) nitric acid. The tube was inserted into a cold home-made block digester, gradually raising the temperature up to 130°C and maintaining this temperature for the whole time of mineralization (2 h). After cooling, the digest was filtered through Whatman N. 541 filter paper, evaporated almost to dryness, and the soluble salts dissolved in 100 mL of ammonia+ammonium chloride buffer solution (pH 9.0). All solutions were prepared with deionized water. Ammonia+ammonium chloride buffer solution (pH=9.0) was prepared by mixing appropriate amounts of 1 M hydrochloric acid (50 mL) and 1 M ammonia (72.4 mL) solutions. Aqueous stock solutions of As(III), Se(IV), Cu(II), Pb(II), Cd(II), Zn(II) and Mn(II) were prepared by dilution of the respective standard 1 g/L solutions. Voltammetric measurements were carried out with an AMEL (Milan, Italy) Model 433 multipolarograph. The voltammetric cell was kept at 20°C. The solutions were de-gassed with pure nitrogen for 15 min prior to the measurements, while a nitrogen blanket was maintained above the solution during the analysis. The solutions were de-gassed for 2 min after each standard addition. The detection limit for each element was around 10^{-9} M. The experimental conditions of the voltammetric analytical procedure and the experimental peak potentials are as follow: deposition potential: -1.050 V and -1.600V, final potential: -1.65 V, and -0.200 V; deposition time: 270s and 240s for As(III),Se(IV) by DPCSV and for Cu(II), Pb(II),Cd(II), Zn(II), Mn(II) by DPASV. Delay time, before the potential sweep was 10 s; potential scan rate was 10 mV/s and amplitude of pulse superposed 50 mV. Pulse duration was 0.065 s and pulse repetition 0.250 s. Stirring rate was 600 rpm. Aqueous reference solutions were used to determine the analytical calibration functions for all the elements by differential pulse cathodic and anodic stripping voltammetry.

Achtenberg and Braungardt also used stripping voltammetry for the determination of trace metal speciation and in-situ measurements of trace metal distributions in marine waters.[9] They used a voltammetric analyzer, a three-electrode cell (working electrode, reference electrode and counter electrode) and a computer for automated measurements and data acquisition. An Ag/AgCl/KCl was used as reference electrode and a platinum wire or a carbon rod as counter electrode. PVC samplers with a Teflon inner lining and a PTFE tap and silicone seals with the volume of 10–20 L and were used. The seawater was then filtered using acid-cleaned membrane filters (typical 0.4 mm polycarbonate, 47 mm diameter) fitted into acid-cleaned filtration units (made from FEP, polysulfone or polyethylene), and then stored in acid-cleaned high density polyethylene sample bottles. Acid cleaning of sample bottles were performed by overnight cleaning with hot detergent, followed by a 1-week soak in 6M HCl and subsequently a 1-week soak in 2 M HNO_3. In between the soaks, the bottles were rinsed with copious amounts of de-ionized water. Prior to use, the bottles were filled with de-ionized water, acidified to pH 2 with quartz-distilled acid and stored in two resealable polyethylene bags. Filtered discrete seawater samples were acidified with ultra-clean quartz-distilled acid, prior to ship-board or land-based analysis (typically samples are acidified to pH 2 in case of subsequent voltammetric analysis). All sample handling should took place in a class-100 laminar flow hood, which was situated in a clean container (supplied with particle-free clean air). Alkali metals presented in seawater cannot interfere in trace metal analysis, but in many cases increase the sensitivity of the voltammetric methods due to their role as electrolyte. The reduction in the sample handling minimizes the risk of sample contamination and allows automation of the instrumentation.

A common treatment of acidified samples is the application of UV-digestion prior to total dissolved trace metal analysis by stripping voltammetry. The UV light breaks down surfactants, which could interfere with the analysis by adsorbing onto the HMDE or MFE during the preconcentration step and therefore, hinder the passage to the electrode of metal cations (ASV) or of metal–AdCSV ligand complexes. In addition, the UV-digestion breaks down metal-complexing organic ligands, which occur naturally in seawater. In order to aid the breakdown, 10 mM H_2O_2 (final concentration) can be added to the sample prior to UV-digestion.

Some other example of using voltammetric techniques for monitoring the heavy metals in various water samples are listed in Table 12.2.

TABLE 12.2 Some Recent Example of Voltammetric Technique Used in Analysis of Important Environmental Cations

Ion	Sample	Voltammetric method	Detection Limit	Ref.
Cd(II)	Waste, surface and domestic water	Differential pulse anodic stripping voltammetry with graphene-modified platinum electrode	10 ppb	10
Cd(II)	Sea water	Square wave anodic stripping voltammetry with silver amalgamated microwire electrode	12 pM	11
Co(II)	River water	Square wave adsorptive Stripping voltammetry with integrated planar metal-film electrode	0.09 ppb	12
Co(II)	River water	Square wave adsorptive stripping voltammetry with rotating-disc bismuth-film electrode	70 ppt	13
Cr(III)	Natural Water	Square wave adsorptive cathodic stripping voltammetric with an improved bismuth film electrode	2.0×10^{-10} M	14
Cr(III) Cr(VI)	River water	Square wave adsorptive cathodic stripping voltammetry	Cr(III): 1.12 nM Cr(IV): 1.40 nM	15
Cr(IV)	Surface water	Differential pulse and normal pulse catalytic adsorptive stripping voltammetry with mercury film-modified silver solid amalgam annular band electrode	0.05 nM	16
Cu(II)	Drinking water	Square wave anodic stripping voltammetry with self-assembly of alkyl functionalized graphene oxide on a metal substrate	2.7 μM	17
Cu(II)	Lake water	Differential pulse voltammetry with naringenin-modified glassy carbon electrode	1.0×10^{-12} M	18
Hg(II)	River water	Ultrasensitive stripping voltammetric with a graphene-based nanocomposite film	6 ppt	19
Hg(II)	Tap water	Differential pulse stripping voltammetry with glassy carbon electrode modified with gold nanoparticles	0.05 pM	20

TABLE 12.2   *(Continued)*

Ion	Sample	Voltammetric method	Detection Limit	Ref.
Ni(II)	River waters	Square wave adsorptive cathodic stripping voltammetry with solid bismuth vibrating electrode	0.6 ppb	21
Ni(II)	River water	Square wave adsorptive stripping voltarnmetry with integrated microfabricated electrode	100 ng L^{-1}	22
Pb(II)	River and tap water	Adsorptive stripping voltammetric with complex onto a hanging mercury drop electrode	5.1×10^{-10} M	23
Pb(II)	Waste water	Differential pulse anodic stripping voltammetry with covalent binderless bulk modified electrode with modification of graphitic carbon with 4-amino salicylic acid	0.9 nM	24
Zn(II)	Sea water	Square wave adsorptive stripping voltammetry with 8-hydroxyquinoline (oxine) and complexing agent with hanging mercury drop electrode	0.05 ppb	25
Zn(II)	Sea water	Square wave stripping voltammetry with tin film/gold nanoparticles/gold microelectrode	5 ppb	26
Zn(II)	Sea water	Differential pulse cathodic stripping voltammetry with adsorptive accumulation of the complexes of Pb(II), Zn(II), and Cu(II) ions with dopamine onto hanging mercury drop electrode	0.25 ppb	27

Compounds of arsenic (III,V) are mainly presented as anions in natural water, in the dissolved state and in suspension. Stripping voltammetry is one of the most important methods for determining arsenic because of its high sensitivity, relatively simple implementation, and the possibility of identifying the valence forms of elements without preseparation. Many works were devoted to the determination of arsenic by stripping voltammetry on mercury, platinum, and carbon containing electrodes previously plated with gold [28–30].

Khustenko et al. [31] determined arsenic in drinking and natural water by stripping voltammetry (SV). A TA-07 universal voltammetric analyzer with a three-electrode cell (Tekhnoanalit, Russia) was used. Silver-silver chloride electrodes (1 M KCl) were used as a reference electrode and an auxiliary electrode. The working electrode was a serial carbon electrode based on a soot-polyethylene composite (Tekhnoanalit, Russia); the working surface was a face that was 4 mm in diameter, which was premodified with gold. Gold was plated electrochemically at 0 V from a 1g/L solution of $AuCl_3$ for 30 s. The electrode has stable characteristics and does not require additional preparation before experiments. To renew the working electrode surface, a thin layer (0.3–0.5 mm) was ground off after the analysis of 200–300 samples. Anodic voltammograms of arsenic were recorded in a supporting electrolyte of 0.04 M Na_2SO_3. The potential of the anodic peak of arsenic was –0.25 V (the electrochemical processes was measured the concentration of arsenic(III) because only arsenic(III) was electrochemically active under experimental conditions.). Complete reduction of arsenic(V) to arsenic(III) occurred upon two methods (1) ozonation and (2) UV irradiation (To minimize the losses and simplify the procedure, arsenic(III) was oxidized by the ozonation or UV irradiation of the solution in the presence of sulfite, chloride, or hydroxide ions, which increased the oxidation efficiency.). The samples were analyzed within 1 h after collection; if it was impossible, the samples were acidified by HCl to pH 4 and stored at 5°C for a month. A 1- to 10-mL portion of the test water was placed in a quartz beaker and diluted with water. Then, one of the following was chosen: (1) A half-milliliter of a 1 M NaOH solution or 0.2 mL of a saturated Na_2SO_3 was added; then, the ozone was bubbled for 1–2 min; (2) A half-milliliter of a 1 M NaOH solution was added; then, the solution was UV irradiated for 2 min upon stirring. The beaker with the sample was placed in an analyzer, and 0.2 mL of a saturated Na_2SO_3 was added. The electrodes were immersed into the solution, and UV irradiation was performed for 5 min upon stirring; after that, 0.02 mL of 0.1 M EDTA was added. The accumulation occurred at –1.6 V under stirring for 30–60 s; then the potential was decreased to –0.6 V. The stirring was switched off, and 5 s later, a voltammogram was recorded from –0.6 V to 0.1 V at a potential sweep rate of 150 mV/s. The solution was mixed by vibrations of the working electrode. Under the same conditions, the voltammogram of the sample was recorded after the addition of arsenic(III), and the concentration of arsenic in the sample was calculated. The arsenic concentration was found by the standard addition method. The anodic dissolution current of As(III) depends linearly on the concentration to 0.05 mg/L at an accumulation time of 60 s. The maximum concentration determinable without sample dilution at the minimum acceptable accumulation time (3–5 s) was 1 mg/L; therefore, the arsenic concentration in the studied solutions did not exceed 1 mg/L.

An indirect determination method was used for sulfide determination in water sample through anodic stripping voltammetry (ASV) [32]. ASV was performed on an Autolab PGSTAT 302 (Metrohm China Ltd.) instrument. Three-electrode system was used for the electrochemical experiment, containing a bismuth-film glassy carbon electrode (3 mm diameter for bare glassy carbon electrode), a saturated calomel reference electrode (SCE) and a platinum wire counter electrode. The measurement was based on the determination of residual Cd^{2+} after reacting with S^{2-}. The principle for sulfide determination by ASV is based on the selective reaction between Cd^{2+} and S^{2-} to form CdS precipitate, and the residual Cd^{2+} can be determinated by ASV using the sensitive response of BiFE to Cd^{2+}. Under the optimal experimental conditions (0.1 M pH 4.5 NaAc-HAc, C_{Cd2+} = 3.6 × 10^{-6} M, deposition potential, E_d = –1.2 V, and reaction time t_R = 120 s), the determination of S^{2-} can be achieved in the range of (0.7–5.0) × 10^{-6} M with a detection limit of 2.1 × 10^{-7} M. the measurement is done as follow: at first, 25 mL of 0.1 M pH 4.5 NaAc-HAc containing certain amount of Cd^{2+} was added into an electrolyte cell and the linear sweep curve was recorded between –1.0 and –0.5 V after deposition 120 s under the preconcentration potential of –1.2 V under stirring condition. Then a certain amount of S^{2-} was added into above solution, and the linear sweep curve was recorded again under the same conditions. After sampling, lake water and wastewater samples were filtered through a 0.45 μm membrane immediately and determined at once.

Another example of indirect determination of sulfide by ASV can be seen in the work of Huang et al. [33] Autolab PGSTAT 302 (Metrohm China Ltd.) instrument was used for ASV. A three electrodes cell consisting of a mercury-film glassy carbon working electrode, a saturated calomel electrode (SCE) and a platinum wire auxiliary electrode was used for electrochemical measurements. The proposed method has been successfully applied to determination of sulfide in synthetic wastewater, lake water, beverage, spring water and real wastewater samples.

The potentiometric methods besides potentiostatic methods were also used for analysis of heavy metals in water samples. It should be noted that their detection limits are not as low as the stripping methods.

An example is determination of zinc in electroplating and battery waste waters [34]. A 1.0 mL of each sample was taken and diluted to 50.0 mL by acetic acid/ sodium acetate buffer (pH 4.0) and distilled water. Then, zinc content of the sample solutions was determined by a zinc selective electrode using the calibration method. A corning ion analyzer 250 pH/mV meter was used for the potential measurements at 25.0°C. The emf observations were made relative to a double-junction saturated calomel electrode (SCE, Philips), with its chamber filled with an ammonium nitrate solution. The PVC membranes were prepared according to the following general procedure. The required amounts of the membrane ingredients (e.g., 30 mg PVC, 64 mg NPOE, 2 mg KTpClPB and 4 mg L) were mixed thoroughly and dissolved in 3 mL of dry THF. The resulting mixture was transferred into a glass dish of 2 cm in diameter. The solvent was then evaporated slowly up to the point that an oily concentrated mixture was created. A Pyrex tube (3–5 mm in top) was dipped into the oily mixture for about 5 s, so that a transparent film of about 0.3 mm thickness was formed. The tube was then removed from the mixture and kept at room temperature for about 12 h. Then, the tube was filled with an internal filling solution (1.0×10^{-3} M of zinc chloride). The electrode was finally conditioned for 12 h by soaking in 1.0×10^{-2} M solution of $ZnCl_2$. A silver-silver chloride electrode was used as an internal reference electrode. All electromotive force measurements were conducted with the following cell assembly:

Ag–AgCl | | internal solution (1×10^{-3} M $ZnCl_2$) | PVC membrane | test solution | | Hg_2Cl_2, KCl saturated

The activities were calculated according to Debye–Hückel procedure.

Table 12.3 lists some reports on potentiometric determination of cationic pollutants in various water samples.

TABLE 12.3 Some Reported Potentiometric Methods Used for Analysis of Cationic Pollutant in Water Samples

Pollutant	Sample	Type of Electrode	Detection limit	Ref.
Cadmium	Water, medicinal plants and soil samples	PVC membrane electrode	3.6×10^{-8} M	35
Cadmium	Tap, river and waste water	Membrane with solid contact electrode	1×10^{-8} M	36
Cadmium	Water and in waste water	PVC membrane electrode	8.4×10^{-7} M	37
Cadmium	Well water	Modified carbon paste electrode	1.8×10^{-7} M	38
Cadmium	Tap, river and waste water	PVC membrane electrode	8×10^{-7} M	39
Cadmium	Different waters	PVC membrane electrode	4.37×10^{-8} M	40
Cadmium	Tap, well, industrial, spring water	PVC membrane electrode	5.0×10^{-8} M	41
Cadmium	Tap, and waste water	PVC membrane electrode	1.0×10^{-7} M	42
Cadmium	Waste water	PVC membrane electrode	9.0×10^{-6} M	43
Cobalt	Natural water	PVC-membrane electrode	6.1×10^{-8} M	44
Cobalt	Wastewaters	Solid contact electrode	1.5×10^{-6} M	45
Cobalt	Waste waters	PVC membrane electrode	2.0×10^{-7} M	46
Cobalt	Wastewater and tap water	PVC membrane electrode	1.0×10^{-6} M	47

TABLE 12.3   *(Continued)*

Pollutant	Sample	Type of Electrode	Detection limit	Ref.
Cobalt	Wastewater	PVC membrane electrode	$6.3 \times 1^{0}\text{--}^{6}$ M	48
Chromium	Industrial waste water	poly(vinyl chloride) membrane electrode	8.6×10^{-8} M	49
Chromium	Waste water	nano-composite carbon paste electrode	7.0×10^{-8}	50
Chromium	Water	PVC membrane electrode	1.8×10^{-6}M	51
Chromium	Wastewaters	PVC membrane electrode	2.0×10^{-7} M	52
Chromium	Water samples	PVC-membrane electrodes polymeric membrane and coated glassy carbon	1.5×10^{-6} M for PVC and 2.0×10^{-7} M for Glassy electrodes	53
Chromium	Wastewater	PVC membranes electrode	3.35×10^{-7}M	54
Chromium	Electroplating bath solutions	Coated wire ion-selective electrode	1.0×10^{-7} M	55
Copper	Drinking water	Carbon Paste Electrode Modified With Multi-walled Carbon Nanotubes	1.1×10^{-6} M	56
Copper	River water	PVC membrane electrode	7.2×10^{-8} M	57
Copper	Tap, river, well and waste water	PVC membrane electrode	2.5×10^{-7} M	58
Copper	Water samples	PVC membrane electrode	6.3×10^{-7} M	59
Copper	Sea water	PVC membrane electrode	4×10^{-6} M	60
Mercury	Tab water, sea water	PVC membrane electrode	10^{-10} M	61
Mercury	Water	PVC membrane electrode	5.0×10^{-6} M	62
Mercury	Water	Modified carbon paste electrode	1.5×10^{-7}M	63
Mercury	Wastewater	PVC membrane electrode	2.4×10^{-6} M	64
Mercury	Wastewater	Carbon paste electrode	1.0×10^{-7} M	65

TABLE 12.3 *(Continued)*

Pollutant	Sample	Type of Electrode	Detection limit	Ref.
Nickle	Waste water of industrial	PVC membrane electrode	6.7×10^{-7} M	66
Nickle	Waste, river and well water	PVC membrane electrode	1.6×10^{-6} M	67
Nickle	Wastewater	PVC membrane electrode	1.8×10^{-6} M	68
Nickle	Well, tap, river, laboratory water	PVC membrane electrode	8×10^{-8} M	69
Nickle	Tap, river, laboratory water	PVC membrane electrode	6.0×10^{-8} M	70
Nickle	Edible oil and wastewater	Coated graphite PVC-membrane electrode	4.0×10^{-8} M	71
Nickle	Wastewater	PVC membrane electrode	8.0×10^{-6} M	72
Nickle	Wastewater	PVC membrane electrode	9×10^{-6} M	73
Lead	Mineral water	PVC membrane coated wire electrode	9.0×10^{-7} M	74
Lead	River and waste water	PVC membrane electrode	8.0×10^{-8} M	75
Lead	Water and in waste water	Carbon paste electrode	2.51×10^{-9} M	76
Lead	Tap water and river water	PVC membrane electrode with solid contact	4.3×10^{-9} M	77
Lead	Tap and river water	PVC membrane electrode	6.31×10^{-7} M	78
Zinc	Water	Polymeric membrane electrode and coated graphite electrode	3.3×10^{-7} M for PVC and 7.9×10^{-8} M for graphite electrodes	79
Zinc	Sea and tap water	PVC membrane electrode	1.1×10^{-7} M	80
Zinc	Wastewater of industrial	PVC membrane electrode	6.3×10^{-7} M	81
Zinc	Wastewater of industry	PVC membrane electrode	8.5×10^{-7} M	82

An example of determination of anions by potentiometric methods is measurements of chromate ions in wastewater samples of chromium electroplating samples.[83] A glass cell where CrO_4^{2-} carbon paste electrode was used consisting of an R684 model Analion Ag/AgCl double junction electrode as a reference electrode. A Corning ion analyzer 250 pH/mV meter was used for the potential measurements at $25.0\pm0.1°C$. The electrochemical cell can be represented as follows:

Ag, AgCl(s), KCl (3 M) ∥ sample solution ∣ nano-composite carbon paste electrode

Calibration graph was drawn by plotting the potential, E, versus the logarithm of chromate ion concentration. Before the monitoring studies, the pH of the waste water should be adjusted at about 7.0 with sodium hydroxide. The indicator electrode is a nanocomposite carbon paste electrode. The nano-composite electrode contained 5% multiwalled carbon nanotube (MWCNT), 67% graphite, 3% nano-silica and 15% room temperature ionic liquid (RTIL), 1-n-butyl-3-methylimidazolium tetrafluoroborate [bmim]BF_4 and 10% of a sensing material. It exhibited the best performance with a Nernstian response (-29.6 ± 0.2 mV/decade) toward chromate ions in a dynamic concentration range of 1.0×10^{-7}–1.0×10^{-2} M and detection limit of 7.0×10^{-8} M. The sensor response was found to be invariable in pH range from of 6.5 to 10.5. The electrode had relatively short response time (20 s).

Table 12.4 shows some examples on anion selective potentiometric electrodes applied for water analysis.

TABLE 12.4 Some Reported Electrochemical Methods Used for the Analysis of Some Pollutant Anions in Water Sample

Anion	Sample	Type of Electrode	Detection Limit	Ref.
Arsenate	Nautral water	membrane electrode	1×10^{-8} M	84
Arsenate	Water	membrane electrode	1×10^{-6} M	85
Arsenate	Water and wastewater	zeolite-modified carbon-paste electrode	5.0×10^{-8} M	86
Arsenate	Drinking and ground water	silica gel membrane electrode	4×10^{-7} M	87
Bromide	Tap water	PVC membrane	1.4×10^{-6} M	88
Bromide	Tap water	Modified carbon paste electrode	4.0×10^{-6} M	89
Chloride	Drinking water	PVC membrane	1.0×10^{-8} M	90
Cynide	Spring water	Coated-wire electrode	3.2×10^{-7} M	91
Cynide	Spring water	Carbon paste electrode	9×10^{-6} M	92

TABLE 12.4 *(Continued)*

Anion	Sample	Type of Electrode	Detection Limit	Ref.
Cynide	Some exhausted electroplating bath samples	PVC membrane electrode	5.8×10^{-6} M	93
Nitrate	Waste water	PVC membrane electrode	3.9×10^{-5} M	94
Nitrate	Fresh water	Coated-wire electrode	5.0×10^{-6} M	95
Nitrate	Rain water	PVC membrane electrode	1×10^{-5} M	96
Nitrite	Drinking Water	PVC membrane electrode	8.0×10^{-7} M	97
Nitrite	Drinking and polluted water	PVC membrane and coated graphite	8.0×10^{-7} M for PVC and 2.0×10^{-8} M for graphite electrode	98
Perchlorate	Tap water and river water	PVC Membrane electrode	4.0×10^{-7} M	99
Perchlorate	Tap water	PVC membrane electrode	4.0×10^{-7} M	100
Perchlorate	River, drinking and sludgy water	PVC membrane electrode	5.0×10^{-7} M	101
Perchlorate	Mineral water	Polymeric membrane electrodes and coated glassy carbon electrodes	1.0×10^{-6} M for PVC and 9.0×10^{-7} M for Glassy electrode	102
Perchlorate	Drinking water	PVC membrane	8.0×10^{-7} M	103
Perchlorate	Tap water	Polymeric membrane and coated glassy carbon	5.0×10^{-6} M for PVC and 7.0×10^{-7} M for Glassy electrode	104
Perchlorate	Wastewater	PVC membrane electrode	2.0×10^{-7} M	105
Phosphate	Environmental sample	Solid state membrane	1.0×10^{-6} M	106
Phosphate	Fertilizers	Carbon-paste	1.28×10^{-5} M	107
Phosphate	Waste waters and fertilisers	Cobalt-wire	1.0×10^{-4} M	108
Posphate	River water	Membrane electrode	1.0×10^{-6} M	109

TABLE 12.4   *(Continued)*

Anion	Sample	Type of Electrode	Detection Limit	Ref.
monohy-drogen phosphate	Waste water	Nano-composite carbon paste	$7.9 \times 10^{+7}$ M	110
monohy-drogen phosphate	Waste water	PVC membrane	6.0×10^{-8} M	111
Thiocya-nide	River water	PVC-polymeric membrane electrode and coated graphite electrode	2.2×10^{-7} M for PVC 6.7×10^{-8} M for Graphite electrode	112
Thiocya-nide	Tap water and river water	PVC membrane electrodes	8.33×10^{-7} M	113
Thiocya-nide	Wastewater	PVC membrane electrodes	$0.1–7.0 \times 10^{-6}$ M $0.1–4.51 \times 10^{-5}$ M and $0.1–4.16 \times 10^{-5}$ M	114

12.5.2 ORGANIC POLLUTANTS IN WATER

Organic water pollutants which can be analyzed by electrochemical methods include detergents, pesticides, insecticides and herbicides. Here, some examples of pesticides analysis using electroanalytical methods have been presented.

El-Shahawi et al. [115] determined organochlorine pesticides namely alachlor (ALC) and chlorfenvinphos (CHL) by differential pulse cathodic stripping voltammetry (DPCSV). Two pesticides were electroactive compounds. The method was applied for the analysis of trace concentrations of ALC and CHL in fresh and marine water (Atlantic and Red Sea) and sediment samples and food stuffs. A Metrohm 757 VA trace analyzer and 747 VA stand (Basel, Switzerland) were used for recording the cyclic, linear and differential pulse cathodic stripping voltammetry.

A three-compartment borosilicate (Metrohm) voltammetric electrochemical cell (10 mL) and hanging mercury drop electrode (HMDE, drop surface area 5 mm²) as a working electrode, double-junction Ag/AgCl,(3 M) KCl, as a reference and platinum wire (BAS model MW-1032) as counter electrode were used. Platinum (Pt, surface area 2 mm²) and gold (Au, surface area 2 mm²) were also used as working electrodes. The general procedures were preceded as follows: An accurate volume (10 mL) of an aqueous solution containing B–R buffer as supporting electrolyte of pH 2–3 was placed in the cell. The solution was stirred and purged with nitrogen gas for 10 min before recording the voltammogram. The stirrer was then stopped and after 10 s quiescence time, the background voltammogram of the supporting electrolyte was recorded by applying a negative going potential scan from 0 to –1.5 V vs. Ag/AgCl at a deposition potential of -0.35 V, accumulation time of 660 s and 750 s; scan rate of 50 mV/s and pulse amplitude of 50 mV. After recording the voltammogram of the blank solution, an accurate concentration (8.35×10^{-8}–11.1×10^{-8} M) of the chlorinated pesticide was placed into the electrochemical cell. The solution was stirred and purged with nitrogen gas for 5 min and the stirrer was then stopped. After 10 s quiescence time, the voltammogram of the pesticide was finally recorded by applying a negative going potential scan from 0.0 to –1.5 V vs. Ag/AgCl under the same experimental conditions as for the blank set up. Sea water samples collected and then were filtered by 0.45 mm cellulose membrane filter and stored in LDPE sample bottles. The standard addition method was used as follows: known volumes (1.0–2.0 mL) of sample extract which its pH adjusted to 2–3 were transferred into the electrochemical cell. The peak current displayed by the test solution before and after addition of various volumes of the standard ALC or CHL pesticide was measured. The change in the peak current was then recorded and used for determining both pesticides.

A new method using differential pulse adsorptive stripping voltammetry for the determination of atrazine (ATZ) in natural water samples using a bismuth film electrode (BiFE) has been proposed by Figueiredo-Filho et al., recently.[116]

Bismuth film electrodes (BiFEs) have recently been used as alternatives to mercury electrodes (DME, HMDE, MFE) due to their very low toxicity and general similarity in electrochemical properties. An additional advantage of this environmentally friendly electrode (BiFE), when compared to mercury electrodes, is the lower interference of dissolved oxygen. The electrochemical measurements were realized using a potentiostat/galvanostat PGSTAT-30 (Autolab, Eco Chemie, Netherlands) driven by GPES 4.9 software (Eco Chemie). All electrochemical experiments were carried out in a three-electrode single compartment glass cell made of Pyrex. Bismuth film on a copper electrode of 4 mm diameter was used as working electrode, Ag/AgCl (3.0 M KCl) as reference electrode and a platinum foil as counter electrode. The copper electrode was mechanically polished with silicon carbide papers (600, 1200 and 1500 grit SiC paper), rinsed with water, dried in air and inserted into the electrochemical cell to avoid oxidation of the surface. The electrodeposition of bismuth film onto the copper electrode was carried out by the chronoamperometry technique applying a deposition potential of -0.18 V for 200 s and 20 mL of following solution. The electrodeposition solution was prepared with 0.02 M $Bi(NO_3)_3.5H_2O$ and 0.M sodium citrate in 1.5 M HCl. DPAdS voltammograms were obtained in a potential range from –0.4 to –1.2 V vs. Ag/AgCl (3.0 M KCl) with an accumulation time of 210 s, a scan rate of 10 mV/ s, a modulation amplitude of 100 mV and a modulation time of 40 ms.

Table 12.5 lists some example of using voltammetric techniques for analysis of some pesticides.

TABLE 12.5 Some Recent Example of Voltammetric Technique Used in Analysis of Important Environmental Pesticides/Insecticides

Pesticides/Insec-ticide	Sample	Method	Detection limit	Ref.
Atrazine	River water	Square wave voltam-metry	10 nM	117
Methiocarb	River water	Square wave voltam-metric	0.45 mg L^{-1}	118
Organophosphorus	Natural water	Differential pulse voltammetry	0.06 μM	119
Thiamethoxam	River water	Square Wave Voltam-metric	0.25 μg mL^{-1}	120
Diafenthiuron	Natural water and soil	Square wave cathodic stripping voltammetric	9.1 mu g L^{-1}	121
Clothianidin	Tap water	Square wave adsorp-tive stripping voltam-metry	8.6×10^{-9} M	122
Clothianidin (CLO) Nitenpyram (NTT) Thiacloprid (TCL)	River water	Square wave voltam-metry	CLO: 0.52 μg mL^{-1} NTT: 0.18 μg mL^{-1} TCL: 0.27 μg mL^{-1}	123
Organophosphate	Underground and seawater	Differential pulse voltammetric	0.05 ppb	124
Organochlorine	Water	Differential pulse strip-ping voltammetry	0.01 μg l^{-1}	125
Cyromazine	River and tap water	Square wave stripping voltammetry	0.12 μg mL^{-1}	126
Clothianidin	Tap and river water	Cathodic stripping square wave voltam-metry	2.0×10^{-8} M	127
Organic Chlori-nated Pesticides	Marine Water	Differential pulse cathodic stripping voltammetry	7.4×10^{-9} M	128

The number of potentiometric electrodes reported for pesticides is low. A good example in this field is a work of AbuShawish et al. in 2012 [129]. In this work, a carbon paste electrode for diquat dibromide (Dq.2Br) pesticide was prepared. The electrode was applied to the potentiometric determination of diquat ions in water and urine samples. The electrode is based on the ion pair, namely, diquatphosphotungstate dissolved in 2-nitrophenyloctyl ether (2-NPOE) as pasting liquid with 1.0% Na-TPB as an additive. The modified electrode showed a near-Nernstian slope of 30.8 mV over the concentration range of 3.8×10^{-6} to 1.0×10^{-3} M with the limit of detection 9.0×10^{-7} M over the pH range of 4.5–9.5. The electrode exhibits good selectivity for Dq cations with respect to a number of inorganic cations.

A highly sensitive disposable screen-printed butyrylcholine potentiometric sensor, based on heptakis (2,3,6-tri-o-methyl)-beta-cyclodextrin (beta-CD) as ionophore has also been reported [130] for butyrylcholinesterase activity monitoring. The proposed electrode is a homemade printing carbon ink including beta-CD, anionic sites, and plasticizer. The fabricated sensor showed Nernstian responses in the range of 10^{-6} to 10^{-2} M with detection limit of 8×10^{-7} M, fast response time (about 1.6s) and adequate shelf-life of about 6 months.

Another example for potentiometry of pesticide is the work of Ristori et al. in 1996 [131]. He designed an enzyme-based sensor for the determination of organophosphate pesticides. It is based on the potentiometric determination of the inhibiting properties of the pesticides on the acetylcholinesterase activity. The enzyme was deposited onto commercially available membranes placed on the surface of an ion sensitive field effect transistor. The usage of this kind of membranes improves the behavior of the system with respect to the application of polymeric matrices chemically anchored to the surface of the sensor. The results showed a good sensitivity for the pesticide and the possibility of using the sensor for a preliminary screening of different pollutants on water samples simply by changing the sensitive membrane.

12.6   ANALYSIS OF POLLUTANTS IN SOIL SAMPLES

Environmental monitoring and speciation of inorganic and organic compounds in soil are of great importance for ecological assessments and for finding the relationship between plant and soil. Soil properties can affect the plant life. Thus, composition of soil and soil solution including concentration of heavy and toxic metals, their speciation, and distribution as well as other organic compounds should be studied.

12.6.1   INORGANIC POLLUTANTS IN SOIL

In most of agricultural lands, knowing the contents of heavy metals in soils or even soil solutions is of great importance.

A novel bismuth modified hybrid binder carbon paste electrode with square-wave anodic stripping voltammetry (SWASV) was used for simultaneous determination of lead and cadmium in soil extract of agricultural samples.[132] With the electrochemically deposited bismuth film on carbon paste electrode, the developed electrode exhibited well-defined and separate stripping peaks for cadmium and lead. Voltammetric measurements were carried out using a CHI660D electrochemical workstation (Chenhua Instrumental Corporation, Shanghai, China). The electrochemical cell was assembled with a conventional three-electrode system: an Ag/AgCl (saturated KCl) reference electrode, a platinum wire counter electrode and the prepared carbon paste working electrodes. A PC controlled magnetic stirrer was used to stir the solution during the deposition and cleaning step. All electrochemical experiments were carried out at room temperature. Soil samples were collected form an agriculture department in china, which is near a chemical industrial area. Soil sample was dried in an oven at 60°C for two hour. Then, the sample was grinded in a pestle and mortar, and further sieved by a 200 μm sieve. A portion (1 g) of soil sample was placed in an extraction tube with 40 mL of 0.11M acetic acid (pH 2.8) added and shaked for 16 h at room temperature. Then the mixture was centrifuged for phase separation, and the aqueous phase was filtered with a membrane (0.2 μm pore size). After these processes, the heavy metal ions of the water and acid soluble fraction in soil are extracted. Before the measurement, the pH of the extract solutions was adjusted to 4.5 by 0.11M NaOH solution. SWASV measurements were performed in 0.11 M acetate buffer solutions in the presence of 300 μgL^{-1} Bi(III) and appropriate target metals. The deposition potential of −1.2V was first applied to the working electrode for 120s under stirring conditions. Then the stirring was stopped, and after a 10s equilibration period, the SWASV potential scan was carried out from −1.2V to +0.3V (square wave amplitude, 25 mV; potential step, 5 mV; frequency, 25 Hz). Prior to the next measurement, a clean step at potential of +0.3V was applied for 30s. All experiments procedures were carried out in the presence of dissolved oxygen in the test solutions. Under the optimal conditions, the linear range for both metal ions was from 1 to 90 μgL^{-1}, and the detection limit was 0.12 μgL^{-1} for cadmium and 0.25 μgL^{-1} for lead, respectively.

Nedeltcheva et al.[133] determined the amount of mobile forms (because Plants can only absorb mobile form of these elements) of some heavy metals (Zn, Pb, Cd and Cu) in extracts form soil samples by an appropriate combination of anodic and cathodic stripping voltammetry. The voltammetric analyzes were carried out by a Metrohm Model 646VAProcessor and Model 647VAStand. A hanging drop mercury electrode was used as a working electrode, and a Ag/AgCl electrode as a reference electrode and a carbon electrode as auxiliary electrode. The base electrolyte solution was a mixture of 0.01M HCl and 0.10M NaCl. The four metals were extracted with 1.0M ammonium nitrate. Each soil sample (about 20 g) was treated with 50 mL of the extract for 2 h. The solid phase was then removed by filtering through a "dry" filter. The solution obtained was preserved by adding 0.5 mL nitric acid (d = 1.40 g/cm^3). No microwave digestion of soil extracts was necessary. A 15 mL of the supporting electrolyte solution was poured into the electrochemical cell. Oxygen was removed by bubbling pure nitrogen through the solution for about 10 min. The peak currents of the analytes in the blank were measured. Then a volume of 0.5 or 1.0 mL of the sample extract was added to the cell. Oxygen was removed again and peak currents were measured. The procedure for recording the analytical signals of Zn, Cd, Pb and Cu in the blank and in the sample solutions, and measuring the peak currents was as follow. Every analysis required three mercury drops: on the first one, zinc was determined; on the second, cadmium and lead; on the third, copper was determined. Zinc, lead and cadmium were determined by conventional differential-pulse anodic stripping voltammetry (DPASV). For copper determination, adsorptive differential-pulse cathodic stripping voltammetry with amalgamation using chloride ions as a complexing agent was applied. Zinc ions were accumulated into a mercury drop at a potential of −1.2V with stirring for 60 s. After a rest period of 20 s, the potential was swept in positive direction to the value of −0.7V. The anodic peak of zinc was recorded by differential-pulse voltammetry with pulse amplitude of 50 mV, pulse duration of 0.6 s and potential sweep of 10 mV/s. A new mercury drop was formed, and the ions of Cu, Pb and Cd were accumulated on it at a potential of −0.8 V with stirring for 60 or 120 s. After a rest period of 20 s the potential was swept in positive direction to the value of −0.2 V (when anodic copper peak was recorded, the end potential value was +0.05 V). Oxidation currents of the three metals were recorded by differential-pulse voltammetry with pulse amplitude of 50mV, pulse duration of 0.6 s and potential sweep of 10 mV/s. A potential of −0.5 V was set to the working electrode, and the copper ions were accumulated on the mercury drop with stirring. Accumulation times of 1, 2 or 3 min were set, thus, a peak height of 10 to 30 nA could be obtained. After an equilibrium period of 20 s, the potential was suddenly changed to +50 mV and swept in negative direction with pulse amplitude of −50 mV, pulse duration of 0.2 s and potential sweep of 40 mV/s. The metal contents were quantified by the standard addition method with three additions for each metal. The values obtained for the curve slopes were used to calculate the metal quantities in the sample solution and in the blank. The difference between the metal contents in sample and blank solutions gives the analytes contents in the extract.

Another example is the work of Jaklová Dytrtová et al. [134]. They used differential pulse anodic stripping voltammetry to determine and specify Cd, Cu, and Pb in soil solution. The ionic contents of elements in the soil solution were determined by the DPASV. The total soil solution contents of Cd, Pb and Cu were determined in 10^{-2} M HNO_3 as base electrolyte. Also, the ionic content of these elements in the soil solution were measured. DPASV measurements in 10^{-3} M $NaClO_4$ were used for determination of Cd^{2+}, Pb^{2+} and Cu^{2+} concentrations in the soil solution at original pH (C_{ion}), which is suitable for assessment of these cations bioavailability in plants. An aliquot of the dried and powdered soil was weighed to 1 mg into a borosilicate glass test-tube and decomposed in a mixture of oxidizing gases ($O_2+O_3+NO_x$) at 400°C for 14 hours in a Dry Mode Mineralizer Apion (Tessek, Czech Republic). The ash obtained from soil was decomposed in a mixture of HNO_3 and HF, evaporated to dryness at 160°C, and dissolved in aqua regia (HNO_3+3HCl). The measurement was carried out in three electrodes connected with a hanging mercury drop electrode (HMDE) as a working electrode, Ag/AgCl as reference electrode and a platinum electrode as the auxiliary electrode, on a PC-controlled EcoTribo Polarograph voltammetric analyzer (Polaro-Sensors, Prague, Czech Republic) with Polar

4 software. The accumulation time was 360 s and the accumulation potential was −1200 mV for Cd and Pb and −800 mV for Cu.

In another work, a sample of soils polluted by oil was analyzed using the potentiometric method.1-Phenyl-2-(2-hydroxyphenylhydrazo)butane-1,3-dione (as selectophore) was used as an effective ionophore for copper-selective poly(vinyl) chloride (PVC) membrane electrodes.[135] Optimization of the composition of the membrane and of the conditions of the analysis was performed, and under the optimized conditions the electrode has a detection limit of 6.30×10–[7]M Cu(II) at pH 4.0 with response time 10 s and displays a linear EMF versus log[Cu^{2+}] response over the concentration range 2.0×10^{-6} to 5.0×10^{-3}M Cu(II) with a Nernstian slope of 28.80±0.11 mV/decade over the pH range of 3.0–8.0. A portion (2.000 g) of soil sample was placed in a glassy carbon casserole and dissolved in a mixture of HF (35%, 38.00 mL), HCl (33%, 24.00 mL), and HNO_3 (65%, 8.00 mL). The obtained paste was treated with 12.00.16.00 mL of conc. HNO_3 at 60–70°C to distil off HF. The obtained residue was dissolved in distilled water, filtered off and diluted to 50.0 mL with water.

All EMF measurements were carried out using the following assembly:

Ag–AgCl|KCl (3M)|internal solution, 1.0×10^{-3}M $Cu(NO_3)_2 \times 2.5H2O$| PVC membrane|test solution|Hg–Hg_2Cl_2,

KCl (saturated).

All the EMF observations were made with a 177DMM (Keithley) microvoltmeter.

Table 12.6 lists some example of cationic pollutant analysis in soil samples by potentiometric method.

TABLE 12.6 Some Reported Electrochemical Methods Used for Analysis of Cationic Pollutents in Soil Samples by Potentiometric Electrodes

Pollutant	Sample	Type of Electrode	Detection limit	Ref.
Cobalt	Water, soil, pharmaceutical samples and medicinal plants	PVC-membrane and Coated Graphite electrodes	7.0×10^{-9} M	136
Lead	Soils	PVC membrane electrode	4.0×10^{-6} to 1.0×10^{-2} M	137
Lead	Mineral rock	Disposable carbon composite PVC-based membrane	3.2×10^{-7}M	138
Lead	Water, black tea, hot and black pepper	Microsensor with a membrane composition	6.0×10^{-10} M	139
Lead	Bath solutions, effluent waters, alloy and battery waste samples	Coated-wire lead ion-selective electrode	6×10^{-7} M	140

TABLE 12.6 *(Continued)*

Pollutant	Sample	Type of Electrode	Detection limit	Ref.
Lead	Mineral rocks and wastewater	Coated-wire electrode	2×10^{-6} M	141
Zinc	Soil and industrial samples	PVC membrane electrode	7.9×10^{-8} M	142
Zinc	Alloy samples	PVC membrane electrode	2.0×10^{-7} M	143
Zinc	Rock materials	PVC membrane electrode	7.0×10^{-6} M	144

12.6.2 ORGANIC SPECIES

Agricultural herbicides are triazines which are environmental pollutants in soils and waters. They are used widely and due to the mobility and solubility, they can be widely distributed in water and can also be strongly adsorbed into soils.

An extraction-anodic adsorptive stripping voltammetric procedure using microwave-assisted solvent extraction and a gold ultramicroelectrode was developed for determining the pesticide ametryn in soil samples.[145] The method is based on the use of acetonitrile as extraction solvent and on controlled adsorptive accumulation of the herbicide at the potential of 0.50 V ($_{vs.}$ Ag/AgCl) in Britton-Robinson buffer (pH 3.3). Soil sample extracts were analyzed directly after drying and redissolution with the supporting electrolyte but without other pretreatment. An AUTOLAB potentiostat/galvanostat, model PSTAT 10, coupled with an ECD module (it is possible to perform measurements at extremely low currents up to 10^{-11} A) from EcoChemie, controlled by a PC through the Model GPES3 software, was used for all electrochemical measurements. The voltammetric measurements were done by a gold ultramicroelectrode, an Ag/AgCl reference electrode, and a cylindrical carbon counter electrode. Samples from each soil type were thoroughly mixed to ensure homogeneity. After air-drying and sieving to a grain size of 2 mm, the soils samples were stored at 4°C. The working gold ultramicroelectrode was inserted in a 2.5–5 mL aliquot of a soil sample residue redissolved with Britton-Robinson buffer at pH=3.3. The preconcentration was accomplished in quiescent solutions at an optimal potential of 0.50 V for a selected deposition time. Following the anodic potential scan, a conditioning potential of 1.0 V was applied to the ultramicroelectrode for 30 s. The square-wave parameters used (except where otherwise stated) were: frequency 50 Hz, amplitude 30 mV; staircase step 3 mV. Voltammetric quantifications were achieved by the standard additions method.

The multiple square-wave voltammetry (MSWV) with gold microelectrode was used to establish an electroanalytical procedure for the determination of the paraquat and diquat pesticides in river sediment samples.[146]

SWV and MSWV were carried out with a PGZ 402 Voltalab potentiostat of the Radiometer Analytical coupled to Voltameter 5.06 software. An Ag-AgCl electrode was used as the reference electrode while the working electrodes were lab-made and constructed with gold micro wires. All measurements were carried out under ambient conditions. Before each experiment, a stream of N_2 was passed through the solution for 10 min. The working electrode was placed in the measuring cell filled with 10 mL of a Na_2SO_4 0.1 M solution containing a known concentration of pesticide. The optimization of the analytical procedure for MSWV was carried out. For both pesticides, two reduction peaks, at about -0.70 V and around -1.00 V *vs.* Ag/AgCl KCl 3.00 M, with profile of the totally reversible redox process, were observed. The best conditions to reduce paraquat and diquat were a pH of 6.0, a frequency of 250 s^{-1}, a scan increment 2 mV, square-wave amplitude of 50 mV and pulse number of 8 pulses of potential in each step of staircase of potential. After the optimization of voltammetric parameters, analytical curves were obtained in pure electrolyte by the standard addition method for the two pesticides. Sediments samples were collected in the River. The samples were collected according to the procedure defined by CRC from two different points of the river, and city. Aliquots of 0.0 (blank), 1.93, 3.86, 7.70 and 12.86 mg/mL of paraquat and 0.0 (blank), 2.59, 3.45, 10.35, of diquat were added into Erlenmeyer flasks, containing 2.0 g of river sediment samples and 20 mL of a solution of the 0.1 M Na_2SO_4 were sealed and shaken at 25°C. After 24 h of agitation, the samples were transferred for 50 mL centrifuge tubes and centrifuged for 20 minutes at 15,000 rpm. From the supernatant, 10 mL aliquots were transferred into electrochemical cells and analyzed by MSWV allied to Au-ME using optimized voltammetric parameters. Also, the electroanalytical procedure proposed was applied for the determination of adsorption isotherms of pesticides on river sediments samples collected.

12.7   ANALYSIS OF POLLUTANTS IN ATMOSPHERE

The concentration of some toxic gases produced from the combustion processes (e.g., power stations, mobile vehicles, etc.) should be monitored in order to reduce the harmful effect on human health and the damages from acid rain. There are some reports on determination of some gas molecules by electrochemical methods.

One of the examples is amperometric determination of nitrogen dioxide gas by on PAn/Au/Nafion® electrode [147].

The NO_2 gas sensor was assembled in terms of a divided cell with prepared PAn/Au/Nafion® as a working electrode, a platinum wire as counter electrode and Ag/AgCl/3 M NaCl solution as a reference electrode. The cell was divided with PAn/Au/Nafion® prepared in this work. 0.5 M H_2SO_4 was fed into the counter chamber and was used as H$^+$ and water source for ion conduction within Nafion film. The working chamber was fed with test gas and the electrochemical reaction occurred at the gas±solid electrode (PAn/Au/ Nafion®) interface. An Au O-ring was contacted with PAn/Au/Nafion® and used as a current collector. Two rubber O-rings were placed at both the sides of PAn/Au/ Nafion® to serve as gaskets for preventing the leakage of gas and aqueous solution. The concentration of nitrogen dioxide and the gas flow rate were controlled by a mass flow-rate controller (Sierra 902C). An electrochemical analyzer (BAS 100B) was used for obtaining the current±potential relationships of the amperometric NO_2 gas sensor.

Another analytical method has been developed for the determination of total gaseous selenium in the atmosphere by honeycomb denuder collection followed by differential pulse cathodic stripping voltammetry (DPCSV) measurement.[148] Gaseous selenium was collected in a denuder coating solution containing 2% HNO_3 and 2% glycerine. The soluble product, selenious acid, was then extracted by water for DPCSV analysis. The collection efficiency for gaseous selenium was 99.1% at a flow rate of 1 l min-1 for 3 h. Excellent linearity in DPCSV was maintained up to Se concentration of 40 ng/mL. This was equivalent to a working concentration of 220 ng/m of selenium in the atmosphere.

The DPCSV detection system consisted of a platinum wire auxiliary electrode, a saturated calomel reference electrode, and a silver working electrode. All potential measurements were reported in volts versus SCE. The surface of the silver working electrode ($f1$ mm, 1.57 mm area) was polished to a mirror finish with aluminum oxide film sheets before use, and then ultrasonically washed with water to remove any aluminum oxide. A clean, dry honeycomb denuder was placed on a clean plastic tray. A total of 10 mL of coating solution (2% HNO3/2% glycerine) was added by pipette to the tray. The plastic cap of the tray immediately was capped. Gently invert and reverse the denuder 10 times, and then rotate about 120° along its axis to insure that all tubes in the denuder were coated as completely as possible. Remove the plastic cap and pour out the residual coating solution. Then the denuder placed on the tray was dried in a clean air-drying system. Another denuder also was coated by the above procedure. The two coated denuders were placed into honeycomb denuder/filter pack sampler and the gaseous selenium in the atmosphere was collected by pump at the flow rate of 1 L/min for 3 h. After collection, the sampler was disassembled and honeycomb denuders were extracted twice with 10 mL of fresh sub-boiling distilled water. Then the extraction solution was transferred quantitatively to a 50 mL volumetric flask. A total of 125 μL of 12 M HC1 and 65 μL of 15 M HNO_3 were added and the solution was diluted to the mark with the above distilled water. To minimize contamination of the denuders, coating, drying and extraction were performed inside a clean air positive pressure hood. After the solution was mixed thoroughly, content of Se(IV) was measured by DPCSV. Four milliliters of aqueous extract containing Se(IV) was placed in the voltammetric cell (25×40 mm). After the solution was deaerated with nitrogen for 2 min, preelectrolysis was carried out at −0.350 V for 20 min with stirring the solution. Then the electrode was removed, rinsed with water, and transferred to another voltammetric cell containing 4 mL of a 2.0 M sodium hydroxide solution. After 10 s, the electrode potential was scanned from −0.350 to −1.20 V with a scan speed of 40 mV/s. A stripping voltammogram was recorded in the differential pulse mode, using a pulse amplitude of 100 mV, a pulse duration of 50 ms, a pulse repetition time of 100 ms, and a sensitivity (A/V) of 5×10^{-5}.

ACKNOWLEDGMENT

The authors are acknowledged the research council of University of Tehran for financial support of this work.

KEYWORDS

- **Classification of Samples**
- **Pollutants in Atmosphere**
- **Pollutants in Soil**
- **Pollutants in Water**
- **Sample Preparations**
- **Step-By-Step Measurement Guide**

REFERENCES

1. Ganjali, M. R., Norouzi, P., & Faridbod, F. (2010). Research Signpost Transworld Research Network, Electrochemical Sensors, Include 11 Chapters.
2. Ganjali, M. R., Norouzi, P., & Rezapour, M. (2006). Encyclopedia of Sensors: Potentiometric Ion Sensors, Potentiometric Ion Sensors, Am. Sci. Publisher (ASP), 8, 197–288.
3. Namieśnik, J., & Górecki T. (2001). Preparation of Environmental Samples for the Determination of Trace Constituents, *Polish J. Environ. Studies*, 10, 77–84.
4. Yokoi, K., Yamaguchi, A., Mizumachi, M., & Koide, T. (1995). Direct Determination of Trace Concentrations of Lead in Fresh Water Samples by Adsorptive Cathodic Stripping Voltammetry of a Lead-Calcein Blue Complex, *Anal. Chim Acta.*, 316, 363–369.
5. Hutton, E. A., Ogorevc, B., Hocevar, S. B., & Smyth, M. R. (2006). Bismuth Film Microelectrode for Direct Voltammetric Measurement of Trace Cobalt and Nickel in Some Simulated and Real Body Fluid Samples, *Anal Chim Acta.*, 557, 57–63.
6. Echeandía, J. S. (2011). Direct Simultaneous Determination of Co, Cu, Fe, Ni and V in Pore Waters by Means of Adsorptive Cathodic Stripping Voltammetry with Mixed Ligands. Talanta, 85, 506–512.
7. Grabarczyk, M., & Koper, A. (2012). Direct Determination of Cadmium Traces in Natural Water by Adsorptive Stripping Voltammetry in the Presence of Cupferron as a Chelating Agent. Electroanalysis, 24, 33–36.
8. Locatelli, C., & Torsi, G. (2001). Voltammetric Trace Metal Determinations by Cathodic and Anodic Stripping Voltammetry in Environmental Matrices in the presence of Mutual Interference, *J. Electroanal. Chem.*, 509, 80–89.

9. Achterberg, E. P., & Braungardt, C. (1999). Stripping Voltammetry for the Determination of Trace Metal Speciation and In-situ measurements of Trace Metal Distributions in Marine Waters, Analysis, Chim, Acta, *400*, 381–397.

10. Tang, F. J., Zhang, F; Jin, Q. H., & Zhao, J. L. (2013). Determination of Trace Cadmium and Lead in Water Based on Graphene-Modified Platinum Electrode Sensor, *Chinese J. Anal Chem.*, *41*, 278–282

11. Bi, Z. S., Salaun, P., & van den Berg, C. M. G. (2013). Determination of Lead and Cadmium in Seawater using a vibrating Silver Amalgam Microwire Electrode, *Anal Chim Acta, 769*, 56–64.

12. Kokkinos, C., Economou, A., & Koupparis, M. (2009). Determination of Trace Cobalt (II) by Adsorptive Stripping Voltammetry on Disposable Microfabricated Electrochemical Cells with Integrated Planar Metal-film Electrodes, Talanta, *77*, 1137–1142.

13. Morfobos, M., Economou, A., & Voulgaropoulos, A. (2004). Simultaneous Determination of Nickel (II) and Cobalt (II) by Square Wave Adsorptive Stripping Voltammetry on a Rotating Disc Bismuth Film Electrode, *Anal Chim Acta*, *519*, 57–64.

14. Zhang, Q., Zhong, S. W., Su, J. L., Li, X. J., & Zou, H. (2013). Determination of Trace Chromium by Square-Wave Adsorptive Cathodic Stripping Voltammetry at an Improved Bismuth Film Electrode, *J Electrochem Soc.*, *160*, H237-H242.

15. Jorge, E. O., Rocha, M. M., Fonseca, I. T. E., & Neto, M. M. M. (2010). Studies on the Stripping Voltammetric Determination and Speciation of Chromium at a Rotating-Disc Bismuth Film Electrode, Talanta, *81*, 556–564.

16. Bas, B., Bugajna, A., Jakubowska, M., & Niewiara, E. (2012). Normal Pulse Voltammetric Determination of Subnanomolar Concentrations of Chromium (VI) with Continuous Wavelet Transformation, *Electroanalysis*, *24*, 2157–2164.

17. Zhang, W., Wei, J., Zhu, H. J., Zhang, K., Ma, F., Mei, Q. S., Zhang, Z. P., & Wang, S. H. (2012). Self-Assembled Multilayer of Alkyl Graphene Oxide for Highly Selective Detection of Copper (II) based on Anodic Stripping Voltammetry, *J. Mater* Chem., *22*, 22631–22636.

18. Mulazimoglu, I. E. (2012). Electrochemical Determination of Copper (II) Ions at Naringenin-Modified Glassy Carbon Electrode: Application in Lake Water Sample, Desalination and Water Treatment, *44*, 161–167.

19. Gong, J. M., Zhou, T., Song, D. D., & Zhang, L. Z. (2010). Monodispersed Au Nanoparticles Decorated Graphene as an Enhanced Sensing Platform for Ultrasensitive Stripping Voltammetric Detection of Mercury (II), Sensor, Actuat B-Chem, *150*, 491–497.

20. Behzad, M., Asgari, M., Shamsipur, M., & Maragheh, M. G. (2013). Impedimetric and Stripping Voltammetric Detection of Sub-Nanomolar Amounts of Mercury at a Gold Nanoparticle modified Glassy Carbon Electrode, *J. Electroanal Soc., 160*, B31–B36.

21. Alves, G. M. S., Magalhaes, J. M. C. S., & Soares, H. M. V. M. (2013). Simultaneous Determination of Nickel and Cobalt using a Solid Bismuth Vibrating Electrode by Adsorptive Cathodic Stripping Voltammetry, *Electroanalysis*, *25*, 1247–1255.

22. Kokkinos, C., Economou, A., Raptis, I., & Speliotis, T. (2008). Disposable Mercury-Free Cell-on-a-Chip Devices with Integrated Microfabricated Electrodes for the Determination of Trace Nickel (II) by Adsorptive Stripping Voltammetry, *Anal Chim Act*a, *622*, 111–118.

23. Grabarczyk, M. (2013). Sensitive Adsorptive Stripping Voltammetric Method for Direct Determination of Trace Concentration of Lead in the Presence of Cupferron in Natural Water Samples, *Int. J. Environ. Anal Chem*, *93*, 1008–1018.

24. Kempegowda, R. G., & Malingappa, P. A. (2012). Binderless, Covalently Bulk Modified Electrochemical Sensor: Application to Simultaneous Determination of Lead and Cadmium at Trace Level, *Anal. Chim Acta*, *728*, 9–17.

25. Arancibia, V., Zuniga, M., Zuniga, M. C., Segura, R., & Esteban, M. (2010). Optimization of Experimental parameters in the Determination of Zinc in Seawater by Adsorptive stripping Voltammetry, *J. Brazil Chem. Soc, 21*, 255–261.

26. Wang, J. F., Bian, C., Tong, J. H., Sun, J. Z., & Xia, S. H. (2012). Simultaneous detection of Copper, Lead and Zinc on Tin Film/Gold Nanoparticles/Gold Microelectrode by Square Wave stripping Voltammetry, Electroanalysis, *24*, 1783–1790.

27. Rajabi, M., Asghari, A., & Mousavi, H. Z. (2010). Trace Amounts Determination of Lead, Zinc and Copper by Adsorptive Stripping Voltammetry in the Presence of Dopamine. *J. Anal. Chem., 65*, 511–517.

28. Rahman, M. R., Okajima, T., & Ohsaka, T. (2010). Selective Detection of AS (III) at the Au (III)-like Polycrystalline Gold Electrode, *Anal Chem, 82*, 9169–9176.

29. Hassan, S. S., Sirajuddin; Solangi, A. R., Kazia, T. G., Kalhoro, M. S., Junejo, Y., Tagar, Z. A., & Kalwar, N. H. (2012). Nafion Stabilized Ibuprofen Gold Nanostructures Modified Screen Printed Electrode as Arsenic (III) sensor. *J. Electroanal. Chem., 682*, 77–82.

30. Nagaoka, Y., Ivandini, T. A., Yamada, D., Fujita, S., Yamanuki, M., & Einaga, Y. (2010). Selective Detection of As(V) with High Sensitivity by as Deposited Boron-Doped Diamond Electrodes, *Chem. Lett.*, 39, 1055–1057.

31. Khustenko, L. A., Tolmacheva, T. P., & Nazarov, B. F. (2009). A Rapid Method of Sample Preparation for determining Arsenic in Water by Stripping, *Voltammetry Journal of Analytical Chemistry, 64(11)*, 1136–1140.

32. Huang, D. Q., Xu, B. L., Tang, J., Yang, L. L., Yang, Z. B., & Bi1, S. P. (2012). Bismuth Film Electrodes for Indirect Determination of Sulfide Ion in Water Samples at Trace Level by Anodic Stripping Voltammetry. *Int. J. Electrochem. Sci., 7*, 2860–2873.

33. Huang, D., Xu, B., Tang, J., Luo, J., Chen, L., Yang, L., Yang, Z., & Shuping, B. (2010). Indirect Determination of Sulfide Ions in Water Samples at Trace Level, by Anodic Stripping Voltammetry using Mercury Film Electrode, *Anal. Met., 2*, 154–158.

34. Hosseini, M., Abkenar, S. D., Ganjali, M. R., & Faridbod, F. (2011). Determination of Zinc (II) Ions in Waste Water Samples by a Novel Zinc Sensor Based on a New Synthesized Schiff's Base, *Mat. Sci. Eng. C., 31*, 428–433.

35. Singh, A. K., Jain, A. K., Upadhyay, A., Thomas, K. R. J., & Singh, P. (2013). Electroanalytical Performance of Cd(II) Selective Sensor Based on PVC Membranes of 5,5-(5,5-(Benzo c 1,2,5 Thiadiazole-4,7-diyl)Bis(thiophene-5,2-diyl))bi s(N1,N1,N3,N3-tetraphenylbenzene-1,3-diamine). *Int. J. Environ. Anal Chem., 93*, 813–827.

36. Wardak, C. A. (2012). Comparative Study of Cadmium Ion-Selective Electrodes with Solid and Liquid Inner Contact, *Electroanalysis, 24*, 85–90.

37. Ensafi, A. A., Meghdadi, S., & Sedighi, S. (2009). Sensitive Cadmium Potentiometric Sensor Based on 4-Hydroxy Salophen as a Fast Tool for Water Samples Analysis, Desalination, *242*, 336–345.

38. Abbastabar-Ahangar, H., Shirzadmehr, A., Marjani, K., Khoshsafar, H., Chaloosi, M., & Mohammadi, L. (2009). Ion-Selective Carbon Paste Electrode based on New Tripodal Ligand for Determination of Cadmium (II). *J. Incl. Phenom. Macro, 63*, 287–293.

39. Rezaei, B., Meghdadi, S., & Zarandi, R. F. (2008). A Fast Response Cadmium-Selective Polymeric Membrane Electrode Based on N, N'-(4-methyl-1, 2-phenylene) Diquinoline-2-Carboxamide as a New Neutral Carrier. *J. Hazard. Mater, 153*, 179–186.

40. Singh, A. K., Mehtab, S., Singh, U. R., & Aggarwal, V. (2007). Comparative Studies of Tridentate Sulfur and Nitrogen-Containing Ligands as Ionophores for Construction of Cadmium Ion-Selective Membrane Sensors. *Electroanalysis, 19*, 1213–1221.

41. Gupta, V. K., Singh, A. K., & Gupta, B. (2007). Schiff Bases as Cadmium (II) Selective Ionophores in Polymeric Membrane Electrodes, *Anal Chim. Acta, 583*, 340–348.

42. Shamsipur, M., & Mashhadizadeh, M. H. (2001). Cadmium Ion-Selective Electrode Based on Tetrathia-12-Crown-4. Talanta, *53*, 1065–1071.
43. Javanbakht, M., Shabani-Kia, A., Darvich, M. R., Ganjali, M. R., & Shamsipur, M. (2000). *Cadmium (II)-Selective Membrane Electrode based on a Synthesized Tetrol Compound, Anal Chim. Acta, 408*, 75–81.
44. Rofouei, M. K., Mohammadi, M., Khodadadian, M., Jalalvand, A. R., & Beiza, A. (2012). Cobalt (II)-Selective Membrane Electrode Based on N, N '-Di (thiazol-2-yl) Formimidamide. *Int. J. Environ. Anal. Chem, 92*, 665–675.
45. Wardak, C. (2008). Cobalt (II) Ion-Selective Electrode with Solid Contact, Cent Eur. *J. Chem., 6*, 607–612.
46. Zamani, H. A., Ganjali, M. R., Norouzi, P., Tajarodi, A., & Hanifehpour, Y. (2007). Fabrication of a Cobalt (II) PVC-Membrane Sensor Based on N-(Antipyridynil)-N'-(2-methoxyphenyl) Thiourea. *J. Chil. Chem. Soc., 52*, 1332–1337.
47. Zamani, H. A., Ganjali, M. R., Norouzi, P., & Adib, M. (2007). Cobalt (II) Ion Detection in Electroplating Wastewater by a New Cobalt Ion-Selective Electrode based on N'- 1-(2-thienyl) Ethylidene-2-Furohydrazide, *Sens. Lett, 5*, 522–527.
48. Singh, A. K., Singh, R. P., & Saxena, P. (2006). Cobalt (II)-Selective Electrode Based on a Newly Synthesized Macrocyclic Compound. Sensor, *Actuat B-Chem., 114*, 578–583.
49. Zamani, H. A., & Sahebnasagh, S. (2013). Potentiometric Detection of Cr^{3+} Ions in Solution by Chromium (III) Electrochemical Sensor Based on Diethyl 2-Phthalimidomalonate Doped in Polymeric Membrane. Int. *J. Electrochem. Sci., 8*, 3708–3720.
50. Ganjali, M. R., Eshraghi, M. H., Ghadimi, S., Moosavi, S. M., Hosseini, M., Haji-Hashemi, H., & Norouzi, P. (2011). Novel Chromate Sensor Based on MWCNTs/Nanosilica/Ionic Liouid/Eu Complex/Graphite as a New Nano-Composite and Its Application for Determination of Chromate Ion Concentration in Waste Water of Chromium Electroplating. *Int. J. Electrochem. Sci., 6*, 739–748.
51. Abu-Shawish, H. M., Saadeh, S. M., Hartani, K., & Dalloul, H. M. A. (2009). Comparative Study of Chromium (III) Ion-Selective Electrodes Based on N, N-Bis (salicylidene)-o-Phenylenediaminatechromium (III). *J. Iran. Chem. Soc., 6*, 729–737.
52. Abedi, M. R., Zamani, H. A., Ganjali, M. R., & Norouzi, P. (2007). Cr (III) Ion-Selective Membrane Sensor Based on 1,3-Diamino-2-Hydroxypropane-N,N,N',N'- Tetraacetic Acid. *Sens. Lett, 5*, 516–521.
53. Shamsipur, M., Soleymanpour, A., Akhond, M., Sharghi, H., & Sarvari, M. H. (2005). Highly selective Chromium (III) PVC-Membrane Electrodes Based on Some Recently Synthesized Schiff's Bases. *Electroanalysis, 17*, 776–782.
54. Choi, Y. W., Minoura, N., & Moon, S. H. (2005). Potentiometric Cr (VI) Selective Electrode Based on Novel Ionophore-Immobilized PVC Membranes. Talanta, *66*, 1254–1263.
55. Sil, A., Ijeri, V. S., & Srivastava, A. K. (2004). Coated Wire Chromium (III) Ion-Selective Electrode Based on Azamacrocycles, *Anal. Bioanal Chem., 378*, 1666–1669.
56. Soleimani, M., & Afshar, M. G. (2013). Potentiometric Sensor for Trace Level Analysis of Copper Based on Carbon Paste Electrode Modified With Multi-Walled Carbon Nanotubes. *Int. J. Electrochem. Sci., 8*, 8719–8729.
57. Tomar, P. K., Chandra, S., Singh, I., Kumar, A., Malik, A., & Singh, A. (2011). Development of a New Copper (II) Ion-Selective PVC Membrane Electrode Based on Tris (2-Benzimidazolylmethyl) Amine, *J. Indian Chem. Soc., 88*, 1739–1744.
58. Ghanei-Motlagh, M., Taher, M. A., Saheb, V., Fayazi, M., & Sheikhshoaie, I. (2011). Theoretical and Practical Investigations of Copper Ion Selective Electrode with Polymeric Membrane Based on N, N'-(2, 2-Dimethylpropane-1, 3-diyl)-Bis (Dihydroxyacetophenone), *Electrochim Acta., 56*, 5376–5385.

59. Sadeghi, S., & Jahani, M. (2009). New Copper (II) Ion-Selective Membrane Electrode based on Erythromycin Ethyl Succinate as a Neutral Ionophore, *Anal. Lett*, *42*, 2026–2040.

60. Buzuk, M., Brinic, S., Generalic, E., & Bralic, M. (2009). Copper (II) Ion Selective PVC Membrane Electrode based on S, S '-Bis (2-aminophenyl) Ethanebis (thioate), *Croat Chem, Acta*, *82*, 801–806.

61. Khan, A. A., & Paquiza, L. (2011). Analysis of Mercury Ions in Effluents using Potentiometric Sensor Based on Nanocomposite Cation Exchanger Polyaniline-Zirconium Titanium Phosphate, *Desalination*, *272*, 278–285.

62. Patel, B., Kumar, A., & Menon, S. K. (2009). Mercury Selective Membrane Electrode Based on Dithio Derivatized Macrotricyclic Compound, *J. Incl. Phenom. Macro Chem.*, *64*, 101–108.

63. Abu-Shawish, H. M. A. (2009). Mercury (II) Selective Sensor Based on N, N '-Bis (salicylaldehyde)-Phenylenediamine as Neutral Carrier for Potentiometric Analysis in Water Samples. *J. Hazard. Mater*, *167*, 602–608.

64. Othman, A. M. (2006). Potentiometric Determination of Mercury (II) Using a Tribromomercurate-Rhodamine BPVC Membrane Sensor, *Int. J. Environ. Anal Chem.*, *86*, 367–379.

65. Mashhadizadeh, M. H., Talakesh, M., Peste, M., Momeni, A., Hamidian, H., & Majum, M. A. (2006). Novel Modified Carbon Paste Electrode for Potentiometric Determination of Mercury (II) Ion, *Electroanalysis*, *18*, 2174–2179.

66. Zamani, H. A., Masrournia, M., Rostame-Faroge, M., Ganjali, M. R., & Behmadi, H. (2008). Construction of Nickel (II) PVC Membrane Electrochemical Sensor Based on 5-Methoxy-5, 6-Diphenyl-4, 5 Dihydro-3(2H)-Pyridazinethione as a Novel Ionophore, *Sens. Lett, 6,* 759–764.

67. Yari, A., Azizi, S., & Kakanejadifard, A. (2006). An Electrochemical Ni (II)-Selective Sensor-Based on a newly Synthesized Dioxime Derivative as a Neutral Ionophore, *Sensor Actuat B-Chem., 119*, 167–173.

68. Kumar, K. G., Poduval, R., John, S., & Augustine, P. A. (2006). PVC Plasticized Membrane Sensor for Nickel Ions, *Microchim. Acta*, *156*, 283–287

69. Mashhadizadeh, M. H., Sheikhshoaie, I., & Saeid-Nia, S. (2003). Nickel (II)-Selective Membrane Potentiometric Sensor using a Recently Synthesized Schiff Base as Neutral Carrier, *Sensor, Actuat B-Chem., 94*, 241–246.

70. Mashhadizadeh, M. H., & Momeni, A. (2003). Nickel (II) Selective Membrane Potentiometric Sensor Using a Recently Synthesized Mercapto Compound as Neutral Carrier. *Talanta*, *59*, 47–53.

71. Ganjali, M. R., Hosseini, M., Salavati-Niasari, M., Poursaberi, T., Shamsipur, M., Javanbakht, M., & Hashemi, O. R. (2002). Nickel ion-Selective Coated Graphite PVC-Membrane Electrode based on Benzylbis (thiosemicarbazone), *Electroanalysis*, *14*, 526–531.

72. Ganjali, M. R., Hosseini, S. M., Javanbakht, M., & Hashemi, O. R. (2000). Nickel (II) Ion-Selective Electrode based on 2, 5-Thiophenyl Bis (5-tert-butyl-1, 3-benzoxazole), *Anal Lett*, *33*, 3139–3152.

73. Ganjali, M. R., Fathi, M. R., Rahmani, H., & Pirelahi, H. (2000). Nickel (II) Ion-Selective Electrode based on 2-Methyl-4-(4-methoxy phenyl)-2, 6-Diphenyl-2H-Thiopyran, *Electroanalysis*, *12*, 1138–1142.

74. Soleymanpour, A., Shafaatian, B., Kor, K., & Hasaninejad, A. R. (2012). Coated Wire Lead (II)-Selective Electrode based on a Schiff Base Ionophore for Low Concentration Measurements, *Monatsh Chem.*, *143*, 181–188.

75. Mazloum-Ardakani, M., Safari, J., Pourhakkak, P., & Sheikh-Mohseni, M. A. (2012). Determination of Lead (II) Ion by Highly Selective and Sensitive Lead (II) Membrane Electrode based on 2-(((E)-2-((E)-1-(2-hydroxyphenyl) methyliden) hydrazono) metyl) Phenol, *Int. J. Environ. Anal Chem.*, *92*, 1638–1649.

76. Afkhami, A., Madrakian, T., Shirzadmehr, A., & Bagheri, H. (2012). Tabatabaee, M. A Selective Sensor for Nanolevel Detection of Lead (II) in Hazardous Wastes using Ionic-Liquid/ Schiff Base/MWCNTs/Nanosilica as a Highly Sensitive Composite, Ionics, *18*, 881–889.

77. Wardak, C. (2011). A Highly Selective Lead-Sensitive Electrode with Solid Contact based on Ionic Liquid. *J. Hazard. Mater*, *186*, 1131–1135.

78. Huang, M. R., Rao, X. W., Li, X. G., & Ding, Y. B. (2011). Lead Ion-Selective Electrodes based on Polyphenylenediamine as Unique Solid Ionophores. *Talanta*, *85*, 1575–1584.

79. Singh, P., Singh, A. K., & Jain, A. K. (2011). Electrochemical Sensors for the Determination of Zn2+ Ions Based on Pendant Armed Macrocyclic Ligand, *Electrochim. Acta*, *56*, 5386–5395.

80. Gholivand, M. B., Shahlaei, M., & Pourhossein, A. (2009). New Zn (II)-Selective Potentiometric Sensor Based on 3-Hydroxy-2-Naphthoic Hydrazide, Sens. Lett, *7*, 119–125.

81. Zamani, H. A., Ganjali, M. R., & Pooyamanesh, M. J. (2006). Zinc(II) PVC-Based Membrane Sensor Based on 5,6-Benzo-4,7,13,16,21,24-Hexaoxa-1,10-Diazabicyclo 8,8,8 Hexacos-5-ene, *J. Brazil. Chem. Soc.*, *17*, 149–155.

82. Ganjali, M. R., Zamani, H. A., Norouzi, P., Adib, M., Rezapour, M., & Aceedy, M. (2005). Zn²⁺ PVC-Based Membrane Sensor Based on 3- (2-furylmethylene) Amino -2-Thioxo-1, 3-Thiazolidin-4-One. B. *Kor. Chem. Soc.*, *26*, 579–584.

83. Ganjali, M. R., Rafiei Sarmazdeh, Z., Poursaberi, T., Shahtaheri, S. J., & Norouzi P. (2012). Dichromate Ion Selective Sensor Based on Functionalized SBA-15/ Ionic Liquid/MWCNTs/ Graphite. *Int. J. Electrochem. Sci.*, *7*, 1908–1916.

84. Khan, A. A., & Baig, U. (2013). Preparation of New Polymethylmethacrylate-Silica Gel Anion Exchange Composite Fibers and Its Application in making Membrane Electrode for the Determination of As (V), *Desalination, 319,* 10–17.

85. Khan, A. A., & Baig, U. (2012). Polyacrylonitrile-Based Organic-Inorganic Composite Anion-Exchange Membranes: Preparation, Characterization and Its Application in Making Ion-Selective Membrane Electrode for Determination of As (V). *Desalination, 289,* 21–26.

86. Mazloum-Ardakani, M., Karimi, M. A., Mashhadizadeh, M. H., Pesteh, M., Azimi, M. S., & Kazemian, H. (2007). Potentiometric Determination of Monohydrogen Arsenate by Zeolite-modified Carbon-Paste Electrode. *Int. J. Environ. Anal Chem.*, *87*, 285–294.

87. Rodriguez, J. A., Barrado, E., Vega, M., Prieto, F., & Lima, J. (2005). Construction and Evaluation of As (V) Selective Electrodes Based on Iron Oxyhydroxide, Embedded in Silica Gel Membrane, *Anal. Chim Acta, 539,* 229–236.

88. Singh, A. K., Mehtab, S., & Saxena, P. A. (2006). Bromide Selective Polymeric Membrane Electrode Based on Zn (II) Macrocyclic Complex. *Talanta*, *69*, 1143–1148.

89. Shamsipur, M., Ershad, S., Samadi, N., Moghimi, A., & Aghabozorg, H. (2005). A Novel Chemically Modified Carbon Paste Electrode based on a New Mercury (II) Complex for Selective Potentiometric Determination of Bromide Ion, *J. Solid State Electrochem*, *9*, 788–793.

90. Gupta, V. K., Goyal, R. N., & Sharma, R. A. (2009). Chloride Selective Potentiometric Sensor Based on a Newly Synthesized Hydrogen Bonding Anion Receptor, *Electrochim Acta*, *54*, 4216–4222.

91. Alizadeh, N., Teymourian, H., Aghamohammadi, M., Meghdadi, S., & Amirnasr, M. (2007). Highly Selective Cyanide Coated-Wire Electrode Based on a Recently Synthesized Co (II) Complex with the N, N '-bis(2-quinolinecarboxamido)-1,2-Benzene applying Batch and Flow Injection Analysis Techniques, *Ieee Sens. J.*, *7*, 1727–1734.

92. Abbaspour, A., Asadi, M., Ghaffarinejad, A., & Safaei, E. (2005). A Selective Modified Carbon Paste Electrode for Determination of Cyanide Using Tetra-3, 4-Pyridinoporphyrazinato cobalt(II), Talanta, *66*, 931–936.

93. Hassan, S. S. M., Marzouk, S. A. M., Mohamed, A. H. K., & Badawy, N. M. (2004). Novel Dicyanoargentate Polymeric Membrane Sensors for Selective Determination of Cyanide Ions, *Electroanalysis, 16*, 298–303.

94. Gupta, V. K., Singh, L. P., Chandra, S., Kumar, S., Singh, R., & Sethi, B. (2011). Anion Recognition Through Amide-Based Dendritic Molecule: A Poly (vinyl chloride) Based Sensor for Nitrate Ion, Talanta, 85, 970–974.

95. Mazloum-Ardakani, M., Dastanpour, A., & Salavati-Niasari, M. A. (2004). Highly Selective Nitrate Electrode Based on a Tetramethyl Cyclotetra-Decanato-Nickel (II) Complex, *J. Electroanal Chem.*, 568, 1–6.

96. Hara, H., & Izumiyama, F. (1997). Continuous-Flow Determination System Based on Null-Point Potentiometry Using a Nitrate Ion-Selective Membrane, *Anal Chim Acta., 355*, 211–216.

97. Ganjali, M. R., Shirvani-Arani, S., Norouzi, P., Rezapour, M., & Salavati-Niasari, M. (2004). Novel Nitrite Membrane Sensor based on Cobalt (II) Salophen for Selective Monitoring of Nitrite Ions in Biological Samples, *Microchim Acta, 146*, 35–41.

98. Shamsipur, M., Javanbakht, M., Hassaninejad, A. R., Sharghi, H., Ganjali, M. R., & Mousavi, M. F. (2003). Highly Selective PVC-Membrane Electrodes based on Three Derivatives of (Tetraphenylporphyrinato) Cobalt (III) Acetate for Determination of Trace Amounts of Nitrite Ion, *Electroanalysis, 15*, 1251–1259.

99. Nezamzadeh-Ejhieh, A., & Badri, A. (2011). Application of Surfactant Modified Zeolite Membrane Electrode towards Potentiometric Determination of Perchlorate, *J. Electroanal. Chem., 660*, 71–79.

100. Gholamian, F., Sheikh-Mohseni, M. A., & Salavati-Niasari, M. (2011). Highly Selective Determination of Perchlorate by a Novel Potentiometric Sensor Based on a Synthesized Complex of Copper, *Mater. Sci. Eng. C., 31*, 1688–1691.

101. Shokrollahi, A., Ghaedi, M., Rajabi, H. R., & Kianfar, A. H. (2009). Highly Selective Perchlorate Membrane Electrode Based on Cobalt (III) Schiff Base as a Neutral Carrier, *Chinese J. Chem., 27*, 258–266.

102. Soleymanpour, A., Hanifi, A., & Kyanfar, A. H. (2008). Polymeric Membrane and Solid Contact Electrodes Based on Schiff Base Complexes of Co (III) for Potentiometric Determination of Perchlorate Ions, *B. Kor. Chem. Soc., 29*, 1774–1780.

103. Mazloum-Ardakani, M., Jalayer, M., Naeimi, H., Zare, H. R., & Moradi, L. (2005). Perchlorate-Selective Membrane Electrode based on a New Complex of Uranil, *Anal Bioanal Chem, 381*, 1186–1192.

104. Shamsipur, M., Soleymanpour, A., Akhond, M., Sharghi, H., & Hasaninejad, A. R. (2003). Perchlorate Selective Membrane Electrodes based on a Phosphorus (V)-Tetraphenylporphyrin Complex Sensor Actuat. *B-Chem., 89*, 9–14.

105. Ganjali, M. R., Yousefi, M., Poursaberi, T., Naji, L., Salavati-Niasari, M., & Shamsipur, M. (2003). Highly Selective and Sensitive Perchlorate Sensors Based on Some Recently Synthesized Ni (II)-Hexaazacyclotetradecane Complexes, *Electroanalysis, 15*, 1476–1480.

106. Tafesse, F., & Enemchukwu, M. (2011). Fabrication of New Solid state Phosphate Selective Electrodes for Environmental Monitoring, *Talanta, 83*, 1491–1495.

107. Ejhieh, A. N., & Masoudipour, N. (2010). Application of a New Potentiometric Method for Determination of Phosphate Based on a Surfactant Modified Zeolite Carbon-Paste Electrode (SMZ-CPE*), Anal Chim Acta, 658*, 68–74.

108. DeMarco, R., Pejcic, B., & Chen, Z. L. (1998). Flow Injection Potentiometric Determination of Phosphate in Waste Waters and Fertilisers using a Cobalt Wire Ion-Selective Electrode, *Analyst, 123*, 1635–1640.

109. Hara, H., & Kusu, S. (1992). Continous-Flow Determination of Phosphate using a Lead Ion-Selective Electrode, *Anal Chim Acta., 261*, 411–417.

110. Norouzi, P., Ganjali, M. R., Faridbod, F., Shahtaheri, S. J., & Zamani, H. A. (2012). Electrochemical Anion Sensor for Monohydrogen Phosphate based on Nano-Composite Carbon Paste, *Int. J. Electrochem. Sci.*, *7*, 2633–2642.

111. Ganjali, M. R., Norouzi, P., Ghomi, M., & Salavati-Niasari, M. (2006). Highly Selective and Sensitive Monohydrogen Phosphate Membrane Sensor based on Molybdenum Acetylacetonate, *Anal Chim Acta, 567*, 196–201.

112. Singh, P., & Singh, A. K. (2011). Determination of Thiocyanate Ions at Nanolevel in Real Samples Using Coated Graphite Electrode based on Synthesised Macrocyclic Zn (II) Complex, *Anal Bioanal Chem.*, *400*, 2261–2269

113. Badri, A., & Pouladsaz, P. (2011). Highly Selective and Sensitive Thiocyanate PVC Membrane Electrodes based on Modified Zeolite ZSM-5, *Int. J. Electrochem. Sci.*, *6*, 3178–3195.

114. Beheshti, S. S., Sohbat, F., & Amini, M. K. (2010). A Manganese Porphyrin-Based Sensor for Flow-Injection Potentiometric determination of Thiocyanate, *J. Porphyr Phthalocya*, *14*, 158–165.

115. El-Shahawi, M. S., Hamza, A., Bashammakh, A. S., Al-Sibaai, A. A., & Al-Saggafa, W. T. (2011). Analysis of Some Selected Persistent Organic Chlorinated, Pesticides in Marine Water and Food Stuffs by Differential Pulse-Cathodic Stripping Voltammetry. *Electroanalysis*, *23*, 1175–1185.

116. Figueiredo-Filho, L. C. S., Azzi, D. C., Janegitz, B. C., & Fatibello-Filho, O. (2012). Determination of Atrazine in Natural Water Samples by Differential Pulse Adsorptive Stripping Voltammetry using a Bismuth Film Electrode, *Electroanalysis*, *24*, 303–308.

117. Svorc, L., Rievaj, M., & Bustin, D. (2013). Green Electrochemical Sensor for Environmental Monitoring of Pesticides: Determination of Atrazine in River Waters Using a Boron-Doped Diamond Electrode, Sens. *Actuator B-Chem.*, *181*, 294–300.

118. Inam, R., & Bilgin, C. (2013). Square Wave Voltammetric Determination of Methiocarb Insecticide Based on Multiwall Carbon Nanotube Paste Electrode. *J. Appl. Electrochem.*, *43*, 425–432.

119. Yang, L. Y., Liu, C. Y., Yang, L. N., & You, J. (2012). A Study of Preparation and Performance of a Dichlorvos Electrochemical Sensor Based on Molecularly Imprinted Technique *Anal. Lett*, *45*, 1036–1044.

120. Putek, M., Guzsvany, V., Tasic, B., Zarebski, J., & Bobrowski, A. (2012). Renewable Silver-Amalgam Film Electrode for Rapid Square Wave Voltammetric Determination of Thiamethoxam Insecticide in Selected Samples, *Electroanalysis*, *24*, 2258–2266.

121. Inam, R., & Tekalp, F. (2012). Square Wave Voltammetric Determination of Diafenthiuron and its Application to Water, Soil and Insecticide Formulation, *Int. J. Environ. Anal Chem.*, *92*, 85–95.

122. Guziejewski, D., Skrzypek, S., & Ciesielski, W. (2012). Application of Catalytic Hydrogen Evolution in the Presence of Neonicotinoid Insecticide Clothianidin, *Food Anal Method*, *5*, 373–380.

123. Brycht, M., Vajdle, O., Zbiljic, J., Papp, Z., Guzsvany, V., & Skrzypek, S. (2012). Renewable Silver-Amalgam Film Electrode for Direct Cathodic SWV Determination of Clothianidin, Nitenpyram and Thiacloprid Neonicotinoid Insecticides Reducible in a Fairly Negative Potential Range, *Int. J. Electrochem* Sci., *7*, 10652–10665.

124. Wu, S., Lan, X. Q., Cui, L. J., Zhang, L. H., Tao, S. Y., Wang, H. N., Han, M., Liu, Z. G., & Meng, C. G. (2011). Application of Graphene for Preconcentration and Highly Sensitive Stripping Voltammetric Analysis of Organophosphate Pesticide, *Anal. Chim Acta.*, *699*, 170–176.

125. Sundari, P. L. A., & Manisankar, P. (2011). Development of Ultrasensitive Surfactants Doped Poly (3, 4-ethylenedioxythiophene)/Multiwalled Carbon Nanotube Sensor for the detection of Pyrethroids and an Organochlorine Pesticide, *J. Appl. Electrochem.*, *41*, 29–37.

126. Mercan, H., Inam, R., & Aboul-Enein, H. Y. (2011). Square Wave Adsorptive Stripping Voltammetric Determination of Cyromazine Insecticide with Multi-Walled Carbon Nanotube Paste Electrode, *Anal Lett*, *44*, 1392–1404.

127. Guziejewski, D., Skrzypek, S., Luczak, A., & Ciesielski, W. (2011). Cathodic Stripping Voltammetry of Clothianidin: Application to Environmental Studies. *Collect. Czech Chem. Commun.*, *76*, 131–142.

128. El-Shahawi, M. S., Hamza, A., Bashammakh, A. S., Al-Sibaai, A. A., & Al-Saggaf, W. T. (2011). Analysis of Some Selected Persistent Organic Chlorinated Pesticides in Marine Water and Food Stuffs by Differential Pulse-Cathodic Stripping Voltammetry, *Electroanalysis*, *23*, 1175–1185.

129. Abu Shawish, H. M., Abu Ghalwa, N., Hamada, M., & Basheer, A. H. (2012). Modified Carbon paste Electrode for Potentiometric Dibromide Pesticide in Water and Urine Samples, Mater Sci & Engin C, *32*, 140–145.

130. Khaled, E., Hassan, H. N. A., Mohamed, G. G., Ragab, F. A., & Seleim, A. E. A. (2010). Disposable Potentiometric Sensors for Monitoring Cholinesterase activity, *Talanta*, *83*, 357–363.

131. Ristori, C., DelCarlo, C., Martini, M., Barbaro, A., & Ancarani, A. (1996). Potentiometric Detection of Pesticides in Water Samples, *Anal Chim Acta*, *325*, 151–160.

132. Wang, Z., Liu, G., Zhang, L., & Wang, H. A. (2012). Bismuth Modified Hybrid Binder Carbon Paste Electrode for Electrochemical Stripping Detection of Trace Heavy Metals in Soil. *Int. J. Electrochem. Sci.*, *7*, 12326–12339.

133. Nedeltcheva, T., Atanassova, M., Dimitrov, J., & Stanislavov, L. (2005). Determination of Mobile Form Contents of Zn, Cd, Pb and Cu in Soil Extracts by Combined Stripping Voltammetry Anal, Chim Acta, *528*, 143–146.

134. Dytrtová, J. J., Šestáková, I., Jakl, M., Száková, J., Miholová, D., & Tlustoš, P. (2008). The Use of Differential Pulse Anodic Stripping Voltammetry and Diffusive Gradient in Thin Films for Heavy Metals Speciation in Soil Solution, *Cent Eur. J. Chem.*, *6*, 71–79.

135. Kopylovich, M. N., Mahmudov, K. T., & Pombeiro, A. J. L. (2011). Poly (vinyl) Chloride Membrane Copper-Selective Electrode Based on 1-Phenyl-2-(2-hydroxyphenylhydrazo) Butane-1, 3-Dione, *J. Hazard Mater*, *186*, 1154–1162.

136. Bandi, K. R., Singh, A. K., Jain, K. A. K., & Gupta, V. K. (2011). Electroanalytical Studies on Cobalt (II) Ion-Selective Sensor of Polymeric Membrane Electrode and Coated Graphite Electrode Based on N2O2 Salen Ligands. *Electroanalysis*, *23*, 2839–2850.

137. Wilson, D., Arada, M. D., Alegret, S., & Del Valle, M. (2010). Lead (II) Ion Selective Electrodes with PVC Membranes Based on Two Bis-Thioureas as Ionophores: 1, 3-Bis (N '-Benzoylthioureido) Benzene and 1, 3-Bis (N '-furoylthioureido) Benzene, *J. Hazard. Mater*, *181*, 140–146.

138. Abbaspour, A., Mirahmadi, E., Khalafi-Nejad, A., & Babamohammadi, S. (2010). A Highly Selective and Sensitive Disposable Carbon Composite PVC-Based Membrane for Determination of Lead Ion in Environmental Samples, *J. Hazard Mater*, *174*, 656–661.

139. Faridbod, F., Ganjali, M. R., Larijani, B., Hosseini, M., Alizadeh, K., & Norouzi, P. (2009). Highly Selective and Sensitive Asymmetric Lead Microsensor Based on 5, 5, Dithiobis(2-nitrobenzoic acid) as an Excellent Hydrophobic Neutral Carrier for Nano Level Monitoring of Lead in Real Samples, *Int. J. Electrochem. Sci.*, *4*, 1528–1540.

140. Bhat, V. S., Ijeri, V. S., & Srivastava, A. K. (2004). Coated Wire Lead (II) Selective Potentiometric Sensor Based on 4-Tert-Butylcalix 6 Arene. *Sensor Actuat B-Chem*, *99*, 98–105.

141. Mazloum-Ardakani, M., Ensafi, A. A., Naeimi, H., Dastanpour, A., & Shamlli, A. (2003). Highly Selective Lead (II) Coated-Wire Electrode Based on a New Schiff Base, Sensor *Actuat B-Chem*, *96*, 441–445.

142. Mizani, F., & Ziaeiha, M. (2012). Design and Construction of High-Sensitive and Selective Zinc (II) Electrochemical Membrane Sensor Based on N, N-Bis (2hydroxy-4-metoxybenzaldehyde)-2,6-Di Amino pyridine. *Int. J. Electrochem. Sci.*, *7*, 7770–7783.

143. Gholivand, M. B., & Mozaffari, Y. (2003). PVC based Bis (2-nitrophenyl) Disulfide sensor for Zinc Ions, Talanta, *59*, 399–407.

144. Saleh, M. B., & Gaber, A. A. A. (2001). Novel Zinc Ion-Selective Membrane Electrode Based on Sulipride Drug. *Electroanalysis*, *13*, 104–108.

145. Tavares, O., Morais, S., Paý´ga, P., & Delerue-Matos, C. (2005). Determination of Ametryn in Soils Via Microwave-Assisted Solvent Extraction Coupled to Anodic Stripping Voltammetry with a Gold Ultramicroelectrode, *Anal. Bioanal Chem*, *382*, 477–484.

146. Souza, D. D., da Silva, M. R. C., & Machado, S. A. S. (2006). The Employ of Multiple Voltammetric Pulses for the Study of the Adsorption of Bipyridilium Pesticides in River Sediment, *Electroanalysis*, *18*, 2305–2313.

147. Do, J. S., & Chang, W. B. (2001). Amperometric Nitrogen Dioxide Gas Sensor: Preparation of PAn/Au/SPE and Sensing Behavior. *Sens. Actuators B.*, *72*, 101–107.

148. Zhang, B., Xu, H., & Yu, J. C. (2002). Determination of Total Gaseous Selenium in Atmosphere by Honeycomb Denuder/Differential Pulse Cathodic Stripping Voltammetry, *Talanta*, *57*, 323–331.

CHAPTER 13

HEAVY METAL DETERMINATION IN ENVIRONMENTAL SAMPLES

NIL OZBEK[1], ASLI BAYSAL[2*], and SULEYMAN AKMAN[1]

[1]Istanbul Technical University, Science and Letters Faculty, Chemistry Department 34469 Maslak Istanbul-Turkey

[2]T.C. Istanbul Aydin University, Health Services Vocational School of Higher Education, 34295 Sefakoy Kucukcekmece - Istanbul, Turkey
*E-mail: baysalas@itu.edu.tr

CONTENTS

13.1 INTRODUCTION

Heavy metals can be listed as the transition metals, some metalloids, lanthanides and actinides [1, 2]. Mostly heavy metals, metalloids and their compounds have been associated with contamination and potential toxicity or ecotoxicity. Lots of references and associations have not made any exact definition for heavy metals. Physicians for Social Responsibility (PSR) defines heavy metals as a metallic chemically element which is of a relatively high density and toxic or poisonous at low concentrations [3]. The Environment Protection Agency (EPA) has listed common heavy metals, which are the main source of pollution, especially in the environment as; As, Cd, Cr, Cu, Hg, Ni, Pb and Zn. Also Ag, Al, Au, Be, Bi, Co, Fe, Hg, Mn, Pt, Se, Sn, Tl, U and V cause health issues and should be treated carefully [4]. The main sources of heavy metals can be listed as industrial processes, mining, medicine, textile, glass and ceramic industries, dying, agriculture and medical processes and, and so on. In Table 13.1, some of the heavy metals and their related health effects are given [4–10].

TABLE 13.1   Some Heavy Metals and Related Diseases [4–10]

Disease	Analyte
Dermal	Arsenic, Nickel, Selenium
Gastrointestinal	Arsenic Beryllium, Cadmium, Copper, Mercury, Lead, Tin, Thallium, Vanadium, Zinc
Hepatic	Arsenic, Copper, Manganese, Thallium
Neurological	Arsenic, Aluminum, Mercury, Manganese, Lead, Thallium
Cardiovascular	Cadmium, Cobalt, Iron, Manganese, Lead, Nickel, Vanadium
Respiratory	Aluminum, Beryllium, Cobalt, Chromium, Manganese, Zinc

Since the 20th century and the beginning the industrial revolution, heavy/trace metals have spread across the environment and influenced living beings and ecosystems. In order to prevent this spread, governments control all kinds of industrial activities which include heavy metals and set some regulations about limit values of all kinds of environmental samples such as air, water, sediment, soil and plants. Because of these regulations and regular controls, quantitative analysis of heavy/trace elements has become very important and the field of quantitative analysis is developing.

13.2 HEAVY METAL ANALYSIS IN ENVIRONMENTAL SAMPLES

Our earth has been contaminated by pollutants from various sources. Air has been exposed to smoke from industries, lakes and rivers have been used as natural waste disposal places, soils and sediments are directly affected by the discharge of waste from industries, urban and agricultural areas. These forms of contamination directly affect ecosystems or human health [11].

Metals and their compounds are generally found in a wide concentration range in the environment, such as urban areas caused by industrial and vehicle exhaust [4]. Different commissions have set some regulations for environmental sample types. For example, the European Commission has set some standards for air samples. According to the EU Commission, humans can be affected by exposure to the air pollutants in ambient air. As a result, the EU has developed extensive standards and objectives of legislation for a number of air pollutants [12]. Table 13.2 summarizes the limit values of heavy metals in air.

TABLE 13.2   Limit Values of Some Heavy Metals in Air Set by European Commission

Pollutant	Concentration	Averaging period
Lead	$0.5\ \mu g/m^3$	1 year
Arsenic	$6\ ng/m^3$	1 year
Cadmium	$5\ ng/m^3$	1 year
Nickel	$20\ ng/m^3$	1 year

Water samples can be found in different kinds of environment. They can be classified as sea water, lake/river water, urban water and waste water. For all different water samples, there are different regulations set by different groups. The most restrictive general regulation about water samples is the Council Directive 98/83/EC on the quality of water intended for human consumption. In Table 13.3, there are limit values set by that Directive [13].

For sea water, Frontier GeoSciences Inc (FGS), which is an advanced research and analytical laboratory specializing in the determination and characterization of trace metals in the environment, reports limit values as in Table 13.3 [14].

TABLE 13.3   Metal Concentrations in Water for Human Consumption [13] and Seawater [14]

	Metal Concentration in Water for Human Consumption[13]	Metal Concentration in Seawater (μL^{-1}) [14]
Aluminum	-	10
Antimony	$5\ \mu g/L$	0.03
Arsenic	$10\mu g/L$	0.05
Barium	-	5
Beryllium	-	0.01

TABLE 13.3 *(Continued)*

	Metal Concentration in Water for Human Consumption[13]	Metal Concentration in Seawater (μL^{-1}) [14]
Boron	1 mg/l	-
Cadmium	5 µg/L	0.005
Chromium	50 µg/L	0.08
Cobalt	-	0.02
Copper	2 mg/l	0.02
Cyanide	50 µg/L	-
Iron		5
Lead	1.5 mg/l	0.01
Mercury	1 µg/L	0.00015
Nickel	20 µg/L	0.02
Selenium	10 µg/L	0.1
Silver	-	0.005
Thallium	-	0.01
Tin	-	5
Uranium	-	0.1
Vanadium	-	0.01
Zinc	-	0.8

Unfortunately there are no limit values for plants but there is also Commission Regulation 1881/2006 on setting maximum levels for certain contamination in food stuffs. In the metals section there are limit values for some heavy metals like lead, cadmium, tin, etc. in many food samples [15].

13.3 SPECTROSCOPIC TECHNIQUES

Determination of trace/heavy metals in environmental samples can be performed by various techniques. In this chapter atomic absorption spectroscopy (AAS), inductively coupled plasma- optical emission spectroscopy (ICP-OES), inductively coupled plasma-mass spectrometry (ICP-MS), microwave plasma atomic emission (MP-AES), anodic stripping voltammetry (ASV) and laser induced breakdown spectroscopy (LIBS) will be discussed.

13.3.1 ATOMIC ABSORPTION SPECTROMETRY (AAS)

Atomic Absorption Spectrometry (AAS) is used for the quantitative determination of metals and metalloids, for approximately 70 elements in all matrices. AAS is based on atomization of the analyte(s) in samples and measurement of the absorption by analyte atoms in the gas phase. The atomization can be performed using a flame and graphite furnace. Each atomization technique has its advantages and disadvantages. The comparison of atomization techniques is given in Table 13.4.

TABLE 13.4 Comparison of Several Atomization Techniques in AAS [16]

	FAAS*	GFAAS**	Hydride Generation AAS	Cold Vapor AAS
Elements	68 different elements	50 different elements	As, Se, Sb, Bi, Pb, Sn	Hg
Limit of Detection	low	high	high	high
Precision	high	low	low	low
Interferences	moderate	low	high	high
Analysis Time	short	long	moderate	moderate
Sample Preparation	easy	easy	moderate	moderate
Needed Operation Skills	low	low	low	low
Operation Costs	expensive	moderate	moderate	moderate
Sample Form	liquid	liquid-solid	liquid	liquid

*Flame AAS.
**Graphite furnace (or electrothermal) AAS.

In Flame Atomic Absorption Spectrometry (FAAS), the sample is aspirated to the flame. Two types of flame (i) nitrous oxide/acetylene flame and (ii) air/acetylene flame are used. The type of flame and the ratio of fuel to oxidant gases, as well as the observation height are chosen according to the thermal stability of the analyte and its possible compounds formed with flame concomitants. For thermal refractory elements; high temperature nitrous oxide/acetylene flame is chosen. A comparison of flame types in FAAS is shown in Table 13.5 [17–19].

TABLE 13.5	Comparison of Flame Type in FAAS [17–19]

	Air/Acetylene Flame	Nitrous Oxide/Acetylene Flame
Temperature	2300°C	3000°C
Elements	Sb, Bi, Cd, Ca, Ce, Cr, Co, Cu, Au, Ir, Fe, Pb, Li, Mg, Mn, Ni, Pd, Pt, K, Rh, Ru, Ag, Na, Sr, Ta, Sn, Zn	Al, Ba, Mo, Os, Rh, Si, Th, Ti, V
Time	10–15 seconds per element	10–15 seconds per element
Sample Form	Liquid	Liquid

In Electrothermal or Graphite Furnace Atomic Absorption Spectrometry (ETA-AS/GFAAS), samples are injected into a graphite tube as 5 to 50 μL and atomized by applying a time-temperature program. Since the analyte atoms are volatilized into a much smaller volume compared to flame, the higher density of atoms resulted in a much lower limit of detection at ppb level or lower. Atomization occurs in main four steps;

1. Drying (evaporation of solvent);
2. Pyrolysis (removal of matrix constituents);
3. Atomization (generation of free gaseous atoms of the analyte);
4. Cleaning (removal of residuals at high temperature).

Generally samples are introduced into the furnace as liquids but there are also some solid sampling instruments. Each analysis cycle takes almost 3 to 4 min. More than 50 elements can be analyzed by graphite furnace atomic absorption spectrometry [18].

Hydride generation atomic absorption spectrometry is a technique for some metalloid elements such as arsenic, antimony, selenium as well as tin, bismuth and lead. In order to generate hydride, sodium borohydride is added to the sample in acidic media in a generator chamber. The volatile hydride of the analyte is then transported to the atomizer by inert gas. This method decreases the limit of detection (LOD) 10–100 times [18–20].

Cold vapor atomization technique (CV) is used for the determination of mercury which is the only element to have enough vapor pressure at room temperature. This element's determination is essential in environmental samples. To facilitate that, mercury in the sample should be converted into Hg^{2+}. Then, Hg^{2+} is reduced with tin(II)chloride or borohydride to produce elemental mercury. With an inert gas, the produced elemental mercury is swept along into a long-pass absorption tube. The concentration is determined by the absorbance of this gas at 253.7 nm. The detection limit of this method is around ppb range. Organic mercury compounds can cause some problems since they cannot be reduced to the element by sodium tetrahydroborate, and particularly not by stannous chloride. Therefore before determination, it is advisable to apply an appropriate digestion method [18].

AAS techniques have some drawbacks. Since a specific hollow cathode lamp (HCL) is used for each element, determination should be done one by one (sequentially and adjusting the optimum conditions for each element), which makes qualitative analysis impractical. Moreover, nonmetals cannot be determined because their atomic absorption wavelengths are in the far UV range which is not suitable for analysis due to the absorption of air components.

Samples can be introduced into AAS using solid, solution and slurry forms. All sampling techniques have their own advantages and disadvantages. Also, if the concentration of the analyte is too low to be determined directly and/or interference due to the matrix cannot be eliminated, the use of a separation/enrichment procedure prior to the measurement step is compulsory. A literature survey on the determination of metal concentrations in environmental samples using different sample introduction ways and separation/preconcentration technique is presented in Table 13.6.

TABLE 13.6   Some Literature from Last Decade Performed by AAS About Environmental Samples

Sample	Technique	Elements	Method details	Reference
Air	GFAAS	Pd	Wet digestion/solid sampling	21
Air	GFAAS	Ag	Solid sampling	22
Air	GFAAS	Hg	Solid sampling	23
Air	GFAAS	Si	Wet digestion	24
Air	GFAAS	As, Cd, Cr, Pb	Wet digestion	
Sea Water	GFAAS	Cd	Pd-modified	25
Sea, Waste Water	GFAAS	Pb	Coprecipitation of lead with cobalt/pyrrolidine dithiocarbamate complex (Co(PDC)(2))/Slurry sampling	26
Sea Water	GFAAS	Pb	Ethylene glycol dimethacrylate methacrylic acid copolymer (EGDMA-MA) treated with ammonium pyrolidine dithiocarbamate (APDC)/Slurry sampling	27
Sea Water	GFAAS	Cd	Hollow fiber supported liquid membrane preconcentration	28
Tap, River, Sea Water	GFAAS	Se	Liquid–liquid microextraction (DLLME)	29
Sea Water	GFAAS	Cu, Cd, Pb	Coprecipitation with aluminum hydroxide	30
Sea Water	GFAAS	Cd, Pb	Metal complex formation with 8-hydroxyquinoline/ on-line coupling of ion chromatography	31

TABLE 13.6   *(Continued)*

Sample	Technique	Elements	Method details	Reference
Sea Water	GFAAS	Cd	Electrodeposition on the graphite electrode/ wet digestion	32
Sea Water	GFAAS	Co, Ni, Cu, Cd	A water-in-oil type emulsion containing 8-quinolinol/ ultrasonic irradiation	33
Sea Water	GFAAS	Pb, Cu, Mn	Preconcentration with polyvinylpyrrolidinone (PVP)/ dissolved with water	34
Sea Water	GFAAS	Bi, Pb, Ni	Solid-phase extraction with silica gel modified with 3-aminopropyltriethoxysilane filled in a syringe	35
River Water	GFAAS	Pb	Extraction of Pb–dithizone chelate with coacervates made up of lauric acid	36
River Water	FAAS	Cr	Cloud point extraction (CPE) system/ using 1-(2-pyridilazo)-2-naphtol (PAN) in a surfactant solution (Triton X-114)	37
River Water	Hydride Generation	As	L-cysteine and the total inorganic arsenic is sorbed onto activated alumina in the acid form in a mini-column coupled to a FI-HG AAS system	38
Lake, River Water	GFAAS	As	Cloud point extraction with surfactant-rich phase in the nonionic surfactant octylphenoxypolyethoxyethanol (Triton X-114)	39
Lake Water	FAAS	Cr	Ytterbium(III) hydroxide coprecipitation	40
River, Lake, Tap Water	GFAAS	Pd	Liquid–liquid microextraction-diethyldithiocarbamate (DDTC) was used as a chelating agent, and carbon tetrachloride and ethanol were selected as extraction and dispersive solvent	41
Lake, Tap Water	GFAAS	Pb	Continuous flow microextraction (CFME)/ 1-phenyl-3-methyl-4-benzoyl-5-pyrazolone (PMBP) dissolved in benzene is injected into a glass chamber by a microsyringe and held at the outlet tip of a PTFE connecting tube	42

TABLE 13.6 *(Continued)*

Sample	Technique	Elements	Method details	Reference
Lake, Well Water	Cold Vapor	Hg	Solid phase extraction mini-column with a neutral extractant Cyanex 923	43
River, Lake, Tap Water	Hydride Generation	As	On-line sequential insertion system	44
Lake, River, Tap Water	GFAAS	Cr	Cloud point extraction (CPE)/ surfactant *p*-octyl polyethyl-eneglycolphenyether (Triton X-100), the complex of Cr(VI) with dibromophenylfluorone (Br-PF) could enter surfactant-rich phase	45
Sea, Waste Water	FAAS	Au	Solid-phase extraction/ tetrahep-tylammonium bromide (THA$^+$. Br$^-$) immobilized polyurethane foams (PUFs)	46
Waste Water	Cold Vapor	Hg	Addition of a sodium hypochlo-rite solution to reduce a sulfide interference	47
Waste Water	GFAAS	Cr	Speciation with liquid anion exchange by Amberlite LA-2 (LAES)	48
Soil, Sediments	GFAAS	V	Slurry sampling/ using modifiers of the iridium (Ir) and carbide-forming elements: tungsten (W) and niobium (Nb) deposited on the graphite tube	49
Soil, Sediments	GFAAS	Sb	Mixed permanent modifiers Ir/ Nb and Ir/W / Slurry sampling prepared in 4% hydroflu-oric acid and 6% suspension of polytetrafluoroethylene	50
Soil	GFAAS	Cd, Cr, Cu, Pb, Zn	Chemical modifiers (Pd/ Mg(NO$_3$)$_2$ for Cd, Pb and Zn, NH$_4$F for Cu) / Solid sampling	51
Soil, Sediments	GFAAS	Cd, Pb	Mixed permanent modifiers niobium (Nb)/iridium (Ir) and tungsten (W)/iridium (Ir)/ Slurry sampling prepared with 5% HNO$_3$	52
Soils	FAAS	Fe, Cu, Mn, Zn	Soil extraction/ High continuum source	53
Soil, Sludge	ETAAS/ FAAS	Cd, Cu, Ni, Pb, Zn	Extraction with 0.05moll^{-1} EDTA and 0.43moll^{-1} acetic acid	54

TABLE 13.6 *(Continued)*

Sample	Technique	Elements	Method details	Reference
Soil, Sediment, Sludge	ETAAS	Sn, Ti	Slurry sampling prepared with 50% (v/v) concentrated hydrofluoric acid/ matrix modifers palladium (30μg) and ammonium dihydrogen phosphate (7% w/v)	55
Soil, Sediment	ETAAS	Au, Ag	Slurry sampling prepared with 3% v/v concentrated nitric acid, 50% v/v concentrated hydrogen peroxide and 25% v/v concentrated hydrofluoric acid for silver and concentrated hydrofluoric acid for gold	56
Soil, Plant	ETAAS/ FAAS	Fe, Cu, Mn, Zn, Ni, Cr, Pb, Co, Cd	Wet ashing and microwave digestion	57
Plant	FAAS	Mo	Extraction with solidified floating organic drop microextraction and preconcentration with 8-Hydroxyquinoline (8-HQ)	58
Plant	GFAAS	Pb	Solid sampling/ matrix modifier using Pd(NO(3))(2), Pd/ Mg(NO(3))(2), NH(4)H(2) PO(4) and the W-coated platform	59
Plant	GFAAS	Cr	Solid sampling	60
Plant	FAAS	Zn	Cloud point extraction with complexed with 2-methyl-8-hydroxyquinoline (quinaldine) and 1-(2-pyridylazo)-2-naphthol (PAN) separately and entrapped in a non-ionic surfactant Triton X-114	61
Plant	FAAS	Cu	direct solid analysis by flame atomic absorption spectrometry	62
Plant	GFAAS	Cd, Pb	Dynamic ultrasound-assisted extraction with dilute nitric acid	63
Sediments	Hydride Generation/ GFAAS	As	Slurry sampling/ prepared with aqua regia and hydrofluoric acid in an ultrasonic bath	64
Marine Sediments	GFAAS	Tl	High-resolution continuum-source atomic absorption spectrometer/ using ammonium nitrate as a modifier, ruthenium as a permanent modifier	65

TABLE 13.6 *(Continued)*

Sample	Technique	Elements	Method details	Reference
Lake Sediments	FAAS	Cr, Co, Ni, Cu, Zn, Cd, Pb, Mn, Fe	A four-stage sequential extraction procedure: Acetic acid, hydroxyl ammonium chloride, hydrogen peroxide plus ammonium acetate, and aqua regia	66
Sediment	Hydride Generation	As	Slurry pretreated by ultrasonic agitation and microwave assisted extraction/ using l-cysteine as an efficient pre-reduction reagent	67
Lake, River Sediment	ETAAS	Bi	Flow injection (FI) on-line sorption separation and preconcentration procedure (diethyldithiophosphate complex of bismuth is formed in 0.5–4% (v/v) HNO_3 and adsorbed onto the inner walls of a PTFE knotted reactor/ eluted with 30% (v/v) HCl)	68
Marine Sediment	GFAAS	Hg	Slurry sampling, homogenized with Triton X-100, Viscalex HV30 and glycerol	69

In Table 13.7, there is some literature from last decade about determination metal concentration in environmental samples.

TABLE 13.7 Some Investigations About Various Sample Type and Analytes in the Literature from Last Decade Performed by ICP-OES About Environmental Samples

Sample	Elements	Reference
Air	Ba, Cu, La, Ni, Cr, Fe, Mn	71
Air	Al, Ba, Ca, Cr, Cu, Fe, Mg, Mn, Pb, Se, Ti, Zn	72
Air	Al, As, Cd, Cr, Cu, Fe, Mn, Ni, Pb, Ti, V, Zn	73
Tap water, River water	Cu	74
Seawater, Well Water, Tap water, Lake water	V	75
Tap water, Well Water, Sea water	Cu	76
Sea water	Cu, Ni, Pb, Zn	77
Tap water	Al, Be, Bi, Ce, Cr, Cu, Dy, Er, Eu, Fe, Ga, Gd, Hf, Ho, Ir, La, Lu, Nd, Pr, Ru, Sc, Sm, Sn, Ta, Tb, Te, Th, Ti, Tm, U, V, Y, Yb, Zn, Zr	78

TABLE 13.7   *(Continued)*

Sample	Elements	Reference
Lake water	V	79
Sediment	Al, As, Co, Cr, Cu, Fe, Mn, Ni, Pb, Zn	80
Sediment	Se, As	81
Sediment	Pd	82
Sediment	Mg, Al, Cr, Mn, Fe, Co, Ni, Ca, Zn, Cd, Pb	83
Soil, Sediment	Cd, Co, Cr, Cu, Mn, Ni, Pb, and Zn	84
Soil	As, Sb	85
Plant	Al, Ba, Cd, Co, Cr, Cu, Fe, Hg, Mn, Ni, Se, Sn, Sr, V, Zn	86
Plant	Cd, Pb, Co, Ni, Cu	87

In Table 13.8, there is some literature from last decade about determination metal concentration in environmental samples.

TABLE 13.8   Some Literature from Last Decade Performed by ICP-MS About Environmental Samples

Sample	Element	Reference
Air	Ti, V, Cr, Mn, Co, Ni, Cu, Zn, As, Mo, Sn, Sb, Ba, Tl, Pb	88
Air	Pt, Pd, Rh	89
Air	Fe, Zn, Mn, Cu, Pb, V, Ti, Ni, As, Cr, Sr	90
Air	Na, Mg, Al, K, Ti, V, Mn, Fe, Co, Ni, Cu, Zn, As, Se, Rb, Sr, Cd, Mo, Sb, Cs, Ba, La, Ce, Pr, Nd, Sm, Gd, Pb, Th, U	91
Air	Pt, Rh	92
Sea Water	V, Mn, Co, Ni, Cu, Zn, Cd and Pb	93
Sea Water	Mn, Fe, Co, Ni, Cu, Zn	94
Sea Water	Tl	95
Sea Water	Pu	96
Sea Water	I	97
Sea Water	Mn, Fe, Co, Ni, Cu, Zn, Cd, Pb	98
Sea Water	Ra	99
Sea Water	Li	100
Sea Water	V, Mn, Co, Ni, Cu, Zn, Mo, Cd, Pb, U	101

TABLE 13.8   *(Continued)*

Sample	Element	Reference
River Water	Gd	102
Waste Water	Cr	103
Waste Water	Pd	104
Sediment	Pu	105
Sediment	Ag, As, Ba, Br, Ca, Cd, Ce, Co, Cr, Cs, Cu, Eu, Fe, Hf, LA, Mn, Mo, Na, Ni, Pb, Rb, Sb, Sc, Sm, Ta, Tb, Th, U, Yb, Zn	106
Sediment	Ti, Mn, Fe, Co, Cu, Zn, Ga, Ge, Nb, Mo, Ag, Cd, U, Bi, Pb, Tl, Re, W, Te, Sb, Sn, In	107
Sediment	U	108
Sediment	Se	109
Soil	Cu, Zn, Pb, sb	110
Soil	Sr	111
Soil	Ag	112
Soil	Ru, Rh, Pd, Re, Os, Ir, Pt	113
Plant	As	114
Plant	Cu, K, Mg, Mn, P, S, B	115
Plant	As	116

13.3.2   *INDUCED COUPLED PLASMA EMISSION*

ICP techniques have highest intensities but they are not free from interference. ICP-OES suffers from spectral interference due to wavelength overlap of different elements. Likewise, the combination of different elements in the torch forms diatomic molecules which give a similar signal as the analyte in mass spectrometer of ICP (16). When you compare ICP techniques with AAS techniques (Table 13.9), the main advantage is their sensitivity. ICP techniques also have the longest linear range when you compare with AAS techniques. If you compare their operation costs, flame atomic absorption is the cheapest one. Also, FAAS has highest precision but lowest sensitivity. It is also the cheapest when you compare the four instruments. Solid sampling can be accomplished by GFAAS and Laser Ablation ICP techniques.

TABLE 13.9   Comparison of ICP and AAS Techniques [70]

	Atomic Absorption Spectrometry		Inductively Coupled Plasma	
	Flame AAS	**Graphite Furnace AAS**	**ICP-AES**	**ICP-MS**
Sample Type	Liquid	Liquid-Solid	Liquid	Liquid Solid (LA-ICP-MS)
Sample Volume	Large	Small	Medium	Small
Detection Limits	Low	Good	Moderate	Excellent
Usage	Easy	Moderate	Easy	Moderate
Sampling Time	Very fast	Slow	Fast	Fast
Operation Cost	Low	Medium	Economic (for many samples/elements)	Economic (for many samples/elements)
Screening Ability	No	No	Yes	Yes

Some of the literature belongs to last decade has been given in Table 13.10.

TABLE 13.10   Some Literature from Last Decade Performed by LIBS on Environmental Samples

Sample	Elements	Reference
Air	Ag, Ba, Cd, Co, Cr, Cu, Hg, V, Mg, Mn, Na, Ni, Pb, Zn	117
Air	Al, Mg, Ca, Na	118
Water	Pb, Si, Ca, Na, Zn, Sn, Al, Cu, Ni, Fe, Mg, Cr	119
Water	Mg, Na	120
Waste Water	Pb, Cu, Cr, Ca, S, Mg, Zn, Ti, Sr, Ni, Si, Fe, Al, Ba, Na, K, Zr	121
Waste Water	Cr	122
Sediment	Al, Ba, C, Ca, Fe, K, Li, Mg, Mn, Na, Si, Sr, Ti	123
Sediment	Fe, Al, Si, Mg, Ca, Ti, Mn, Ba	124
Soil	Al, Ba, Ca, Cu, Fe, Ti	125
Soil	Cr	126
Soil	Cr	127

TABLE 13.10 *(Continued)*

Sample	Elements	Reference
Soil	Al, Ca, Cr, Cu, Fe, Mg, Mn, Pb, Si, Ti, V, Zn	128
Plant	Al, Ca, CN, Co, Cr, Cu, Fe, K, Mg, Mn, Mo, Na, Ni, Pb, Si, Ti, V, Zn	129
Plant	Ca, Cl, Cu, Fe, Mg, Mn,P, N, S, Zn,	130
Plant	Al, Ca, Cu, Fe, Mg, Mn, P, Zn	131
Plant	Al, Ba, Be, Ca, Cl, CN, Co, Cr, Cu, F, Fe, H, K, Li, Na, Mg, Mn, Mo, N, Ni, Rb, Sr, S, Si, Ti, V	132
Plant	As, Ca, Cu, Cr, Na, Zn	133

13.4 ANODIC STRIPPING VOLTAMMETRY (ASV)

Anodic Stripping Voltammetry (ASV) is an analytical technique that specifically detects heavy metals in various matrices. This method proposes sensitivity which is 10–100 times greater than ETAAS for some metals. Since it is so sensitive, there may be no need for a preconcentration step for this method. Also, simultaneous determination of 4 to 6 metals can be accomplished. The ASV technique consists of three steps:

(i) Deposition: electroplating of certain metals in solution onto an electrode which concentrates the metal.

(ii) Rest Period: stirring is stopped and the electrode waits at lower potential in order to let the material distribute evenly.

(iii) Stripping: metals on the electrode are stripped off which generates a current that can be measured (Fig. 13.1). This current is characteristic for each metal and by its magnitude, quantification can be done. The stripping step can be either linear, staircase, square wave, or pulse [134].

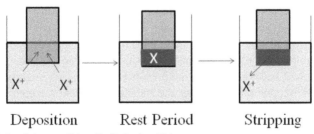

Deposition Rest Period Stripping

FIGURE 13.1 Scheme of Anodic Stripping Voltammetry.

Some of the literature from the last decade is given in Table 13.11.

TABLE 13.11   Some Literature from Last Decade Performed by ASV on Environmental Samples

Sample	Element	Detection Limit	Type of Electrode	References
Water	Cu	0.3 nM	Gold Microelectrode	135
Lake Water, Well Water, Wastewater, Sludge, Soil	Ag, Cu, Pb, Cd, Zn	Not given	Nanocrystalline diamond thin film electrode	136
Waste water, Rain water	Hg	1.1 ng mL^{-1}	Screen-printed Gold Electrodes	137
Water	As(III)	0.005 ppb	Gold Coated Diamond Thin-Film Electrode	138
	As(V)	0.08 ppb		
Soil	Sb	Not given	Hanging mercury drop electrode	139
Soil	Pb, Cd	Not given	Static mercury drop electrode	140
Plant	Sb	Not given	Copper Plant Electrode	141
Plant	TI(I)	0.15 ng mL^{-1}	Hanging mercury drop electrode	142
Plant	Cd	0.010 mg kg^{-1}	Hanging mercury drop electrode	143
	Pb	0.12 mg kg^{-1}		

13.5   PRACTICAL PROCEDURES

The objective of this section is to explain briefly sampling, sample handling, sample pretreatment (digestion/extraction/preconcentration) and main instrumentation tips using standard applications for heavy metal analysis in the different environmental samples by AAS and ICP. The basic procedures for heavy metal analysis mainly include: (i) Sampling (sample collection, preservation, handling); (ii) Sample pretreatment (digestion, if necessary separation/preconcentration); (iii) Analysis (measurement, calibration), including reporting and statistical evaluation. Figure 13.2 shows the sampling and analytical step of the whole environmental analysis.

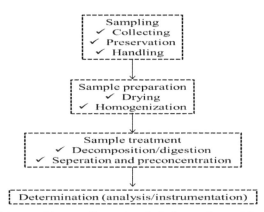

FIGURE 13.2   Scheme of the sampling and analytical procedures of the sample.

The main procedure includes these steps with a sample throughput, cost, ease of use, time, and availability of method. Other important considerations for the procedure are how to deal with interference and improve the preconcentration/separation methods for effective determination.

Interferences come up in the atomic absorption and inductively coupled plasma spectrometries for the heavy metal determination in environmental samples are (i) spectral, (ii) chemical, (iii) physical. To overcome interference, some solutions are summarized in Table 13.12.

TABLE 13.12   Interferences and Their Elimination Methods

Technique		Interferences	Elimination
AAS	Flame	Chemical	Releasing reagent Nitrous-acetylene flame
		Physical	Dilution Matrix matching Method addition
	Graphite (electrothermal)	Spectral	Zeeman background correction
		Chemical/physical	Stabilized temperature platform
ICP	OES	Spectral	Background correction Use alternative lines
		Matrix	Internal addition
	MS	Matrix	Internal addition
		Mass overlap	Interelement correction Dynamic reaction cell uses High mass resolution

When the concentration of the analyte is too low to be determined directly, usage of a separation/preconcentration technique is advised for the determination of heavy metals. Separation is a process in which the components constituting the starting mixture are separated from each other. Pre-concentration is a technique by which the ratio of concentration (or the amount) of trace components to the concentration (or the amount) of macro component is increased. A separation/enrichment method is unavoidably applied prior to the measurement step in all instrumental methods. General separation and preconcentration techniques are explained below:

   i)   *Solvent extraction:* it uses the concept of unique solute distribution ratios between two immiscible solvents. Solvent extraction is widely applied to the processes of metal ion recovery, ranging from aqueous solutions in hydrometallurgical treatment to environmental applications. Numerous organic solvents have been used to remove heavy metals; most of them are, in part, made from petroleum. Recently, solvents such as vegetable oil and organo-phosphorous compounds have also been employed as metal extractants.

   ii) *Precipitation and coprecipitation:* another common technique is the coprecipitation of the analyte with a gathering precipitate and subsequently its transferal (completely or partially) into solution phase using a suitable eluent or dissolving the precipitate completely by means of a solvent for quantification. The literature is full of papers on preconcentration/separation of trace elements by coprecipitation technique using different carriers. In order to coprecipitate various analytes, different gathering precipitates such as naphthalene, manganese dioxide obtained by the reduction of $KMnO4$, cerium(IV) hydroxide, lanthanum hydroxide, yttrium phosphate, 2,3-dihydroxypyridine, yttrium hydroxide, nickel diethyldithiocarbamate, and cobalt diethyldithiocarbamate have been used [26, 30].

   iii) *Volatilization:* it is based on differences in the vapor pressures of some elements (As, Hg, Se, Tl, etc.). This method is used for nonmetallic and amphoteric elements which have high vapor pressures. Hydride generation and halogen volatilization are the most known volatilization methods.

   iv) *Use of the ion exchanger and other sorbents (e.g., solid phase extraction):* is a separation process by which compounds that are dissolved or suspended in a liquid mixture are separated from other compounds in the mixture according to their physical and chemical properties. Solid phase extraction can be used to isolate analytes of interest from a wide variety of matrices. For the separation and preconcentration of heavy metals, the most commonly and practically applied method is the solid phase extraction, either off-line (column or batch) or on-line, which is the collection of the analyte on a sorbent using different materials; numerous inorganic and organic, natural and synthetic materials directly or modified with a chelating group, bacteria, etc. are used as sorbents, all of which have their own advantages and disadvantages [144]. Nowadays, nanomaterials such as nano carbon tubes, nano ZnO, ZrO_2, TiO_2, SiO_2 are commonly and successfully used as sorbents.

An important point of this method is the elution step. Elution causes a loss of time, contamination, high blank values and it highly influences the quality of analytical data such as recovery, precision, enrichment of analytes, etc. Ideally the elution process should be fast and reproducible. Therefore, this step determines the appropriateness of the method including speed, easiness, accuracy (or recovery) and precision [145].

13.5.1 AIR SAMPLES

Air is the most important substances for life. Most toxins are taken in through inhalation. In order to have a less toxic environment, clean air is necessary. Air samples which are important for air quality research can be divided into different types;

1. aerosols: suspensions of small particles;
2. smoke: ash, soot, grit emitted from combustion processes;
3. haze: suspension of small particles causing a reduction in visibility;
4. mist: suspension of droplets;
5. fog: suspension of water droplets.

An important point of air analysis is particle size. The main difference in air analysis is particle size which is particulate matter (PM) (>10 micrometer, <10 micrometer, <2.5 micrometer, etc.).

Samples of suspended particulate matter are collected on a filter (including different particulate size) with various methodologies which are:

1. low volume sampler (having low flow rates);
2. high volume sampler (having high flow rates).

After samples are collected in an appropriate way, the particulate matter of air samples collected on a filter is extracted (digested) with nitric acid for heavy metal determination by atomic absorption spectrometry and ICP techniques. Results are reported in appropriate form [146, 147].

13.5.2 WATER SAMPLES

Water samples are divided as river water, sea water, ground water, tap water and drinking water. Trace or heavy metal concentrations should range within some levels for each of the water samples not to cause any toxicity for the environment or humans. Water samples need less pretreatment and can be analyzed directly. Heavy metals are mostly determined by atomic absorption spectrometry and ICP techniques according to the sensitivity of the metal.

Water samples are taken into bottles by an appropriate method explained in the sampling section. Samples are filtered and preserved with nitric acid until analysis. If the water samples include a heavy matrix like sea water, waste water, etc., the samples are mostly digested with concentrated nitric acid and then analyzed by AAS or ICP methods. If the water samples are clean or do not have any salts, like tap water or drinking water, the samples are analyzed directly by AAS or ICP methods without any pretreatment [148, 149].

13.5.3 SOIL, SEDIMENTS, SLUDGE, AND DUST SAMPLES

Heavy metal determination for solid samples can be classified as hard in contrast to water samples, which can be analyzed directly or with minimal sample treatment. Solid samples in environmental analysis can be soil, sediment, sludge and dust. Solid samples have great importance because of the biogeochemical cycling of nutrients and pollutants.

Heavy metals in soil, sediment, sludge and dust are mostly determined by atomic absorption spectrometry and inductively coupled plasma techniques after acid digestion. AAS and ICP methods are the standard instrumentation for most of the standard methods like EPA. In addition to these techniques, anodic stripping voltammetry method (ASV) can be used.

The procedure can include:

1. Sampling: Sampling of the solids is difficult because of the heterogeneity and dimensions (depth/width) of the media. The sampling program should be carefully planned for this reason. Samplings of the solids are given extensively in the sampling section. Success of the analysis is complete sampling which also includes drying and homogenization.
2. Digestion (extraction) of the sample: heavy metals in solids are digested in acid solution and analyzed by AAS and ICP. To determine metals in solids, different acid digestion can be used with various acid ratios, digestion and duration etc., which are as follows:

- Nitric acid digestion: for some complex matrixes nitric acid digestion may be not effective but if you are not able or willing or is unsafe to use hydrofluoric acid-perchloric acid, you can use nitric acid. There are many variations which can be used for nitric acid digestion, such as nitric acid+hydrogen peroxide, nitric acid+sulfuric acid or nitric acid+hydrochloric acid (aqua regia).
- Hydrofluoric acid+perchloric acid: this is the most effective digestion or extraction mixture for metal analysis but it is extremely dangerous and hazardous for the environment. Aqua regia has replaced hydrofluoric acid+perchloric acid in most procedures for safety reasons.

After digestion is completed, the digested samples are analyzed by AAS and ICP [150, 151].

13.6 CONCLUSION

There is a number of spectrometric methods for the measurement of heavy/trace metals in environmental samples; in general, none of the methods is completely free of problems. Every method has its own advantages and drawbacks with respect to time, cost, easiness, sensitivity, accuracy, interference, precision, etc. Therefore, the most appropriate method should be chosen considering the capabilities and insufficiencies of the methods.

For example, while AAS has been proved to be a popular or most appropriate method for the analysis of metals for environmental samples, ICP techniques offer the option of sensitive simultaneous multi element analysis of an environmental sample with real advantages in the rates of sample analysis.

KEYWORDS

- **Air Samples**
- **Anodic Stripping Voltammetry**
- **Heavy Metals in Environment**
- **Practical Procedures**
- **Sediments**
- **Sludge and Dust Samples**
- **Soil**
- **Spectroscopic Techniques**
- **Water Samples**

REFERENCES

1. Morris, C. (1992). Dictionary of Science and Technology, Academic Press: San Diego, CA.
2. Duffus, J. H. (2002). "Heavy Metals", a Meaningless Term? Pure Appl. Chem. *74*, 793–807.
3. Physicians for Social Responsibility. Heavy metals, http://www.psr.org/environment-and-health/confronting-toxics/heavy-metals, (accessed Aug 12, 2013).
4. Athar, M., & Yohra, S. B. (2001). Heavy Metals and Environment, New Age International Limited Publishers: Noida.
5. United States Environmental Protection Agency (EPA) in Arsenic Rule Benefit Analysis, August 9, 2001.
6. Toxic Subtances Portal, Agency for Toxic Substances& Disease Registry. U. S. Department of Health and Human Services. http://www.atsdr.cdc.gov (Accessed July 30, 2013).
7. Ballatori, N. (2000). Molecular Mechanisms of Hepatic Metal Transport, In Molecular Biology and Toxicology of Metals, Zalups RK, *Koropatnick J*, Ed. Taylor & Francis: New York, 346–381.

8. Levantine, A., & Almeyda, J. (1973). Drug Induced Changes in Pigmentation. *British Journal of Dermatology*, *89*, 105–112.

9. Nellson, G. (2013). Occupational Respiratory Diseases in the South African mining Industry, Global Health Action, *6*, 19520.

10. Iron Disorders Institute. http://www.irondisorders.org/chronic-diseases-affected-by-iron (accessed July 25, 2013).

11. Abida, B., HariKrishna, S., & Khan, I. (2009). Analysis of Heavy Metals in Water, Sediments and Fish Samples of Madivala Lakes of Bangalore, Karnataka *Int. J. Chem.* Tech. Res. *1*, 245–249.

12. Air Quality Standarts. European Comission, http://ec.europa.eu/environment/air/quality/standards.htm accessed June 20, 2013).

13. Council Directive, 98/83/EC of 3 November 1998 on the Quality of Water Intended for Human Consumption http://eurlex.europa.eu/LexUriServ/LexUriServ.do?uri=OJ:L:1998:330:0032:0054:EN: PDF Accessed June 20, 2013).

14. http://www.frontiergs.com/files/Measure_Trace_Metals_in_Seawater.pdf (accessed June 27, 2013).

15. COMMISSION REGULATION (EC) No 1881/2006 of 19 December 2006 Setting Maximum Levels for Certain Contaminants in Foodstuffs. http://eurlex.europa.eu/LexUriServ/LexUriServ.do? uri=CONSLEG:2006R1881:20100701:EN: PDF (accessed June 27, 2013).

16. Baysal, A., Ozbek, N., & Akman, S. (2013). Determination of Trace Metals in Waste Water and their Removal Processes, Waste Water-Treatment Technologies and Recent Analytical Developments, Prof. Fernando Sebastián García Einschlag Ed. Intech: Rijeka, 146–171.

17. Eaton, A. D., Clesceri, L. S., Rice, E. W., & Greenberg, A. E. (2005). Standard Methods for the examination of Water and Waste Water, 21st ed., New York.

18. Skoog, D. A., Holler, F. J., & Nieman, T. A. (1998). Principles of Instrumental Analysis Saunders College Pub: Philadelphia.

19. Csuros, C., & Csuros, M. (2002). Environmental Sampling and Analysis for Metals, CRC, Press.

20. Dedina, J., & Tsalev, D. L. (1995). Hydride Generation Atomic Absorption Spectrometry, Wiley: New York.

21. Atilgan, S., Akman, S., Baysal, A., Bakircioglu, Y., Szigeti, T., Óvári, M., & Záray, G. (2012). Monitoring of Pd in Airborne Particulates by Solid Sampling High Resolution Continuum Source Electrothermal Atomic Absorption Spectrometry Spectrochim, Acta B, *70*, 33–38.

22. Araujo, R. G. O., Vignola, F., Castilho, I. N. B., Welz, B., Goreti, R. Vale, M., Smichowski, P., Ferreira, S. L. C., & Becker-Ross, H. (2011). Determination of Silver in Airborne Particulate Matter Collected on Glass Fiber Filters Using High-Resolution Continuum Source Graphite Furnace Atomic Absorption Spectrometry and Direct Solid Sampling Spectrochim. Acta B, *66*, 378–382

23. Araujo, R. G. O., Vignola, F., Castilho, I. N. B., Borges, D. L. G., Welz, B., Goreti, R., Vale, M., Smichowski, P., Ferreira, S. L. C., & Becker-Ross, H. (2011). Determination of Mercury in Airborne Particulate Matter Collected on Glass Fiber Filters using High Resolution Continuum Source Graphite Furnace Atomic Absorption Spectrometry and Direct Solid Sampling Spectrochim. Acta B, *66*, 378–382.

24. Mukhtar, A., & Limbeck, A. (2009). A New Approach for the Determination of Silicon in Airborne Particulate Matter using Electrothermal Atomic Absorption Spectrometry Anal. Chim Acta, *646*, 17–22

25. Konečná, M., Komárek, J., & Trnková, L. (2008). Determination of Cd by Electrothermal Atomic Absorption Spectrometry after Electrodeposition on a Graphite Probe Modified with Palladium Spectrochim, Acta B, *63*, 700–703

26. Baysal, A., Akman, S., & Calisir, F. A. (2008). Novel Slurry Sampling Analysis of Lead in different Water Samples by Electrothermal Atomic Absorption Spectrometry after Coprecipitated with Cobalt/Pyrrolidine Dithiocarbamate Complex, *J. Hazard.* Mater *158*, 454–459

27. Baysal, A., Tokman, N., Akman, S., & Ozeroglu, C. (2008). Slurry Analysis after Lead Collection on a Sorbent and Its Determination by Electrothermal Atomic Absorption Spectrometry, *J. Hazard* Mater 150, 804–808

28. Peng, J. F., Liu, R., Liu, J., He, B., Hu, X., & Jiang, G. (2007). Ultrasensitive Determination of Cadmium in Seawater by Hollow Fiber Supported Liquid Membrane Extraction Coupled with Graphite Furnace Atomic Absorption Spectrometry, Spectrochim. Acta B, *62*, 499–503.

29. Bidari, A., Jahromi, E. Z., Assadi, Y., & Hosseini, M. R. M. (2007). Monitoring of Selenium in Water Samples using Dispersive Liquid–Liquid Microextraction Followed by Iridium modified Tube Graphite Furnace Atomic Absorption Spectrometry, *Microchemicals J. 87*, 6–12.

30. Doner, G., & Ege, A. (2005). Determination of Copper, Cadmium and Lead in Seawater and Mineral Water by Flame Atomic Absorption Spectrometry after Coprecipitation with Aluminum Hydroxide Anal Chim Acta, 547, 14–17.

31. Ceccarini, A., Cecchini, I., & Fuoco, R. (2005). Determination of Trace Elements in Seawater Samples by Online Column Extraction/Graphite Furnace Atomic Absorption Spectrometry *Microchem J. 79*, 21–24.

32. Knápek, J., Komárek, J., & Krásenský, P. (2005). Determination of Cadmium by Electrothermal Atomic Absorption Spectrometry using Electrochemical Separation in a Microcell Spectrochim, Acta B, *60*, 393–398.

33. Matsumiya, H., Kageyama, T., & Hiraide, M. (2004). Multielement Preconcentration of Trace Heavy Metals in Seawater with an Emulsion Containing 8-Quinolinol for Graphite-Furnace Atomic Absorption Spectrometry, Anal.Chim. Acta, *507*, 205–209.

34. Tokman, N., Akman, S., & Ozeroglu, C. (2004). Determination of Lead, Copper and Manganese by Graphite Furnace Atomic absorption Spectrometry after Separation/Concentration Using a Water-Soluble Polymer Talanta, *63*, 699–703.

35. Tokman, N., Akman, S., & Ozcan, M. (2003). Solid-Phase Extraction of Bismuth, Lead and Nickel from Seawater Using Silica Gel Modified with 3-Aminopropyltriethoxysilane Filled in a Syringe Prior to their Determination by Graphite Furnace Atomic Absorption Spectrometry Talanta, 59, 201–205.

36. Hagarová, I., Bujdoš, M., Matúš, P., & Kubová, J. (2013). Coacervative Extraction of Trace Lead from Natural Waters Prior to Its Determination by Electrothermal Atomic Absorption Spectrometry Spectrochim, Acta B. Doi: 10.1016/j.sab.2013.03.010.

37. Matos, G. D., Dos Reis, E. B., Costa, A. C. S., & Ferreira, S. L. C. (2009). Speciation of Chromium in River Water Samples Contaminated with Leather Effluents by Flame Atomic Absorption Spectrometry after Separation/Preconcentration by Cloud Point Extraction *Microchem. J. 92*, 135–139.

38. Bortoleto, G. G., & Cadore, S. (2005). Determination of Total Inorganic Arsenic in Water Using On-Line Pre-Concentration and Hydride-Generation Atomic Absorption Spectrometry Talanta, *67*, 169–174.

39. Tang, A., Ding, G., & Yan, X. (2005). Cloud Point Extraction for the Determination of As (III) in Water Samples by Electrothermal Atomic Absorption Spectrometry Talanta, *67*, 942–946.

40. Duran, A., Tuzen, M., & Soylak, M. (2011). Speciation of Cr (III) and Cr (VI) in Geological and Water Samples by Ytterbium (III) Hydroxide Coprecipitation System and Atomic Absorption Spectrometry Food Chem. Toxicol. *49*, 1633–1637.

41. Liang, P., Zhao, E., & Li, F. (2009). Dispersive Liquid-Liquid Microextraction Preconcentration of Palladium in Water Samples and Determination by Graphite Furnace Atomic Absorption Spectrometry Talanta, *77*, 1854–1857.

42. Cao, J., Liang, P., & Liu, R. (2008). Determination of Trace Lead in Water Samples by Continuous Flow Microextraction Combined with Graphite Furnace Atomic Absorption Spectrometry *J. Hazard.* Mater *152*, 910–914.

43. Duan, T., Song, X., Xu, J., Guo, P., Chen, H., & Li, H. (2006). Determination of Hg (II) in Waters by On-Line Preconcentration Using Cyanex 923 as a Sorbent Cold Vapor Atomic Absorption Spectrometry Spectrochim, Acta B, *61*, 1069–1073.

44. Anthemidis, A. N., Zachariadis, G. A., & Stratis, J. A. (2005). Determination of Arsenic (III) and Total Inorganic Arsenic in Water Samples using an On-Line Sequential Insertion System and Hydride Generation Atomic Absorption Spectrometry Anal, Chim. Acta, *547*, 237–242.

45. Zhu, X., Hu, B., Jiang, Z., & Li, M. (2005). Cloud Point Extraction for Speciation of Chromium in Water Samples by Electrothermal Atomic Absorption Spectrometry Water Res. *39*, 589–595.

46. El-Shahawi, M. S., Bashammakh, A. S., Al-Sibaai, A. A., Orief, M. I., & Al-Shareef, F. M. (2011). Solid Phase Preconcentration and Determination of Trace Concentrations of Total Gold (I) and/or (III) in Sea and Wastewater by Ion Pairing Impregnated Polyurethane Foam Packed Column Prior Flame Atomic Absorption Spectrometry *Int. J. Miner.* Proces *100*, 110–115.

47. Kagaya, S., Kuroda, Y., Serikawa, Y., & Hasegawa, K. (2004). Rapid Determination of Total Mercury in Treated Waste Water by Cold Vapor Atomic Absorption Spectrometry in Alkaline Medium with Sodium Hypochlorite Solution Talanta, *64*, 554–557

48. Stasinakis, A. S., Thomaidis, N. S., & Lekkas, T. D. (2003). Speciation of Chromium in Wastewater and Sludge by Extraction with Liquid Anion Exchanger Amberlite LA-2 and Electrothermal Atomic Absorption Spectrometry Anal Chim Acta, *478*, 119–127.

49. Dobrowolski, R., Adamczyk, A., & Otto, M. (2013). Determination of Vanadium in Soils and Sediments by the Slurry Sampling Graphite Furnace Atomic Absorption Spectrometry using Permanent Modifiers, Talanta, *113*, 19–25.

50. Dobrowolski, R., Adamczyk, A., Otto, M., & Dobrzyńska, J. (2011). Determination of Antimony in Sediments and Soils by Slurry Sampling Graphite Furnace Atomic Absorption Spectrometry using a Permanent Chemical Modifier Spectrochim, Acta B, *66*, 493–499.

51. Török, P., & Žemberyová, M. (2011). A Study of the Direct Determination of Cd, Cr, Cu, Pb and Zn in Certified Reference Materials of Soils by Solid Sampling Electrothermal Atomic Absorption Spectrometry Spectrochim, Acta B, *66*, 93–97.

52. Dobrowolski, R., Adamczyk, A., & Otto, M. (2012). Comparison of Action of Mixed Permanent Chemical Modifiers for Cadmium and Lead Determination in Sediments and Soils by Slurry Sampling Graphite Furnace Atomic Absorption Spectrometry Talanta, *82*, 1325–1331.

53. Raposo, Jr. J. L., Ruella de Oliveira, S., Caldas, N. M., & Gomes Neto, J. A. (2008). Evaluation of Alternate Lines of Fe for Sequential Multi-Element Determination of Cu, Fe, Mn and Zn in Soil Extracts by High-Resolution Continuum Source Flame Atomic Absorption Spectrometry Anal Chim Acta, *627*, 198–202.

54. Žemberyová, M., Barteková, J., Závadská, M., & Šišoláková, M. (2007). Determination of Bioavailable Fractions of Zn, Cu, Ni, Pb and Cd in Soils and Sludges by Atomic Absorption Spectrometry Talanta, *71*, 1661–1668.

55. López-Garcia, I., Arnau Jerez, I., Campillo, N., & Hernández-Córdoba, M. (2004). Determination of Tin and Titanium in Soils, Sediments and Sludges using Electrothermal Atomic Absorption Spectrometry with Slurry Sample Introduction Talanta, *62*, 413–419.

56. López- Garcia, I., Campillo, N., Arnau Jerez, I., & Hernández -Córdoba, M. (2003). Slurry sampling for the Determination of Silver and Gold in Soils and Sediments using Electrothermal Atomic absorption Spectrometry, Spectrochim, *Acta B, 58,* 1715–1721.
57. Tüzen, M. (2003). Determination of Heavy Metals in Soil, Mushroom and Plant Samples by Atomic Absorption Spectrometry, *Microchem J. 74,* 289–297.
58. Oviedo, J. A., Fialho, L. L., & Nóbrega, J. A. (2013). Determination of Molybdenum in Plants by Vortex-Assisted Emulsification Solidified Floating Organic Drop Microextraction and Flame Atomic Absorption Spectrometry Spectrochim. Acta B, *86,* 142–145.
59. Rêgo, J. F., Virgilio, A., Nóbrega, J. A., & Gomes Neto, J. A. (2012). Determination of Lead in Medicinal Plants by High-Resolution Continuum Source Graphite Furnace Atomic Absorption Spectrometry using Direct Solid Sampling, Talanta, *100,* 21–26.
60. Virgilio, A., Nóbrega, J. A., Rêgo, J. F., & Gomes Neto, J. A. (2012). Evaluation of Solid Sampling High-Resolution Continuum Source Graphite Furnace Atomic Absorption Spectrometry for Direct Determination of Chromium in Medicinal Plants Spectrochim, Acta B, *78,* 58–61.
61. Kolachi, N. F., Kazi, T. G., Khan, S., Wadhwa, S. K., Baig, J. A., Afridi, H. I., Shah, A. Q., & Shah, F. (2011). Multivariate Optimization of Cloud Point Extraction Procedure for Zinc Determination in Aqueous Extracts of Medicinal Plants by Flame Atomic Absorption Spectrometry Food Chem, Toxicol *49,* 2548–2556.
62. De Moraes Flores, E. M., Fleig Saidelles, A. P., De Moraes Flores, E. L., Mesko, M. F., Pedroso, M. P., Dressler, V. L., Bittencourt, C. F., & Da Costa, A. B. (2004). Determination of Copper in Medicinal Plants used as Dietary supplements by Atomic absorption Spectrometry with Direct Flame Solid Analysis, *Microchem J. 77,* 113–118.
63. Ruiz-Jiménez, J., Luque- Garcìa J. L., & Luque de Castro, M. D. (2003). Dynamic Ultrasound Assisted Extraction of Cadmium and Lead from Plants Prior to Electrothermal Atomic Absorption Spectrometry Anal. Chim Acta, *480,* 231–237.
64. Vieira, A., Welz, B., & Curtius, A. J. (2002). Determination of Arsenic in Sediments, Coal and Fly Ash Slurries after Ultrasonic Treatment by Hydride Generation Atomic Absorption Spectrometry and Trapping in an Iridium Treated Graphite Tube Spectrochim, Acta B, *57,* 2057–2067.
65. Welz, B., Vale, M. G. R., Silva, M. M., Becker-Ross, H., Huang, M., Florek, S., & Heitmann, U. (2002). Investigation of Interferences in the Determination of Thallium in Marine Sediment Reference Materials Using High-Resolution Continuum-Source Atomic Absorption Spectrometry and Electrothermal Atomization Spectrochim, Acta B, *57,* 1043–1055.
66. Tokalioğlu, Ş., Kartal, Ş., & Elçi, L. (2000). Determination of Heavy Metals and their Speciation in Lake Sediments by Flame Atomic Absorption Spectrometry after a Four-Stage Sequential Extraction Procedure Anal. Chim Acta, *413,* 33–40.
67. Mierzwa, J., & Dobrowolski, R. (1999). Slurry Sampling Hydride Generation Atomic Absorption Spectrometry for the Determination of Extractable/Soluble as in Sediment Samples Spectrochim, Acta B, *53,* 117–122.
68. Ivanova, E., Yan, X. P., & Adams, F. (1997). Determination of Bismuth in Cod Muscle, Lake and River Sediment by Flow Injection On-Line Sorption Preconcentration in a Knotted Reactor Coupled with Electrothermal Atomic Absorption Spectrometry Anal. Chim Acta, *354,* 7–13.
69. Bermejo-Barrera, P., Moreda-Piñeiro, J., Moreda-Piñeiro, A., & Bermejo-Barrera, A. Palladium as a Chemical Modifier for the Determination of Mercury in Marine Sediment Slurries by Electrothermal Atomization Atomic Absorption Spectrometry Anal. Chim Acta *296,* 2, 10 October.
70. www.thermo.com.

71. Limbeck, A., Wagner, C., Lendl, B., & Mukhtar, A. (2012). Determination of Water Soluble Trace Metals in Airborne Particulate Matter Using a Dynamic Extraction Procedure with Online Inductively Coupled Plasma Optical Emission Spectrometric Detection, Anal. Chim Acta, *750*, 111–119.

72. Dos Santos, M., Gómez, D., Dawidowski, L. Gautier, E., & Smichowski, P. (2009). Determination of Water-Soluble and Insoluble Compounds in Size Classified Airborne Particulate Matter, *Microchem J. 91*, 133–139.

73. Smichowski, P., Marrero, J., & Gómez, D. (2005). Inductively Coupled Plasma Optical Emission Spectrometric Determination of Trace Element in PM10 Airborne Particulate Matter Collected in an Industrial Area of Argentina, *Microchem. J.* 80, 9–17.

74. Escudero, L. A., Cerutti, S., Olsina, R. A., Salonia, J. A., & Gasquez, J. A. (2012). Factorial Design Optimization of Experimental Variables in the On-Line Separation/Preconcentration of Copper in Water Samples Using Solid Phase Extraction and ICP-OES Determination, *J. Hazard,* Mater. 0, *183*, 218–223.

75. Xiong, C., Qin, Y., & Hu, B. (2010). Online Separation/Preconcentration of V (IV)/V (V) in Environmental Water Samples with CTAB-Modified Alkyl Silica Microcolumn and their Determination by Inductively Coupled Plasma Optical Emission Spectrometry, *J. Hazard* Mater, *178*, 164–170.

76. Faraji, M., Yamini, Y., & Shariati, S. (2009). Application of Cotton as a Solid Phase Extraction Sorbent for On-Line Preconcentration of Copper in Water Samples Prior to Inductively Coupled Plasma Optical Emission Spectrometry Determination, *J. Hazard* Mater *166*, 1383–1388.

77. Otero-Romaní, J., Moreda-Piñeiro, A., Bermejo-Barrera, P., & Martin-Esteban, A. (2009). Inductively Coupled Plasma-Optical Emission Spectrometry/Mass Spectrometry for the Determination of Cu, Ni, Pb and Zn in Seawater after Ionic Imprinted Polymer based Solid Phase Extraction, Talanta, *79*, 723–729.

78. Karbasi, M. H., Jahanparast, B., Shamsipur, M., & Hassan, J. (2009). Simultaneous Trace Multielement Determination by ICP-OES after Solid Phase Extraction with Modified Octadecyl Silica Gel, *J. Hazard* Mater *170*, 151–155.

79. Wu, Y., Jiang, Z., & Hu, B. (2005). Speciation of Vanadium in Water with Quinine Modified Resin Micro-Column Separation/Preconcentration and their Determination by Fluorination Assisted Electrothermal Vaporization (FETV)–Inductively Coupled Plasma Optical Emission Spectrometry (ICP-OES) Talanta, *67*, 854–861.

80. Chand, V., & Prasad, S. (2013). ICP-OES Assessment of Heavy Metal Contamination in Tropical Marine Sediments: A Comparative Study of Two Digestion Techniques *Microchem. J.* 111, 53–61.

81. Da Luz Lopes, W., Santelli, R. E., Oliveira, E. P., De Fátima, M., De Carvalho, B., & Bezerra, M. A. (2009). Application of Multivariate Techniques in the Optimization of a Procedure for the Determination of Bioavailable Concentrations of Se and as in Estuarine Sediments by ICP OES Using a Concomitant Metals Analyzer as a Hydride Generator Talanta, *79*, 1276–1282.

82. Nakajima, J., Ohno, M., Chikama, K., Seki, T., & Oguma, K. (2009). Determination of Traces of Palladium in Stream Sediment and Auto Catalyst by FI-ICP-OES using On-Line Separation and Preconcentration with Quadrasil TA Talanta, *79*, 1050–1054.

83. Barreto, S. R. G., Nozaki, J., De Oliveira, E., Do Nascimento Filho, V. F., Aragão, P. H. A., Scarminio, I. S., & Barreto, W. (2004). Comparison of Metal Analysis in Sediments Using EDXRF and ICP-OES with the Hcl and Tessie Extraction Methods Talanta, *64*, 345–354

84. Bettinelli, M., Beone, G. M., Spezia, S., & Baffi, C. (2000). Determination of Heavy Metals in Soils and Sediments by Microwave-Assisted Digestion and Inductively Coupled Plasma Optical Emission Spectrometry, Analysis Anal Chim Acta, *424*, 289–296.

85. Tighe, M., Lockwood, P., Wilson, S., & Lisle, L. (2004). Comparison of Digestion Methods for ICP-OES Analysis of a Wide Range of Analytes in Heavy Metal Contaminated Soil Samples with Specific Reference to Arsenic and Antimony Commun.Soil Sci. Plant Anal, *35*, 9–10.

86. Bressy, F. C., Brito, G. B., Barbosa, I. S., Teixeira, L. S. G., & Korn, M. G. A. (2013). Determination of Trace Element Concentrations in Tomato Samples at Different Stages of Maturation by ICP OES and ICP-MS Following Microwave-Assisted Digestion, *Microchem J. 109*, 145–149.

87. Mikuła, B., & Puzio, B. (2007). Determination of Trace Metals by ICP-OES in Plant Materials after Preconcentration of 1,10-Phenanthroline Complexes on Activated Carbon Talanta, *71*, 136–140

88. Suzuki, Y., Sato, H., Hiyoshi, K., & Furuta, N. (2012). Quantitative Real Time Monitoring of Multi-Elements in Airborne Particulates by direct Introduction into an Inductively Coupled Plasma Mass Spectrometer, Spectrochim Acta B, *76*, 133–139.

89. Zereini, F., Alsenz, H., Wiseman, C. L. S., Püttmann, W., Reimer, E., Schleyer, R., Bieber, E., & Wallasch, M. (2012). Platinum Group Elements (Pt, Pd, Rh) in Airborne Particulate Matter in Rural Vs. Urban Areas of Germany: Concentrations and Spatial Patterns of Distribution Sci. Total Environ. *416*, 261–268.

90. Niu, J., Rasmussen, P. E., Wheeler, A., Williams, R., & Chénier, M. (2010). Evaluation of Airborne Particulate Matter and Metals Data in Personal, Indoor and Outdoor Environments using ED-XRF and ICP-MS and Co-Located Duplicate Samples Atmosph. Environ. *44*, 235–245.

91. Kulkarni, P., Chellam, S., Flanagan, J. B., & Jayanty, R. K. M. (2007). Microwave Digestion—ICP-MS for Elemental Analysis in Ambient Airborne Fine Particulate Matter: Rare Earth Elements and Validation Using a Filter Borne Fine Particle Certified Reference, Material Anal. Chim Acta, *599*, 170–176.

92. Gómez, B., Gómez, M., Sanchez, J. L., Fernández, R., & Palacios, M. A. (2001). Platinum and Rhodium Distribution in Airborne Particulate Matter and Road Dust Sci. Total Environ. *269*, 131–144

93. Veguería, S. F. J., Godoy, J. M., de Campos, R. C., & Araújo, G. R. (2013). Trace Element Determination in Seawater by ICP-MS Using Online, Offline and Bath Procedures of Preconcentration and Matrix Elimination, *Microchem J. 106*, 121–128.

94. Lagerström, M. E., Field, M. P., Séguret, M., Fischer, L., Hann, S., & Sherrell, R. M. (2013). Automated Online Flow-Injection ICP-MS Determination of Trace Metals (Mn, Fe, Co, Ni, Cu and Zn) in Open Ocean Seawater, Application to the Geotraces Program, Mar Chem *155*, 71–80.

95. Krasnodębska-Ostręga, B., Sadowska, M., Piotrowska, K., & Wojda, M. (2013). Thallium (III) Determination in the Baltic Seawater Samples by ICP MS after Preconcentration on SGX C18 Modified With DDTC Talanta, *112*, 73–79.

96. Zheng, J., & Yamada, M. (2012). Determination of Plutonium Isotopes in Seawater Reference Materials using Isotope-Dilution ICP-MS, Appl. Radiat Isotopes, *70*, 1944–1948.

97. Zheng, J., Takata, H., Tagami, K., Aono, T., Fujita, K., & Uchida, S. (2012). Rapid Determination of Total Iodine in Japanese Coastal Seawater using SF-ICP-MS Microchem J. 100, 42–47.

98. Biller, D. V., & Bruland, K. W. (2012). Analysis of Mn, Fe, Co, Ni, Cu, Zn, Cd, and Pb in Seawater Using the Nobias-Chelate PA1 Resin and Magnetic Sector Inductively Coupled Plasma Mass Spectrometry (ICP-MS) Mar. Chem. 130–131, 12–20.

99. Bourquin, M., Van Beek, P., Reyss, J. L., Riotte, J., & Freydier, R. (2011). Determination of 226Ra Concentrations in Seawater and suspended Particles (NW Pacific) using MC-ICP-MS, Mar. Chem. 126, 132–138.

100. Choi, M. S., Shin, H. S., & Kil, Y. W. (2010). Precise Determination of Lithium Isotopes in Seawater using MC-ICP-MS, *Microchem J. 95*, 274–278.
101. Rahmi, D., Zhu, Y., Fujimori, E., Umemura, T., & Haraguchi, H. (2007). Multielement Determination of Trace Metals in Seawater by ICP-MS with Aid of Down-Sized Chelating Resin-Packed Minicolumn for Preconcentration Talanta, *72*, 600–606.
102. Hennebrüder, K., Wennrich, R., Mattusch, J., Stärk, H. J., & Engewald, W. (2004). Determination of Gadolinium in River Water by SPE Preconcentration and ICP-MS Talanta, *63*, 309–316.
103. Balarama Krishna, M. V., Chandrasekaran, K., Rao, S. V., Karunasagar, D., & Arunachalam, J. (2005). Speciation of Cr (III) and Cr (VI) in Waters using Immobilized Moss and determination by ICP-MS and FAAS, Talanta, *65*, 135–143.
104. Balarama Krishna, M. V., Ranjit, M., Chandrasekaran, K., Venkateswarlu, G., & Karunasagar, D. (2009). Online Preconcentration and Recovery of Palladium from Waters using Polyaniline (PANI) Loaded in Mini-Column and Determination by ICP-MS; Elimination of Spectral Interferences Talanta, 79, 1454–1463.
105. Liao, H., Zheng, J., Wu, F., Yamada, M., Tan, M., & Chen, J. (2008). Determination of Plutonium Isotopes in Freshwater Lake Sediments by Sector-Field ICP-MS after Separation using Ion-Exchange Chromatography, Appl. Radiat. Isotopes, *66*, 1138–1145.
106. Papaefthymiou, H., Papatheodorou, G., Christodoulou, D., Geraga, M., Moustakli, A., & Kapolos, J. (2010). Elemental Concentrations in Sediments of the Patras Harbour, Greece, using INAA, ICP-MS and AAS, *Microchem J. 96*, 269–276.
107. Dolor, M. K., Helz, G. R., & McDonough, W. F. (2009). Sediment Profiles of Less Commonly Determined Elements Measured by Laser Ablation ICP-MS, Mar. Pollution Bull, *59*, 182–192.
108. Zheng, J., & Yamada, M. (2006). Determination of U-Isotope Ratios in Sediments using ICP-QMS after Sample Cleanup with Anion-Exchange and Extraction Chromatography, Talanta, *68*, 932–939.
109. Pinho, J., Canário, J., Cesário, R., & Vale, C. (2005). A Rapid Acid Digestion Method with ICP-MS Detection for the determination of Selenium in Dry Sediments, Anal. Chim Acta, *551*, 207–212.
110. Koelmel, J., & Amarasiriwardena, D. (2012). Imaging of Metal Bio accumulation in Hay Scented Fern (Dennstaedtia Punctilobula) Rhizomes growing on contaminated Soils by Laser Ablation ICP-MS, Environment Pollution, *168*, 62–70.
111. Feuerstein, J., Boulyga, S. F., Galler, P., Stingeder, G., & Prohaska, T. (2008). Determination of 90Sr in Soil Samples Using Inductively Coupled Plasma Mass Spectrometry Equipped with Dynamic Reaction Cell (ICP-DRC-MS), *J. Environ. Radioactiv 99*, 1764–1769.
112. Guo, W., Hu, S., Zhang, J., & Zhang, H. (2011). Elimination of Oxide Interferences and Determination of Ultra-Trace Silver in Soils by ICP-MS with Ion–Molecule Reactions, Sci. Total Environment, *409*, 2981–2986.
113. Fritsche, J., & Meisel, T. (2004). Determination of Anthropogenic Input of Ru, Rh, Pd, Re, Os, Ir and Pt in Soils along Austrian Motorways by Isotope Dilution ICP-MS. Sci. Total Environ. 325, 145–154.
114. Pell, A., Márquez, A., López-Sánchez, J. F., Rubio, R., Barbero, M., Stegen, S., Queirolo, F., & Díaz-Palma, P. (2013). Occurrence of Arsenic Species in Algae and Freshwater Plants of an Extreme arid Region in Northern Chile, the Loa River Basin, Chemosphere, *90*, 556–564.
115. Wu, B., Chen, Y., & Becker, J. S. (2009). Study of Essential Element Accumulation in the Leaves of a Cu-Tolerant Plant Elsholtzia Splendens after Cu Treatment by Imaging Laser Ablation Inductively Coupled Plasma Mass Spectrometry (LA-ICP-MS), Anal Chim Acta, *633*, 165–172.

116. Jedynak, L., Kowalska, J., Harasimowicz, J., & Golimowski, J. (2009). Speciation Analysis Of Arsenic In Terrestrial Plants From Arsenic Contaminated Area, Sci. Total Environ. *407*, 945–952.

117. Kuhlen, T., Fricke-Begemann, C., Strauss, N., & Noll, R. (2008). Analysis of Size-Classified Fine and Ultrafine Particulate Matter on Substrates with Laser-Induced Breakdown Spectroscopy, Spectrochim, Acta B, *63*, 1171–1176.

118. Carranza, J. E., Fisher, B. T., Yoder, G. D., & Hahn, D. W. (2001). Online Analysis of Ambient Air Aerosols using Laser-Induced Breakdown Spectroscopy, Spectrochim Acta B 1, *6*, 851–864.

119. Fichet, P., Mauchien, P., Wagner, J. F., & Moulin, C. (2001). Quantitative Elemental Determination in Water and Oil by Laser Induced Breakdown Spectroscopy. Anal.Chim. Acta, *429*, 269–278.

120. Charfi, B., & Harith, M. A. (2002). Panoramic Laser-Induced Breakdown Spectrometry of Water, Spectrochim, Acta B, *57*, 1141–1153.

121. Gondal, M. A., & Hussain, T. (2007). Determination of Poisonous Metals in Wastewater Collected from Paint Manufacturing Plant Using Laser-Induced Breakdown Spectroscopy. Talanta, *71*, 73–80.

122. Rai, N. K., & Rai, A. K. (2008). LIBS-an Efficient approach for the Determination of Cr in Industrial Wastewater, *J. Hazard* Material, *150*, 835–838.

123. Lazic, V., Colao, F., Fantoni, R., Spizzichino, V., & Jovićević, S. (2007). Underwater Sediment Analyses by Laser Induced Breakdown Spectroscopy and Calibration Procedure for Fluctuating Plasma Parameters, Spectrochim, Acta B, *62*, 30–39.

124. Barbini, R., Colao, F., Lazic, V., Fantoni, R., Palucci, A., & Angelone, M. (2002). On Board LIBS Analysis of Marine Sediments Collected During the XVI Italian Campaign in Antarctica, Spectrochim Acta B, *57*, 1203–1218

125. ElHaddad, J., Villot Kadri, M., Ismaël, A., Gallou, G., Michel, K., Bruyère, D., Laperche, V., Canioni, L., & Bousquet, B. (2013). Artificial Neural Network for Onsite Quantitative Analysis of Soils using Laser Induced Breakdown Spectroscopy, Spectrochim, Acta B, 79–80, 51–57.

126. Senesi, G. S., Dell'Aglio, M., Gaudiuso, R., De Giacomo, A., Zaccone, C., De Pascale, O., Miano, T. M., & Capitelli, M. (2009). Heavy Metal Concentrations in Soils as Determined by Laser-Induced Breakdown Spectroscopy (LIBS), with Special Emphasis on Chromium, Environ. Resear. *109*, 413–420.

127. Gondal, M. A., Hussain, T., Yamani, Z. H., & Baig, M. A. (2009). On-Line Monitoring of Remediation Process of Chromium Polluted Soil Using LIBS. *J. Hazard. Material*, *163*, 1265–1271.

128. Pontes, M. J., Cortez, J., Galvão R. K., Pasquini, C., Araújo, M. C., Coelho, R. M., Chiba, M. K., De Abreu, M. F., & Madari, B. E. (2009). Classification of Brazilian Soils by using LIBS and Variable Selection in the Wavelet Domain, Anal Chim Acta, *642*, 12–18.

129. Zhang, D. C., Ma, X., Wen, W. Q., Liu, H. P., & Zhang, P. J. (2009). Studies of Laser Induced-Breakdown Spectroscopy of Holly Leaves, *J. Phys. Conf. Ser. 185*, 1–4.

130. Yao, M., Liu, M., Zhao, J., & Huang, L. (2010). Identification of Nutrition Elements in Orange Leaves by Laser Induced Breakdown Spectroscopy Int. Conf. Comp. Electric. Syst. Sci. Eng. 398–401

131. Sun, Q., Tran, M., Smith, B. W., & Winefordner, J. D. (1999). Direct Determination of P, Al, Ca, Cu, Mn, Zn, Mg And Fe in Plant Materials by Laser-Induced Plasma Spectroscopy Can. *J. Anal. Sci. Spectrosc. 44*, 164–170

132. Juvé, V., Portelli, R., Boueri, M., Baudelet, M., & Yu, J. (2008). Space-Resolved Analysis of Trace Elements in Fresh Vegetables using Ultraviolet Nanosecond Laser-Induced Breakdown Spectroscopy Spectrochim, Acta B, *63*, 1047–1053

133. Martin, M. Z., Labbé, N., Rials, T. G., & Wullschleger, S. D. (2005). Analysis of Preservative-Treated Wood by Multivariate Analysis of Laser-Induced Breakdown Spectroscopy Spectra Spectrochim, Acta Part B, *60*, 1179–1185

134. Kounaves, S. P. (2012). Voltammetric Techniques in Handbook of Instrumental Techniques for Analytical Chemistry, Settle, F. Ed., Prentice-Hall: New-Jersey.

135. Zhuang, J., Zhang, L., Lu, W., Shen, D., Zhu, R., & Pan, D. (2011). Determination of Trace Copper in Water Samples by Anodic Stripping Voltammetry at Gold Microelectrode, *Int. J. Electrochem* Sci. *6*, 4690–4699.

136. Sonthalia, P., Mcgaw, E., Show, Y., & Swain, G. M. (2004). Metal Ion Analysis in contaminated Water samples using Anodic Stripping Voltammetry and a Nanocrystalline Diamond Thin-Film Electrode, Analytica Chimica Acta, *522*, 35–44.

137. Bernalte, E., Sánchez, C. M., & Gil, E. P. (2011). Determination of Mercury in Ambient Water Samples by Anodic Stripping Voltammetry on Screen Printed Gold Electrodes, Anal. Chim Acta, *689*, 60–64.

138. Song, Y., & Swain, G. M. (2007). Total Inorganic Arsenic Detection in Real Water Samples Using Anodic Stripping Voltammetry and a Gold-Coated Diamond Thin-Film Electrode, Anal Chim Acta, *12(593)*, 7–12.

139. Use of Anodic Stripping Voltammetry for Determination of Antimony in Soils, Proceedings of Ecopole, Jadwiga OPYDO *2(2)*, (2008).

140. Rúriková, D., & Kudravá M. (2012). Anodic Stripping Voltammetric Determination of Lead and Cadmium in Soil Extracts, Slovak Academy of Sciences, *60*, 22–26.

141. Bond, A. M., Kratsis, S., Michael, O., & Newman, G. (2005). Determination of Antimony (III) and Antimony (V) in Copper Plant Electrolyte by Anodic Stripping Voltammetry, Electroanalysis, *9*, 681–684.

142. Pałdyna, J., Krasnodebska-Ostrega, B., Sadowska, M., & Gołebiewska, J. (2013). Indirect Speciation Analysis of Thallium in Plant Extracts by Anodic Stripping Voltammetry, Electroanalysis, *25*, 1–7

143. Mamani, M. C. V., Aleixo, L. M., De Abreu, M. F., & Rath, S. (2005). Simultaneous Determination of Cadmium and Lead in Medicinal Plants by Anodic Stripping Voltammetry, *J. Pharmaceut Biomed 37*, 709–713

144. Camel, V. (2003). Solid Phase Extraction of Trace Elements: A Review. Spectrochim Acta B., *58*, 1177–1233.

145. Baysal, A., Kahraman, M., & Akman, S. (2009). The Solid Phase Extraction of Lead Using Silver Nanoparticles –Attached to Silica Gel Prior to its Determination by FAAS Current Analytical Chemistry, *5*, 352–357.

146. Environment Protection Agency Office of Environmental Enforcement (OEE), Air Emissions Monitoring Guidance, Note 2, (AG2), Ireland.

147. Technical Standard Operating Procedure, High Volume Indoor Dust Sampling at Residences for Determination of Risk-Based Exposure to Metals, SOP SRC-DUST-01 Revision No: 0, Date: 04/01/04.

148. Environment protection Agency Method, *200*, 9.

149. Technical Standard Operating Procedure Surface Water Sampling, SOP #EH-01. Surface Water Collection (Adapted From ERT/REAC SOP 2013 Rev 1.0).

150. US EPA Environmental Response Team, SOP #2012; Revision 0.0; 11/16/94; U.S. EPA Contract 68-C4–0022.

151. EPA, EPA/600/R-92/128 July (1992) Preparation of Soil Sampling Protocols: Sampling Techniques and Strategies.

CHAPTER 14

DETERMINATION OF ORGANIC POLLUTANTS IN ENVIRONMENTAL SAMPLES

Y. MOLINER-MARTHNEZ, A. ARGENTE-GARCÍA,
R. HERRÁEZ-HERNÁNDEZ, J. VERDÚ-ANDRÉS,
C. MOLINS-LEGUA, and P. CAMPÍNS FALCÓ*

Departamento de Química Analítica, Facultad de Química, Universitat de Valencia C/ Dr. Moliner 50, E46100- Burjassot, Valencia. Spain
*Email: pilar.campins@uv.es

CONTENTS

14.1   INTRODUCTION

The presence of organic pollutants (OPs) in the environment is a topic of major concern. There are several thousand compounds that can be categorized as OPs, and the list is constantly being expanded. OPs have a wide variety of uses and origins (urban, industrial, agricultural), but the most relevant compounds are those synthesized at industrial scale. The impact of some biogenic pollutants, e. g. hormones, is also important. The problem of diffuse pollution caused by industrial, agricultural and human activities has resulted in directives to control the sources of pollution, to contribute to the protection of the environment, and to guarantee the utilization of natural resources. It has long been known that OPs accumulate in the natural environment and in living organisms causing damage to health. As a result, the use of many of these chemicals is now prohibited or, if permitted, environmental quality standards (EQS) have been established [1–3]. In the last decades, important advances to control and to treat conventional organic contaminants generated in human and industrial activities have been achieved. However, new toxic substances called emergent contaminants (ECs), which were not initially recognized as a threat, are frequently detected in the environment. The term EC is somewhat ambiguous, since these contaminants are not necessarily new substances. ECs encompass a diverse group of compounds, including algal and cyanobacterial toxins, brominated and organophosphate flame retardants, plasticizers, hormones and other endocrine disrupters compounds (EDCs), pharmaceuticals and personal-care products (PPCPs), drugs of abuse and their metabolites, disinfection by-products (DBPs), organometallics, nanomaterials, polar pesticides and their degradation/transformation products, perfluorinated compounds (PFCs), and surfactants and their metabolites [4].

The concentration levels, spatial and temporal distribution, and the fate of OPs depend not only on their production rates but also on their physicochemical properties, being volatility and polarity the main features that have to be considered. In the aquatic media polar and semipolar OPs are of particular concern because, owing to their high solubility, large amounts of such compounds can enter into the masses of water and disperse. Moreover, many of them are persistent, so they can easily move through different water compartments. For these reasons, increasing attention is being devoted to control of polar OPs. OPs present a wide variety of chemical structures. In Table 14.1, the main groups of OPs that occur in environmental samples and some representative compounds of each group are listed [5].

With the sensitivity attainable by modern analytical techniques, several OPs can be detected in various environmental compartments, and different studies have been published on the occurrence of these compounds in the aquatic media. River and lake waters have been extensively characterized because they can be highly polluted by industrial, mining and urban wastewaters, as well as by the surface runoff of agricultural areas. In the last years attention is also being devoted to sea water and, especially, to ground water, as ground water is the main source of public drinking water supplies. These studies have provided evidence that some OPs can be considered ubiquitous contaminants of aquatic media. This is illustrated in Fig. 14.1, which shows the frequency of detection in ground and surface water, as illustrative matrices, of some representative OPs [6–10]. These results have been derived from a series of studies that involved samples from different European countries, as well as from monitoring programs developed in the US. It should be remarked that the frequency of detection depends on the limits of detection (LODs) of the analytical methods applied to analyze water. The LODs attainable with modern sample treatment techniques and instrumentation, especially gas chromatography (GC) and high performance liquid chromatography (HPLC or LC) coupled to mass spectrometry (MS), are typically of low to sub ng.L^{-1}. As a result, frequencies of detection near of 100% are observed for some OPs. Compounds such as caffeine, perfluorooctanoic acid (PFOA), ^{1}H benzothiazol, atrazine and its degradation product desethylatrazine, and nonylphenolethoxilatescarboxilates (NPEC) are detected in a vast majority of the samples analyzed. Other compounds that are frequently found in surface waters are diethyltoluamide (DEET), methylbenzothiazol, perfluorooctanoic sulfonate (PFOS), simazine, carbamazepine, propazine, 2,4-dinitrophenol, Bisphenol A (BPA), triclosan, sulfamethoxazol, and many others which are not shown in Fig. 14.1 such as cholesterol and coprostanol. In general terms, the compounds most frequently detected in surface water are also the pollutants most frequently found in ground water, although ground water is less polluted. It is interesting to note that sometimes the degradation products are more frequently detected than the parent compounds. This is the case of the pesticides atrazine and terbutylazine and their respective desethyl- degradation products.

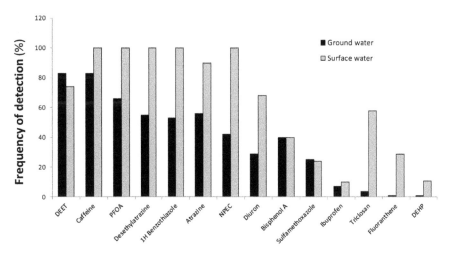

FIGURE 14.1 Frequency of detection in ground and surface water of some representative OPs according with recent studies [6–10].

TABLE 14.1 Groups and Representative Examples of OPs that Can Be Found in Environmental Samples

Group	Examples
Polar pesticides	Atrazine*, simazine, terbutylazine, chlorphenvinphos, chlorpyriphos*, fenitrotion, diuron*, linuron, glyphosphate, mecoprop, dichloprop, alachlor, endosulfan, carbaryl
Plasticizers	Diethylphthalate, dibutylphthalate, DEHP*
	BPA
Steroids and hormones	Cholesterol, coprostanol, testosterone, estriol, 17β-estradiol and estriolderivatives
Surfactants and metabolites	Linear alkylbenzenesulphonates,
	Nonylphenol*, octylphenol, nonylphenolethoxilatescarboxilates (NPEC), nonylphenolpolyethoxilates*
	Quaternary ammonium compounds
Perfluorinated compounds	Perfluorooctanoic acid (PFOA), perfluorooctanoic sulfonate (PFOS)*
Benzotriazoles	Benzotriazole, tolyltriazole, xylyltriazole
Benzothiazoles	¹H Benzothiazole, 2-hydroxybenzothiazol
PPCPs	
Antibiotics	Sulfamethoxazole, ciprofloxacin, tetracycline's
Prescribed and non-prescribed drugs	Atenolol, carbamazepine, gemfibrozil, diclofenac, ketoprofen, ibuprofen, acetaminophen, furosemide, codeine, caffeine
Organic UV filters	
Antimicrobial agentes	Benzophenones
Insect repellant	Triclosan, triclocarban
Musk fragrances	DEET
	Galaxolide
Water disinfection by-products	Bromoform, chloromethane, bromochloromethane
Halomethanes and other transformation products	Haloaldehydes, HAAs, halonitromethanes, nitrosamines
PAHs	Anthracene*, naphtalene*, fluoranthene, pyrene, benzo[a]pyrene
Complexing agents	Ethylenediaminetetraacetic acid (EDTA), nitrilotriacetic acid (NTA),diethylene triamine pentaacetic acid (DTPA)
PBDEs	Deca-, octa-, penta-brominated diphenyl esters*
Other (solvents, additives, etc)	Chloroalkanes*, 1,4-dioxane, chlorophenol and nitrophenol derivatives, PCBs, dioxins*, etc.

The asterisk denotes compounds included in the list of priority contaminants of the EU.

On the other hand, OPs found at the highest frequencies are not always those found at the highest concentrations. For example, according to Loos et al. [7], PFOA is one of the most common compounds in ground water (66% frequency) but its average concentration was only 3 ng/L, whereas some other compounds were detected in the same samples at concentration levels in the µg range. This was the case of nonylphenol, BPA, simazine and carbamazepine [7].

The concentrations reported in a vast majority of studies for the most important OPs are generally in the sub µg/L levels, and rarely exceed the regulated values. However, multiple OPs are often found in a given sample, thus revealing a high pollution degree. For a particular aquatic system, the concentration levels are related to the density of population and the economic activities involved in the area (indeed, the climate and geography are also determinant). In addition, wastewater treatments plants (WWTPs) are not totally effective in removing them. As a result, concentrations exceeding the maximum legislated levels have been reported, especially in raw wastewater. A good example is contamination by pesticides. There are numerous studies dealing with the presence of pesticides in samples of areas with an intensive agriculture activity, but contamination of water by these compounds in urban areas can be also of high impact, and even in ground water concentrations >1 µg/L have been reported for some pesticides. For example, the maximum concentration reported for dichloprop in Eq. (7) was 3.2 µ/L, whereas the limit fixed by the EU for individual pesticides is 0.1 µg/L. Other pesticides found at concentrations >0.1 µg/L in some of the samples tested in such study were atrazine, terbutylazine, desethylatrazine, desethyltherbutylazine and diuron. Similar results have been reported for river water and water of WWTPs. It should be remarked that most of these compounds are among the most frequently detected either in ground and surface waters (Fig. 14.1).

Detergent metabolites derived from nonylphenol are also important pollutants taking into account the abundance and the concentration levels at which they are found. The mean concentrations of these OPs are generally in the 0.1–1.0 µg/L range, but values as high as 20 µg/L have been reported. Other relevant OPs in terms of concentration are BPA and cholesterol. Although the average concentrations for both compounds are typically < 0.1 µg/L, much higher concentration values have been reported in raw wastewater. As regards pharmaceuticals, a vast majority of these compounds are present at concentrations in the 1–1000 ng/L range, but there are significant exceptions. This is the case of widely prescribed drugs such as carbamazepine, sulfamethoxazole, codeine and ketoprofen. High concentration levels of caffeine have been also detected in surface waters of areas with a high density of population. The concentrations of other OPs such as perfluorinated derivatives, phthalates, artificial fragrances or hormones are typically in the low ng/L level.

The reliable quantification of OPs in environmental samples such as waters, sediments, atmosphere and biota, has appeared as a huge challenge to environmental analytical chemists. On the other hand, cost-effective methods should be developed.

It is our objective to show the current performance in the quantification of OPs in the mentioned matrices, and to highlight some recent advances in the area.

14.2 GAS CHROMATOGRAPHY

GC was the first chromatographic separation technique applied to the analysis of OPs, and still it has not lost any of its eminence. CG has been traditionally applied for the determination of various families of environmental contaminants, including regulated pollutants such as volatile organic compounds (VOCs), polycyclic aromatic hydrocarbons (PAHs), polychlorinated biphenyls (PCBs), polychlorinated dibenzo-*p*-dioxins (PCDDs), polychlorinated dibenzo-furans (PCDFs), pesticides, polybrominated diphenyl esters (PBDEs) and, more recently, nonregulated pollutants such as some EDCs, PPCPs, brominated flame retardants (BFRs), steroid hormones, synthetic musk compounds and alkyl phenols. In this chapter a short overview is presented on environmental analysis of the different families of compounds. Table 14.2 summarizes some important characteristics of GC based methods, including sample treatments.

TABLE 14.2 Pre-treatment Procedures and CG Analysis of Representative Organic Pollutants

Family	Sample treatment	Sample type	Sample amount	Column	Detector
VOCs	Direct injection [12]	Water samples	250 µL	Restrictor (de-activated silica capillary column of 0.3 m × 0.1 mm)	GC-FID [12]
	LLE, P&T, HS- SPME [13, 14], close loop stripping analysis, SBSE[11]	Water samples	Several mL (1 to 10 mL)	DB-130 m × 0.32 mm (i.d.) × 5 µm column) (polydiphenylsiloxane, PDMS), 30 m × 0.25 mm (i.d.) × 0.25 µm film thickness	GC-ECD [17], FID [14, 16], PID [18], MS [19, 20]
HAAs	Derivatization LLE-HS [27], HS-SPME [28], SMPD [29], HF-LPME [30]	Water samples	Several mL (40 mL) (LLE) 10 mL [27, 30] (LLE-HS/ HF-PLME), 3 mL (SPMD) [29]	TRB-5 (phenyl-methylpolydiphe-nylsiloxane) 30 m × 0.25 mm (i.d.) × 0.25 µm film thickness	GC-ECD [28] o MS [27]

PAHs	LLE [31], SPE [32], SPME [33], LLMicroextraction [34], USAEME [35]	Water samples	2 mL to 10 mL (SPME) [33]	Non polar stationary phases (methylpolysiloxane, phenyl methyl polysiloxane)	GC-FID o MS
			5 mL HF-LPME [34]		
			From 10 mL (USAEME) to 200 mL LLE or SPE		
C h l o r i - nated com- pounds	LLE, SPE [38], SPME [39], DLLME [40], USAEME [41], PLE [42. 43], SFE [44, 45], Bead injection lab- on-valve [46], SBSE [47], HF-LPME [46], pressurized solvent extraction [49]	Bird liver [39]	Liver- 1 g [39]	DB-5-MS capillary column (length, 30 m; column (i.d.), 0.25 mm; film thickness, 0.25 μm)	GC-MS (SIM)
		Serum [38]	1 mL serum [38]		
		Water samples			GC-ITD (MS/MS) [38]
			From 10 mL (USAEME [41] or HF-LPME [46]) to 200 mL (LLE o SPE)	VF-5MS (Fac-torFour™, 55 m, 0.25 mm (i.d.), 0.25 μm film thickness)	
		Sediments [43]			
		Seaweed samples			
			0.5–1 g [43]		GC-ECD
			0.5 g [45]		GC-ECD
				non polar (phenyl methyl polysiloxane) and semi-polar (cyanopropyl methyl polysiloxane)	
				HT-5 (25 m × 0.22 mm, 0.10 μm 5%-phenyl-polycarboran-siloxane,	
				BP5 capillary column (30 m × 0.25 mm (i.d.), 0.25 μm film)	
B r o m i - nated com- p o u n d s (PBDEs)	Soxhlet [52], ASE [53], UAE [54], MAE [52], SFE [55], SPE [56], MSPD [57, 58]	Mussels [57]	MSPD 0.1 g	Non-polar (e.g. DB-5, TBR-5)	GC-ECD o MS
			MAE 2–3 g;	30 m DB-5 column (0.10 μm film thickness, 0.25 mm (i.d.))	HRGC/ HRMS
		Soil or fish [52]	SPE-2–3g; ASE 2–3 g		
			10 mL milk [59]		GC–NCI– MS
		Milk mother [59]		A HP-5 ms (30 m × 0.25 mm (i.d.), 0.25 μm film thickness)	

Pesticides	LLE [62], Soxhet [62], SPE [63], SPME [64]	Sediments [62]	LLE – 10 g	Organohalogen compounds (OCPs)- non stationary phases (DB1 or DB 5)	OCPs and OPPs (ECD[64] and NPD)
		Tissue [62]	Soxhlet 2–5 g		Orgnano-tin, OPPs and OSPs (FID)
		Berries [63]	SPE 10 g	OPPs- semipolar stationary phases (OV-17, OV-1701)	GC-MS[62,63]
		Water [64]	SPME 4 mL		
				More polar compounds- polar stationary phases (DB-Wax)	
ECs	Derivatization (polar compounds) – on line [67]	Water [67]	SPE-500 mL [67]	Non-polar DB-35, HP-5MS, DB5-MS	GC-MS [67, 69]
	SPE [68], SPME, LPME, lyophilization [69]	Water [68]	SPE-100 mL [68]	DB-5MS capillary column (30 m × 0.25 mm i.d., 0.25 μm film, connected to 2 m of deacti-vated fused-silica per-column	GC-IT-(EI) MS/MS [68]
				VF-5 ms capil-lary column (5% diphenyl 95% dimethylsiloxane), 30 m × 0.25 mm i.d., 0.25 μm	

14.2.1 REGULATED POLLUTANTS

14.2.1.1 VOCS

Nowadays, GC plays an important role in the identification and quantification of VOCs in the environment [11]. VOCs are reported in the range ng/L–μg/L, and these concentrations can be determined by injecting large sample volumes [12]. Alterna-tively, sample enrichment techniques, such as purge and trap methods (P&T), head-space (HS) and HS-solid-phase microextraction (SMPE), liquid-liquid extraction (LLE), liquid-phase microextraction (LPME), close loop stripping analysis, and stir-bar sorptive extraction (SBSE) have been applied (see Table 14.2). The application of HS-SPME is one of the best options because it takes less time for sample precon-centration and also because the required sensitivity can be reached [13]. Recently, Campíns-Falcó et al. [14] have developed a method for the determination of VOCs in water samples (river and sea water) based on SPME extraction and GC with flame ionization detection (FID). As observed in Fig. 14.2, the HS-SPME approach offered improved selectivity and sensitivity with respect to the direct injection into the GC equipment. In the direct injection method, the most volatile analytes overlapped with the solvent peak (methanol) and the resolution for the most retained analytes was poorer. By working under optimized conditions very low LODs can be obtained (from 0.005 to 0.02 µg/L). A wide variety of columns have been used for separation of VOCs [15] as shown in Table 14.2. Dimethylsiloxane stationary phases are preferred for separation of VOCs in water samples. Conventional procedures for water samples are usually carried out with classical detectors such as FID [16], electron-capture detector (ECD) [17] and photoionization detector (PID) [18]. GC-MS plays an important role in the identification and quantification of VOCs in environmental samples. Several ionization techniques are also used, although the electron ionization (EI) is the most popular one [11]. A low-resolution ion trap mass spectrometer has also been applied in Ref. [19]. A few applications are described using tandem MS for identification and quantification of VOCs [20].

14.2.1.2   DISINFECTION BY-PRODUCTS (DBPS)

Another important group of volatile analytes in water are DBPs. Extensive research has focused on the formation of DBPs in water samples, with emphasis in trihalomethanes (THMs) and, more recently, in haloacetic acids (HAAs) which are nonvolatile chlorinated by-products. Table 14.2 shows the most used procedures. An HS-SPME-GC-ECD procedure has been proposed in Ref. [21] for determination of halogenated compounds generated in water disinfection by chlorination. In this case, an ECD detector was employed providing good results for water and air samples. The LODs obtained were lower than those reported in the literature by using P&T-GC-MS or by SPME-GC-MS [22]. Although some applications have been reported for the direct determination of HAAs in waters by reverse-phase [23] and ion [24] LC and capillary electrophoresis [25], most of the analytical methods reported to date involved GC with ECD or MS. A prior derivatization step is necessary because of their low volatility and high polarity (see Table 14.2). Fisher esterification recommended by the US Environmental Protection Agency (EPA) [26] is the most common reaction used in the derivatization step. Gallego et al. [27] have proposed a simultaneous LLE/methylation by HS-GC-MS. This procedure reduces the time required for sample treatment and the consumption of organic solvents, and also minimizes the degradation of the HAAs during the whole procedure. A similar procedure has been recently developed by Hammami et al. [28] by performing a dry derivatization and HS-SPME combined with GC-ECD. Several alternatives such as single-drop microextraction (SDME) with in-microvial derivatization [29], or in situ derivatization and hollow fiber LPME (HF-LPME) [30] have been also proposed.

14.2.1.3 POLYCYCLIC AROMATIC HYDROCARBONS (PAHS)

PAHs are environmental pollutants of both natural and anthropogenic origin. They can be found in air, water, soil, sediments and food matrices. These compounds consist of fused aromatic rings, and do not contain heteroatoms or carry substituents. Naphthalene is the simplest compound as it only has two aromatic fused rings, but other PAHs may contain up to seven or even more rings. GC was first option applied to the analysis of PAHs in the early 1960s, and it is still one of the standard methods for the determination of these compounds in environmental matrices. However, PAHs with more than 24 carbon atoms cannot be analyzed by GC because of their lack of volatility. The analysis of PAHs entails an extraction or preconcentration step (Table 14.2). LLE is probably the most commonly used method for the extraction of PAHs from water samples [31]. Alternative techniques such as solid-phase extraction (SPE) [32], SPME [33] and different modes of LPME [34] and ultrasound-assisted emulsification microextraction (USAEME) [35] have been explored in order to establish efficient and economical sample preparation methods (see Table 14.2). Ozcan et al. [36] compared different procedures and demonstrated that the USEAME-based approach showed efficiencies comparable to those obtained with by applying LLE. In general, nonpolar stationary phases, such as methylpolysiloxane or phenylpolysiloxane, are suitable for separation these compounds [37]. FID is adequate for sensitive detection, but GC-MS offers grater selectivity through the application of selected ion monitoring (SIM), using EI and chemical ionization (CI) modes.

14.2.1.4 CHLORINATED COMPOUNDS

At present, GC is still the most frequently used technique for the analysis of organic halogenated contaminants such as PCDDs, PCDFs, PCBs and organohalogen compounds (OCPs) in the environment. Determination of PCBs at trace levels, is usually performed by GC combined with a preconcentration step such as LLE and SPE [38], SPME [39], dispersive liquid-liquid microextraction (DLLME) [40], USAEME [41], pressurized liquid extraction (PLE) [42, 43] supercritical fluid extraction (SFE) [44, 45], bead injection with a miniaturized lab-on-valve device [46], SBSE [47], dynamic HF-LPME [48] and ultrasound-assisted PLE [49]. These pretreatment methods have been applied to the enrichment and determination of PCBs. In general, the separation is performed using long capillary columns (50 m) with nonpolar (phenylmethylpolysiloxane) and semipolar (cyanopropylmethylpolysiloxane) stationary phases, temperatures from $100°C$ to $280–320°C$ and splitless or on column injection. ECD and MS are the most commonly used options for detecting halogenated compounds in environmental samples. Although ECD generally offers high sensitivity at low cost, the use of MS is continually increasing. Negative CI-MS is extremely sensitive for highly chlorinated compounds, although PCBs with less than four chlorine atoms cannot be measured with this technique [50].

14.2.1.5 BROMINATED COMPOUNDS

PBDEs are considered an emerging class of contaminants. Increasing concentrations levels of PBDEs observed in indoor environments (house dust), human tissues as well as environmental samples (water, soil, sediments, biota, etc.) have resulted in an increasing interest in the analysis of PBDEs. GC is often applied for PBDEs analysis [51] because the polarity and vapor pressure of these compounds. Due to the complexity of the matrix in most methods a complex extraction is employed. Soxhlet extraction [52], accelerated solvent extraction (ASE) [53], ultrasonic assisted extraction (UAE) [54], microwave assisted extraction (MAE) [52], SFE [55] and SPE are the extraction techniques commonly used for PBDEs analysis in environmental and biota samples [56] (see Table 14.2). Matrix-solid phase dispersion (MSPD) was used by Campíns-Falcó et al. [57] for the determination of PBDEs in mussels. OCPs were determined together with PBDEs, and three different extraction procedures based on MSPD followed by a cleanup step and GC-ECD were tested and compared. The best results were obtained with a miniaturized MSPD assembly involving a blend of 0.1 g of sample, 0.4 g of C_{18} phase and 0.1 g of Florisil as cosorbent. A volume of acetonitrile as low as 1.2 mL was sufficient to elute the compounds of interest. After dilution with water, the extract was processed by solid-phase microextraction (SPME) to concentrate the analytes, avoiding solvent evaporation and possible loss of volatile compounds. The benefits of miniaturization are evident, but in some cases of heterogeneous specimens, aliquots of prehomogenized sample are too small that they may not ensure representativeness unless a treatment involving intimate mixing (e.g., freeze drying) is used. Figure 14.3 shows the diagram of the procedure employed compared with conventional MSPD (58). The LODs obtained were in the range of 3–7.1 ng/g, and hexachlorobenzene was found in the tested mussels samples at levels of (2.2±0.2), (2.9±0.3) and (1.8±0.2) ng/g. The injection systems employed are: split, splitless, and the programmed temperature vaporizing (PTV) injector. The temperatures of the port and column often require special attention, due to the possible degradation of brominated congeners of decabromodiphenyl ether. The selection of the column is also crucial. PBDEs are mainly separated using nonpolar stationary phases (e.g., DB-5), 30–50 m nonpolar or semipolar phases and diameters <0.25 mm (59–60). ECD and MS are the detection systems most commonly applied. However, ECD can be used at high concentration levels (ng/g). GC-GC coupled to MS with time of flight analyzer has been reported to overcome some coelution problems encountered in the analysis of PBDEs congeners [61].

14.2.1.6 PESTICIDES

The most important classes of pesticides analyzed by GC are OCPs, organophosphorous (OPPs), organonitrogen compounds and organosulfur (OSPs) compounds, as well as organotin pesticides. As listed in Table 14.2, the extraction procedures typically used for cleanup and preconcentration are LLE [62], SPE [63] and SPME

[64]. More than 60% of these pesticides and/or their metabolites are amenable to GC. The selection of the stationary phase depends on the features of the pesticides to be separated. For example, for OCPs non polar stationary phases such as DB-1 or DB 5 are usually employed (see Table 14.2). For the separation of more polar compounds such as OPPs, semipolar stationary phases (OV-17 and OV-1701) are preferred. Polar stationary phases, e.g. DB-Wax are employed for polar compounds such as methamidophos. Usually splitless injection is preferred for the analysis of pesticides, although on-column and PTV have also been used [65]. Large samples volumes using PTV or an autoloop interface have been used for the analysis of pesticides in water samples. For detection, ECD and nitrogen phosphorous detector (NPD) are quite popular in the quantification of OCPs and OPPs residues, respectively, and FID for organotin, OPPs and OSPs compounds. Alternatively, MS is a universal detector employed not only for the quantification but also for the identification of the majority of pesticides in complex matrix samples [66]. In the full-scan MS method, all ions produced could be employed in confirmation and quantitation of the target analyte, allowing high confidence in the results. However, the LODs of MS detectors are higher than those obtainable with ECD and NPD. So far, the improvement of preconcentration methods is absolutely necessary to reach good sensitivity with MS detectors.

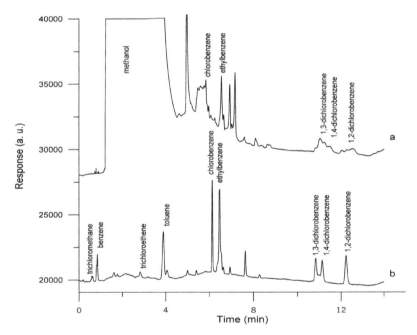

FIGURE 14.2   GC-FID Chromatograms corresponding to (a) the direct injection of 2 μL of mixture of VOCs in methanol; (b) HS-SPME (sample volume, 10 mL, adsorption time, 10 min, and desorption time, 2 min). Concentration of each VOC, 1 g mL^{-1}.

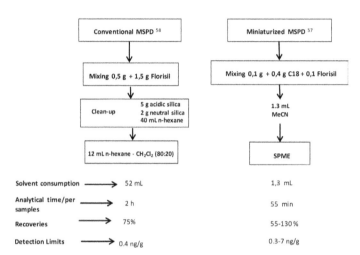

FIGURE 14.3   A comparative schedule of the MSPD and miniaturized MSPD.

14.2.2   NON-REGULATED COMPOUNDS

In general, the analysis of pharmaceuticals is performed by chromatography combined to MS or to MS/MS. LC-MS/MS is becoming more commonly used for this purpose. GC-MS in SIM mode or GC-MS/MS is a very good alternative for the analysis of these compounds. However, prior to GC-MS analysis, derivatization of polar pharmaceutical compounds is necessary, often involving toxic and/or carcinogenic derivatization reagents. An on-line derivatization in the injection port using a large volume (10 μL) sample is an alternative [67]. Another option was that proposed by Gómez et al. [68], who developed a multiresidue method based on the employment of SPE with Oasis HLB sorbent followed by GC-MS/MS for the determination of a group of 10 acidic and neutral pharmaceutical and related compounds in wastewater, avoiding the tedious and critical step of derivatization. Overall, SPE with different sorbents (RP-C$_{18}$, LiChrolut, Oasis) is the extraction technique mainly employed (see Table 14.2). A few articles deal with the use of tandem MS for identification and quantification of pharmaceutical compounds. Most of articles use GC-MS with derivatization for the most polar compounds. Although the LODs reached by LC-MS/MS are slightly higher than those obtained by GC-MS methods, LC-MS offers some advantages in terms of versatility and sample preparation requirements.

As for the others ECs, sample preparation is the most critical step in the determination of PPCPs [69]. SPE is the method of choice as it is particularly well suited to multiresidue analysis, including compounds with a wide range of polarities and physicochemical properties [70]. Other procedures such as SPME and semipermeable membrane devices (SPMDs) are becoming alternatives in the analysis of these compounds. An on fiber derivatization reaction has been developed by combining on-line derivatization with SPME for polar species [71]. Guitar et al. [72] have proposed a method for the simultaneous determination of PPCPs, fecal steroids and phenolic endocrine disrupters by GC-MS and PTV-GC-MS, and limited applicability of the PTV inlet was reported for environmental samples, affording only a modest improvement in chromatographic signal-to noise ratios. The choice between GC and LC is generally based on the physicochemical properties of the target compounds. LC-MS is usually employed to determine more polar and less volatile compounds. GC-MS and GC-MS/MS provide the suitable sensitivity and the selectivity for the identification and quantification of GC-amenable PPCPs at trace levels. For multiresidue procedures, it has been found that harsh chemicals or high derivatization temperature may result in thermal breakdown or transformation of underivatizable parent compounds, so it has been proposed to split the sample into two fractions prior to GC-MS analysis: half of the sample is submitted to derivatization for analysis of polar compound and the other half is directly analyzed.

Figure 14.4 summarizes the main extraction techniques used for EDCs during last five years. As it can be deduced SPE is the technique most commonly used but the number of publications that involve SPME is increasing. Moliner-Martínez et al. [73] proposed a quick and simple procedure to determine nonylphenol, BPA and 4-tert-octylphenol in WWTPs effluents by using SPME with GC-MS. This procedure was put into practice in three sampling campaigns for controlling WWTPs effluents along the region of Comunitat Valenciana (Spain) held in 2006, 2007 and 2008. Technical nonylphenol was found in several sampling points, while 4-nonylphenol was below its LOD in 98% of the sampling points. BPA concentrations were generally below its LODs, and only in 1% of the studied samples the concentration was between the LOD and the limit of quantification (LOQ). Figure 14.5 shows the GC-MS chromatograms corresponding to a real sample.

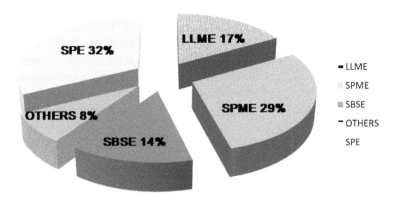

FIGURE 14.4   Graphic of the main extraction techniques employed for EDCs in last five years.

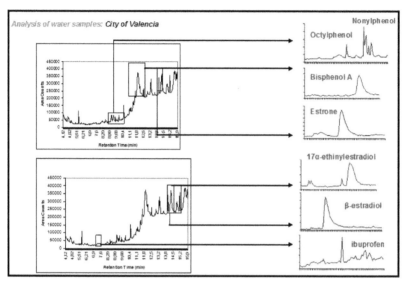

FIGURE 14.5 TIC (scan range from 100 to 300 m/z) and SIM chromatograms for a real water sample providing target.

14.3 LIQUID CHROMATOGRAPHY

We have selected organic pollutants (legislated and emergent) given in Table 14.3 in order to show the state of art of the LC methods for environmental samples.

14.3.1 ALIPHATIC AMINES

Short-chain aliphatic amines can be found in many different matrices such as natural and waste waters, industrial raw materials, or gaseous emissions from waste incineration and WWTPs. The combination of SPME and chemical derivatization for aliphatic amines was firstly reported by Pawliszyn et al. [74]. The authors demonstrated that the transformation of the analytes into less polar derivatives reduced considerably the LODs over direct SPME, using GC for separation. In Ref. [75] on-fiber derivatization and LC was proposed, using the manifold described in Fig. 14.6. In order to achieve a better understanding of the processes involved, the results were compared with those obtained by processing directly the samples after a conventional solution derivatization. The proposed SPME/on-fiber derivatization method can be considered a simple alternative to those procedures which use LLE or SPE, often combined with chemical derivatization, to reach the required sensitivity. It can be also an alternative to SPME based methods which use fiber coatings specially designed for short-chain aliphatic amines. Unlike those methods, the aforementioned approach does not require the synthesis of new coatings but only the saturation of commercially available fibers with the reagent, thus resulting in a very simple and rapid alternative. The method is also useful for air samples [76]. In Ref. [77] a microscale method was proposed that makes unnecessary any form of analyte enrichment.

14.3.2   TRIAZINES

Triazine herbicides and some of their transformation products are considered one of the most important classes of chemical pollutants owing to their widespread use and toxicity. These compounds are applied in agriculture as selective pre and postemergence weed control of corn, wheat, barley, sorghum and sugar cane, but they are also widely used for nonagricultural uses (railways and roadside verges). Triazine herbicides are relatively soluble and stable in water. The determination of triazines in water samples has been described using a variety of techniques to preconcentrate and purify the analytes such as LLE, SPE, on-line SPE and SPME. Off-line LLE and SPE are the procedures which require the largest sample volumes. SPME-based procedures usually require volumes 100 times lower in order to achieve adequate LODs. SPME has been extensively used for the determination of triazines in combination to GC. Successful results have been reported for the analysis of a variety of pesticides including triazines such as atrazine, simazine, propazine, terbuthylazine, simetryn, desethylatrazine, prometon, trietazine, terbutylazine, prometryn, terbutryn and ametryn, using ECD, thermionic specific detection and MS. In the screening analysis of triazines the LOD is the most important factor to be considered in the selection of an appropriate method for each type of water. In Ref. [78] several LC methods coupled to SPME were tested. In-tube SPME (IT-SPME) with a packed column and capillary LC (CapLC) (Table 14.3) was a quite fast and automated procedure. The on-fiber SPME method was the most time consuming method and it had the lowest degree of automation (see Table 14.3). The retention times were quite similar in all SPME-based methods tested in (78). The cost of the analysis was also evaluated. IT-SPME with packed column method incorporates a precolumn and a switching valve that increases the price of the equipment (Fig. 14.7). However, the procedure based in IT-SPME with an open column only requires a capillary GC column connected to the injection valve (Fig. 14.8). In the on-fiber SPME procedure coupled to conventional LC a SPME-HPLC interface is required (Fig. 14.6), although the capLC system is more expensive than the conventional LC system. The IT-SPME with an open column-capillary chromatography approach provided low LODs and allowed the analysis of the majority of environmental water samples.

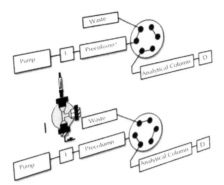

FIGURE 14.6   Configuration of the SPME interfaced to HPLC (I interface).

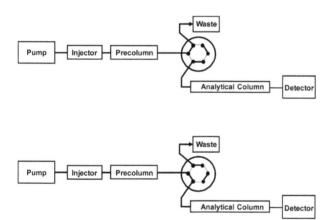

FIGURE 14.7   Configuration for column switching CapLC.

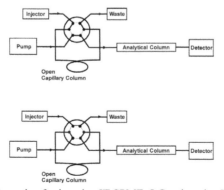

FIGURE 14.8   Configuration for in valve IT-SPME- LC or in valve IT-SPME CapLC.

TABLE 14.3 Selected Procedures for the Analysis the Main Organic Pollutants by LC

Family	LC characteristics	Pre-treatment	Matrix	Observations	Reference
Aliphatic amines (primary, secondary and tertiary)	1. Conventional quaternary pump. SPME-HPLC interface and high-pressure six-port valve. Fluorescence detector Lichrospher 100 RP$_{18}$ 125 × 4 mm i.d, 5 μm and Hypersil C$_{18}$ 20×2.1 mm i.d., 30 μm	SPME with CW-TR fiber loaded with 9-fluorenylmethyl chloroformate (FMOC) reagent 25 mL of sample 35 min	Tap, river and waste waters	See Figure 6 for configuration. On-fiber derivatization Chromatographic time=10 min LODs 5–250 μg/L	[75]
		o-phthalaldehyde/N-acetyl-L-cysteine (OPA-NAC) solution derivatization	Air		[76]
	2. CapLC with a Zorbax SB C$_{18}$, 150 mm × 0.5 mm i.d., 5 μm and UV detection	100 μL of sample 5 min	Fog water from leaf surfaces	LOD 12 mg/m^3 Chromatographic time=15 min LODs 8–50 μg/L	[77]
Triazines	1. Like aliphatic amines (1) but without derivatization and with UV detector 2. IT-SPME and CapLC with a Zorbax SB C$_{18}$, 150 × 0.5 mm i.d., 5 μm: a) packed capillary column (Zorbax SB C$_{18}$, 35 × 0.5 mm i.d., 5 μm), injection valve and automatic switching micro-valve and b) TRB5 (300x0.25 mm i.d., 0.25 μm film thickness) as injection loop. UV detector	1. Filtration of samples through 0.45 μm nylon membranes. 25 mL of sample 45 min 2a) 100 μL of sample on-line, 5 min 2b) 1000 μL of sample on-line, 1 min	Field, river, irrigation, ground, waste, industrial waste, drinking, water-treatment plant and pond waters	See Fig. 14.7 for configuration 2a) and Fig. 14.8 for configuration 2b). Chromatographic time=15 min LODs: system set-up 1: 25–125 μg/L; system set-up 2a): 0.025–0.25 μg/L; system set-up 2b): 0.1–0.5 μg/L	[78]

Organophos-phorus pesticides	1. Like triazines system set-up 2b)	Filtration of samples through 0.45 µm nylon membranes 1000 µL of sample on-line, 1 min	Field, river, ground, irriga-tion, 22 effluents of water-treat-ment plants	See Fig. 14.8 Chromatographic time=20 min Absolute recoveries obtained by TRB5 IT-SPME between 0.136% for para-thion to 43% for fonofos.	[79]
	2. IT-SPME with TRB-5 (35 cm, 0.32 mm i.d, 3 µm film thickness) and CapLC with Onyx monolith (150 mm × 0.2 mm i.d.) column. DAD detector.	MSPE with 50mg of silica supported Fe_3O_4 nanoparticles for 20 mL of sample. Isolation of sorbent with a Nd disk magnet, add 500 µL of methanol, extract the analytes and add 1 mL of water (1.5 mL processed on-line); 15 min.	Waste-waters	Good selectivity LODs for legislated organophosphorus: clorfenvinphos and clorpiriphos, 1 and 0.1 µg/L, respec-tively.	[80, 81]
	3. IT-SPME and magnetic IT-SPME with silica sup-ported Fe_3O_4 nanopar-ticles capillary column (60 cm, 0.075 mm i.d, 10 µm) and CapLC with Zorbax SB C_{18} (35 mm × 0.5 mm i.d, 5 µm). DAD detector.	3 mL of sample On-line; 5 min		See Fig. 14.9 for MSPE and Fig. 14.8 for system set-up. Chromatographic time=10 min LODs 0.085 and 0.015 µg/L, respec-tively for IT-SPME (see Fig. 14.10 for system set-up)	
			Waste-waters	0.05 and 0.01 µg/L, respectively for magnetic IT-SPME (see Fig. 14.10 for system set-up) Chromatographic time=10 min	[81–83]

Phthalates esters	1.IT-SPME with TRB-5 (80 cm, 0.32 mm i.d, 3 μm film thickness) placed between the sample injection loop and the injection needle of the autosampler and LC with Genesis C_{18} (5 cm × 4.6 mm i.d. , 4 μm particle size). UV detector.	Automatic program of the autosampler. 1.5 mL of sample On-line	Waste-waters, drinking water	Total analysis time=20 min LODs: 1 and 2.5 μg/L for dibutyl and DEHP, respectively.	[84]
	2.IT-SPME with TRB-5 (40 cm, 0.32 mm i.d, 3 μm film thickness) and CapLC with Zorbax SB C18 (35 mm × 0.5 mm i.d, 3.5 μm). DAD detector.	Matrix solid-phase dispersion 0.1 g lyophilized sample 2 mL of treated sample was processed on-line; 10 min.	Mus-sels and coastal waters in which they live.	Bioconcentration factors have been discussed. LOD for DEHP: 170 μg/Kg Chromatographic time = 25 min	[88]
Polycyclic Aromatic Hydrocarbons	1. IT-SPME with TRB-5 (70 cm, 0.32 mm i.d, 3 μm film thickness) and LC with Onyx monolithic C_{18} (100 mm × 3 mm i.d, mesopores of 13 nm, macropores of 2μm). Fluorescent detector.	MSPD 0.1 g lyophilized sample 2 mL of treated sample processed on-line. 10 min	Bivalves	LODs ≤ 0.6 μg/Kg (dry weight) Chromatographic time=10 min	[93]
	2. IT-SPME with TRB-35 (70 cm, 0.32 mm i.d, 3 μm film thickness) and LC with LiChrospher PAH (250 mm × 4.6 mm i.d, 5 μm). Fluorescent detector.	Cleaning sorbents for MSPD 0.1 g sample 2 mL of treated sample processed on-line, 10 min.	Sedi-ments	Quantitation at μg/Kg levels of PAHs. Chromatographic time=15 min	[94]
Others: Faecal sterols	IT-SPME with TRB-5 (70 cm, 0.32 mm i.d, 3 μm film thickness) and CapLC with Zirchrom Sachtopore RP-TiO_2 (100 mm × 0.5 mm i.d, 5 μm). DAD detector.	200 μL of sample On-line	Waste-water	time=10 min LODs: 10 and 1.2 μg/L for coprostanol and cholesterol, respectively.	[105]
Phenols	CapLC with Zorbax SB C18 (35 mm × 0.5 mm i.d, 5 μm). Electrochemical detection.	2 μL of sample	Treated waters		[106]

Pharmaceu-ticals	CapLC with Zorbax SB C18 (35 mm × 0.5 mm i.d, 5 µm). MS detector.	MSPE with 30mg of silica supported Fe_3O_4 NPs for 20 mL of sample. Isolation of sorbent with a Nd disk magnet, add 200 µL of methanol, extract the analytes. 10 min 2 µL of sample taken.	Water treatment plants effluents, river water.	Total analysis time=10 min LODs: 1µg/L for PN, o-cresol and 2-chloroPN and 2 µg/L for BPA. Chromatographic time=10 min	[80]
Carbonyl compounds	silica supported Fe_3O_4 nanoparticles capillary column (60 cm, 0.075 mm i.d, 10 µm) and CapLC with Zorbax SB C18 (150 mm × 0.5 mm i.d, 3.5 µm). DAD detector. IT-SPME with TRB-5 (70 cm, 0.32 mm i.d, 3 µm film thickness) and CapLC with Zorbax SB C_{18} (35 mm × 0.5 mm i.d, 5 µm) or LC with Lichrospher RP-$_{18}$ (125 mm × 4 mm i.d., 5 µm). DAD detector.	200 µL of sample On-line 1/8 portion of filter of PM_{10} extracted in 25 mL of water. Derivatization with DNPH 20 min 2–4 mL of derivatized extract. On-line	Waters Air PM_{10}	Chromatographic time=10 min LODs: 50, 150, 50 and 40 µg/L for acetylsalicilic acid, acetoaminophen, diclofenac and ibuprofen. See Figure 9 for MSPE	[82–83] [111]
Perfluori-nated com-pounds	IT-SPME with TRB-35 (43 cm, 0.32 mm i.d, 3 µm film thickness) and CapLC with Zorbax SB C_{18} (35 mm × 0.5 mm i.d, 5 µm). MS detection.	1/8 portion of filter of $PM_{2.5}$ extracted in 25 mL of water. Derivatization with DNPH20 min 2 mL of derivatized extract. On-line	Air $PM_{2.5}$ Water	Chromatographic time=15 min LODs: 5, 5, 2.5, 1.7 and 2 µg/L for acetylsalicilic acid, atenolol acetamino-phen, diclofenac and ibuprofen	[112]

Surfactants	LC–MS/MS with XTerra (150 mm × 2 mm i.d., 3 μm); UHPLC- MS/MS with Zorbax C_{18} column (50 mm × 2.1 mm, 1.8 μm); CapLC-MS with Zorbax SB-C_{18} column (150 mm × 0.5 mm, 3.5 μm)	SPE with Oasis WAX 250 mL of water sample. Evaporation to dryness. Redissolution in water:methanol (80:20, v/v). 30 min. Injection volume 5 μL in LC-MS-MS and UHPLC-MS/MS and 2 μL in CapLC-MS	Waters	See Figure 10 for system set-up Chromatographic time=20 min Screening of carbonyl compounds. LODs: 30–198 ng/L	[113]
					[114]
Multiresidue	Ion-pair IT-SPME with TRB-35 (20 cm, 0.32 mm i.d, 3 μm film thickness and CapLC with Zirchrom Sachtopore RP-TiO$_2$ column (100 mm × 0.3 mm, 5 μm). DAD detection		Waters	Chromatographic time=35 min Screening of carbonyl compounds. LODs: 0.9–8.2 ng/L Chromatographic time: 40, 15, 20 min for LC-MS/MS, UHPLC- MS/MS and CapLC-MS, respectively.	[115]
					[116]
		2 mL of sample with added tetrabutylammonium chloride. On line			
		2 mL of sample.			
	IT-SPME with TRB-35 (43 cm, 0.32 mm i.d, 3 μm film thickness and CapLC withIN-ERTSIL CN-3 column (150x0.5mm i.d., 3 μm). MS detection.	On-line		18 compounds LODs: 0.1–2 ng/L, 0.002–4 ng/L, 0.003–5 ng/L for LC-MS/MS, UHPLC- MS/MS and CapLC-MS, respectively.	[117]
		100 μL of sample On-line			[118]
		4 mL of sample			
	Capillary Column switching chromatography: like triazines 2a). DAD detector	On-line		Chromatographic time=6 min Selective for C12-BAK	
	IT-SPME –CapLC like triazines 2b). DAD detector	4 mL of sample On-line		LOD: 500 ng/L. Chromatographic time=20 min C12, C14, C16 and C18-BAK	
	IT-SPME and CapLC monolithic C_{18} column (150 mm × 0.2 mm i.d.). DAD detector and MS			LODs: 0.1 μg/L Total time=25 min 28 pollutants LODs: 0.025–10 μg/L	
	IT-SPME and UHPLC-MS/MS			LODs: 0.02–0.4 μg/L Total time=25 min 9 pollutants LODs: 5–50 ng/L Total time=25 min 9 pollutants LODs: 25250 ng/L	

14.3.3 OPPS

In Ref. [79] an on-line method for the screening of eight OPPs in water is described, which uses IT-SPME and CapLC (see Table 14.3). The IT-SPME assembly allowed the on-line enrichment of the analytes with the advantages of minimum sample manipulation, high speed, low cost and high sensitivity (the LODs were considerably reduced with respect to those obtained through the direct injection of the analytes). Moreover, under the proposed conditions no matrix effects or interferences due to other pollutants potentially present in water such as PAHs, nonylphenol, organochloro pesticides or PBDEs were observed. The proposed procedure was cost-effective and allowed the identification and quantification of the target compounds in the range of low parts-per-billion. Therefore, it was suitable to control of water quality for OPPs according to the maximum concentration levels established in legislation.

On the other hand, hybrid magnetic nanoparticles (NPs) have been widely used to develop new (micro) extraction techniques. In this context, most applications are based on their use for magnetic solid phase extraction (MSPE) or dispersive micro-SPE. The magnetic behavior of magnetic NPs give rises to the synthesis of adsorbents that can be easily concentrated and separated from sample solutions at low magnetic field gradients (see Fig. 14.9). These methods, however, require several sample pretreatment steps (Table 14.3). In this respect, IT-SPME can overcome the aforementioned limitation as it facilitates sample processing, especially in hyphenated LC techniques. Recently, we reported the use of silica supported Fe_3O_4 magnetic NPs as sorbent phase for IT-SPME (magnetic-IT-SPME) (80, 82, 83). In this approach, a magnetic hybrid material was immobilized on the surface of a bared fused silica capillary column and used as injection loop, providing quantitative extraction efficiencies for some ECs (see Fig. 14.10). In Ref. [81] it was demonstrated the potential of silica supported Fe_3O_4 NPs as adsorbent phase for MSPE and magnetic-IT-SPME to determine OPs in water samples. The results obtained with MSPE-CapLC-diode array detection (DAD) were satisfactory taking into account the maximum allowable concentrations of these pollutants. Additionally, the adsorbent phase has proved to be useful to remove OPs from water samples. However, more favorable results were achieved with magnetic-IT-SPME as pretreatment step yielding extraction efficiencies of 60% and 84% for chlorfenvinphos and chlorpyrifos, respectively. These values can be considered very high for IT-SPME. This approach provides a potential extraction procedure that avoids the pretreatment step and improves sensitivity in water analysis (see Table 14.3).

14.3.4 PHTHALATES

Phthalates esters are synthetic compounds used as polymer additives in the production of plastics, rubber, cellulose and styrene. They are present in many consumer products, such as children toys, cosmetics, personal care products, blood bags, organic solvents, packaging, paper coatings and insecticides. Due to the widespread use of phthalates they are considered as ubiquitous environmental pollutants. They migrate easily from the packaging, bottling material or manufacturing processes into foods, beverages and drinking water, being absorbed into the body. Alkyl phthalate esters, out gassing from the construction materials and process equipment are drawing attention as causes of microcontamination of indoor air. Phthalates can have adverse effects on human health and, although it has not been demonstrated, they can be considered EDCs due to their carcinogenic action. The most widely used techniques for determining phthalates in environmental samples are GC and LC.

GC requires an extensive cleaning to avoid contamination by phthalates. The ubiquitous distribution of phthalates in the environment requires a thorough cleaning in all stages of analysis for all techniques. Table 14.3 shows an automatic procedure for estimating dibutylphthalate and di(2-ethylhexyl)phthalate (DEHP) in waters [84]. Chaler et al. [85] demonstrated that DEHP was the only compound found in all types of biota examined. These authors employed 18 g of sample for digestion (18 h) and subsequent extraction and fractionation, using high volumes of organic solvents and several evaporation steps; the resulting extracts were processed by GC-MS. Huang et al. [86] studied the occurrence of phthalates in sediment and biota by using ASE and GC-MS; 2 g of fish sample was needed. ASE was carried out in 1h and the procedure required an evaporation step. The highest concentration of DEHP in fish samples were found by Ref. [86] in *Liza subviridis* (253.9 mg. Kg^{-1}) and *Oreochromis miloticus niloticus* (129.5 mg/kg). The authors indicated that living habits of fish and physical-chemical properties of phthalates, like octanol/water partition coefficient, K_{ow}, may influence the bioavailability of phthalates in fish. In another procedure proposed by Lin et al. [87] 5 g of biota sample were blended with 15–20 g of prebaked Na_2SO_4, and ground with a mortar and a pestle to a free-flowing powder. The homogenate was placed in a flask, extracted with 50 mL of 1:1 (v/v) dichloromethane/hexane for 10 min, and shaken for 10 min. Once the suspended particles settled, the supernatant was removed. The extraction was repeated two more times with fresh solvent. The combined extracts were concentrated to 5 mL with a gentle stream of nitrogen. The concentrate was quantitatively transferred onto a 350 mm × 10 mm i.d. glass column packed with 15 g of deactivated alumina (15% HPLC water, w/w) and capped with 1–2 cm of anhydrous Na_2SO_4. To prepare samples for analysis by LC-ESI MS, the column was eluted first with 30 mL of hexane, which was discarded, and then with 50 mL of 1:1 (v/v) dichloromethane/ hexane. The organic fraction was evaporated to dryness under a stream of nitrogen. Finally, the residue was reconstituted in 2 mL of methanol and analyzed by LC-ESIMS.

Campíns-Falcó et al. developed in Ref. [88] a miniaturized method based on MSPD for the analysis of DEHP in biota samples by cap-LC coupled to IT-SPME and DAD (see Table 14.3); for MSPD a C_{18} phase was used as the dispersant and acetonitrile-water as eluting solvent. Recovery studies showed that the combination of C_{18}-Florisil® was optimal using low amount of samples (0.1 g) and with low volumes of acetonitrile-water (2.6 mL 1:3.25 v/v). The samples were processed in less than 30 min and no evaporation step was required. The proposed method was applied to the analysis of DEHP in mussels and in the coastal waters from which they were collected. Table 14.4 compares some figures of merit in function of the analytical column used and the sample matrix. Different C_{18} capillary columns, three particulate (with different stationary phase and/or dimensions) and a monolithic column, were evaluated and compared for the estimation of DEHP in water, biota and sediment samples. For analyte enrichment, IT-SPME was coupled on-line to the analytical column, using DAD for detection. Water was processed directly in the chromatographic system by passing the sample (4 mL) through the extractive capillary of the IT-SPME device. Biota (lyophilized mussels) and sediment samples were previously treated by miniaturized MSPD employing 0.1 g of sample; no evaporation step was required. Best chromatographic profiles were obtained for standard solutions of DEPH with the monolithic C_{18} column (Fig. 14.11). Thus, this column was optimum for the quantification of DEHP in water samples. However, for the extracts obtained from biota and sediment samples, better stability and selectivity, respectively, were achieved with C_{18} particulate columns regardless the stationary phase and/or dimensions. The total analysis time for water samples was 15 min, and biota and sediments samples could be characterized in less than 30 min.

The Fourier Transformed infrared spectroscopy-attenuated total reflectance (FTIR-ATR) spectra shown in Fig. 14.12 were obtained from lyophilized mussels, C_{18} solid-phase, MSPD mixture and after-extracted MSPD mixture. The spectrum of mussels is characterized by broad and strong absorption bands in the 3000–3600 cm^{-1} interval (O-H and N-H stretching vibrations) and moderate C-H stretching bands in the interval from 2800–2960 cm^{-1}. A strong C=O band at 1640 cm^{-1} (amide I) and N-H deformation at 1540 cm^{-1} (amide II) are typical bands of proteins. This is in agreement with the composition of mussels: water (84.4%), proteins (10.1%), carbohydrates (3.4%) and lipids (1.9%). All these bands are also present in the spectra corresponding to MSPD and after-extracted MSPD mixtures, but presenting lesser intensities. The proteins present in the samples are probably responsible for the problems of deterioration of the Onyx monolithic C_{18} analytical column used in the analysis of biota. Note that the proteins passed partially to the acetonitrile extract, as after-extracted MSPD mixture presents smaller absorbance than that observed for the MSPD mixture (see **Fig. 14.12**).

The particulate Zorbax SB-C_{18} columns (150 × 0.3 mm id and 5 μm and 150 × 0.5 mm id and 3.5 μm particle diameters) provided suitable separation, working with the solvent composition of the MSPD extract, and using 50 μL of nanopure water and 10 μL of acetonitrile for cleaning the IT-SPME extractive capillary. However, unlike the monolithic column, the Zorbax SB-C_{18} column proposed in Ref. [88] could be used for the repetitive injection of extracts of biota without column deterioration.

Finally, the monolithic column did not provided satisfactory resolution of DEHP in the separation of the extracts obtained for sediment samples, as different matrix compounds eluted at retention times close to that of the analyte (Fig. 14.13). Additional cleaning of the extractive capillary did not improve the selectivity. It should be remarked that a change in the mobile-phase composition (e.g., a reduction in the percentage of acetonitrile) would most probably result into a satisfactory separation of the analyte in this column. However, an eluent of acetonitrile/water (95:5, v/v) was necessary to desorb the DEHP from the extractive capillary coating. No deterioration of the column was observed through repetitive injections, in contrast with the results observed when using the mussel extracts.

FIGURE 14.9   Photography of the magnetic solid phase extraction procedure

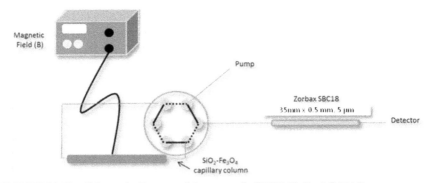

FIGURE 14.10   Schematic diagram of the magnetic-IT-SPME-CapLC-DAD.

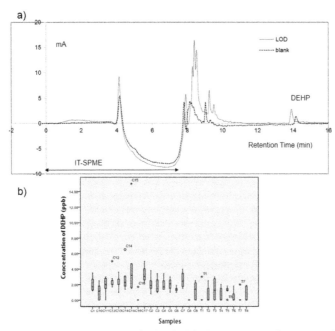

FIGURE 14.11 (a) Chromatograms obtained with the Onyx monolithic C_{18} column under conditions of Table 1 for a blank (pure water) and a standard solution containing 0.25 μg/L of DEHP. For other details, see text. (b) Box plots for the concentrations of DEHP founds in the water samples analyzed. Sea water samples are represented by a C, while transition waters are represented by a T. For other details, see text.

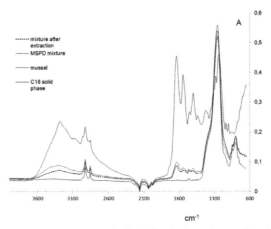

FIGURE 14.12 FTIR-ATR spectra obtained for C18 used as solid-phase in the MSPD procedure, lyophilized mussel, MSPD mixture and MSPD mixture after extraction with acetonitrile. For other details, see text.

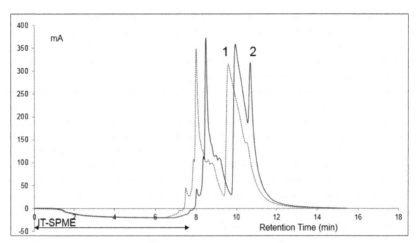

FIGURE 14.13 Chromatograms obtained with the Onyx monolithic C_{18} column under conditions of Table 1 for: (1) a sediment sample, and (2) spiked sediment sample with 120 $\mu g.L^{-1}$ of DEHP. For other details, see text.

14.3.5 PAHS

Because of their high toxicity (these compounds exhibit mutagenic and carcinogenic effects), the EPA has listed 16 PAHs as priority pollutants [89]. PAHs are adsorbed in soils and sediments, being the sediments the most important pool of these pollutants in the marine environment. Sample treatments are required to desorb PAHs from these samples, because they are strongly retained by the organic matter. Once PAHs have been removed from the samples, the most common analytical technique used for their determination is reverse phase LC with fluorescence detection (most of the PAHs present strongly native fluorescence).

For sample treatment, Soxhlet extraction has been traditionally used as it provides good extraction recoveries. However, Soxhlet extraction requires large volumes of organic solvents and long times of analysis. In order to reduce these times, UAE [90] and MAE [91] have been used as alternatives, but in both cases an additional preconcentration step is required. Quick, easy, cheap, effective, rugged and safe method (QuEChERS) has also been applied for the determination of PAHs [92], but it requires the addition of several reagents and sorbents. In addition, all of these methods provide a large volume of organic extract, which usually is not compatible with reversed phase LC conditions. Therefore, evaporation of extracts and reconstitution in another solvent is required. Thus, losses of the most volatile PAHs can occur.

MSPD coupled with IT-SPME has been proposed as a feasible alternative for PAHs analysis in biological samples and sea sediment samples [93, 94] (See Table

14.3). In MSPD the sampled is blended with a solid dispersant in a mortar [95]. This dispersant is usually a SPE supporting material. Sample and dispersant are introduced into an empty SPE column, and PAHs are eluted with an organic compatible solvent (acetonitrile, methanol, isopropanol). After increasing the polarity of the extract by addition of water, the mixture is passed through a capillary column that acts as an injection loop of the chromatographic system (see Fig. 14.8).

For the chromatographic separation of the 16 EPA priority pollutants a C_{18} column can be used, but overlapping is frequently observed; for example, benzo[a] pyrene and indene[1,2,3-cd]pyrene tend to coelut. Specially designed stationary phases have been developed to increase resolution. Isocratic separation can be used if only a few PAHs are going to be separated, but gradient elution is needed for the separation of 16 or more compounds. The detection of PAHs is usually performed by fluorescence (although ultraviolet detection can be also employed, but the sensitivity is lower), because 15 of the 16 EPA priority pollutants present native fluorescence (only acenaphtylene has a low fluorescence quantum yield). However, because maximum of excitation and emission wavelengths do not match for all compounds, several strategies are used. EPA method 8310 uses a single excitation wavelength (280 nm) and measures all emission at wavelengths above 389 nm, but with this methodology selectivity is poor (96). In other cases, to increase the selectivity, a wavelength-programmed fluorescence approach is used. PAHs are typically grouped as listed in Table 14.5, or in a similar scheme, depending on the separation achieved between analytes [97].

TABLE 14.4 Determination of DEHP: Some Figures of Merit for Several Methods for Water, Biota and Sediment in Function of the Analytical Column Used and Sample Matrix

Sample type	Column	IT-SPME conditions	Mobile phase flow rate (µL/min)	Analytical parameters						
				LOQ (µg/L)	LOD (µg/L)	Linearity (y = a + b x)[a]				
						Working concentration range (µg/L)	a ±S$_a$	b ±S$_b$	R^2	
Water	Onyx monolithic C_{18}	4 mL of aqueous standard + 100µL H$_2$O	10	0.75	0.25	0.75–10	5 ± 5	12.5 ± 0.7	0.993	
	Synergi 4u Fusion –RP 80 A	4 mL of aqueous standard + 50µL H$_2$O	7	3	1	3–100	10±2	23 ±1	0.999	
	Zorbax SB-C$_{18}$ (0.5 mm id, 3,5 µm)	4 mL aqueous standard + 50µL H$_2$O	10	3	1	3–100	9±4	30±1	0.997	

TABLE 14.4 *(Continued)*

Biota	Onyx monolithic C_{18}	2 mL standard acetonitrile:H_2O (1:3,25) + 50 μL H_2O+10 μL acetonitrile	10	15	5	15–30	70±40	11± 3	0.993
	Zorbax SB-C_{18}(0,3 mm id, 5 μm)	2 mL standard acetonitrile:H_2O (1:3,25) + 50 μL H_2O+10 μL acetonitrile	7	30	10	30–200	50±60	24 ± 2	0.990
	Zorbax SB-C_{18} (0,5 mm id, 3.5 μm) [18]	2 mL standard acetonitrile:H_2O (1:3,25) + 50 μL H_2O+10 μL acetonitrile	10	30	10	30–200	45±40	29±2	0.991
	Synergi 4u Fusion –RP 80 A	2 mL standard acetonitrile:H_2O (1:3,25) + 50 μL H_2O+10 μL acetonitrile	10	30	10	15–300	50 ±40	26 ±5	0.990
Sedi-ment	Onyx monolithic C_{18}	3 mL standard acetonitrile:H_2O (1:3,25) + 50 μL H_2O	7	15	5	30–120	80 ± 80	13.8± 0,9	0.990
	Zorbax SB-C_{18} (0,5 mm id, 3.5 μm)	3 mL standard acetonitrile:H_2O (1:3,25) + 50 μL H_2O	10	30	10	30–120	70 ± 80	30 ±5	0.990
	Synergi 4u Fusion –RP 80 A,	3 mL standard ACN:H_2O (1:3,25) + 50 μL H_2O	10	15	5	15–100	70 ± 20	29± 7	0.995

[a] expressed in μg.L^{-1};
[b] measured concentration.

TABLE 14.5   Typical Wavelength-Programmed Fluorescence Scheme for the Determination of PAHs

$\lambda_{excitation}$ (nm)	$\lambda_{emission}$ (nm)	Analyte
226	330	Naphthalene, acenaphthene, fluorene
250	380	Phenanthrene, anthracene
270	440	Fluoranthene
250	390	Pyrene, benzo[a]anthracene, chrysene
290	430	Benzo[b] and benzo[k]fluoranthene, benzo[a]pyrene
296	406	Dibenzo[a,h]anthracene, benzo[g,h,i]perylene
290	500	Indeno [1,2,3-c,d]pyrene

The choice of lower excitation wavelengths results in enhanced detectability for naphthalene and acenaphthene [98]. However, the best results are obtained when the detector can measure different excitation or emission wavelengths simultaneously. Under such conditions several signals are obtained from the same chromatographic run, with additional benefits. As an example, 15 PAHs have been detected by using a Hewlett-Packard 1046 Series fluorimetric detector and the program shown in Table 14.6 [94].

The chromatographic signals for a standard sample and a sea sediment sample are presented in Fig. 14.14. Because at least two signals can be used to determine each analyte, additional information can be obtained for most compounds. For example, in the case of benzo[a]anthracene the ratio for peak area is different in the standard and in the sample for wavelengths 375, 425 and 475 nm, showing the presence of an unknown interferent at 375 nm. The existence of such a kind of error cannot be detected when only one signal is available.

TABLE 14.6   Wavelength-programmed fluorescence conditions for the determination of PAHs using different excitation or emission wavelengths simultaneously [94]

Time (min)	$\lambda_{excitation}$ (nm)	$\lambda_{emission}$ (nm)
0	235	335, 375, 425, 475
7.5	265	335, 375, 425, 475
10	265	375, 425, 475, 525

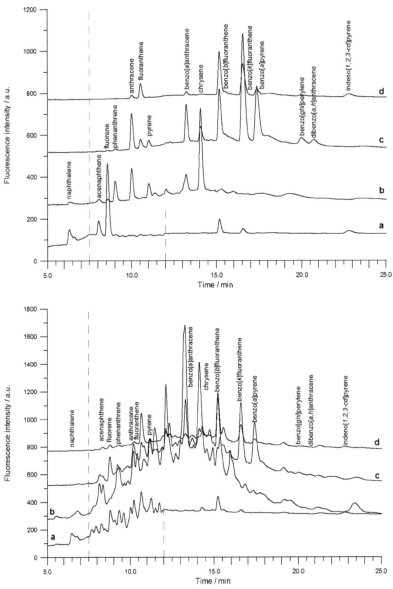

FIGURE 14.14 Chromatograms at four different excitation/emission wavelengths for an standard (top) and a sediment sample (bottom): (a) t=0 min, λ_{exc}=235 nm; λ_{ems}=335 nm; t=7.5 min, λ_{exc}=265 nm; λ_{ems}=335 nm; t=0 min, λ_{exc}=265 nm; λ_{ems}=525 nm; (b) t=0 min, λ_{exc}=235 nm; λ_{ems}=375 nm; t=7.5 min, λ_{exc}=265 nm; λ_{ems}=375 nm; (c) t=0 min, λ_{exc}=235 nm; λ_{ems}=425 nm; t=7.5 min, λ_{exc}=265 nm; λ_{ems}=425 nm; t=0 min, λ_{exc}=235 nm; λ_{ems}=475 nm; t=7.5 min, λ_{exc}=265 nm; λ_{ems}=475 nm.

14.3.6   OTHER FAMILIES

Coprostanol is the major human sterol found in water, and it is produced by micro-bial reduction of cholesterol. The concentration ratio of coprostanol to cholesterol has been proposed to estimate the degree of fecal matter in water samples. Waters having coprostanol to cholesterol concentration ratios greater than 0.2 are consid-ered contaminated by fecal material [99]. The most used technique for the analysis of fecal sterols in environmental waters is GC. To achieve the required sensitivity and selectivity the analytes must be extracted from the samples, concentrated and then derivatized before the chromatographic step. For extraction different strategies have been proposed which involve SFE [100], UAE [101], Soxhlet extraction [102] and LLE [103]. Although good extraction efficiencies and sensitivities are generally achieved, the resulting procedures are very laborious and very time consuming (to-tal analysis times longer than 24 h have been reported [101]). LC is also well-suited for the analysis of organic compounds in water matrices, but only a few procedures using this technique have been proposed for the analysis of sterols, probably be-cause these compounds are less sensitive towards common LC detectors than other classes of pollutants. Piocos et al. [104] proposed a method for measuring copros-tanol in water, which combined SPE and postcolumn derivatization. The method reported in [105] combined IT-SPME for analyte enrichment and CapLC DAD for identification and quantification of fecal sterols coprostanol and cholesterol (see Table 14.3). The reliability of the proposed method was tested by processing several waste water samples.

SPE and LLE combined to LC and different detection modes have been tradi-tionally used for the analysis of phenolic compounds. Although these methodolo-gies provided good LODs the main drawbacks are the high volumes of samples that need to be processed, the large volume solvents and, in some cases, the long analysis times. In Ref. [106] a method that does not require sample treatment is proposed (see Table 14.3). Phenol and o-cresol were selected by their abundant use in the world, BPA since it is a component of polycarbonate plastics and epoxy resins and 2-chloro-phenol because it can be a by-product of phenol in chloride water treat-ment. Phenol and 2-chloro-phenol are listed as EPA priority pollutants, and BPA is under evaluation by the EPA for action plan development owing to its harmful effects as endocrine disruptor.

On the other hand, the synthesis of the material proposed in Ref. [80] for MSPE (see Table 14.3 and Fig. 14.9), has been carried out taking advantage of (i) organ-ic-phase synthesis of Fe_3O_4 NPs, (ii) cethyltrimethylammonium bromide (CTAB) transfer to obtain water soluble CTAB coated Fe_3O_4 NPs, and (iii) Fe_3O_4 NPs sup-ported on SiO_2 matrix by sol gel procedure. The new magnetic material was used to extract and preconcentrate ECs such as acetylsalicylic acid, acetaminophen, di-clofenac and ibuprofen from environmental water samples prior to their analysis by CapLC-MS (Table 14.3). The use of the proposed silica supported Fe_3O_4 magnetic NPs provides surfactant free extracts for the analysis with MS detection without interferences in the ionization step. Magnetic forces offer great advantages in analytical applications as magnetic interactions are not influenced by chemical variables such as pH, concentration or surface charges.

As described above, extraction, desorption and injection can be performed automatically in IT-SPME, allowing a shorten analysis time and better precision and accuracy [107]. The main disadvantage of IT-SPME is the low extraction efficiency due to large breakthrough volume and a small amount of adsorbent phases, although it is possible to obtain good LODs by processing high volumes of sample (up to several mL). Intensive efforts have been devoted to the development of new adsorbent phases to improve the extraction efficiency since 1997, when Eisert and Pawliszyn coupled SPME to LC to develop IT-SPME. Polypyrrolyne [108] and molecular imprinted polymers (MIPs) [109] have been described as capillary coatings that improve these parameters. Monolithic capillary columns have also been proposed [110]. Magnetic-IT-SPME [82, 83] has been developed (see Table 14.3 and Fig. 14.10), taking advantage of magnetic microfluidic principles to improve extraction efficiency of IT-SPME systems. First, the magnetic hybrid material formed by Fe_3O_4NPs supported on SiO_2 was synthesized and immobilized in the surface of a bared fused silica capillary column to obtain a magnetic adsorbent extraction phase. The capillary column was placed inside a magnetic coil which allowed the application of a variable magnetic field. Acetylsalicylic acid, acetaminophen, atenolol, diclofenac and ibuprofen were tested as target analytes. The application of a controlled magnetic field resulted in quantitative extraction efficiencies of the target analytes, giving recoveries in the 70–100% range. These results demonstrated that magnetic forces overcome the low extraction efficiency (10–30%) of IT-SPME systems.

Carbonyl compounds are organic substances mainly formed through an oxidative process (photochemical oxidation, lipid peroxidation or chemical oxidation). For this reason, they are detected in numerous matrices such as atmosphere (air and particulate matter), treated water (disinfection using ozone), biological fluids (urine, plasma, serum), food (wine, beer) and emissions of industrial and WWTPs. GC and LC are the most widely employed separation techniques for the analysis of carbonyl compounds. With respect to sample treatment, SPE and SPME are frequently combined with a derivatization step due to the polarity and reactivity of these compounds. A new device for carbonyl compounds based on coupling on-line and miniaturizing both, sample pretreatment and chromatographic separation, is reported in [111]. Different combinations of IT-SPME and derivatization using 2,4-dinitrophenylhydrazine (DNPH) were examined for mixtures containing 15 carbonyl compounds (aliphatic, aromatic and unsaturated aldehydes and ketones). A screening analysis of aqueous extracts of atmospheric particulate matter with diameter < 10 μm (PM_{10}) was carried out. The possibility of coupling IT-SPME and conventional LC was also tested. Derivatization solution and IT-SPME coupled to capLC provided the best results for carbonyl compounds in atmospheric particulate analysis. The LODs achieved using DAD ranged from 30 to 198 ng/L, improving markedly the LODs reported by conventional SPME–LC–DAD (see Table 14.3).On line IT-SPME-CapLC with MS detection has been applied for the first time to determine carbonyl compounds in particulate matter with diameter <2.5 μm ($PM_{2.5}$) and in water samples at ng/L levels [112]. This methodology has been employed for the chromatographic separation and screening of ten aliphatic aldehydes from formaldehyde to decyladehyde, unsaturated aldehyde of four carbon atoms (crotonaldehyde) and a cyclic ketone (cyclohexanone). This combination provides a remarkable improvement of the LODs compared to existing methods. This improvement has allowed the screening and quantification of carbonyl compounds in water-soluble fractions of $PM_{2.5}$, avoiding the additional evaporation steps needed for concentrating the obtained extract in order to reach ng L^{-1} concentration levels (see Table 14.3). In the IT-SPME step, cleanup and preconcentration are carried out on-line; therefore, the analysis time can be reduced to sampling and extraction of the analytes from the sampling filters. Thus, this methodology not only improves the sensitivity but also simplifies the sample pretreatment, as water soluble fraction of $PM_{2.5}$ can be directly processed into the chromatographic system. This procedure can be useful for increasing the knowledge about $PM_{2.5}$, as well as for analyzing water samples.

PFCs embody a large group of highly stable man-made compounds, which are amphiphilic and consist of a perfluorinated hydrophobic, linear carbon chain attached to one or more hydrophilic heads. PFCs have been widely produced and used since the 1950s for many industrial purposes and consumer-related applications such as protective coatings for carpets and textiles, surfactants, lubricants and food packaging. The strong carbon-fluorine bonds make them resistant to hydrolysis, photolysis, metabolism and biodegradation. Due to their long half-life time and their high bioaccumulativity, PFCs are found widespread over the world in biota, wildlife and humans. In fact, PFOS was added in 2009 to Annex B of the Stockholm Convention on persistent OPs as well as to the Annex III Substances subject to review for possible identification as priority substances or priority hazardous substances of the Directive 2008/105/EC of the European Parliament and Council of 16 December 2008 concerning the EQS in the field water policy. Although humans are exposed to PFCs from a number of sources, food (including drinking water) could be the dominant intake pathway. Toxicological consequences might be the disturbance of the fatty acid metabolism, the affection of the reproductive system and the induction of adverse effects in liver and other tissues. So far, most of the analytical methods to determine PFCs are based on LC-MS or LC-MS/MS preceded by SPE.

In Ref. [113] the analytical performance of HPLC-MS/MS, ultrahigh performance liquid chromatography (UHPLC)-MS/MS and CapLC-MS has been evaluated for the analysis of PFCs in several water samples (see Table 14.3). When comparing the three systems, it was found that HPLC-MS/MS provided the worst linearity, precision and sensitivity. The capabilities of that system were close to the instrumental limits for determining PFCs at the concentrations reported in environmental samples. Furthermore, LC-MS/MS required the longest analysis times.

UHPLC-MS/MS provided the best linearity, shortest analysis time and for most of the compounds the best sensitivity. However, precision was between that obtained by the LC-MS/MS and the CapLC-MS approaches. The precision and sensitivity achieved by CapLC-MS were comparable to those obtained by the UHPLC-MS/MS approach, thus being a viable and economical alternative to determine PFCs in environmental samples. SPE is an appropriate technique to isolate and concentrate PFCs providing recoveries in the 63–104% range which are appropriate for quantification purposes. This preconcentration procedure improved 500 folds the method detection limit and the method quantification limit. The analysis of waters demonstrated the occurrence of several PFCs congeners and the need to develop alternative confirmation methods to eliminate false positives, even when very selective systems such as MS/MS are used. In this sense, a confirmatory study was carried out by analyzing the samples with the UHLC-MS/MS and CapLC-MS procedures.

Cationic surfactants with biocide properties are consumed in thousands tons per year in several applications such as the production of pharmaceutical formulations, disinfection products, swimming pool algaecides and water treatment plants. The toxicity of some cationic surfactants such as benzalkonium chloride (BAK) has been assessed. This is the case of the BAK effective concentration at 10% inhibition (EC_{10}) for aquatic organisms such as some microalgaes and coastal phytoplankton. However, EC_{10} values (between 16 and 33 µg/L) fall within the range of BAK concentrations found for different effluents. Moreover, problems in BAK degradation due to nitrification effect, adsorption on suspended particles and sediments and inhibition of depuration process in WWTPs have been reported. Although cationic surfactants have been included in environmental legislation, regulatory concentrations have not been established. The novelty in Ref. [114] was the use of an ion-pair reagent for the C_{12}-BAK extraction in the capillary coated with polydimethylsiloxane (PDMS) containing 35% diphenylsiloxane monomer (TBR phase), and the use of a capillary reversed phase TiO_2 column for the C_{12}-BAK separation (see Table 14.3). In such a way, two steps of the analysis, sample treatment by IT-SPME and chromatographic separation, are coupled on-line and miniaturized. Obvious advantages such as rapidity and minimization of errors due to manipulation, analysis time and residues were accomplished in the on-line analysis. Miniaturization provided a drastic reduction of solvent and reagents and improves the sensitivity. A mixed mechanism of ion exchange and of titanophilic interaction can be accomplished with an adequate choice of ion-pair reagent. This mechanism could explain the results obtained.

The retention and separation of four homologues of benzalkonium chloride (alkyl C_{12}, C_{14}, $C_{16,}$ C_{18} dimethylbenzylammonium chloride) have been studied in TRB and nitrile capillary phases in [115]. Under the optimized conditions (50% acetonitrile in the processed samples, 35% of diphenyl content of the TRB extraction capillary, capillary length 43 cm and water: methanol 60:40 as replacing solvent), the extraction efficiency was similar for all the homologues with satisfactory reproducibility and independently of the amount and proportion of homologues. Industrial washing waters have been analyzed without previous treatment (see Table 14.3).

The different chemical behavior of OPs lead to analyze them for family groups, and a few procedures have been described in the literature capable to determine them by using a single method. This fact and the low concentrations limits established by the legislation, make difficult their chemical analysis. Thus, the current methods often require an enrichment step, and the final determination is usually carried out by LC or GC in combination with one or more detectors. For these reasons a trend in the environmental field is the development of accurate, automated and sensitive analytical methods that reduce sample handling. GC is mainly used in multiresidue approaches. There are quite a few LC multiresidue procedures and most of them use MS. Several sample pretreatments have been employed for the analysis of organic pollutants in waters, from more traditional techniques such as LLE or SPE to more moderns methodologies such SBSE, DLLME or SPME. Most of the procedures are performed off-line, using high sample volume and a preconcentration step in order to reach low LODs. On-line extraction techniques with sorbents are very useful because the samples can be often introduced with minimal preparation into the systems, and the preconcentration and cleanup steps are easily performed. Therefore, parameters such as time, cost or sample preparation are reduced while others like reproducibility or sensitivity are increased.

In Ref. [116] an on-line analyte multiresidue procedure was proposed without any previous sample treatment which was based on IT-SPME and CapLC with DAD. This system has been compared with a column switching procedure that involve a packed C_{18} column (see Table 14.3). Different configurations have been compared in order to improve detectability. The optimal IT-SPME procedure has been applied to the analysis of waste waters discharged into the Mediterranean Sea (Comunitat Valenciana region, Spain). Samples were collected from 22 different places in three different dates (April, July and November of 2007 and 2008). The proposed method established a suitable protocol to be followed by research and routine laboratories for the simultaneous analysis of 28 pollutant compounds in water samples according to water quality legislation. Monolithic columns have emerged as an alternative to traditional packed-bed columns for high efficiency separations in LC. The main advantages of monolithic columns are good permeability and fast mass transfer, versatile surface chemistry, easy fabrication and fritless design. The advantages of monolithic columns in CapLC have been extensively exploited in the biomedical field, especially in the analysis complex mixtures of peptides for proteome analysis. In principle, the high efficiency and resolution of monolithic capillary columns make them ideal for assessing environmental pollution. In Ref. [117] the potential utility of silica-based particulate and monolithic columns for multiresidue organic pollutants analysis using IT-SPME for on-line enrichment of the analytes and capLC were compared. The columns were evaluated not only in terms of resolution capabilities, but also for their suitability to be used in an on-line system that integrates analytes enrichment, separation and detection. Several compounds of different chemical structure and hydrophobicity were used as model compounds: simazine, atrazine and terbutylazine (triazines), chlorfenvinphos and chlorpyrifos (OPPs), diuron and isoproturon (phenyl ureas) trifluralin (dinitroaniline) and DEHP. Improved LODs were obtained (see Table 14.3).The results shown in Ref. [118] illustrate the advantages of the UHPLC-MS/MS approach (Table 14.3) for on-line analysis.

14.4   OTHER TECHNIQUES, CHEMICAL SENSORS AND BIOSENSORS

In the following section, some representative examples of the applications of sensors to the determination of several pollutants are summarized. Figure 14.15 depicts the state of art of this issue for the examples shown in this section.

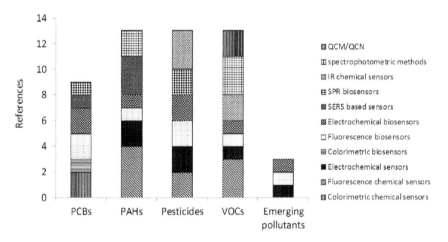

FIGURE 14.15   Summary of the nonchromatographic techniques, chemical sensors and biosensors employed for determining organic pollutants in environmental samples.

14.4.1   PCBS

Recently, chemical sensors based on electrochemical detection have been developed for determining PCBs in environmental samples. Jin et al. described the use of a porous anodic alumina based capacitive sensor for PCBs detection at trace level [119]. Furthermore, cyclodextrin-decorated single walled carbon nanotubes (SWCNTs) with electrochemical impedance detection have been proposed for this purpose [120]. Biochemical sensors have been exploited for PCBs determination in environmental samples. Recently, immunomagnetic electrochemical sensors using magnetic beads have been proposed as immunochemical test for PCBs detection [121, 122]. On the other hand, the determination of PCBs by using sensitive optical sensors has been reported. As an example, Endo et al. developed a micro flow inmunosensor chip based on Co-PCB antibodies immobilized polystyrene beads, which with the fluorogenic substrate produced changes in the fluorescence response as function of PCBs concentration [123]. Most recently, DNA-aptamers to recognize PCBs have been described as a highly sensitive sensor for fast environmental monitoring [124]. Colorimetric biosensors have also been proposed. Gavlasova et al. described a cell based on the formation of a yellow compound for the selective and semiquantitative detection of these compounds [125]. Another possibility to determine PCBs in environmental samples is the use of surface-enhanced Raman scattering (SERS). This analytical tool allows the detection of such compounds at low detection limits and at real-time, being a cost-effective methodology. In these applications, it is mandatory the development of highly sensitive and reproducible SERS substrates. In this sense, the use of nanomaterials is an important issue as they allow achieving these properties. For example, Ag denditric nanostructures have been reported as SERS substrate for the determination of PCBs at trace level [126]. Additionally, nanostructured surface plasmon resonance based biosensor can be an alternative tool for PCBs detection at concentration level ranging from 0.1 to 0.8 ppb [127].

14.4.2   PAHS

The high fluorescence quantum yield of PAHs has been explored as tool for their determination with fluorescence based analytical methods. In the last decade, methods based on laser-induced fluorescence have been proposed for the sensitive determination of PAHs in environmental matrices such as waters [128]. Time resolved fluorescence has also been reported for real time in-situ monitoring of these pollutants [129]. Despite the advantages that these techniques can provide, multivariate analysis is needed to resolve mixtures of PAHs. On the other hand, the development of fluorescence chemical sensors is nowadays an important trend for determining PAHs in environmental samples. Several fluorescence based sensors have been described in the last decade. A fluorescence optosensor have been proposed for the analysis of several water samples [130]. PAHs sensing technologies have also taken advantage of the feasibilities of MIPs to develop fluorescence optosensors useful for water analysis [131,132]. Most recently, quantum dots (QDs) have been employed to develop fluorescence sensors because their optical properties make them especially attractive for fluorescence sensors. For example, QDs modified TiO_2 nanotubes (NTs) have been proposed for the sensitive detection of PAHs [133]. SERS sensing has also demonstrated to be an alternative for determining PAHs in environmental samples. This methodology allows the rapid identification with the high sensitivity necessary for detection at low ppb levels. In the most recently investigations,

NPs based SERS substrates have been proposed for sensing PAHs in environmental samples. Gold nanoparticles based SERS substrates have been reported by Peron et al. [134]. 6-deoxy-(6-thio)-beta-cyclodextrin modified by silver NPs (AgNPs) [135] and thiol-functionalized Fe_3O_4@Ag magnetic NPs have also been proposed [136]. AgNPs aggregates have been used for SERS detection of PAHs with portable Raman spectrometer [137]. The sensing of PAHs in environmental samples has been also carried out by using electrochemical sensors. Recently, grapheme based films have been proposed as promising tools for the electrochemical sensing of these pollutants [138]. Pinyayev et al. reported the applicability of a fluorescence spectroelectrochemical sensor to determine 1-hydroxypyrene with a detection limit of 1×10^{-9} M [139]. Furthermore, many examples of fluorescence and electrochemical biosensors have been published for the determination of PAHs. A bioluminescent sensor was proposed by Valdman and Gutz for naphthalene in air samples. Their results showed satisfactory LODs (20 nM) [140]. Electrochemical biosensors have also been reported to determine PAHs in water samples [141].

14.4.3 PESTICIDES

The determination of pesticides in environmental samples by using nonchromatographic techniques is a dynamic topic within the sensing technology in order to develop in situ devices. Several approaches based on the development of chemical sensors have been recently published. Taking advantage of the microescales technologies combined with the use of AuNPs, Lafleur et al. have published a sensitive sensor for dithiocarbamate in water samples [142]. More recently, Zhao et al. have demonstrated the applicability of the combination of QDs and MIP nanospheres for the fluorescence detection of pesticides at ppb level [143]. IR spectroscopy has also been usefully employed for the analysis of pesticides in environmental samples. Near-IR spectroscopy combined with chemometric analysis has been proposed for the analysis of imazapyr and dimethoato in soil samples [144]. Mid-IR evanescent wave sensor has also been proposed to determine OPPs in river water samples [145]. Portable SERS sensor has been reported by Li et al. for the determination of the mention above pesticides at trace level in water samples [146]. Carbon nanotubes (CNTs) based electrochemical sensors to determine pesticides have also been recently reported. An amperometric sensor using inhibition of enzymes has been described to detect organophosphorous pesticides [147]. Methyl-parathion has been determined with CNTs-Web modified electrodes with a LOD of 1 p.m. [148]. Other recent advances on the development of biosensors for pesticides determination are the development of sensitive fluorescence sensing strategies. Xu et al. have been described a fluorescent polarization immunoassay for the analysis of mixtures of OPPs pesticides [149]. Colorimetric based biosensors have also been proposed for the determination of carbofuran in soil samples [150]. Surface plasmon resonance (SPR) technology has also been described for biosensing of several pesticides. By way of example, chlorpyrifos and carbaryl have been determined in water samples at very low concentration levels [151]. Magneto-optical surface plasmon resonance has been proposed as an alternative sensing strategy [152]. Finally, electrochemical biosensors are also an alternative for pesticides determination owing to the increasing demand of sensitive, cost effective and in-situ analysis for the environmental monitoring. CNTs and magnetic NPs based electrochemical sensors have been reported to determine pesticides in water samples [153, 154].

14.4.4   VOCS

Alternative approaches for the determination of several VOCs have been proposed as an attempt to simplify chromatographic methods. Direct spectrophotometry on PDMS adsorbent was proposed by Lamotte et al. to determine benzene, toluene, ethylbenzene and xylenes (BTEX) in air and water samples [155]. Amino compounds have also been determined in air and water samples by using detector supports combined with spectrophotometry measurements [156, 157]. In the last decade, several optical chemical sensors have been developed for the determination of VOCs in environmental matrices. Recently, Silva et al. proposed the use of a remote optical fiber microsensor for the determination of BTEX in landfill leachate [158]. VOCs have also been determined by using optical sensors using polythiophene as active layer [159]. Optical waveguide sensors have been described for the determination of formaldehyde at low concentration levels [160] and surface acoustic wave have been described for the analysis of toluene and octane in air samples [161]. IR based sensors have been used for the analysis of several volatile compounds in environmental samples. Young et al. reported a thermal desorption FTIR hollow waveguide for the sensitive detection of BTEX in water samples [162]. Near IR combined with multivariate analysis [163] has also been proposed for VOCs determination in environmental samples. Quartz crystal micro and nanobalances have been proposed owing to low cost, rapid response, portability, nonhazardous label-free real-time procedure, and high sensitivity [164, 165]. Another group of sensors widely proposed for VOCs determination in the recent years are electrochemical sensors. As an example, Staginus et al. have proposed a capacitive chemical sensor for determining several VOCs in environmental matrices [166]. A colorimetric biosensor that improves the detection and discrimination of 20 VOCs in air samples has been also described by Lin et al. [167], while Tizzard et al. developed a selective electrochemical biosensor for BTEX determination [168].

14.4.5   ECS

Analytical methods based on chemical sensor and biosensors have also been reported for some ECs such as pharmaceuticals, flame retardants and hormones. Sousa et al. described a potentiometric sensor for monitoring ibuprofen in water samples

[169]. An optical bioassay was published by Tschmelak et al. for the determination of several ECs such as pharmaceuticals, hormones and EDCs in waters [170]. More recently, Lin et al. have developed an electrochemical biosensor for BFRs [171].

14.5   SELECTED PROCEDURES

In this section some selected procedures for the analysis of representative OPs in different matrices are given, which reflect current tendencies in environmental analysis.

14.5.1   DETERMINATION OF PCBS IN WATER USING DLLME AND SUBSEQUENT SEPARATION AND DETECTION BY GC-ECD

The method has been optimized to determine 2,4,4'-trichlorobiphenyl, 2,2,'5,5'-tetrachlorobiphenyl, 2,3,'4,5,5'-pentachlorobiphenyl, 2,3,3,'4,4'-pentachlorobiphenyl, 2,3,4,4,'5-pentachlorobiphenyl, 3,3,'4,4,'5-pentachlorobiphenyl, 2,2,'3,4,4,'5,5'-hexachlorobiphenyl, 2,2,'4,4,'5,5'-hexachlorobiphenyl, 2,2,'3,3,'4,4,'5-heptachlorobiphenyl, and 2,2,'3,4,4,'5,5'-heptachlorobiphenyl in different kinds of water [40].

The analysis entails sample collection in glass bottles and storage in the dark at 4°C until analysis. Samples can be analyzed within 48 h of collection without any previous treatment or filtration. For analysis, aliquots of 5 mL of the samples are placed in a screw cap conical bottom test tube. To remove any organic contamination, the tubes have to be heated at 500°C for 30 min in a furnace before their use. For DLLME, 500 μL of acetone (disperser solvent) and 10 μL of chlorobenzene (extraction solvent) are added to the samples. The mixture is then centrifuged for 2 min at 5000 rpm. The sedimented phase is removed with a 10 μL-microsrynge and transferred to vials for GC.

Aliquots of 0.50 μL of the extracts are processed in a GC chromatograph equipped with a ^{63}Ni ECD and a split/splitless injector. A BPX-5 capillary column (30 m x 0.25 mm (i.d.), 0.25 μm film thicknesses, 95% methyl, 5% phenyl) is used. Ultrapure helium at a flow of 30 mL/min is the carrier. The injector port is held at 300°C and at the split ratio of 1:50. The oven temperature program is as follows: 3 min at 100°C first ramp at 25°C/min to 200°C, second ramp at 5°C/min to 290°C (held for 2 min). The ECD temperature is held at 300°C. Under these conditions, the ten PCBs studied can be satisfactorily resolved in 25 min.

The recoveries of the extraction range from 92–114%, making unnecessary the addition of salt (NaCl) to the samples in order to increase the extraction efficiency. The method provides good linearity in the 0.005–2 μg/L range using external calibration, and the relative standard deviation (RSD) ranged from 4% to 11%. LODs < 0.002 μg/L are obtained for the PCBs assayed. The method provides satisfactory results for the quantification of PCBs in natural waters such as well, river and sea-water. For these waters, no matrix effect is observed.

14.5.2   DETERMINATION OF ALIPHATIC AMINES IN AIR AND WATER BY COMBINING SPME/DERIVATIZATION AND LC

This method involves SMPE with on-fiber derivatization and LC, and it has been optimized for trimethylamine [76]. However, the methodology can be extended to other aliphatic primary and secondary amines.

Water samples are collected in glass bottles and filtered with 0.45 μm nylon membranes after their arrival to the laboratory, and then stored in the dark at 4°C until analysis. Before analysis, the samples are treated with a solution of NaOH (1%, w/v) to adjust the pH to 10.0. Passive sampling is used for air samples.

A SPME assembly with replaceable extraction fibers coated with a Carbowax-Templated Resin phase (50 μm) is used for analyte extraction and derivatization. Fibers are previously immersed in a solution of the derivatization reagent for 5 min. This solution is a mixture of 10 mL of 25 mM 9-fluorenylmethyl chloroformate (FMOC) prepared in acetonitrile and 0.5 mL of 0.05 M borated buffer of pH 9.0. The treated fibers are exposed to the air samples for 15 min or immersed into water samples for 30 min. After sampling, the fiber is protected within the SPME assembly until analysis.

For separation and detection a LC chromatograph equipped with a binary pump, an SPME-LC interface and a fluorimetric detector is used. A precolumn and a six-port switching valve are placed between the SPME-LC interface and the separative column to effect peak compression (see Fig. 14.6). The precolumn (20 mm × 2.1 mm (i.d.)) is dry packed with a Hypersil C_{18}, 30 μm stationary phase. A Lichrosphere 100 RP_{18}, 125 mm × 4.6 mm i.d., 5 μm, column is used for separation. The mobile phase is acetonitrile-water in gradient elution mode, and the flow-rate is 1 mL/min. The signal is recorder at excitation and emission wavelengths of 264 nm and 313 nm, respectively.

After sampling, the SMPE assembly is placed into the interface. For desorption of the target compounds, the interface chamber is filled with 1 mL of acetonitrile; the desorption time is 5 min. Next, the SPME-LC interface is activated and the chromatographic run is started. The analyze is sent to the precolumn with a mobile phase of 100% water. At 0.5 min the switching valve is rotated, so the precolumn and the separative column are connected; the analyte derivative is transferred to the analytical column for separation. The acetonitrile content was increased from 0 at zero min to 60% at 2.5 min, kept constant from 2.5 to 3.5 min, and then increased to 70% at 10 min, and to 100% at 15 min.

The reagent is also reactive towards other aliphatic amines. The chromatographic conditions used in the assay were selected to achieve adequate resolution of the peak of the FMOC-trimethylamine from the derivatives originated by other amines such as ethylamine, *n*-butylamine and *n*-pentylamine. All these derivatives could be separated by modifying the gradient elution program, and therefore, the method could be extended to those aliphatic amines. However, dimethylamine originates the same derivative than trimethylamine, and therefore it is a direct interferent. In water samples, this interference can be easily overcome by adding 5 mL of a solution of dinitrobenzoil chloride (40 mM) before the SPME step. This later reagent is reactive only to primary and secondary amines, and therefore, prevents the formation of the FMOC-dimethylamine derivative.

As regards the analytical performance, the method provides good linearity in the 25–200 mg/m^3 range (25°C, 1.013×10^{-5} Pa) for air, and in the 1.0–10.0 µg/L for water (external calibration). The precision is satisfactory at the concentration levels assayed, with RSDs ranging from 9–14% for air samples, and from 12–23% for water samples. The LODs are 12 mg/m^3 for air, and 0.25 µg/L for water samples. The methodology is useful for the measurement of trimethylamine in polluted environments and occupational atmospheres, as well as in tap water, ground water, river water and sea water.

14.5.3 DETERMINATION OF BTEX IN AIR AND WATER BY DIRECT SPECTROPHOTOMETRY

The method is based on the extraction of the analytes (BTEX) into PDMS blocks and direct spectrophotometry [155].

Water samples are collected in precleaned amber glass bottles; the bottles have to be completely filled to avoid migration of BTEX from the aqueous solution to the HS [14]. Samples have to be stored at 4°C until analysis. For the analysis of air, PDMS is directly exposed to the sample.

The analytes are extracted on PDMs blocks of 6 mm × 10 mm (extracting face) and 2 mm of thickness (optical path). The PDMS blocks are placed in a holder and exposed to air samples for 60 min or immersed into the water samples for 60 min. After sampling, the PDMS block holder is removed and the blocks are immediately placed in a regular fused silica cell (1 cm × 1cm) in the presence of pure water for absorbance measurement. The immersion of the PDMS in water prevents desorption of the analytes during the measurement step. In addition, the refractive index of PDMS is closer to that of water than that of air, which improves the optical transparency of the PDMS phase.

Absorption spectra are registered in the range 240–300 nm. The working signal is obtained by difference between the spectra of the same PDMS block before and after sampling. The absorbance is measured as the difference between the maximum of the band at 262 nm and the extrapolation line of the underlying background.

Calibration graphs are linear up to concentrations of 200 mg/m^3 for both air and water by using the external calibration method, being the sensitivity maxima for the xylenes. The lowest sensitivity is obtained for benzene. The LODs were 3–6 mg/m^3,

100–200 μg/m³ for toluene and 30–60 μg/m³ for o-xylene and p-xylene. In the original research, and although the authors used PDMS designed for chromatographic applications, the sensitivity attained was enough for application to pollution monitoring in water and industrial air, according to the legislated values. Nevertheless, the sensitivity could be improved by using PDMS specially prepared for optical applications.

The method is not as accurate and precise as methods based in GC, but it is relative simple and well-suited alternative for on-site monitoring. Another advantage is that the same PDMS block can be used several times by removing the adsorbed volatile compounds by means of a dry nitrogen stream at ambient temperature.

14.5.4 DETERMINATION OF EMERGING CONTAMINANTS IN WATER USING FE₃O₄ MAGNETIC NPS FOR EXTRACTION AND CAPLC

The method entails MSPE and capLC-MS, and it has been optimized for pharmaceuticals such as acetylsalicilic acid, acetaminophen, diclofenac and ibuprofen. These compounds are among the most frequently prescribed pharmaceuticals and, consequently, they are frequently detected in water, particularly in WWTPs.

The extractive material uses magnetite Fe_3O_4 NPs and cetyltrimethylammonium bromide (CTAB) supported on a silica matrix. For details on how to prepare the magnetic sorbent, see Ref. [80]. It has to be remarked that the sorbent is specially designed to avoid the presence of surfactants in the final extract (surfactants are necessary to assist the extraction of organic compounds in MSPE). The presence of surfactants in the extracts is one of the main drawbacks of MSPE, especially in methods which involve MS, because they can suppress ionization.

The samples are collected in dark glass containers and stored at 4°C until analysis. For analysis, aliquots of the samples (20 mL) are placed into glass vials and mixed with 30 mg of magnetic sorbent material (see Fig. 14.9). After stirring for 10 min, the magnetic sorbent is isolated from the mixture with a Nd disk magnet. The sorbent is air dried to eliminate the excess of water for 30 s. Then the analytes are extracted into 200 μL of methanol which are added to the isolated magnetic sorbent. The methanolic extract is processed into the CapLC system.

The CapLC system consists of a quaternary capillary pump, an injection valve equipped with an internal loop of 2 μL, and an MS detector quipped with an atmospheric pressure ionization source electrospray. The mobile-phase is a mixture acetonitrile-water in gradient elution mode at a flow rate of 10 μL/min. The solvent composition was acetonitrile-water 30:70 (v/v) from zero to 4 min, then the acetonitrile content was increases up to 50% at min 5 and maintained constant over the next 10 min. The MS is operated in negative ion mode, drying glass flow of 4 mL/min, nebulizer pressure of 12 psi and capillary voltage of 3000 V. For quantification purposes SIM is used.

The extraction efficiencies for the tested analytes are quite high, in the 80% -110% range, and independent from the sample matrix. The same sorbent can be used with any treatment for 20 times without losing its extraction capability. The method provides good linearity in the 1–10 µg/L range, using the external calibration approach. The RSD coefficients are <12%, whereas the LODs are 40 ng/L for ibuprofen, 50 ng/L for diclofenac and acetylsalicylic acid, and 150 ng/L for acetaminophen. The method is useful for the analysis of WWTPs effluents.

14.5.5   DETERMINATION OF PAHS IN BIOTA AND SEDIMENTS MICROSAMPLES BY MSPD AND IT-SPME COUPLED TO LC-FUORIMETRIC DETECTION

This methodology combines miniaturized MSPD combined with IT-SPME and LC with fluorescence detection, and it can be applied to quantify several representative PAHs included in the list of priority pollutants of the EPA in biota and marine sediments [93, 94]. A simplified version can be used to monitor PAHs in environmental waters.

Bivalves (in average 40 specimens per sample) are collected in 2-L plastic bottles, previously washed with HNO_3, and transported to the laboratory in a fridge. Once in the laboratory, edible tissues are removed from the shells, homogenized and frozen. Finally, samples are lyophilized and pulverized in a ball mill, and kept at 4°C in glass flask tightly closed until analysis. Marine sediments are collected in glass vials, air dried after their arrival to the laboratory, and kept in glass flask tightly closed at ambient temperature until analysis.

MSPD is used to extract the PAHs from the samples, using a C_{18} phase (40 µm) as dispersant. For conditioning the C_{18} phase, portions of 0.4 g are placed in prefritted SPE tubes, and then the tubes are rinsed with 1 mL of THF and sonicated into the ultrasonic bath for other 10 min. The THF layer is removed by vacuum and the sorbent is washed 1 mL of acetone. Portions of 0.1 g of the samples (lyophilized mussel or marine sediment) are placed in a glass mortar and ground by means of a pestle with 400 mg of conditioned C_{18} phase for 5 min. The resulting powder is then transferred to a 3 mL polypropylene SPE tube with a polyethylene frit (20 µm) placed at the bottom. For biota samples, 0.1 g of Florisil phase (60–100 mesh) are placed at the bottom of the SPE tube before transferring the blended sample in order to retain the fat content. Another frit is then placed at the top of the tube with the aid of a syringe plunger. To desorb the analytes, the cartridges are flushed with 1.2 mL of acetonitrile by applying positive pressure, and the extracts are collected into glass vials.

In this method, no preconcentration of the extracts is affected. Instead, the required sensitivity is reached by processing a large volume of the extracts by the IT-SPME methodology. For this purpose, the stainless steel loop of a six-port high-pressure injection valve is replaced by a TBR-35 capillary column (70 cm length

× 0.32 mm i.d., 3 μm film thickness) (see Fig. 14.8). A volume of the extract as large as necessary is loaded into the capillary until sufficient analyte is extracted. The addition of water to the acetonitrile extracts is necessary for the analytes to be retained into the capillary coating. For this reason, 2.6 mL of water are added to the extracts, and then 3.0 mL of the resulting mixture are introduced into the capillary column by means of a 1.0-mL precision syringe. After sample loading, the sample portion remaining in the capillary is flushed-out with 100 μL of nanopure water, and the valve is rotated, so the analytes are desorbed from the capillary coating and transferred to the separative column by the mobile-phase. The chromatographic run is started when the valve is rotated.

For separation and detection a LC chromatograph equipped with a quaternary pump and a programmable fluorimetric detector is used. The column is a LiChrospher PAH, 250 mm × 4.6 mm i.d., 5 μm column. The initial composition of the mobile phase is water-methanol-acetonitrile 20:55:25 (v/v/v). This composition is kept constant until 8 min, and then, the acetonitrile content is increased to reach 100% at 16 min; 100% acetonitrile is used until the end of the chromatographic run. The mobile-phase flow rate is 0.8 mL/min in the 0–8 min time interval, and then it is linearly increased up to 1.5 mL/min at 16 min; the mobile-phase flow rate is then kept constant until the end of the run (24 min). The detector program used is that of Table 14.6.

No significant differences on the performance of the method have been found between biota and sediment samples. The linearity of the method is suitable in the range 10–100 μg/kg using the external calibration method. The accuracy of the method is also satisfactory. The concentrations found when applying the method to a standard reference material (mussels) and during interlaboratory assays carried out with marine sediments were of 76–151% of the expected concentrations. These results are satisfactory taking into account the low working concentrations. The RSD coefficients were <20%, whereas the LODs were <5 μg/kg. Therefore, the sensitivity is excellent as only 0.1 g of samples is processed, and no solvent evaporation steps are involved. The methodology is also advantageous over classical approaches in terms of time of analysis, instrumental requirements, and consumption of reagents and generation of wastes.

A simplified version can be used for the detection and determination of PAHs in waters at sub μg/L levels. In this case, the untreated samples (up to 4 mL) can be directly loaded in the IT-SPME device for analyte enrichment, and then the analytes are separated and detected under conditions recommended for biota and sediments samples.

14.5.6   DETERMINATION OF HAAS IN WATER

This method is based on the simultaneous derivatization and extraction of HAAs with HF-LPME followed by GC-ECD. It can be applied for the determination of monochloroacetic acid, dichloroacetic acid, trichloroacetic acid, monobromoacetic acid, dibromoacetic acid and bromochloroacetic acid [30].

The method has been validated for bottled drinking water and chlorinated tap water. Tap water samples are collected in screw cap plastic bottles with no head-space, and analyzed immediately after sampling.

Aliquots of 10 mL of the samples (drinking and tap water) are placed in a 20-mL head space vial. In order to increase extraction percentages the samples are made 20% Na_2SO_4 (w/v). Then analytes are derivatized into their methyl esters with acidic methanol and simultaneously extracted, first by adding 1 mL of 98% sulfuric acid and then 1 mL of methanol. For HF-LPME a polypropylene fiber with 600 μm of inner diameter, wall thickness of 200 μm, average pore size of 0.2 μm (70% porosity) and 6.5 cm length is used. The acceptor phase is n-octanol (20 μL) and the extraction is performed in the head space of the vial for 60 min at 55°C. After the extraction time one end of the fiber is detached from the syringe needle, and the extracting solvent is flushed by an air blow into a 0.1 mL micro insert, which is then placed in a 2-mL PTFE/rubber septum crimp GC vial. The extracts are kept at 4°C until analysis.

Separation and detection of the methyl derivatives of HAAs are carried out into a GC equipment with a [63]Ni ECD and a split/splitless injector is port. The column is a DB-1701 capillary column (30 m × 0.32 mm (i.d.), 0.25 μm film thickness, 14% cyanopropylphenyl-86% dimethylpolysiloxane). Ultrapure helium at a flow of 1 mL/min is the carrier. The injector port is held at 270°C and at the split ratio of 1:10; the injection volume is 0.5 μL. The oven temperature was programmed at 40°C for 1 min, and then increased at 15°C/min to 270°C. The detector temperature is held at 300°C. Ultrapure nitrogen is used as make-up gas at 60 mL/min.

The method provides good linearity up to concentrations of 300 μg/L by the external calibration model. The recoveries ranged from 97 to 109%, being the RSD coefficients less than 12%. The LODs for most analytes were <1 μg/L, although the LOD for monochloroacetic acid is18 μg/L. The sensitivity is adequate to control the presence of HAAs in water according to the recommendations of the EPA.

A similar approach (without the derivatization step) can be used to determine THMs up to concentrations of 100 μg/L.

14.5.7 MULTIRESIDUE METHOD FOR PESTICIDES DETERMINATION IN FRUITS

The method entails extraction of the pesticides followed by SPE, and separation and detection by GC-MS, and it has been developed for 88 compounds belonging to the main families of pesticides. Raspberry, strawberry, blueberry and grape samples can be analyzed.

Samples are homogenized and portions of 10 g are placed into 50 mL centrifuge tubes and mixed with 20 mL of acetonitrile. The mixtures are vortexed for 2 min, then added 5 g of NaCl and vortexed for 2 min again, and finally centrifuged for 5 min at 3000 rpm. The supernatant is then transferred to a pear-shaped flask which is placed into a nitrogen evaporator and evaporated at 35°C until near dryness. The extracts are reconstituted with 2 mL of acetonitrile-toluene (3:1, v/v) and subjected to SPE into Envi-Carb cartridges (3 mL, 250 mg) coupled to NH_2-LC cartridges (3 mL, 250 mg). The employment of two phases is necessary owing to the wide variety of analytes included in this multiresidue method. A layer of about 1 cm of anhydrous sodium sulfate is placed into the Envi-Carb cartridge to remove traces of water; the column is then washed with 5 mL of acetonitrile-toluene (3:1, v/v) and attached below the NH_2-LC cartridge before elution. The extracts of the analytes are then loaded into the cartridges and passed at a flow-rate of 1 mL/min. The retained analytes are eluted with 25 mL of acetonitrile-toluene (3:1, v/v) at a flow rate of 2 mL/min. The 5 mL to 25 mL fraction of eluate is collected and evaporated by nitrogen stream at 35°C. The residue is redissolved in 1 mL of acetone and filtered through a 0.45 μm PTFE filter for GC-MS analysis.

For separation and detection a GC chromatograph equipped with a split/splitless injector and a MS detector is used. A DB-1701 capillary column (30 m × 0.25 mm (i.d.), 0.25 μm film thickness) is used, with ultrapure helium at 1 mL/min as the carrier. The temperature of the injector port is 250°C, and 1 μL of the extracts collected after SPE is injected; the injector port is held at 270°C, and the split ratio is 1:10. The oven temperature is programmed at 70°C for 1 min, and then increased at 10°C/min to 130°C, then at 6°C/min to 230°C, and finally 8°C/min to 250°C, and held for 13 min. The MS spectrometer operates in electron ionization mode with an ionizing energy of 70 eV, ion source temperature 230°C, MS quad temperature 150°C, electron multiplier voltage 1750 V when performing SIM, scanning from m/z 50 to 500 at 2.35 s per scan, and delay solvent of 8.5 min.

Analysis is performed in the SIM mode based on the use of one target and two or three qualifier ions. The target and the qualifier abundances are determined by injecting the standards of the individual pesticides under the same chromatographic conditions in full-scan mode with m/z rations from 50 to 500. Pesticides are identified according to their retention times, the target and qualifier ions, and the qualifier to target abundance ratios. Quantification is based on the peak area ratio of the target ion in the sample divided by that in of the external standards (external calibration).

Good linearity is obtained for all the tested pesticides in the 0.05–0.5 mg/Kg range. The overall recoveries were in the 63–137%, whereas the RSD varied from 1% to 19%. Finally, the LODs varied form 0.006–0.05 mg/Kg. No differences in the analytical performance of the method are observed between the different types of fruits analyzed.

14.6   CONCLUSIONS AND REMARKS

Current legislation provides a comprehensive frame for the control of OPs towards which the chemical science and technology have contributed with a large number of solutions. New analytical techniques allow the detection of substances at very low concentration levels, making possible to analyze in greater depth the behavior of ECs that were not possible to detect only a few years ago. Some organic ECs are EDCs, PPCPs, and alkylphenol derivatives.

Recent research efforts are oriented towards the development of sensitive and simplified analytical methodologies which reduce the time of analysis, the use of chemicals and the generation of wastes. This is especially important in environmental analysis as a great number of samples must be processed in order to assure the environmental quality.

The determination of OPs in environmental samples cannot be accomplished without some sample pretreatment because samples are too diluted and too complex. Traditionally, LLE has been used to detect organic micropollutants, but this technique requires large volumes of solvents and it has been superseded by SPE, which uses less solvents and is less time consuming. Nowadays, the general trend is to simplify the sample preparation, to diminish the sample volume needed, the number of off-line steps and the amount of solvents employed. SPME proposed by Pawliszyn is a very important contribution. MSPD is a good option for biota and sediment samples. The determination of OPs is usually carried out by GC or LC, depending on their polarity, volatility and the risk of decomposition at high temperature. The relative low concentrations, high polarity and thermal liability, and their interaction with host of complex environmental matrices, make the analysis of OPs a challenging task.

Monitoring environmental pollution due to OPs requires analytical procedures that provide low LODs and that include several families of compounds as well. The different chemical behavior of the compounds lead to analyze them grouped by families, and no many procedures have been described capable to determine all of them by using a single method. Multiresidue analyzes is based on GC and LC are increasingly common, generally coupled to MS or tandem MS. However, new efforts are necessary because the analytical problem is not resolved. The LOD is the most important factor to select an appropriate method.

Nowadays, most modern pesticides are medium to high polarity substances with thermal stability, making LC the preferred analytical separation technique. Indeed, coupling LC to MS or to MS/MS has become a powerful tool for environmental samples at sub-μg/L levels. DAD detection is also, while other detectors have been less frequently employed. On-line IT-SPME envisages as one of the most useful approaches for sample preparation. IT-SPME is a mode of SPME which typically uses a GC capillary column with a proper coating to extract the analytes. An increased number of papers are focused on proposing new phases. Analytical methodologies have started to take advantage of recent advances in the development of magnetic NPs, hybrid magnetic (nano) materials and magnetic composites, to improve the performance of existing methodologies.

There are different approaches for increasing the sensitivity in LC in order to reduce off-line preconcentration steps. The combination of a reduction of the column diameter (micro, capillary or nano-LC) and/or the particle size is one of them. The injection of a large sample volume via an IT-SPME device is an effective alternative for the analysis of OPs in environmental samples. The study of the applicability of new materials is also an important research field for the development of new procedures or for the improvement of existing ones.

As discussed above, chromatographic techniques are the most widely used techniques for applications in environmental analysis. Nevertheless, there is an increasing number of reports dealing with the development of analytical methods for environmental contaminants using nonchromatographic methods. In particular, recent advances in sensing technologies have shown the potential of chemical sensors and biosensors for environmental analysis. The importance of these new strategies can be explained by the need to develop analytical techniques with rapid response, high throughput of samples, less cost, and the possibility to work remotely, on-line or in-situ. The development of chemical or biosensors can be framed within the concept Green Analytical Chemistry, and it is nowadays supported by the advances in others areas such as nanotechnology and system miniaturization to achieve highly sensitive devices. However, specificity, and stability of the devices, in particular biosensors, are still some limitations that need to be overcome.

ACKNOWLEDGMENTS

The authors are grateful to the Spanish Ministerio de Economía y Competitividad (project CTQ 2011–26760) and to the Generalidad Valenciana (Prometeo Program 2012/045).

KEYWORDS

- Biosensors
- Chemical Sensors
- Gas Chromatography
- Liquid Chromatography
- Non-Regulated Compounds
- Regulated Pollutants
- Selected Procedures

REFERENCES

1. World Health Organization (WHO), (2005). Guidelines for Drinking Water Quality.
2. European Parliament Directive 2008/105/EC of the European Parliament and of the Council of 16 December 2008 on Environmental Quality Standards in the Field of Water Policy. See Also COM (2011). 876 Final.
3. WHO/IPCS Global Assessment of the State-of-the-Science of Endocrine Disruptors, World Health Organization/International Program on Chemical Safety, WHO/PCS/EDC/02.2, (2002).
4. Wille, K., De Brabander, F., De Wulf, E., Van Caeter, P., Janssen, C. R., & Vanhaecke, L. (2012). Coupled Chromatographic and Mass Spectrometric Techniques for the Analysis of emergent Pollutants in the Aquatic Environment, *TRAC, 35*, 87–108.
5. Giger, W. (2009). Hydrophilic and Amphiphilic Water Pollutants: Using Advanced Analytical Methods for Classic and Emerging Contaminants, *Anal. Bioanal Chem. 393*, 37–44.
6. Loos, R., Locoro, G., Comero, S., Contini, S., Schwesig, D., Werres, F., Balsaa, P., Gans, O., Weiss, S., Blaha, L., Bolchi, M., & Gawlik, B. M. (2010). Pan-European survey on the Occurrence of selected Polar Organic Persistent Pollutants in Ground Water, *Water Res, 44*, 415–4126.
7. Loos, R., Locoro, G., & Contini, S. (2010). Occurrence of Polar Organic Contaminants in the Dissolved Water Phase of the Danube River and Its Major Tributaries using SPE-LC-MS[2] *Water Res, 44*, 2325–2335.
8. Kolpin, D. W., Furlong, E. T., Meyer, M. T., Thurman, E. M., Zaugg, S. D., Barber, L. B., & Buxton, H. T. (2002). Pharmaceuticals, Hormones and Other Organic Wastewaters Contaminants in US Streams, 1999–2000: A National Reconnaissance, *Environ. Sci. Technol, 36*, 1202–1211.
9. Kolpin, D. W., Blazer, V. S., Gray, J. L., Focazio, M. J., Young, J. A., Alvarez, D. A., Iwanowicz, L. R., Foreman, W. T., Furlong, E. T., Speiran, G. K., Zaugg, S. D., Hubbard, L. E., Meyer, M. T., Sndstrom, M. W., & Barber, L. B. (2013). Chemical Contaminants in Water and Sediment near Fish Nesting Sites in the Potomac River Basin: Determining Potential Exposures to Smallmouth Bass (Micropterus Dolomieu) *Sci. Total Environ, 443*, 700–716.
10. Köck-Schulmayer, M., Villagrasa, M., López De Alda, M., Céspedes-Sánchez, R., Ventura, F., & Barceló, D. (2013). Occurrence and Behaviour of Pesticides in Wastewater Treatment Plants and Their Environmental Impact *Sci. Total Environ, 458–460*, 466–476.
11. Santos, F. J., & Galcerán, M. T. (2002). The Application of Gas Chromatography to Environmental Analysis *TRAC, 21*, 672–685.
12. Kubinec, R., Adamuščin, J., Jurdáková, H., Foltin, M., Ostrovský, I., Kraus, A., & Soják, L. (2005). Gas Chromatographic Determination of Benzene, Toluene, Ethylbenzene and Xylenes using Flame Ionization Detector in Water Samples with Direct Aqueous Injection Up to 250 ML, *J. Chromatogr. A, 1084*, 90–94.
13. Sridhara-Chary, N., & Fernandez-Alba, A. R. (2012). Determination of Volatile Organic Compounds in Drinking and Environmental Samples, *TRAC, 32*, 60–75.
14. Moliner-Martínez Y., Herráez-Hernandez R., Verdú-Andres J., Campíns-Falcó P., Garrido-Palanca C., & Molins-Legua C, A. (2013). Seco, Study of the Influence of Temperature and Precipitations on the Levels of BTEX in Natural Waters, *J. Hazard Mater*, http://Dx.Doi.Org/10.1016/J.Jhazmat, 2013, 07.037.
15. Poolem, C. F., Qian, J., Kiriden, W., Dekay, C., & Koziol, W. W. (2006). Evaluation of the Separation Characteristics of Application-Specific (Volatile Organic Compounds) Open-Tubular Columns for Gas Chromatography, *J. Chromatogr A, 1134*, 284–290.

16. Ben-Zen Wu., Tien-Zhi, F., Usha, S., Kong-Hwa, C., & Jiunn-Guang, L. (2006). Sampling and Analysis of Volatile Organics Emitted from Wastewater Treatment Plant and Drain System of an Industrial Science Park, *Anal. Chim Acta, 576*, 100–111.

17. Yung-Tsun, L., Bei-Zen W., Hung-Chi N., Hsing-Jung C., Kong-Hwa C., & Jiunn-Guang, L. (2009). Process Sampling Module Coupled with Purge and Trap–GC–FID for in Situ Auto-Monitoring of Volatile Organic Compounds in Wastewater, Talanta, *80*, 903–908.

18. Cavalcante, R. M., De Andrade, M. V. F., Marins, R. V., & Oliveira, L. D. M. (2010). Development of a Headspace-Gas Chromatography (HS-GC-PID-FID) Method for the Determination of Vocs in Environmental Aqueous Matrices: Optimization, Verification and Elimination of Matrix Effect and VOC Distribution on the Fortaleza Coast, Brazil, *Microchem. J.* 96, 337–343.

19. Furtula, V., Davies, J. M., & Mazumder, A. (2004). An Automated Headspace SPME-GC-ITMS Technique for Taste and Odour Compound Identification Water Qual. *Res. J. Canada,* 39, 213–222.

20. Regueiro, J., Lompart, M., Jares, C. G., & Cela, R. (2009). Development of a Solid-Phase Microextraction Gas Chromatography Tandem Mass Spectrometry Method for the Analysis of Chlorinated Toluenes in Environmental Waters, *J. Chromatogr* A, *1216*, 2816–2824.

21. Molins-Legua, C., Moliner-Martínez, Y., Campins-Falcó, P., Herráez-Hernández, R., Verdú-Andrés, J., Tello Lopez, M., & Garrido-Palanca, C. Effect of Disinfection Procedure in the Thms Concentrations in Indoor Swimming Pools, Submitted.

22. Lara-Gonzalo, A., Sánchez-Uría, J. E., Segovia-García, E., & Sanz-Medel, A. (2008). Critical Comparison of Automated Purge and Trap and Solid-phase Microextraction for Routine Determination of Volatile Organic Compounds in Drinking Waters by GC–MS Talanta, *74*, 1455–1462.

23. Kou, D., Wang, X., & Mitra, S. (2004). Supported Liquid Membrane Microextraction with High-performance Liquid Chromatography UV Detection for Monitoring trace Haloacetic acids in Water, *J. Chomatogr* A, *1055*, 63–69.

24. Pall, B., & Barret, L. (2004). Using Ion-Chromatography to Monitor Haloacetic Acids in Drinking Water, A Review of Current Technologies, *J. Chromatogr* A., *1046*, 1–9.

25. Martinez, D., Borrull, F., & Calull, M. (1999). Evaluation of Different Electrolyte Systems and Online Preconcentrations for the Analysis of Haloacetic Acids by Capillary Zone Electrophoresis, *J. Chromatogr* A., *835*, 187–196.

26. EPA Method 552.2, EPA/600/R-95/131 US Environmental protection Agency, Cincinnati, OH (1995).

27. Cardador, M. J., Serrano, A., & Gallego, M. (2008). Simultaneous Liquid–Liquid Microextraction/Methylation for the Determination of Haloacetic Acids in Drinking Waters by Headspace Gas Chromatography, *J. Chromatogr* A, *1209*, 61–69.

28. Hammami, B., & Driss, M. R. (2013). Development of Dry Derivatization and Headspace Solid-Phase Microextraction Technique for the GC-ECD Determination of Haloacetic Acids in Tap Water, *J. Anal* Chem., *68*, 671–679.

29. Mohammad, S., & Akbar-Hajialiakbari, B. A. (2009). Single-Drop Microextraction with In-Microvial Derivatization for the Determination of Haloacetic Acids in Water Samples by Gas Chromatography Mass Spectrometry, *J. Chromatogr* A, *1216*, 1059–1066.

30. Varanusupakul, P., Vora-Adisak, N., & Pulpoka, B. (2007). In Situ Derivatization and Hollow Fiber Membrane Microextraction for Gas Chromatographic Determination of Haloacetic Acids in Water, *Anal Chim Acta*, *598*, 82–86.

31. Tor, A., Cengeloglu, Y., Aydin, M. E., Ersoz, M., Wichmann, H., & Bahadir, M. (2003). An Investigation into the Analytical Methods for the Determination of Selected Chlorinated Phenols in Aqueous Phase Fresen, *Environ. Bull.*, *12*, 732–735.

32. Aydin, M. E., Wichmann, H., & Bahadir, M. (2004). Priority Organic Pollutants in Fresh and Waste Waters of Kenya Turkey, *Fresen Environment Bull, 13*, 118–123.

33. Xianhoa, C., Forsythe, J., & Peterkin, E. (2013). Some Factors affecting SPME Analysis of Pahs in Philadelphia's Urban Waterways, *Water Res 47*, 2331–2340.

34. Magdalini, C., Psillakis, E., Mantazavino, D., Kalogerakis, N., & Tor, A. (2005). Analysis of Polycyclic Aromatic Hydrocarbons in Wastewater Treatment Plant Effluents using Hollow Fibre Liquid-Phase Microextraction, *Chemosphere, 60(5),* 690–698

35. Saleh, A., Faraji, M., Rezaee, M., & Ghambarian, M. (2009). Ultrasound-Assisted Emulsification Microextraction Method Based on Applying Low Density Organic Solvents Followed by Gas Chromatography Analysis for the Determination of Polycyclic Aromatic Hydrocarbons in Water Samples, *J. Chromatogr.* A, *1216*, 6673–6679

36. Ozcan, S., Tor, A., & Aydin, M. E. (2010). Determination of Polycyclic Aromatic Hydrocarbons in Waters by Ultrasound assisted Emulsification-Microextraction and Gas Mass Spectrometry, *Anal. Chim Acta, 665*, 193–199.

37. Wegener, J. W. M., Cofino, W. P., Maier, E. A., & Kramer, G. N. (1999). The Preparation, Testing and Certification of Two Freshwater Sediment Reference Materials for Polycyclic Aromatic Hydrocarbons and Polychlorinated Biphenyls: BCR CRM 535 and CRM 536TRAC, *18*, 14–25.

38. Ramos, J. J., Gómara, B., Fernández, M. A., & González, M. J. (2007). A Simple and Fast Method for the Simultaneous Determination of Polychlorinated Biphenyls and Polybrominated Diphenyl Ethers in Small Volumes of Human Serum, *J. Chromatogr* A, *1152*, 124–129.

39. Lambropoulou, D. A., Konstantinou, I. K., & Albanis, T. A. (2006). Sample Pretreatment Method for the Determination of Polychlorinated Biphenyls in Bird Livers Using Ultrasonic Extraction Followed by Headspace Solid-Phase Microextraction and Gas Chromatography Mass Spectrometry, *J. Chromatogr.* A, *1124*, 97–105.

40. Rezaei, F., Bidari, A., Birjandi, A. P., Hosseini, M. R. M., & Assadi, Y. (2008). Development of a Dispersive Liquid–Liquid Microextraction Method for the Determination of Polychlorinated Biphenyls in Water, *J. Hazard Mater 158*, 621–627.

41. Ozcan, S., Tor, A., & Aydin, M. E. (2009). Determination of Selected Polychlorinated Biphenyls in Water Samples by Ultrasound-Assisted Emulsification-Microextraction and Gas Chromatography-Mass-Selective Detection. *Anal Chim Acta, 647*, 182–188.

42. Haglund, P., Sporring, S., Wiberg, K., & Bjolrklund, E. (2007). Shape selective Extraction of Pcbs and Dioxins from Fish and Fish Oil using in Cell Carbon Fractionation Pressurized Liquid Extraction, *Anal Chem 79*, 2945–2951.

43. Josefsson, S., Westbom, R., Mathiasson, L., & Björklund, E. (2006). Evaluation of PLE Exhaustiveness for the Extraction of PCBs from Sediments and the Influence of Sediment Characteristics, *Anal Chim Acta, 560*, 94–102.

44. Bavel, B., Jalremo, M., Karlsson, L., & Lindström, G. (1996). Development of a Solid Phase Carbon Trap for Simultaneous Determination of PCDDs, PCDFs, PCBs and Pesticides in Environmental Samples by Using LC/SPE, *Anal Chem., 68*, 1279–1283.

45. PunínCrespo, M. O., & Lage Yusty, M. A. (2005). Comparison of Supercritical Fluid Extraction and Soxhlet Extraction for the Determination of PCBs in Seaweed Samples Chemosphere, *59*, 1407–1413.

46. Quintana, J. B., Boonjob, W., Miró, M., & Cerdá, V. (2009). Online Coupling of Bead Injection Lab on Valve Analysis to Gas Chromatography (BI-LOV-GC): Application to the Determination of Trace Levels of Polychlorinated Byphenyls (PCBs) in Solid Waste Leachate Samples. *Anal Chem 81*, 4822–4830.

47. Popp, P., Keil, P., Montero, L., & Rückert, M. (2005). Optimized Method for Determination of 25 Polychlorinated Biphenyls in Natural Waters by Use of Stir Bar Sorptive Extraction

Followed by Termodesorption Gas Chromatography Mass Spectrometry *J. Chromatogr. A*, *1071*, 155–162.

48. Li, G., Zhang, L., & Zhang, Z. (2008). Determination of Polychlorinated Biphenyls in Water using Dynamic Hollow Fiber Liquid Phase Microextraction and Gas Chromatography Mass Spectrometry, *J. Chromatogr* A, *1204*, 119–122.

49. Rocco, G., Toledo, C., Ahumada, I., Sepúlveda, B., Cañete, A., & Richter, J. P. (2008). Determination of Polychlorinated Biphenyls in Biosolids using Continuous Ultrasound-Assisted Pressurized Solvent Extraction and Gas Chromatography Mass Spectrometry, *J. Chromatogr* A, *1193*, 32–41.

50. Wester, P. G., De Boer, J., & Brinkman, U. A. Th. (1996). Determination of Polychlorinated Terphenyls in Aquatic Biota and Sediment with Gas Chromatography Mass Spectrometry using Negative Chemical Ionization Environ, Sci. Technol. *30*, 473–480.

51. Tadeo, J. L., Sanchez-Brunete, C., & Miguel, E. (2009). Determination of Polybrominated Diphenyl Ethers in Human Hair by Gas Chromatography Mass Spectrometry Talanta, *78*, 138–143.

52. Wang, P., Zhang, W., Wang, Y., Wang, T., Li, X., Ding, L., & Jiang, G. (2010). Evaluation of Soxhlet Extraction Accelerated Solvent Extraction and Microwave-Assisted Extraction for the Determination of Polychlorinated Biphenyls and Polybrominated Diphenyl Ethers in Soil and Fish Samples *Anal. Chim Acta*, *663*, 43–48.

53. Samara, F., Tsai, C. W., & Aga, D. S. (2006). Determination of Potential Sources of Pcbs and Pbdes in Sediments of the Niagara River Environ. *Pollut*, *139*, 489–497.

54. Sahangmuganathan, D., Megharaj, M., Chen, Z., & Naidu, R. (2011). Polybrominated Diphenyl Ethers (Pbdes) in Marine Foodstuffs in Australia: Residue Levels and Contamination Status of Pbdes, *Mar. Pollut. Bull. 63*, 154–159.

55. Hyotylainen, T., & Hartonene, K. (2002). Determination of Brominated Flame Retardants in Environmental Samples TRAC, *21*, 3–29.

56. Krol, S., Zabiegala, B., & Namiensnik, J. (2012). Pbdes in Environmental Samples: Sampling and Analysis *Talanta*, *93*, 1–17.

57. Moliner-Martinez, Y., Campins-Falco, P., Molins-Legua, C., Segovia-Martinez, L., & Seco-Torrecillas, A. (2009). Miniturized Matrix Solid Phase Dispersion Procedure and Solid phase Microextraction for the Anaysis of Organochlorinated Pesticides and Polibrominated Diphenylethers in Biota Samples by Chromatography Electron Capture Detection, *J. Chromatogr. A*, *1216*, 6741–6745.

58. Martínez, A., Ramil, M., Montes, R., Hernanz, D., Rubí, E., Rodríguez I., & Cela Torrijos, R. (2005). Development of a Matrix Solid-Phase Dispersion Method for the Screening of Polybrominated Diphenyl Ethers and Polychlorinated Biphenyls in Biota Samples Using Gas Chromatography with Electron-Capture Detection *J. Chromatogr. A*. *1072*, 83–91.

59. Lacorte, S., & Guillamon, M. (2008). Validation of a Pressurized Solvent Extraction and GC–NCI–MS Method for the Low Level Determination of 40 Polybrominated Diphenyl Ethers in Mothers' Milk, *Chemosphere*, *73*, 70–75.

60. Covaci, A., Voorspoels, S., & De Boer, J. (2003). Determination of Brominated Flame Retardants with Emphasis on Polybrominated Diphenyl Ethers (Pbdes) in Environmental and Human Samples. A Review Environ. Int. *29*, 735–756.

61. Wang, D., & X. Li, Q. (2010). Application of Mass Spectrometry in the Analysis of 311 Polybrominated Diphenyl Ethers Mass Spectrom, Rev. *29*, 737–775.

62. Metcalfe, T. L., & Metcalfe, C. D. (1997). The Trophodynamics of PCBs, Including Mono- and Non-Ortho Congeners, in the Food Web of North-Central Lake Ontario. *Sci. Total Environ.*, *201*, 245–272.

63. Yang, X., Zhang, H., Liu, Y., Wang, J., Zhang, C. Y., Dong, A. J., Zhao, H. T., Sun, C. H., & Cui, J. (2011). Multiresidue Method for Determination of 88 Pesticides in Berry Fruits us-

ing Solid-Phase Extraction and Gas Chromatography-Mass Spectrometry Food *Chem., 127,* 855–865.

64. Raposo Júnior, J. L., Ré-Poppi, N. (2007). Determination of Organochlorine Pesticides in Ground Water Samples Using Solid-Phase Microextraction by Gas Chromatography Electron Capture Detection Talanta, *72,* 1833–1841.

65. Engewlad, W., Teske, J., & Efer, J. (1999). Programmed Temperature Vaporisers based Large Volume Injection in Capillary Gas Chromatography, *J. Chromatogr.* A, *842,* 143–161.

66. Tahboub, Y. R., Zaater, M. F., & Al-Talla, Z. A. (2005). Determination of the Limits of Identification and Quantification of Selected Organochlorine and Organophosphorous Pesticides Residues in Surface Water by Full Scan Gas Chromatography/Mass Spectrometry, *J. Chromatogr* A, *1098,* 150–155.

67. Wan-Ching, L., Hsin-Chang, C., & Wang-Hsien, D. (2005). Determination of Pharmaceutical Residues in Waters by Solid-Phase Extraction and Large Volume Online Derivatization with Gas Chromatography Mass Spectrometry, *J. Chromatogr* A, *1065,* 279–285.

68. Gomez, M. J., Agüera, A., Hurtado, J., Mocholi, F., & Fernandez-Alba, A. R. (2007). Simultaneous Analysis of Neutral and Acidic Pharmaceuticals as Well as Related Compounds by Gas Chromatography Tandem Mass Spectrometry in Wastewater Talanta, *73,* 314–320.

69. Fatta, D., Nikolaou, A., & Meriç, S. (2007). Analytical Methods for Tracing Pharmaceutical Residues In Water and Wastewater, TRAC, *26,* 515–533.

70. Pichon, V. (2000). Solid-Phase Extraction for Multiresidue Analysis of Organic Contaminants in Water, *J. Chromatogr* A, *885,* 195–215.

71. Canosa, P., Rodrigez, I., Rubi, E., Bollain, M. H., & Cela, R. (2007). Determination of Parabens and Triclosan in Indoor Dust Using Matrix Solid-Phase Dispersion and Gas Chromatography with Tandem Mass Spectrometry, *Anal Chem., 79,* 1675–1681.

72. Guitar, C., & Raedman, J. W. (2010). Critical Evaluation of the Determination of Pharmaceuticals, Personal Care Products, Phenolic Endocrine Disrupters and Fecal Steroids by GC/MS and PTV-GC/MS in Environmental Watersanal, *Chim Acta., 658,* 32–40.

73. Moliner,-Martinez, Y., Pastor-Carbonell, J. M., Bouzas, A., Seco, A., Abargues, M. R., & Campins-Falcó, P. (2013). Guidelines for Alkylphenols Estimation as Alkylphenol Polyethoxiylates Pollution Indicator in Wastewater Treatment Plant Effluents, *Anal Methods, 5,* 2209–2217.

74. Pan, L., Chong, J. M., & Pawliszyn, J. (1997). Determination of Amines in Air and Water using Derivatization Combined with Solid Phase Microextraction, *J. Chromatogr* A, *773,* 249–60.

75. Herráez-Hernández, R., Cháfer-Pericás, C., Verdú-Andrés, J., & Campíns-Falcó, P. (2006). An Evaluation of Solid-Phase Microextraction for Aliphatic Amines using Derivatization with 9-Fluorenylmethyl Chloroformate and Liquid Chromatography, *J. Chromatogr* A, *1104,* 40–46.

76. Cháfer-Pericás, C., Campíns-Falcó, P., & Herráez-Hernández, R. (2006). Comparative Study of the Determination of Trimethylamine in Water and Air by Combining Liquid Chromatography and Solid Phase Microextraction with on-Fiber Derivatization, Talanta, *69,* 716–723.

77. Moliner-Martínez, Y., Herráez-Hernández, R., & Campíns-Falcó, P. A. (2007). Microanalytical Method for Ammonium and Short-Chain Primary Aliphatic Amines using Precolumn Derivatization and Capillary Liquid Chromatography, *J. Chromatogr.* A, *1164,* 329–333.

78. Cháfer-Pericás, C., Herráez-Hernández, R., & Campíns-Falcó, P. (2006). On-Fibre Solid-Phase Microextraction Coupled to Conventional Liquid Chromatography Versus In-Tube Solid-Phase Microextraction Coupled to Capillary Liquid Chromatography for the Screening Analysis of Triazines in Water Samples *J. Chromatogr.* A, *1125,* 159–171.

79. Cháfer-Pericás, C., Herráez-Hernández, R., & Campíns-Falcó, P. (2007). In-Tube Solid-Phase Microextraction-Capillary Liquid Chromatography as a Solution for the Screening

Analysis of Organophosphorus Pesticides in Untreated Environmental Water Samples, *J. Chromatogr.* A, *1142*, 10–21.

80. Moliner-Martinez, Y., Ribera, A., Coronado, E., & Campíns-Falcó, P. (2011). Preconcentration of Emerging Contaminants in Environmental Water Samples by Using Silica Supported Fe_3O_4 Magnetic Nanoparticles for Improving Mass Detection in Capillary Liquid Chromatography, *J. Chromatogr.* A, *1218*, 2276–2283.

81. Moliner-Martinez, Y., Vitta, Y., Prima-García, H., Gónzalez-Fuenzalida, R. A., Ribera, A., Campíns-Falcó, P., & Coronado, E. (2013). Silica Supported Fe_3O_4 Magnetic Nanoparticles for Magnetic Solid-Phase Extraction and Magnetic In-Tube Solid-Phase Microextraction: Application to Organophosphorous Compounds Anal. Bioanal Chem, DOI 10.1007/S00216–013–7379-Y.

82. Campíns-Falcó, P., Coronado, E., Moliner-Martinez, Y., Ribera, A., & Prima-Garcia, H. (2011). Patent Application P00823, Spain, International Patent in Progress.

83. Moliner-Martínez, Y., Prima-Garcia, H., Ribera, A., Coronado, E., & Campíns-Falcó, P. (2012). Magnetic In-Tube Solid phase Microextraction, Anal Chem *84*, 7233−7234.

84. Cháfer-Pericás, C., Campíns-Falcó, P., & Prieto-Blanco, M. C. (2008). Automatic Method for the Estimation of Dibuthyl and Di-2-Ethylhexil Phthalate in Environmental Water Samples Based on In-Tube SPME Coupled to Fast Liquid Chromatography, Anal. Chim Acta, *610*, 268–273.

85. Chaler, R., Cantón, L., Vaquero, M., & Grimalt, J. O. (2004). Identification and Quantification of N-Octyl Esters of Alkanoic and Hexanedioic Acids and Phthalates as Urban Wastewater Markers in Biota and Sediments from Estuarine Areas, *J. Chromatogr* A, *1046*, 203–210.

86. Huang, P. C., Tien, C. J., Sun, Y. M., Hsieh, C. Y., & Lee, C. C. (2008). Occurrence of Phthalates in Sediment and Biota: Relationship to Aquatic Factors and the Biota-Sediment Accumulation Factor Chemosphere, *73*, 539–544.

87. Lin, Z. P., Ikonomou, M. G., Jing, H., Mackintosh, C., & Gobas, F. A. P. C. (2003). Determination of Phthalates Ester Congeneres and Mixtures by LC/ESI-MS in Sediment and Biota and Urbanized Marine Inlet. *Environ. Sci. Technol*, *37*, 2100–2108.

88. Muñoz-Ortuño, M., Moliner-Martínez, Y., Cogollos-Costa, S., Herráez-Hernández, R., & Campíns-Falcó, P. A. (2012). Miniaturized Method for Estimating Di(2-Ethylhexyl)Phthalate in Bivalves as Bioindicators *J. Chromatogr.* A, *1260*, 169–173.

89. USEPA, Office of the Federal Registration (OFR). (1982). Appendix A: Priority Pollutants Fed Reg *47*, 52309.

90. Yebra-Pimentel, I., Martínez-Carballo, E., Regueiro, J., & Simal-Gándara, J. (2013). The Potential of Solvent-Minimized Extraction Methods in the Determination of Polycyclic Aromatic Hydrocarbons in Fish Oils Food Chem, *139*, 1036–1043.

91. Guo, L., & Lee, H. K. (2013). Microwave Assisted Extraction Combined with Solvent Bar Microextraction for One-Step Solvent-Minimized Extraction, Cleanup and Preconcentration of Polycyclic Aromatic Hydrocarbons in Soil Samples, *J. Chromatogr.* A, *1286*, 9–15.

92. Sadowska-Rociek, A., Surma, M., & Cieslik, E. (2013). Application of Quechers Method for Simultaneous Determination of Pesticide Residues and Pahs in Fresh Herbs Bull. Environ. Contam Toxicol, *90*, 508–513.

93. Campíns-Falcó, P., Verdú-Andrés, J., Sevillano-Cabeza, A., Molins-Legua, C., & Herráez-Hernández, R. (2008). New Micromethod Combining Miniaturized Matrix Solid-Phase Dispersion and In-Tube In-Valve Solid-Phase Microextraction for Estimating Polycyclic Aromatic Hydrocarbons in Bivalves, *J. Chromatogr.* A, *1211*, 13–21.

94. Moliner-Martínez, Y., González-Fuenzalida, R. A., Herráez-Hernández, R., Campíns-Falcó, P., & Verdú-Andrés, J. (2012). Cleaning Sorbents used in Matrix Solid-Phase Dispersion with Sonication: Application to the Estimation of Polycyclic Aromatic Hydrocarbons at Ng/G Levels in Marine Sediments *J. Chromatogr.* A, *1263*, 43–50.

95. Barker, J. (2000). Matrix Solid-Phase Dispersion *J. Chromatogr* A, *885*, 115–127.

96. United States Environmental Protection Agency Determination of Polycyclic Aromatic Hydrocarbons in Groundwater and Wastes Method 8310, EPA Environmental Monitoring Systems Laboratory, Office of Research and Development, Cincinnati, OH, 1986.

97. Bocanegra-Salazar, M., Ortiz-Pérez, D., Cámara, C., & Sanz-Landaluze, J. (2010). Miniaturisated Method for the Analysis of Polycyclic Aromatic Hydrocarbons in Leaf Samples, *J. Chromatogr* A, *1217*, 3567–3574.

98. Kibby, J., & Russell, G. Application Note SI-02107. Analysis of Polycyclic Aromatic Hydrocarbons using Time-Programmed Fluorescence Detection with the Varian 920-LC and Pursuit™ 3 PAH Column, Varian Inc.

99. Readman, J. W., Fillman, G., Tolosa, I., Bartocci, J., & Mee, L. D. (2005). The Use of Steroid Markers to assess Sewage Contamination of the Black Sea, Mar. Pollution Bull, *50*, 310–318.

100. Grimalt, J. O., Fernández, P., Bayona, J. M., & Albaiges, J. (1990). Assessment of Fecal Sterols and Ketones as Indicators of Urban Sewage Inputs to Coastal Waters, Environ Sci. Technol., *24*, 357–363.

101. Jayasinghe, L. Y., Marriott, P. L., Carpenter, P. D., & Nichols, P. D. (1998). Application of Pentafluorophenyldimethylsilyl Derivatization for Gas Chromatography-Electron Capture detection of Supercritically Extracted Sterols, *J. Chromatogr* A, *809*,109–120.

102. Isobe, K. O., Tarao, M., Zakaria, M. P., Chiem, N. H., Minh, L. Y., & Takada, H. (2002). Quantitative Application of Fecal Sterols Using Gas Chromatography-Mass Spectrometry to Investigate Fecal Pollution in Tropical Waters: Western Malaysia and Mekong Delta, Vietnam Environ. Sci. Technology, *36*, 4497–4507.

103. Shah, V. K. G., Dunstan, H., & Taylor, W. (2006). An Efficient Diethyl Ether-Based Soxhlet Protocol to Quantify Faecal Sterols from catchment Waters, *J. Chromatogr* A, *1108,* 111–115.

104. Piocos, E. A., & De La Cruz, A. A. (2000). .Solid Phase Extraction and High Performance Liquid Chromatography with Photodiode Array Detection of Chemical Indicators of Human Fecal Contamination in Water, *J. Liq Chromatogr Relat Technol.*, *23*, 1281–1291.

104. Gilli, G., Rovere, R., Traversi, D., Schilirò, T., & Pignata, C. (2006). Faecal Sterols Determination in Wastewater and Surface Water, *J. Chromatogr* B, *843*, 120–124.

105. Moliner-Martínez, Y., Herráez-Hernández, R., Molins-Legua, C., & Campíns-Falcó, P. (2010). Improving Analysis of Apolar Organic Compounds by the Use of a Capillary Titania-based Column, Application to the direct Determination of Faecal Sterols Colesterol and Coprostanol in Waste water Samples, *J. Chromatogr* A, *1217*, 4682–4687.

106. Segovia-Martínez, L., Moliner Martínez, Y., & Campíns Falcó, P. A. (2010). Direct Capillary Liquid Chromatography with Electrochemical Detection Method for Determination of Phenols in Water Samples, *J. Chromatogr* A, *1217*, 7926–7930.

107. Eisert, R., & Pawliszyn, J. (1997). Automated In-Tube Solid-Phase Micro Extraction Coupled to High-Performance Liquid Chromatography, Anal Chem, *69*, 3140–3147.

108. Wu, J., Lord, H. L., & Pawliszyn, J. (2001). Determination of Stimulants in Human Urine and Hair Samples by Polypyrrole Coated Capillary In-Tube Solid Phase Microextraction Coupled with Liquid Chromatography-Electrospray Mass Spectrometry Talanta, *54*, 655–672.

109. Mullet, W. M., Martin, P., & Pawliszyn, J. (2001). Intube Molecularly Imprinted Polymer Solid-Phase Microextraction for the Selective Determination of Propranololanal, Chem., *73*, 2383–2389.

110. Zhang, M., Wei, F., Feng, Y. Q., Nie, J., & Feng, Y. Q. (2006). Novel Polymer Monolith Microextraction Using a Poly(Methacrylic Acid-Ethylene Glycol Dimethacrylate) Monolith and Its Application to Simultaneous Analysis of Several Angiotensin II Receptor Antagonists in Human Urine by Capillary Zone Electrophoresis, *J. Chromatgr.* A, *1102*, 294–301

111. Prieto-Blanco, M. C., López-Mahía, P., & Campíns-Falcó, P. (2011). Online Analysis of Carbonyl Compounds with Derivatization in Aqueous Extractsof Atmospheric Particulate PM_{10} by Intube Solid-Phase Microextraction Coupled to Capillary Liquid Chromatography, *J. Chromatogr.* A, *1218*, 4834–4839.

112. Prieto-Blanco, M. C., Moliner-Martinez, Y., López-Mahía, P., & Campíns-Falcó, P. (2013). Determination of Carbonyl Compounds in Particulate Matter $PM_{2.5}$ by Intube Solid-Phase Microextraction Coupled to Capillary Liquid Chromatography/Mass Spectrometry Talanta, *115*, 876–880.

113. Onghena, M., Moliner-Martinez, Y., Picó, Y., Campíns-Falcó, P., & Barceló D. (2012). Analysis of 18 Perfluorinated Compounds in River Waters: Comparison of High Performance Liquid Chromatography-Tandem Mass Spectrometry, Ultra-High Performance Liquid Chromatography Tandem Mass Spectrometry and Capillary Liquid Chromatography Mass Spectrometry, *J. Chromatogr.* A, *1244*, 88–97.

114. Prieto-Blanco, M. C., Moliner-Martínez, Y., López-Mahía, P., & Campíns-Falcó, P. (2012). Ion-Pair In-Tube Solid-Phase Microextraction and Capillary Liquid Chromatography using a Titania-Based Column: Application to the Specific Lauralkonium Chloride Determination in Water *J. Chromatogr.* A, *1248*, 55–59.

115. Prieto-Blanco, M. C., Moliner-Martínez, Y., & Campíns-Falcó, P. (2013). Combining Poly (Dimethyldiphenylsiloxane) and Nitrile Phases for Improving the Separation and Quantitation of Benzalkonium Chloride Homologues: In-Tube Solid Phase Microextraction-Capillary Liquid Chromatography-Diode Array Detection-Mass Spectrometry for Analysing Industrial Samples *J. Chromatogr.* A, *1297*, 226–230.

116. Campíns-Falcó, P., Verdú-Andrés, J., Sevillano-Cabeza, A., Herráez-Hernández, R., Molins-Legua, C., & Moliner-Martinez, Y. (2010). In-Tube Solid Phase Microextraction Coupled by in Valve Mode to Capillary LC-DAD, Improving Detectabilily to Multiresidue Organic Pollutants Analysis in Several Whole Waters, *J. Chromatogr.* A, *1217*, 2695–2702.

117. Moliner-Martinez, Y., Molins-Legua, C., Verdú-Andrés, J., Herráez-Hernández, R., & Campíns-Falcó, P. (2011). Advantages of Monolithic Over Particulate Columns for Multiresidue Analysis of Organic Pollutants by In-Tube Solid-Phase Microextraction Coupled to Capillary Liquid Chromatography, *J. Chromatogr.* A, *1218*, 6256–6262.

118. Masiá, A., Moliner-Martinez, Y., Muñoz-Ortuño, M., Pico, Y., & Campíns-Falcó, P. (2013). Multiresidue Analysis of Organic Pollutants by Intube Solid Phase Microextraction Coupled to Ultra-High Pressure Liquid Chromatography Electrospray-Tandem Mass Spectrometry, *J. Chromatogr.* A, *1306*, 1–11.

119. Jin, Z., Liu, J., Li, M., Kong, L., & Liu, J. A. (2011). Novel Porous Anodic Alumina Based Capacitive Sensor Towards Trace Detection Sens Actuat. B, *157*, 641–646.

120. Wei, Y., Kong, L. T., Yang, R., Wang, L., Liu, J. H., & Huang, X. J. (2011). Electrochemical Impedance Determination of Polychlorinated Biphenyl using a Pyrenecyclodextrin Decorated Single-Walled Carbon Nanotube, Hybrid Chem Commun., *47*, 5340–5342.

121. Centi, S., Laschi, S., Fránek, M., & Mascini. M. A. (2005). Disposable Inmunomagnetic Electrochemical Sensor Based on Functionalised Magnetic Beads and Carbon-Based Screen-Printed Electrodes (Spces) for the Detection of Polychlorinated Biphenyls (PCBs) Anal. Chim Acta, *538*, 205–212.

122. Lin, Y. Y., Liu, G., Wai, C. M., & Lin, Y. (2008). Biolectrochemical Inmunoassay of Polychlorinated Biphenyls Anal, Chim. Acta, *612*, 23–28.

123. Endo, T., Okuyama, A., Matsubara, Y., Nishi, K., Kobayashi, M., Yamamura, S., Morita, Y., Takamura, Y., Mizukami, H., & Tamiya, E. (2005). Fluorescence based Assay with Enzime Amplification on a Micro-Flow Immunosensor Chip for Monitoring Coplanar Polychlorinated Biphenyls Anal. Chim Acta, *531*, 7–13.

124. Xu, S. M., Yuan, H., Chen, S. P., Xu, A., Wang, J., & Wu, L. J. (2012). Selection of DNA Aptamers against Polychlorinated Biphenyls as a Potential Biorecognition Element for Environmental Analysis Anal, Biochem, *423*, 195–201.
125. Gavlasova, P., Kuncova, Kochankova, L., & Mackova, M. (2008). Int. Whole Cell Biosensors for Polychlorinated Biphenyl Analysis Based on Optical Detection, Int. Biodeterior Biodegra, *62*, 304–312.
126. Yang, Y., & Meng, G. (2010). Ag Denditric Nanostructures for Rapid Detection of Polychlorinated Biphenyls Based on Surface Enhanced Raman Scattering Effect, *J. Appl. Phys. 107*, 044315-1–5.
127. Hong, S., Kang, T., Moon, J., Oh, S., & Yi, J. (2006). Highly Responsive Sensor on a Nanostructured surface via Self-assembly of a Biomolecule with an Evanescent Wave Technique, *J. Nanosci. Nanotechnol, 6*, 3604–3607.
128. Levinson, J., Sluszny, C., Yasman, Y., Bulatov, V., & Schechter, I. (2005). Detector for Particulate Polycyclic Aromatic Hydrocarbons in Water, Anal Bioanal Chem., *381*, 1584–1591.
129. Selli, E., Zaccaria, C., Sens, F., Tomasi, G., & Bidoglio, G. (2004). Application of Multi-Way Models to the Time-Resolved Fluorescence of Polycyclic Aromatic Hydrocarbons Mixtures in Water Water Res., *38*, 2269–2276.
130. Fernandez-Sanchez, J. F., Carretero, A. S., Cruces-Blanco, C., & Fernandez-Gutierrez, A. (2004). Highly Sensitive and Selective Fluorescence Optosensor to Detect and Quantify Benzoa Pyrene in Water samples, Anal Chim Acta, *506*, 1–7.
131. Valero-Navarro, A., Salinas-Castillo, A., Fernandez-Sanchez, J. F., Segura Carretero, A., Mallavia, R., & Fernandez-Gutierrez, A. (2009). The Development of a MIP-Optosensor for the detection of Monoamine Naphtalenes in drinking Water Biosens, Bioelectron, *24*, 2305–2311.
132. Sanchez-Barragan, I., Costa-Fernandez, J. M., Pereiro, R., Sanz-Medel, A., Salinas, A., Segura, A., Fernandez-Gutierrez, A., Ballesteros, A., & Gonzalez, J. M. (2005). Molecular Imprinted Polymers Based on Iodinated Monomers for Selective Room-Temperature Phosphorescence Optosensing of Fluoranthene in Water, Anal Chem, *77*, 7005–7011.
133. Yang, L. X., Chen, B. B., Luo, S. L., Li, J. X., Liu, R. H., & Cai, Q. Y. (2010). Sensitive Detection of Polycyclic Aromatic Hydrocarbons Using Cdte Quantum Dot-Modified Tio_2 nanotube Array Trough Fluorescence Resonance Energy Transfer, Environ. Sci. Technol., *44*, 7884–7889.
134. Peron, O., Rinnert, E., Toury, T., De La Chapelle, M. L., & Compere, C. (2011). Quantitative SERS Sensors for Environmental Analysis of Naphthalene Analyst, *136*, 1018–1022.
135. Xie, Y. F., Wang, X., Han, X. X., Xue, X. X., Ji, W., Qi, Z. H., Liu, J. Q., Zhao, B., & Ozaki, Y. (2010). Sensing of Polycyclic Aromatic Hydrocarbons with Cyclodextrin inclusion Complexes on Silver Nanoparticles by Surface enhanced Raman Scattering Analyst, *135*, 1389–1394.
136. Du, J. J., & Jing, C. Y. (2011). Preparation of Thiol Modified Fe_3O_4@Ag Magnetic SERS Probe for Pahs Detection and Identification, *J. Phys. Chem.*, *115*, 17829–17835.
137. Jiang, X. H., Lai, Y. C., Yang, M., Yang, H., Jiang, W., & Zhan, J. H. (2012). Silver Nanoparticle Aggregates on Copper Foil for Reliable Quantitative SERS Analysis of Polycyclic Aromatic Hydrocarbons with a Portable Raman Spectrometer Analyst, *137*, 3995–4000.
138. Zhu, G. B., Wu.L., Zhang, X., Liu, W., Zhang, X. H., & Chen, J. H. (2013). A New Dual-Signalling Electrochemical Sensing Strategy Based on Competitive Hostguest Interaction of a Beta-Cyclodextrin/Poly (N-Acetylaniline)/Graphene-Modified Electrode: Sensitive Electrochemical Determination of Organic Pollutants, Chem. Eur. J. *19*, 6368–6373.
139. Pinayayev, T. S., Seliskar, C. J., & Heineman, W. R. (2010). Fluorescence Spectroelectrochemical Sensor for 1-Hydroxypyrene, Anal. Chem, *82*, 9743–9748.

140. Valdman, E., & Gutz, I. G. R. (2008). Biolumininescent Sensor for Naphtalenes in Air: Cell Immobilization and Evaluation with a Dynamic Standard Atmosphere Generator Sens. Actuat. B, *133*, 656–663.

141. Moore, E. J., Kreuzer, M. P., Pravda, M., & Guilbault, G. G. (2004). Development of a Rapid Single-Drop analysis Biosensor for Screening of Phenanthrene in Water Samples, Electronanal, *16*, 1653–1659.

142. Lafleur, J. P., Senkbeil, S., Jensen, T. G., & Kutter, J. P. (2012). Gold Nanoparticles-Based Optical Microfluidic Sensors for Analysis of Environmental Pollutants Lab. Chip, *12*, 4651–4656.

143. Zhao, Y. Y., Ma, Y. X., Li, H., & Wang, L. Y. (2012). Composite Qds@MIP Nanospheres for Specific Recognition and Direct Fluorescent Quantification of Pesticides in Aqueous Media, Anal Chem. *84*, 386–395.

144. Soto Barajas, M., Gonzalez Martin, I., Hernandez Hierro, J. M., Prado, B., Hidalgo, C., & Etchevers, J. (2012). NIR Spectroscopy to Identify and Quantify Imazapyr in soil, Anal Methods, *4*, 2764–2771.

145. Janotta, M., Karlowatz, M., Vogt, F., & Mizaikoff, B. (2003). Sol-Gel Based Mid-Infrared Evanescent Wave Sensors for Detection of Organophosphate Pesticides in Aqueous Solution Anal. Chim Acta, *496*, 339–348.

146. Li, D., Li, D. W., Fossey, J. S., & Long, Y. T. (2010). Portable Surface Enhanced Raman Scattering Sensor for Rapid Detection of Aniline and Phenol Derivates by On-Site Electrostatic Preconcentration Anal. Chem, *82*, 9299–9305.

147. Jha, N., & Ramaprabhu, S. (2010). Carbon-Nanotube-Polymer based Nanocomposite as Electrode Material for the Detection of Paraoxon, J. Nanosci. Nanotechnol, *10*, 2798–2802.

148. Musameh, M., Notivoli, M. R., Hickey, M., Huynh, C. P., Hawkins, S. C., Yousef, J. M., & Kyratzis, I. L. (2013). Carbon Nanotube-Web Modified Electrodes for Ultrasensitive Detection of Organophosphates Pesticides, Electrochim. Acta, *101*, 209–215.

149. Xu, Z. L., Wang, Q., Lei, H. T., Eremi, S. A., Shen, Y. D., Wang, H., Beier, R. C., Yang, J. Y., Maksimova, K. A., & Sun, Y. M. (2011). A Simple, Rapid and High Troughoutput Fluorescence Polarization Immunoassay for Simultaneous Detection of Organophosphorous Pesticides in Vegetable and Environmental Samples, Anal Chim Acta, *708*, 123–129.

150. Duford, D. A., Xi, Y. Q., & Salin, E. D. (2013). Enzyme Inhibition based Determination of Pesticide Residues in Vegetable and Soil in Centrifugal Microfluidic Devices, Anal Chem *85*, 7834–7841.

151. Mauriz, E., Calle, A., Abad, A., Montoya, A., Hildebrandt, A., Barceló, D., & Lechuga, L. M. (2006). Determination of Carbaryl in Natural Water Samples by a surface Plasmon Resonance Flow-Through Immunosensor Biosens, *Bioelectron*, *21*, 2129–2136.

152. Tran, H. V., Yougnia, R., Reisberg, S., Piro, B., Serradji, N., Nguyen, T. D., Tran, L. D., Dong, C. Z., & Pham, M. C. (2012). A Label Free Electrochemical Immunosensor for Direct, Signal-on and Sensitive Pesticide Detection Biosens, *Bioelectron*, *31*, 62–68.

153. Chen, H. D., Zuo, X. L., Su, S., Tang, Z. Z., Wu, A. B., Song, S. P., Zhang, D. B., & Fan, C. H. (2008). An Electrochemical Sensor for Pesticide Assays Based on Carbon Nanotube Enhanced Acetylcholinesterasae Activity Analyst, *133*, 1182–1186.

154. Braham, Y., Barhoumi, H., & Maaref, A. (2013). Urease Capacitive Biosensors Using Functionalized Magnetic Nanoaprticles for Atrazine Pesticide Detection in Environmental Samples Anal. Methods, *5*, 4898–4904.

155. Lamotte, M., Fornier De Violet, P., Garrigues, P., & Hardy, M. (2002). Evaluation of the Possibility of Detecting Benzenic pollutants by Direct Spectrophotometry on PDMS Solid Adsorbent, Anal Bioanal, Chem. *372*, 169–173.

156. Moliner-Martinez, Y., & Campins-Falcó, P. (2005). Detector Supports: Application to Aliphatic Amines in Wastewater Talanta, 217–222.

157. Moliner-Martinez, Y., Campins-Falcó, P., & Herráez-Hernández. (2004). A Method for the Determination of Dimethylamine in Air by Collection on Solid Support Sorbent with the Subsequent Derivatization and Spectrophotometric, Analysis *J. Chromatogr* A, *1059*, 17–24.

158. Silva, L. I. B., Panteleitchouk, A. V., Freitas, A. C., Rocha-Santos, T. A. P., & Duarte, A. C. (2009). Microscale Optical Fiber sensor for B-TEX Monitoring in Landfill Leachate, Anal Methods, *1*, 100–107.

159. Goncalves, V. C., & Balogh, D. T. (2012). Optical Chemical Sensors using Polythiophene Derivates as Active Layer for Detection of Volatile Organic Compounds, Sens. Actuat.B, *162*, 307–312.

160. Nizamidin, P., Yimit, A., Abdurrahman, A., & Itoh, K. (2013). Formaldehyde Gas Sensor Based on Silver-and Ytrrium-Codoped Lithium Iron Phosphate Thin Film Optical Waveguide Sens. Actuat. B, *176*, 460–466.

161. Sayago, I., Fernandez, M. J., Fontecha, J. L., Horillo, M. C., Vera, C., Obieta, I., & Bustero, I. (2012). New Sensitive Layers for Surface Acoustic Wave Gas Sensors Based on Polymer and Carbon Nanotube Composite Sens. Actuat. B, *175*, 67–72.

162. Young, C. R., Menegazzo, N., Riley, A. E., Brons, C. H., Dibanzo, F. P., Givens, J. L., Martin, J. L., Disko, M. M., & Mizaikoff, B. (2011). Infrared Hollow Waveguide Sensors for Simultaneous Gas phase detection of Benzene, Toluene, and Xylenes in Field Environments, Anal Chem, *83*, 6141–6147.

163. Lima, K. M. G., Raimundo, I. M., & Pimentel, M. F. (2011). Simultaneous Determination of BTX and Total Hydrocarbons in Water Employing near Infrared Spectroscopy and Multivariate Calibration, Sens. Actuat. B, *160*, 691–697.

164. Pejcic, B., Myers, M., Ranwala, N., Boyd, L., Baker, M., & Ross, A. (2011). Modifying the response of a Polymer based Quartz Crystal Microbalance Hydrocarbon Sensor with Functionalized Carbon Nanotubes Talanta, *85*, 1648–1657.

165. Mirmohseni, A., Abdollahi, H., & Rostamizadeh, K. (2007). Analysis of Transcient Response of Single Quartz Crystal Nanobalance for determination of Volatile Organic Compounds, Sens Actuat B, 121, 365–371.

166. Staginus, J., Aerts, I. M., Chang, Z. Y., Meijer, G. C. M., De Smet, L. C. P. M., & Sudholter, E. J. R. (2013). Capacitive Response of PDMS-Coated IDE Platforms Exposed to Aqueous Solutions Containing Volatile Organic Compounds Sens. Actuat. B, *184*, 130–142.

167. Lin, H. W., Jang, M., & Suslick, K. S. (2011). Preoxidation for Colorimetric Sensor Array Detection of Vocsj, Am. Chem. Soc. 133, 16786–16789.

168. Tizzard, A. C., & Lloyd-Jones, G. (2007). Bacterial Oxygenases: in Vivo Enzyme Biosensors for Organic Pollutants Biosens, Bioelectron 22, 2400–2407.

169. Sousa, T. F. A., Amorim, C. G., Montenegro, M. C. B. S. M., & Araujo, A. N. (2013). Cyclodextrin Based Potenciometric Sensor for Determination of Ibuprofen in Pharmaceuticals and Waters, Sens. Actuat. B, *176*, 660–666.

170. Tschmelak, J., Proll, G., & Gauglitz, G. (2005). Optical Biosensor for Pharmaceutical, Antibiotics, Hormones, Endocrine Disrupting Chemicals and Pesticides in Water: Assay Optimization Process for Estrone as Example Talanta, *65*, 313–323.

171. Lin, M., Liu, Y., Chen, X., Fei, S., Ni, C., Fang, Y., Liu, C., & Cai, Q. (2013). Poly (Dopamine) Coated Gold Nanocluster Functionalized Electrochemical Immunosensor for Brominated Flame Retardants Using Multienzyme Labeling Carbon Hollow Nanochains as Signal Amplifiers Biosens Bioelectron. *45*, 82–88.

CHAPTER 15

ANALYTICAL PYROLYSIS PRINCIPLES AND APPLICATIONS TO ENVIRONMENTAL SCIENCE

MICHAEL A. KRUGE

Earth and Environmental Studies Department Montclair State University, Montclair, New Jersey, 07043, USA; E-mail: krugem@mail.montclair.edu

CONTENTS

15.1   INTRODUCTION

Pyrolysis is the heating of organic substances in an inert, oxygen-free atmosphere, thereby avoiding combustion. When performed on a large scale, pyrolysis is involved in industrial processes as diverse as the manufacture of coke from coal and the conversion of biomass into biofuels. In contrast, *analytical pyrolysis* is a laboratory procedure in which small amounts of organic materials undergo thermal treatment, the products of which are subsequently quantified and/or characterized, for example, by gas chromatography. The pyrolysis may be performed "off-line" or "on-line." In the off-line case, pyrolysis occurs in stand-alone reactor. The pyrolysis products are then extracted or trapped manually prior to further evaluation by chromatographic or other means. In on-line methods, the pyrolysis reactor is coupled directly to the analytical system, be it the injector of a gas chromatograph or a detector such as a flame ionization device or a mass spectrometer, with the pyrolyzate swept along its course by inert carrier gas. In some cases, a trapping mechanism such as cryofocusing is employed, which can permit the use of multiple detection or analytical systems. On-line methods typically only require milligram or even submilligram quantities of sample. Samples may be analyzed with little pretreatment, thereby minimizing the use of hazardous solvents in the spirit of environmentally conscious "green chemistry."

Types of samples suitable for analytical pyrolysis include, for example, petroleum source rocks, sediments, soils, biological materials, and artificial polymers. The method is appropriate for macromolecular organic materials of many types, be they "geopolymers" such as kerogen, asphaltenes and humic substances, biopolymers such as proteins and lignin, and manufactured plastics. Such materials are not directly amenable to gas chromatography, so the thermal treatment opens an alternate avenue for molecular analysis.

This chapter primarily focuses on the use of analytical pyrolysis for the chemical characterization of organic contaminants in environmental media, particularly sediments and soils, but also air and water. It presents a variety of instrumental configurations, by which pyrolysis microreactors are directly coupled to detection systems, with or without intervening chromatographic separation of the pyrolyzate. In each instance the chapter provides examples of environmental applications. The pyrolysis terminology employed herein conforms to the IUPAC recommendations [1].

15.2   PYROLYSIS-DIRECT DETECTION

In pyrolysis-direct detection systems, the pyrolysis furnace is coupled to one or more of the detectors typically used in gas chromatographic (GC) systems, but without the intervening GC column. The rapidity of the procedure is its advantage, particularly if bulk characterization is desired. In this configuration, a flame ionization detector (FID), a thermal conductivity detector (TCD), or a mass spectrometer (MS) is commonly employed (Fig. 15.1). The temperature of the pyrolysis furnace may be programmed to gradually increase, allowing the detector to monitor the thermal evolution of the sample.

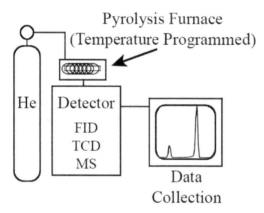

FIGURE 15.1 Simplified schematic diagram of a temperature-programmed pyrolysis-direct detection system.

Pyrolysis systems, be they direct detection or with a coupled chromatograph, may be operated at subpyrolytic temperatures to affect a *thermodesorption* or "thermal extraction" of the sample, by which volatile materials are liberated from the matrix. The sample may then be heated further, at a higher temperature appropriate for true pyrolysis (*stepwise pyrolysis*). Note that if the stepwise approach is not taken and only the higher (true pyrolysis) temperature is employed, the result may likely be a mixture of thermally desorbed and pyrolysis products.

15.2.1 ROCK-EVAL PYROLYSIS

The Rock-Eval pyrolysis system [2–4] is a direct detection instrument widely used in the petroleum industry, most often for the evaluation of petroleum source rock potential. It is an automated device, suitable for the rapid, bulk analysis of multiple samples. Temperature programming is employed to achieve both thermodesorption and stepwise pyrolysis, calibrated quantitatively by external standards. Milligram quantities of crushed rock are placed in a crucible, which is then robotically introduced into a furnace preheated typically to 300°C. After being held at the initial temperature for several minutes, the furnace is heated to the final temperature at a prescribed rate (e.g., to 550°C at 25°C min^{-1}) where it remains for several minutes before returning to the starting temperature (Fig. 15.2A). During the initial isothermal and subsequent temperature ramp stages, the resulting effluent is delivered to an FID, typically producing two peaks on the pyrogram (Fig. 15.2B). The first peak (S1) corresponds to the yield of thermally desorbed "free hydrocarbons," while the second peak (S2) is due to the true pyrolysis products, that is, from the high temperature cracking of the kerogen. Carbon dioxide produced during the early portion of the program is trapped and diverted automatically by valves to a thermal conductivity detector (TCD), registering as the S3 peak. The trap is closed at a sufficiently low temperature to exclude CO_2 evolving from the break-down of carbonate minerals, so that the S3 peak is interpreted to represent pyrolytic CO_2 arising from oxygen-bearing functional groups in the organic matter. By using the primary S1, S2, and S3 parameters, along with standard ratios employing them, petroleum source rock richness, quality (the likelihood to generate oil versus gas), and maturity can be readily ascertained [2–4].

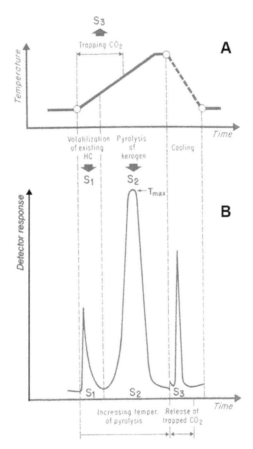

FIGURE 15.2   Typical Rock-Eval temperature program and resulting pyrogram. (Modified from Ref. 3. Used with permission.)

While not commonly employed in environmental research, the Rock-Eval instrument has proven itself to be valuable when it has. The method permitted inferences about nutrient inputs and marine productivity in a paleoenvironmental study of Plio-Pleistocene sapropelic sediments in the western Mediterranean Sea [5]. It has also been used to study carbon cycling in modern mangrove sediments [6] and to investigate the association of organic matter with trace metal pollutants in sediments [7, 8].

The Rock-Eval S1 and S2 parameters are useful for initial screening of sediment samples suspected of contamination. For example, organic-rich surface sediment samples from westernmost Lake Ontario, Canada (in this case, defined as those with total organic carbon (TOC) contents in excess of about 2%) show Rock-Eval S2 values above 3 mg pyrolysis yield/g sediment (Fig. 15.3). The sources of organic matter at this location include spilled petroleum and coal, as well as "natural" materials such as aquatic algae, the growth of which was likely enhanced by anthropogenic nutrient inputs [9]. A more extreme organic enrichment was in evidence at the site of an urban sewage sludge spill on the Mediterranean Sea floor off the coast of Barcelona, Spain [10]. The affected sediment is clearly distinguished by Rock-Eval S1 and S2 values above 2 and 4 mg/g, respectively (Fig. 15.3).

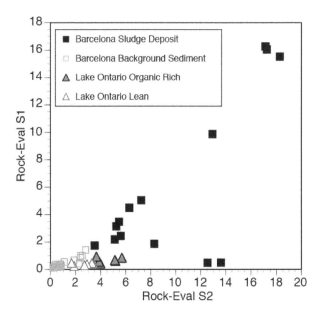

FIGURE 15.3 Cross-plot of the primary Rock-Eval S1 (mg thermally-desorbed products per gram of sample) and S2 (mg pyrolysis products per gram of sample) parameters as recorded by a flame ionization detector (FID). This illustrates the application of this petroleum prospecting technique to the study of contaminated sediments (data from Refs. 9 and 10).

The Rock-Eval hydrogen and oxygen indices are among the most widely employed standard parameters produced by this instrument. They are essentially the ratios of S2 to TOC and S3 to TOC, respectively. (Conveniently, the Rock-Eval instrument is commonly configured to also determine TOC.) In the classical interpretation used in petroleum source rock studies, kerogen types I, II and III can be readily recognized on cross-plots of the two indices [2–4] (Fig. 15.4). However, these parameters have also proven their utility in screening samples for contamination. The organic-rich sediments and sludge deposits from Lake Ontario and Barcelona mentioned above are clearly distinguished from organic-lean and uncontaminated background sediments by elevated hydrogen index values [9, 10] (Fig. 15.4). While it is interesting to note that organic matter in modern sediments is enriched in oxygen compared to ancient kerogen, the kerogen type designations are neither relevant nor necessary in contamination studies.

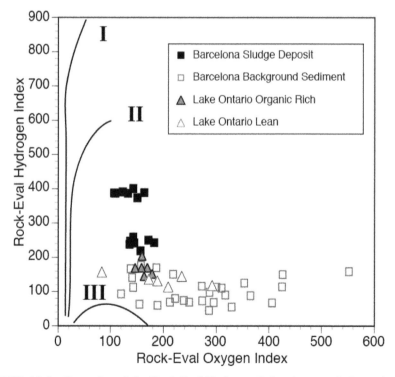

FIGURE 15.4   Cross-plot of the Rock-Eval Hydrogen Index (mg pyrolysis products/g organic carbon) and Oxygen Index (mg pyrolytic CO_2/g organic carbon) for the same samples shown in Figure 15-3. These samples plot outside the classic kerogen type ranges (I, II, III) established by Espitalié and others [2] commonly used in petroleum exploration and shown here for reference (data from Refs. 9 and 10).

15.2.2 *PYROLYSIS-MASS SPECTROMETRY*

Another often-used pyrolysis-direct detection configuration employs a mass spectrometer as the detector [11], providing the researcher with some molecular information without sacrificing rapidity of analysis. The Py-MS method has been actively employed since the 1970's and early 1980's [12–14], often in chemotaxonomic studies of microorganisms [15, 16] and in classification of industrial polymers with forensic applications [17]. The resulting mass spectrometric data are typically subjected to multivariate analysis to aid interpretation. A study of the process of peat formation was an early environmental application of Py-MS [18]. Remmler et al. used the similar technique of thermogravimetry/mass spectrometry to observe the desorption of individual polycyclic aromatic hydrocarbons (PAHs) from petrochemical plant sludges and contaminated soils as a function of analysis temperature [19]. As part of a characterization of coal wastewaters, Pörschmann and co-workers analyzed fulvic and humic acids isolated from groundwater by Py-MS [20].

High-resolution mass spectrometry provides an enhanced variant of the Py-MS technique. As with standard Py-MS, it has been used effectively in chemotaxonomic studies of bacteria, with attendant multivariate analysis of the MS data [21]. Field ionization mass spectrometry (FIMS) provides yet another approach, with a "softer" ionization that avoids the molecular fragmentation characteristic of standard electron impact mass spectrometers. As such, a FIMS spectrum of a complex mixture such as a pyrolyzate consists largely of the molecular ions of the constituent compounds (the *molecular ion* being the mass spectral ion indicative of the compound's molecular weight). Py-FIMS systems have been used effectively in detailed molecular investigations of dissolved organic matter in natural waters (Fig. 15.5A) and soil organic matter [22–24]. In a recent advance, pyrolyzers have been coupled with a metastable ion time-of-flight mass spectrometer (Py-MAS-TOF-MS) for the analysis of microbes and microbial lipids in medical [25] and soil [26] studies. The MAS permits better control over ionization and fragmentation than conventional electron ionization, while the TOF-MS has sensitive, rapid spectral acquisition [25]. A low power Py-TOF-MS instrument has been developed with the intention of making it sufficiently robust to travel to the moon to perform lunar soil analyzes [27].

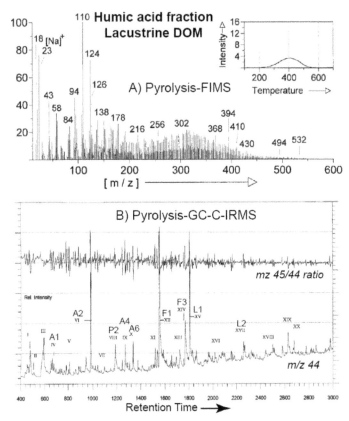

FIGURE 15.5 Results from the pyrolysis of the humic acid fraction of dissolved organic matter (DOM) in lake water. A) Field ionization mass spectrum of the pyrolyzate with associated profile of pyrolysis yield as a function of temperature. B) The m/z 44 mass chromatogram (lower trace) and corresponding m/z 45/44 ratio from pyrolysis-GC-stable carbon isotope ratio mass spectrometry (C-IRMS). See Table 1 for peak identification. (Modified from Ref. 22. Used with permission.)

15.3 PYROLYSIS-GAS CHROMATOGRAPHY

Analytical pyrolysis is most commonly employed in conjunction with a gas chromatograph (Py-GC), suitable for a wide variety of investigations (e.g., forensic, art appraisal, archeological) of textiles, paints, inks, and biopolymers [28]. In such a system, an in-line pyrolyzer is directly coupled to the GC injector (Fig. 15.6). The chromatograph is commonly equipped with an flame ionization detector (sensitive to hydrocarbons) or a mass spectrometer (discussed separately in Section 15.4). Other devices, such as the sulfur-sensitive flame photometric detector (FPD) and the atomic emission detector (AED), are used less frequently. There are a variety of microscale pyrolyzers available, most notably those with a furnace, with an inductively heated filament (Curie point), and with a resistance coil or ribbon, each with its particular advantages and disadvantages [29, 30].

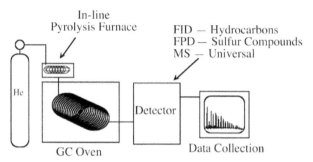

FIGURE 15.6 Simplified schematic diagram of an analytical pyrolysis-gas chromatography system. Different detectors may be coupled to the system, including for example, a flame ionization detector (Py-GC-FID), a mass spectrometer (Py-GC/MS), and more rarely, a flame photometric detector (Py-GC-FPD).

In 1954, Davison and colleagues were perhaps the earliest to advocate Py-GC as a means for polymer analysis [31]. By 1970, Giraud was applying the method for petroleum source rock characterization [32]. In 1979, Irwin provided a review of the technique and its applications in the first article published in the newly created *Journal of Analytical and Applied* Pyrolysis [33]. At about the same time, Gutteridge compiled a review of Py-GC usage in microbial chemotaxonomy [12] and Dembicki with co-workers described Py-GC methods developed at a major petroleum company for source rock evaluation [34].

The Py-GC technique has proven itself to be useful in evaluating environmental contamination. In an early such study, Whelan et al. [35] subjected contaminated marine sediments to thermodesorption (TD) and stepwise pyrolysis-GC. Peak 1 in Fig. 15.7A denotes the desorbed products liberated from the sediments at about 135°C, subsequently injected automatically into a GC-FID. The resulting chromatogram (Fig. 15.7B) reveals a limited number of identifiable *n*-alkanes and monoaromatic hydrocarbons, as well as what appears to be the chromatographic hump ("unresolved complex mixture" or UCM) characteristic of biodegraded oil. (Refer to Table 15.1 for peak identification.) The sediment sample was then heated to a true pyrolysis temperature of 690°C, producing Peak 2 (Fig. 15.7A) composed of a mixture of aromatic and saturate hydrocarbons lighter than those in Peak 1, without the UCM hump (Fig. 15.7C). Note that Peaks 1 and 2 in this study (Fig. 15.7A) correspond approximately to the Rock-Eval S1 and S2 peaks (Fig. 15.2). The stepwise Py-GC approach is not as rapid as the Rock-Eval method, but some limited molecular data are now available.

TABLE 15.1   Peak identification codes for chromatographic figures. Also, +: n-alkanes, ^: n-alk-1-enes, IS: internal standards, X: phthalate contaminants introduced during sample handling. Mass spectral base peaks and/or molecular ions are given

Peak	Compound or isomer group	m/z	Peak	Compound or isomer group	m/z
Aromatic compounds			H4	hopane	191
A1	benzene	78	H5	C_{31} hopanes	191
A2	toluene	92	H6	C_{32} hopanes	191
A3	ethylbenzene	106	Isoprenoids		
A4	1,3 & 1,4-dimethylbenzenes	106	I1	norpristane	71
A5	1,2-dimethylbenzene	106	I2	pristane	71
A6	styrene	104	I3	prist-1-ene	69
A7	C_3-alkylbenzene	120	I4	prist-2-ene	69
A8	methylstyrene isomer	118	I5	phytane	71
A9	indene	116	I6	neophytadiene	68
A10	methylindene	130	I7	phyta-1,3(E)-diene	82
A11	naphthalene	128	I8	phytol	71
A12	2-methylnaphthalene	142	I9	unidentified isoprenoid	68
A13	1-methylnaphthalene	142	Methoxyphenols (lignin markers)		
A14	biphenyl	154	L1	guaiacol	124
A15	dimethylnaphthalenes	156	L2	vinylguaiacol	150
A16	acenaphthylene	152	Organonitrogen compounds		
A17	acenaphthene	154	N1	pyridine	79
A18	methylbiphenyl	168	N2	pyrrole	67
A19	dibenzofuran	168	N3	methylpyridine	93
A20	trimethylnaphthalenes	170	N4	methylpyrrole	80
A21	fluorene	166	N5	C_2-alkylpyrrole	94
A22	tetramethylnaphthalenes	184	N6	alkyl-alkylidene amine	98
A23	phenanthrene	178	N7	benzonitrile	103
A24	anthracene	178	N8	alkyl-alkylidene amine?	98
A25	methylphenanthrenes	192	N9	alkyl-alkylidene amine	140
A26	phenylnaphthalene	204	N10	benzoacetonitrile	117
A27	dimethylphenanthrenes	206	N11	piperidinone	98
A28	fluoranthene	202	N12	benzenepropanenitrile	131
A29	pyrene	202	N13	quinoline	129

A30	trimethylphenahthrenes	220	N14	indole	117
A31	retene	234	N15	methylindole	131
A32	methylpyrene isomers	216	N16	indole dione	147
A33	dimethylpyrene isomers	230	N17	diketodipyrrole	186
A34	benzo[a]anthracene	228	N18	diketopiperazine (Pro-?)	70
A35	chrysene	228	N19	carbazole	167
A36	methylchrysene isomers	242	N20	n-hexadecanitrile	110
A37	dimethyl chrysene	256	N21	n-hexadecanamine, N,N-di-methyl	58
A38	benzo[b]fluoranthene	252	N22	n-octadecanitrile	110
A39	benzo[j]fluoranthene	252	N23	n-octadecanamine, N,N-di-methyl	58
A40	benzo[k]fluoranthene	252	N24	n-hexadecamide	59
A41	benzo[e]pyrene	252	Polysaccharide markers		
A42	benzo[a]pyrene	252	P1	cyclopentenone	82
A43	perylene	252	P2	furancarboxaldehyde	95
A44	indeno[1,2,3-cd]pyrene	276	P3	methylcyclopentenone	96
A45	benzo[ghi]perylene	276	P4	methylfuranone	98
Carboxylic acids			P5	methylfurancarboxaldehyde	110
C1	C_{14} alkanoic acid	73	Sulfur compounds		
C2	hexadecenoic acid	69	S1	dibenzothiophene	184
C3	C_{16} alkanoic acid	73	S2	methyldibenzothiophenes	198
Linear alkylbenzenes (LABs)			S3	dimethyldibenzothiophenes	212
D1	5-phenyldecane	91	S4	trimethyldibenzothiophenes	226
D2	4-phenyldecane	91	S5	benzonaphthothiophene	234
D3	3-phenyldecane	91	S6	elemental sulfur (S_8)	256
D4	1-phenylnonane	91	Steroids		
D5	2-phenyldecane	91	S1	cholestene	215
D6	6-phenylundecane	91	S2	cholestene	215
D7	5-phenylundecane	91	S3	cholestene	215
D8	4-phenylundecane	91	S4	cholestene	215
D9	3-phenylundecane	91	S5	5α(H) cholestane (20R)	217
D10	1-phenyldecane	91	S6	C_{27} steradiene	215
D11	2-phenylundecane	91	S7	methylcholestene	215

D12	6-phenyldodecane	91		S8	methylcholestene	215
D13	5-phenyldodecane	91		S9	methylcholestene	215
D14	4-phenyldodecane	91		S10	methylcholestene	215
D15	3-phenyldodecane	91		S11	5α(H) methylcholestane (20R)	217
D16	1-phenylundecane	91		S12	ethylcholestene	215
D17	2-phenyldodecane	91		S13	ethylcholestene	215
D18	6-phenyltridecane	91		S14	ethylcholestene	215
D19	5-phenyltridecane	91		S15	5α(H) ethylcholestane (20R)	217
D20	4-phenyltridecane	91		S16	ethylcholestene	215
D21	3-phenyltridecdane	91		S17	coprostanol	388
D22	1-phenyldodecane	91		S18	C_{29} steradiene	215
D23	2-phenyltridecdane	91		S19	C_{27} stanone	386
D24	1-phenyltridecdane	91		S20	24-methylsteratetraene	378
Phenolic compounds				S21	24-methylsteratriene	380
F1	phenol	94		S22	24-methylsteradiene	382
F2	2-methylphenol	108		Tricyclic terpanes		
F3	4- & 3-methylphenols	108		T1	C_{23} tricyclic terpane	191
F4	4-ethylphenol	107		T2	C_{24} tricyclic terpane	191
F5	vinylphenol	120		T3	C_{25} tricyclic terpane	191
Hopanes				T4	C_{26} tricyclic terpane	191
H1	18α(H)-trisnorhopane (Ts)	191		T5	C_{28} tricyclic terpanes	191
H2	17α(H)-trisnorhopane (Tm)	191		T6	C_{29} tricyclic terpanes	191
H3	norhopane	191				

Hala undertook a systematic study of contaminated soils in a small abandoned oil field [36], comparing residual oil floating on water in a decrepit wooden holding tank with soil samples collected in the vicinity. He performed Py-GC analysis of solid samples in conjunction with standard GC/MS characterization of the whole oil and solvent extracts of the soil. The conventional extract results showed unsurprisingly that the oil in the tank and contaminated soils was biodegraded, with prominent isoprenoid alkanes and the chromatographic hump due to the unresolved complex mixture of hydrocarbons. The distributions of steranes and terpanes in the tank oil and contaminated soil extracts confirmed their common origin. The extract of the unaffected soil presented almost exclusively the odd carbon-numbered long-chain *n*-alkanes characteristic of natural land plant material [36]. The Py-GC results provided a complementary view (Fig. 15.8). Rather than pyrolyze the whole oil, only its asphaltene fraction was used. This fraction is more resistant to biodegradation and, although a solid, it is nonetheless amenable to microscale pyrolysis. Upon pyrolysis, the oil asphaltene yielded a series of n-alkanes from C_4 to at least C_{27}, marked with $+$ signs in Fig. 15.8A. The chromatographic resolution here is higher than that depicted in Fig. 15.7, so it is apparent that the n-alkanes constitute the second peak in a couplet. The first peak in each pair is the corresponding n-alk-1-ene. This a characteristic of pyrolyzates of aliphatic-rich macromolecules, a phenomenon most readily seen when pyrolyzing artificial polyethylene [28]. Monoaromatic and isoprenoid hydrocarbons are also apparent (Fig. 15.8A). The soil samples were pyrolyzed simply after drying, without solvent extraction. In the strict sense, they yielded a mixture of thermally desorbed compounds and true pyrolysis products. The pyrolyzate of contaminated soil collected near the tank shows a distribution

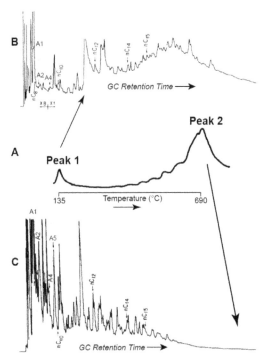

FIGURE 15.7 An early application of thermodesorption- and pyrolysis-GC to the characterization of contaminated sediments. A) FID pyrogram showing product evolution as a function of increasing temperature. The first peak corresponds to thermodesorption products while the second is due to the true pyrolysis products. B) GC trace of the thermodesorbed materials. C) GC trace of the pyrolysis products. See Table 1 for peak identification. (Modified from Ref. 35. Used with permission.)

of *n*-alkenes and *n*-alkanes very similar to that of the oil asphaltene (Fig. 15.8A, B), validating the assumption that asphaltene pyrolyzates would be useful for environmental forensic fingerprinting. In addition, the soil yielded a prominent UCM hump, likely indicative of thermally desorbed degraded oil. Aromatic hydrocarbons are relatively more abundant than in the oil asphaltene. Simple phenolic compounds are also apparent in the soil pyrolyzate, a feature not seen in that of the asphaltene. The pyrolyzate of the uncontaminated soil consists primarily simple monoaromatic hydrocarbons and phenols (Fig. 15.8C), likely produced from natural organic matter in the soil. As such, its Py-GC trace is clearly distinguishable from that of the contaminated soil (Fig. 15.8B). The phenolic compounds and at least a portion of the monoaromatic hydrocarbons produced by the contaminated sample are likely also due to admixed natural soil organic matter. Hala found that simple ratios of C_2-alkylbenzenes (peaks A3, A4, A5) to toluene (A2) and to phenol (F1) correlated positively with the degree of oil contamination of the soil, as determined quantitatively by solvent extraction [36].

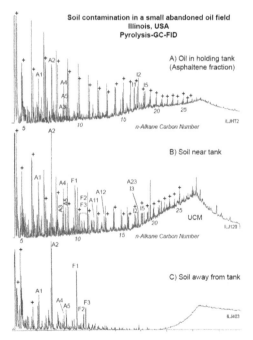

FIGURE 15.8 Application of pyrolysis-GC-FID to the study of contaminated soil in a small abandoned oil field, central Illinois, USA. A) Asphaltenes of biodegraded oil residues in a holding tank. B) Dried, unextracted soil collected 7 m away from tank. UCM: chromatographically unresolved complex mixture of hydrocarbons characteristic of biodegraded petroleum. C) Dried, unextracted soil collected 17 m away from tank. Coil pyrolysis at 610 °C for 20 s (data from Ref. 36).

15.4 PYROLYSIS-GAS CHROMATOGRAPHY/MASS SPECTROMETRY

While Py-GC-FID alone was shown to be capable of distinguishing between clean and petroleum-contaminated soils by visual inspection, the complexity of the pyrolyzates precludes much more than rudimentary identification of individual compounds (Figs. 15.7 and 15.8). By reference to external or internal standards or by the recognition of obvious chromatographic elution patterns, a limited number of compounds might be identified on FID traces. The ability of a standard electron impact mass spectrometer to identify unknown compounds is clearly an advantage and thus pyrolysis-gas chromatography/mass spectrometry (Py-GC/MS) has become the most widely used analytical pyrolysis method in environmental studies. Identification of compounds shown in Fig. 15.8 was in fact facilitated by the reanalysis of a limited number of samples by Py-GC/MS [36].

Py-GC/MS was used in kerogen studies as early as 1975 [37]. In one of the earliest reported environmental applications of Py-GC/MS, de Leeuw et al. [38] used individual ion chromatograms to identify a large number of aromatic compounds in a contaminated soil sample (Fig 15.9). For example, benzo[a]anthracene and chrysene (peaks A34, A35) are readily apparent on the m/z 228 trace, while barely visible on the total ion current (TIC) chromatogram above. Environmental researchers continued to use both Py-GC and Py-GC/MS, for example, in studies of natural organic matter and petroleum contaminated soils [39] and of urban atmospheric contamination residues on surfaces of buildings [40]. However, Py-GC/MS became the preferred technique, particularly as smaller bench-top mass spectrometers became generally available.

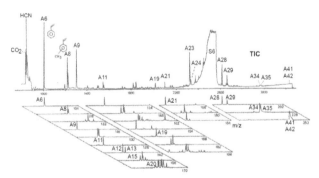

FIGURE 15.9 An early example of the use of pyrolysis-GC/MS for the analysis of contaminated sediments. The partial mass chromatograms below the TIC trace display the distributions of compounds of interest. See Table 1 for peak identification. (Modified from Ref. 38. Used with permission.)

The method has been widely used in studies of soil humus [24, 41, 42] and dissolved organic matter in natural waters [23, 43]. Py-GC/MS was employed to investigate binding of pollutants such as PAHs and petrochemical plant sludges to soil organic matter [19, 44], urban air pollutants on architectural patinas [45], and brown coal dust on various sediment size fractions [46].

Thermodesorption and stepwise pyrolysis-GC/MS proved effective in the characterization of heavily contaminated fluvial sediments from northern Indiana, USA [47]. Selected ion monitoring of the molecular ions of dibenzothiophene and C_1 to C_3-alkyldibenzothiophenes (peaks S1, S2, S3, S4) produced a chromatogram comparable to the pyrolysis-GC-FPD trace of the same sample (Fig. 15.10A, B). With a study of atmospheric sulfate particulate matter as a rare exception [48], the sulfur-selective FPD is seldom employed with pyrolysis and therefore the flexibility of a mass spectrometer is particularly attractive. In addition to the relatively abundant thermally desorbed thiophenes, the total ion current trace reveals a series of 3- and 4-ring PAHs, notably the methylphenanthrenes, pyrene, and chrysene (peaks A25, A29, and A35 in Fig. 15.10C). After thermodesorption at 310°C for 20 seconds in a coil pyrolyzer, the sample was heated to the true pyrolysis temperature of 610°C, also for 20 seconds. These stepwise pyrolysis products show a shift to relatively more high molecular weight PAHs, in particular the 4- and 5-ring (peaks A35, A36, A39, A41, A42 in Fig. 15.10D). It is likely that these are in fact still thermally desorbed but required higher volatilization temperatures, as was also observed via TD-MS of sludge-contaminated soils [19].

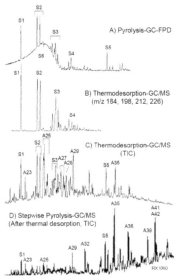

FIGURE 15.10 Characterization of aliquots of the same polluted fluvial sediment (West Branch of the Grand Calumet River, Indiana, USA) by a variety of techniques. The retention time range shown in these examples is approximately that of n-C16 to n-C30. See Table 1 for peak identification. A) Distribution of sulfur compounds as revealed by pyrolysis-GC using sulfur-selective flame photometric detection (FPD). B) The (alkyl)dibenzothiophene distribution in the 310 °C thermal desorption product as seen on a mass chromatogram of their summed molecular ions. C) The total ion current trace of the same thermodesorption product showing a predominance of polycyclic aromatic hydrocarbons and thioarenes. D) TIC trace of the products of stepwise pyrolysis at 610 °C after thermodesorption showing a predominance of larger (4- and 5-ring) PAHs (data from Ref. 47).

With the understanding that it would likely yield a mixture of thermally desorbed and pyrolysis products, a single analytical run heating the sample at a high temperature (e.g., 610°C for 20 s) may still be advantageous in terms of time and cost savings. This approach effectively detected organic contaminants in harbor sediments from Connecticut, USA (Fig. 15.11). The target analytes included acyclic alkanes and alkenes, 2- to 6-ring PAHs, petroleum biomarkers (hopanes), and n-alkylnitriles, as well as low molecular weight monoaromatic hydrocarbons, phenolic and nitrogen compounds. After exploratory runs in full scan mode, the analysis was repeated with the mass spectrometer programmed to use only the base peaks of the compounds of interest in selected ion monitoring (SIM) mode, resulting in improved sensitivity and signal-to-noise. The total ion current trace shown is in fact a summation of these purposefully selected ions. A similar approach was employed in experiments with surface sediments from western Lake Ontario, Canada, but with a fast GC temperature ramp of 20°C min^{-1}. This shortened the run time considerably, to just 22 minutes, compared to the 80 minutes required for the Connecticut analysis done at 5°C min^{-1}. This is advantageous if rapid screening for known contaminants is required. As in the previous example, the low molecular weight compounds generated by pyrolysis of the natural sedimentary organic matter are predominant (Fig. 15.12A). The acyclic hydrocarbon distribution is more clearly evident on the trace of the sum of m/z 69 and 71, although the n-alkane/alkene doublets are poorly resolved due to the fast GC program (Fig. 15.12B). The main objective, however, was the detection of the PAHs and the 4- to 6-ring parent compounds are indeed readily apparent on their composite mass chromatogram (Fig. 15.12C). This trace was assembled by joining the four constituent mass chromatograms (m/z 202, 228, 252, and 276) end-to-end with no overlap and thus provides a compact graphical means to summarize the PAH data.

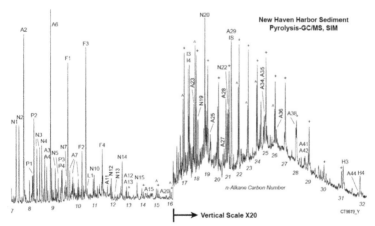

FIGURE 15.11 Total ion current chromatogram showing pyrolysis products obtained from a harbor sediment sample (New Haven (CT), USA). Data were collected in selected ion monitoring mode employing molecular ions or base peaks of compounds of interest, as determined by prior full scan analysis. See Table 1 for peak identification.

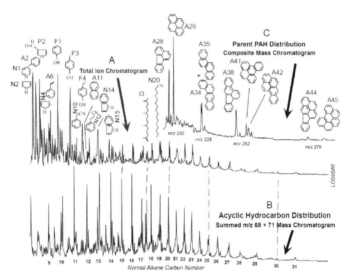

FIGURE 15.12 Application of pyrolysis-mass chromatography to the study of organic matter in Lake Ontario surface sediments. Data were collected using selected ion monitoring (SIM) of molecular ions and base peaks of major as well as trace constituents known to be present from previous full-scan analyses. The data were collected with a "fast" GC temperature program of 20 °C/min. on a 25 m HP-1 column, such that only 22 minutes of run time were needed. See Table 1 for peak identification. A) Total ion current trace of the selected ions. B) Distribution of normal and isoprenoid alkanes and alk-1-enes on a summed m/z 69 + 71 mass chromatogram. C) Distribution of 4-, 5-, and 6-ring parent PAHs as seen on a composite mass chromatogram of their respective molecular ions (m/z 202, 228, 252, 276). The traces of these four individual ions are linked end-to-end, without overlap or summation.

Munson provided a general overview summarizing environmental applications of analytical pyrolysis, with particular attention to Py-GC/MS [49]. Fabbri and others used Py-GC/MS to detect polystyrene, polyvinyl chloride, and other manufactured polymer contamination in the urban lagoonal sediments [50–52]. Work on plastics in environment can be furthered by consulting reference works on their pyrolytic behavior [53, 54]. Kruge and others investigated contamination in Barcelona, Spain harbor sediments [55], while Faure and co-workers used the method to study industrially contaminated fluvial sediments [56, 57].

Mass chromatography is also advantageous when applied to thermodesorption data, formalized as the U.S. Environmental Protection Agency's Method 8275A [58]. The temperature conditions required to effectively thermally desorb hydrocarbon contaminants from petroleum sludges and river sediments have been well documented [19, 56]. In experiments on contaminated fluvial sediment from the Passaic River (New Jersey, USA), thermally desorbed products were analyzed by GC/MS in SIM mode selecting the base peaks of common petroleum biomarkers and polycyclic aromatic compounds (Fig. 15.13A). Mass chromatography permits the visualization of the distributions of the key compounds within each class. The prominent isoprenoids pristane and phytane (peaks I2, I5) and the unresolved complex mixture of hydrocarbons (UCM) seen on the m/z 71 trace are the hallmarks of biodegraded petroleum (Fig. 15.13B). Tricyclic terpane and hopane biomarkers on the m/z 191 mass chromatogram provide further evidence of petroleum contamination (Fig. 15.13C). The alkylnaphthalene and (alkyl)dibenzothiophene isomer clusters are well-resolved on the composite mass chromatogram constructed from their respective molecular ions (Fig. 15.13D). The same is true for the alkylated phenanthrene/anthracene, pyrene/fluoranthene, and chrysene/benzo[a]anthracene series (Figs. 15–13E, F, G) as well as for the parent pentaaromatic hydrocarbons (Fig. 15–13H).

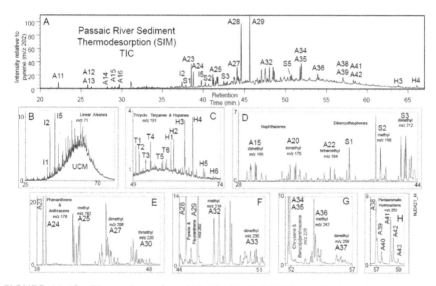

FIGURE 15.13 Thermodesorption (310 °C, 20 s) GC/MS results from the analysis of contaminated fluvial sediment by selected ion monitoring (Passaic River, Kearny (NJ), USA), using molecular ion or base peaks for compounds of interest chosen after prior full scan analyses. See Table 1 for peak and isomer group identification. A) Total ion current trace summing all ions employed. B) m/z 71 trace showing acyclic alkanes and the "unresolved complex mixture" (UCM). C) m/z 191 trace showing the distribution of tricyclic terpanes and hopanes. D) Composite mass chromatogram (m/z 156, 170, 184, 198, 212) showing the distributions of alkylnaphthalenes and (alkyl)dibenzothiophenes. E) Composite mass chromatogram (m/z 178, 192, 206, 220) showing the distribution of (alkyl)phenanthrenes and anthracenes. F) Composite mass chromatogram (m/z 202, 216, 230) showing the distribution of fluoranthene, pyrene, and alkylpyrene isomers. G) Composite mass chromatogram (m/z 228, 242, 256) showing the distribution of benzo[a]anthracene, chrysene, and alkylchrysene isomers. H) m/z 252 mass chromatogram showing the distribution of pentaaromatic hydrocarbons. Composite mass chromatograms display the data for the indicated ions end-to-end, without summation or overlap.

15.4.1   QUANTITATION AND MULTIVARIATE ANALYSIS OF PY-GC/MS DATA

Bar graphs of quantitated chromatographic peak values provide a simpler form of data presentation, particularly useful for a visual comparison of results from several samples. When used for environmental forensic purposes, the members of an isomer cluster (e.g., the dimethylphenanthrenes, marked A27 in Fig. 15.13E) are often summed as a single value upon quantitation [59, 60]. To prepare data for the graphic display, the chromatographic peaks of interest are quantitated using their respective mass chromatograms (e.g., Figs. 15.13B–H). Correction factors (computed from full mass spectra of reference compounds) are then applied for each peak and values normalized for each sample. While environmental forensics practitioners normally use standard solvent extraction methods, data from TD-GC/MS are also amenable to simplified graphical display (Fig. 15.14, Table 15.2). In this figure sample A is the same as the one presented chromatographically in Fig. 15.13, while sample B is sediment collected several kilometers downstream in the same river. This fingerprinting procedure permits ready comparison between the thermal extracts of the two samples, revealing, for example, the relatively higher petroleum biomarker and lower alkylphenanthrene concentrations in sample B.

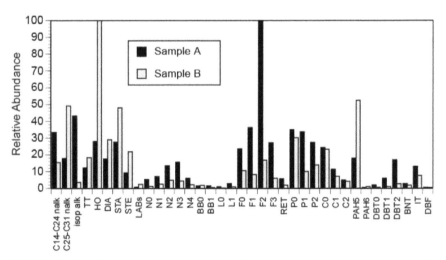

FIGURE 15.14 Bar graph comparing the distribution of saturate and aromatic thermodesorption products in two contaminated Passaic River sediment samples. Sample A: Kearny (NJ), same as in Figure 15-13. Sample B: Newark (NJ). See Table 2 for compound group codes.

TABLE 15.2   Compound codes for Figure 15–14

Code	Compound or isomer group	m/z
C14-C24 nalk	C_{14} to C_{24} *n*-alkanes	71
C25-C31 nalk	C_{25} to C_{31} *n*-alkanes	71
isop alk	Isoprenoid alkanes	71
TT	Tricyclic terpanes	191
HO	Hopanes	191
DIA	Diasteranes	217
STA	Steranes	217
STE	Sterenes	215
LABs	Linear alkylbenzenes	91
N0	Naphthalene	128
N1	Methylnaphthalenes	142
N2	Dimethylnaphthalenes	156
N3	Trimethylnaphthalenes	170
N4	Tetramethylnaphthalenes	184
BB0	Biphenyl	154
BB1	Methylbiphenyls	168
L0	Fluorene	166
L1	Methylfluorenes	180
F0	Phenanthrene & anthracene	178
F1	Methyl phenanthrenes & anthracenes	192
F2	C_2-phenanthrenes & anthracenes	206
F3	C_3-phenanthrenes & anthracenes	220
RET	Retene	232
P0	Pyrene & fluoranthene	202
P1	Methylpyrene & isomers	216
P2	Dimethylpyrene & isomers	230

TABLE 15.2   *(Continued)*

Code	Compound or isomer group	m/z
C0	Chrysene & benzo[*a*]anthracene	228
C1	Methylchrysene & isomers	242
C2	Dimethylchrysene & isomers	256
PAH5	Pentaaromatic hydrocarbons	252
PAH6	Hexaaromatic hydrocarbons	276, 278
DBT0	Dibenzothiophene	184
DBT1	Methyldibenzothiophenes	198
DBT2	Dimethyldibenzothiophenes	212
BNT	Benzonaphthothiophene	234
IT	Isoprenoid thiophenes	308
DBF	Dibenzofuran	168

The various pyrolysis and thermodesorption methods discussed in this chapter are most suitable for inexpensive, rapid sample screening and comparative fingerprinting. However, quantitative analysis is also possible if appropriate internal or external standards are employed, at least for estimation purposes. A long (5 m) sediment core from the Passaic River was subsampled and analyzed by Py-GC/MS, after the addition of a measured amount of perdeuterated pyrene. It was then possible to estimate concentrations of pyrene (thermally desorbed during the single-step run at 610°C) which were found to range between 2 and 30 µg/g sediment, generally increasing with depth (Fig. 15.15). These values are overall less than those achieved using the U. S. Environmental Protection Agency's standard solvent extraction-based methods, but the depth trend is the same (Fig. 15.15). The pyrolysis data were acquired on single samples, whereas the solvent work was performed on composite samples from the same core, in some cases different from those used for pyrolysis. This likely accounts for some of the discrepancy. Buco and others found comparable solvent extraction and pyrolysis quantitation results for parent PAHs in study employing contaminated soil and certified reference materials [61].

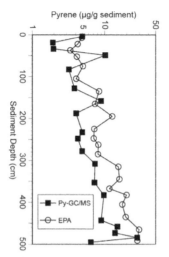

FIGURE 15.15 Comparison of quantitative results for pyrene concentrations in sediments produced using the standard U.S. Environmental Protection Agency methods for PAH determination and by Py-GC/MS with an internal standard (decadeuteropyrene). Note logarithmic scale. Data are from Lower Passaic River sediment core 7A (data from Ref. 64).

Environmental forensics practitioners often employ multivariate methods when interpreting data produced by classical solvent extraction methods [62, 63]. With the large molecular datasets generated by Py- and TD-GC/MS (e.g., Figs. 15.13 and 15.14), such an approach is equally advantageous. As with the extract data, multi-variate methods such as principal components analysis (PCA) can be restricted, if desired, to a suite of compounds of particular interest (such as the PAHs or petro-leum biomarkers) detected in the pyrolyzate. In addition, the pyrolysis products of natural organic matter present in the sediment, such as lignin marker and organoni-trogen compounds (Table 15.1), can be included to monitor background environ-mental conditions at the site of investigation. As an example, the Py-GC/MS dataset from the Passaic River sediment core mentioned above (Fig. 15.15) was subjected to PCA. A total of 138 individual compounds and isomer groups (including PAHs, petroleum biomarkers, and pyrolysis products of natural organic matter) were quan-titated, normalized, and scaled by taking the square root to dampen wide variations in magnitude. The resulting first principal component accounts for 63% while the second accounts for an additional 21% of the variance in a dataset of 138 variables. Having a total of 84% of the variance in only two composite variables permits a visualization of the essential trends on a simple two-dimensional graph (Fig. 15.16). In this case, high positive values of the first principal component correspond to a greater preponderance of natural organic matter and petroleum contamination in the samples, while negative values indicate relatively greater importance of parent polycyclic aromatic compounds. Although the second principal component is less significant than the first, it adds nuance to the interpretation, with positive values indicative of higher PAH content, while on the negative side it points to a greater influence of terrestrial plant matter. These PCA results reveal a clear stratification within the 5 m long core (Fig. 15.16). The lower portion of the core shows a shows a strong PAH contamination, interpreted to be the legacy of a manufactured gas plant formerly located on the adjacent river bank. A terrestrial vegetation signature is evident in the upper portion of the core, likely reflecting changes in watershed ecology after the decline of the area's heavy industry [64].

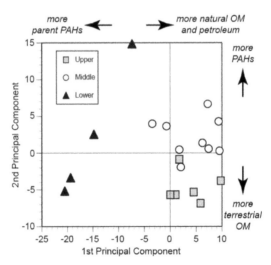

FIGURE 15.16   First two principal components from the multivariate analysis of the Py-GC/MS data from Lower Passaic River (New Jersey, USA) core 7A. Data input included 138 individual compounds and isomer groups from the Py-GC/MS analysis of 19 samples taken from the 5 m long sediment core. The PCA analysis indicates compositional differences between samples taken from the upper, middle and lower segments of the core and serves as an example of environmental geochemical and forensic insights that can be derived from Py-GC/MS results (data from Ref. 64).

15.4.2   VGI INDEX

As noted above, natural sedimentary organic matter may derive from terrestrial plant debris washed into bodies of water via runoff or from aquatic organisms such as algae living in the water column. Distinguishing between such allochthonous and autochthonous materials can be important in environmental and sedimentological studies, for which indicators such as the molar C/N and stable carbon isotope ratios are commonly employed [65, 66]. The Vinylguaiacol Indole Index (VGII or "Veggie" Index) was recently proposed as an additional parameter employing Py-

GC/MS data [67]. Vinylguaiacol is a methoxyphenol and one of the most abundant pyrolysis products of lignin [68], while the organonitrogen compound indole is produced upon pyrolysis of algae and bacteria [69]. Both of these compounds are frequently detected in the sediment pyrolyzates and may be quantitated using their mass spectral base peaks of m/z 150 and 117, respectively. The index is computed as the simple ratio of vinylguaiacol (VG) to the sum vinylguaiacol plus indole (I) (i.e., VG/(VG+I)) using the m/z 117 and 150 quantitation results directly, without applying mass spectral response factors [67]. End member samples, such as green algae (collected in Newark Bay, New Jersey) and pine wood have VGII values of nearly 0 and 1 respectively. As examples, sediment from Newark Bay has a low VGII of 0.34 indicating a predominance of aquatic organic matter, while a Passaic River bank sediment sample collected several kilometers upstream from Newark Bay has a higher VGII of 0.67 due to a greater terrestrial plant input (Fig. 15.17).

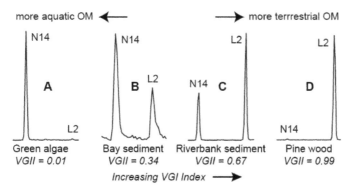

FIGURE 15.17   Examples of summed partial m/z 117 + 150 mass chromatograms from four pyrolyzates illustrating the rationale for the Vinylguaiacol Indole Index (VGII or "Veggie" Index), which assesses the relative contributions of terrestrial and aquatic organic matter to sediments. The VGI Index increases from 0 to 1 with increasing relative amounts of terrestrial organic matter. A) Green algae (*Ulva* sp.?) from Newark Bay, New Jersey, USA representing the aquatic end member. B) Newark Bay sediment showing a predominance of aquatic organic matter. C) Passaic River (New Jersey) bank sediment with the terrestrial component predominant. D) Pine wood (*Pinus strobus* twig) representing the terrestrial end member.

15.4.3   PYROLYTIC MARKER COMPOUNDS FOR ALGAL BLOOMS AND SEWAGE

The particulate organic matter in suspended sediment is readily amenable to Py-GC/MS analysis [43, 70–72]. Eutrophic conditions in a Serbian tributary of the Danube River precipitated a diatom bloom. The results of the pyrolysis of the suspended particulate matter collected during this bloom [73] provide an example of the insights obtainable via this approach (Fig. 15.18). With protein-derived nitrogen compounds such as indole and methylindole (peaks N14, N15) strongly predominant over lignin marker compounds (too small to be visible on this total ion current trace), the sample has an unambiguously aquatic fingerprint and a correspondingly very low VGII of 0.01. Other distinctive compounds contributing to the algal signature include diketopiperazines (the two N18 peaks), C_{14} and C_{16} fatty acids (C1, C2, C3), phytadienes, phytol, and other isoprenoids (peaks I5-I9), and 24-methylsterenes ($20–$22). 24-Methylcholestadienol is a biological marker for diatoms [74] and was detected in the solvent extract of this sample. The steradienes and -trienes are the pyrolysis products of this distinctive diatom marker compound [73].

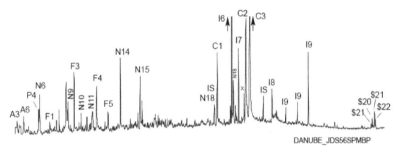

FIGURE 15.18 Full scan, total ion current chromatogram showing results of the pyrolysis (610 °C, 20 sec.) of suspended sediments in the Velika Morava River (Serbia), a tributary of the Danube River. See Table 1 for peak identification. The phytadienes and fatty acids provide evidence of algal blooms in the water column, likely involving diatoms, as indicated by the strong relative contributions of C28 steroids. (Adapted from Ref. 73. Used with permission.)

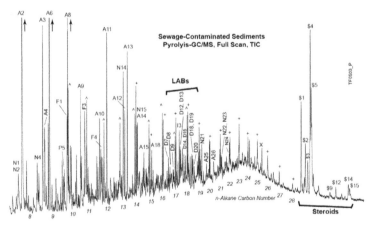

FIGURE 15.19 Full scan, total ion current chromatogram showing the results of the pyrolysis (610 °C, 20 sec.) of sewage sludge sampled offshore Barcelona (Spain). See Table 1 for peak identification. Note prominent linear alkyl benzenes (LABs) and sewage-related steroids. (Adapted from Ref. 10. Used with permission.).

The Rock-Eval pyrolysis results for the sewage spill off the coast of Barcelona, Spain were presented in Section 15.2.1 above and in Figs. 15.3 and 15.4. The pyrolyzate of a sewage sludge sample from this site is notable for its relatively abundant organonitrogen compounds (Fig. 15.19) and a correspondingly low VGII of 0.15. Even more remarkable are the series of linear alkylbenzenes (LABs, peaks D7–D20) and steroids ($1–$15) on this full-scan chromatogram, useful as markers for sewage contamination of sediments [10]. In detail, the full series of C_{15}–C_{19} LAB isomer groups can be seen on the m/z 91 trace of this sample's pyrolyzate (Fig. 15.20). These compounds are markers for alkylbenzene sulfonate surfactants and are characteristic of urban wastewater streams. In this case, they are likely to have been thermally desorbed at the 610°C temperature employed in this experiment [10]. The sterol markers are of even greater utility in detecting sewage contamination. While the steroids are sufficiently abundant in this pyrolyzate to appear prominent on the total ion current trace (Figs. 15.19 and 15.21A), mass chromatography permits an examination of their distribution in detail. Since the majority of these steroids are monounsaturated, the m/z 215 trace is perhaps the most useful (Fig. 15.21B). Their molecular ions (m/z 370, 384, 398, Fig. 15.21E) indicate that these are C_{27}, C_{28}, and C_{29} sterenes and that the C_{27} are the most abundant while the C_{28} are the least. The C_{27}–C_{29} steranes are also present, in about the same relative carbon number proportions as the sterenes (Fig. 15.21C). Since the 5α(H) (20R) stereoisomers strongly dominate in each carbon number (peaks $5, $11, $15) petroleum contamination is precluded. Coprostanol is the primary sterol sewage marker and is a C_{27} compound with a molecular ion of m/z 388 (Fig. 15.21D). While its minor presence in this pyrolyzate is likely due to thermodesorption, the dominant C_{27} sterenes are interpreted to be largely its pyrolysis products [10]. The m/z 316, 330, and 344 fragment ions are indicative of the C_{27}–C_{29} sterenes with the double bond at the C-2 position ($2, $3, $8, $9, $13, $14 in Fig. 15.21F). C_{27} and C_{29} steradienes are also present (Fig. 15.21G). Using this steroid distribution as a guide, particularly the C27 > C29 > C28 sterene pattern seen most readily on a m/z 215 chromatogram, Py-GC/MS can be useful for the rapid detection of sewage contamination in sediments.

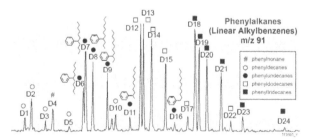

FIGURE 15.20 Distribution of wastewater-derived linear alkylbenzenes (LABs) as seen on a partial m/z 91 chromatogram using selected ion monitoring data from the pyrolysis (610 °C, 20 sec.) of sewage sludge of the same sample shown in Figure 15-19. (Adapted from Ref. 10. Used with permission.)

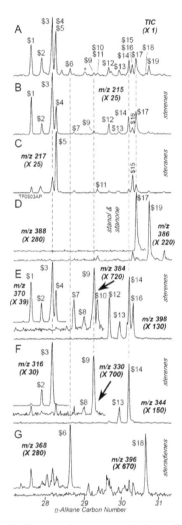

FIGURE 15.21   Complex distribution of sewage-derived steroids as seen on partial mass chromatograms (full scan) from the pyrolysis (610 °C, 20 sec.) of the same sample shown in Figure 15-19. Monounsaturated sterenes are the most abundant pyrolysis products, with C27 > C29 > C28. See Table 1 for peak identification. (Adapted from Ref. 10. Used with permission.)

15.4.4   ANALYTICAL PYROLYSIS OF AIRBORNE PARTICULATE MATTER

The organic components in urban airborne particulate matter are amenable to Py-GC/MS analysis. For example, dry particulate matter from a sampling device in

Lanzhou, China was pyrolyzed directly, without further preparation. The total ion current trace displays a series of *n*-alkanes and mono- to pentaaromatic hydrocarbons (Fig. 15.22A). The sample was reanalyzed in selected ion monitoring mode and the resulting *n*-alkane distribution together with the presence of norpristane, pristane and phytane (Peaks I1, I2, I5) indicate the presence of unburned fossil fuels (Fig. 15.22B). There is an odd over even *n*-alkane predominance in the C_{27} to C_{31} range, indicating admixed terrestrial plant waxes. The parent 2- to 5-ring PAHs are readily seen on a composite mass chromatogram constructed of their molecular ions (Fig. 15.22C). The methylphenanthrenes (A25) are also shown and since they are relatively much less abundant than the parent compound phenanthrene (A23), it is likely that the PAHs are mostly combustion-derived. Once again, the Py-GC/MS method provides a rapid means for the screening of environmental samples.

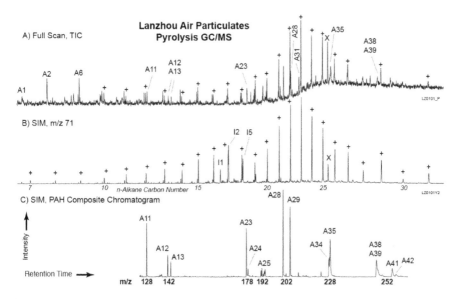

FIGURE 15.22 Distribution of the principal hydrocarbons detected in the pyrolyzate (610 °C, 20 sec.) of urban airborne particulate matter (Lanzhou, China). See Table 1 for peak identification. A) Total ion current trace, full scan analysis. B) m/z 71 mass chromatogram, SIM analysis. C) Composite mass chromatogram showing PAH distributions. In the composite trace, the indicated mass chromatograms are linked end-to-end, without summation or overlap.

15.4.5 ANALYTICAL PYROLYSIS OF SPILLED PETROLEUM

In the wake of the 2010 *Deepwater Horizon* disaster in the Gulf of Mexico, large quantities of spilled oil came ashore. A relatively fresh tarball, collected within hours

of landfall during the early days of the crisis, displays a partially intact series of normal and isoprenoid alkanes atop a prominent UCM hump (Fig. 15.23A). Although
a pyrolysis temperature of 610°C was used for this experiment, it is likely that the
yield was mostly the result of thermodesorption. The sample is obviously partly
biodegraded, which occurred during its transport on marine currents from wellhead
to shore. A second sample was collected after at least six months of exposure on
the beach and displays a more severe degree of degradation (Fig. 15.23B). The *n*-
alkanes are missing, although the isoprenoids (peaks I1, I2, I5) and hopane (H4) are
still present. Relative to the fresher sample, the UCM has evidently lost some of
its lower molecular weight components. Again it appears that most of the material
was thermally desorbed, swamping any true pyrolysis products that may have been
generated from the oil's asphaltene fraction. Analytical pyrolysis provides a novel
means for the characterization of these unusual samples.

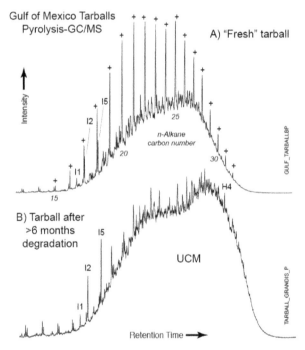

FIGURE 15.23 Total ion current traces, full scan, of pyrolyzates (610 °C, 20 sec.) of
beach tarballs collected on the Gulf of Mexico coast after the Deepwater Horizon disaster in
2010. See Table 1 for peak identification; note prominent hump produced by the "unresolved
complex mixture" of compounds (UCM). A) "Fresh" tarball collected quickly after the spill
reached the shore. B) Tarball degraded after 6 or more months of exposure on beach.

15.5 THERMOCHEMOLYSIS-GAS CHROMATOGRAPHY/MASS SPECTROMETRY

In lieu of dry heating for standard thermodesorption and pyrolysis-GC/MS, the pyrolyzer may be used as a thermochemolysis reactor, in which the sample is heated in the presence of reagents, particularly tetramethylammonium hydroxide (TMAH). This reaction hydrolyzes the sample and methylates polar products yielding individual methyl esters. The products of thermochemolysis and pyrolysis experiments performed on the same samples can appear quite different [42], permitting enhanced interpretation as each method can reveal different aspects of the sample's nature.

Poerschmann and others successfully applied both conventional pyrolysis- and thermochemolysis-GC/MS to the characterization of PAH-contaminated sediment [75, 76]. Deshmukh and co-workers also applied both techniques to characterize a suite of contaminated and noncontaminated sediment samples, but found that the most interesting results illuminated the nature of the associated biogenic organic matter [77]. In a similar approach, Mansuy and others used thermochemolysis-GC/MS to evaluate the humic fraction of polluted river sediments, effectively distinguishing between natural and anthropogenic organic matter sources [78].

15.6 PYROLYSIS WITH OTHER DETECTION SYSTEMS

In lieu of the standard mass spectrometer, a Py-GC system has been coupled to a combustion-isotope ratio mass spectrometer (Py-GC-C-IRMS) and used to characterize lacustrine dissolved organic matter [22]. The results of pyrolysis-FIMS (see Section 15.2.2) and Py-GC-C-IRMS of the same sample show a predominance of higher plant-derived material (Fig. 15.5). The m/z 44 trace (Fig. 15.5B) corresponds to the CO_2 produced as the eluates exiting the GC column are combusted and is functionally equivalent to a standard GC-FID chromatogram. The m/z 45/44 ratio trace corresponds to the $^{13}CO_2/^{12}CO_2$ ratio of the analytes reaching the detector, permitting the computation of the $\partial^{13}C$ values of individual compounds, in turn permitting source inferences.

A portable pyrolysis-gas chromatography-ion mobility spectrometer (Py-GC-IMS) was designed for field deployment to detect potentially hazardous microbes in the environment with military and public safety applications [79]. A seldom-employed but interesting configuration combines pyrolysis with an atomic emission detector (Py-GC-AED), which can be tuned to detect elements of interest, such as carbon, as well as the heteroatoms oxygen, sulfur, nitrogen and chlorine [57]. The AED carbon channel produces results similar to an FID or full-scan total ion current MS trace, while the sulfur channel output resembles that of a sulfur-selective FPD [20, 57].

15.7 CONCLUSIONS

Over the past half century, analytical pyrolysis has proven itself to be an effective means for the semiquantitative characterization of complex macromolecular organic substances. It has been demonstrated that instruments such as Py-FID, Py-MS, and in particular, Py-GC/MS can provide valuable geochemical insights when applied to a wide variety of problems in environmental science. The more widespread use of analytical pyrolysis methods in the evaluation of environmental pollution is recommended, because of their relatively low cost and information-rich results.

KEYWORDS

- **Contamination**
- **Environmental Chemistry**
- **Environmental Forensics**
- **Petroleum Spill**
- **Pollution**
- **Polycyclic Aromatic Hydrocarbons (PAHs)**
- **Pyrolysis-Gas Chromatography/Mass Spectrometry (Py-GC/MS)**
- **Rock-Eval Pyrolysis**
- **Sewage**
- **Thermodesorption**

REFERENCES

1. Uden, P. C. (1993). Nomenclature and Terminology for Analytical Pyrolysis (IUPAC Recommendations 1993). *Pure and Applied Chemistry, 65(11)*, 2405–2409.
2. Espitalié, J., Laporte, J. L., Madec, M., Marquis, F., Leplat, P., Paulet, J., & Boutefeu, A. (1977). Méthode Rapide de Caractérisation des Roches Mères, de Leur Potentiel Pétrolier et de Leur Degré d'Évolution, *Oil & Gas Science and Technology Rev. IFP, 32(1)*, 23–42.
3. Tissot, B. P., & Welte, D. H. (1984). *Petroleum Formation and Occurrence.* 2nd Ed., Springer: 702 P.
4. Peters, K. E. (1986). Guidelines for Evaluating Petroleum Source Rock Using Programmed Pyrolysis, *AAPG Bulletin, 70(3)*, 318–329.
5. Meyers, P. A., & Doose, H. (1999). Sources, Preservation, and Thermal Maturity of Organic Matter in Pliocene-Pleistocene Organic-Carbon-Rich Sediments of the Western Mediterranean Sea, In *Proceedings of the Ocean Drilling Program, Scientific Results*, Zahn, R., Comas, M. C., Klaus, A., Eds. 383–390.
6. Marchand, C., Lallier-Vergès, E., Disnar, J. R., & Kéravis, D. (2008). Organic Carbon Sources and Transformations in Mangrove Sediments: A Rock-Eval Pyrolysis Approach, *Organic Geochemistry, 39, (4)*, 408–421.

7. Stern, G. A., Sanei, H., Roach, P., Delaronde, J., & Outridge, P. M. (2009). Historical Inter-related Variations of Mercury and Aquatic Organic Matter in Lake Sediment Cores from a Subarctic Lake in Yukon, Canada: Further Evidence Toward the Algal-Mercury Scavenging Hypothesis, *Environmental Science & Technology, 43(20)*, 7684–7690.

8. Duan, D., Ran, Y., Cheng, H., Chen, J. A., & Wan, G., Contamination Trends of Trace Metals and Coupling with Algal Productivity in Sediment Cores in Pearl River Delta, South China, *Chemosphere*, (In Press).

9. Kruge, M. A., Mukhopadhyay, P. K., & Lewis, C. F. M. (1998). A Molecular Evaluation of Contaminants and Natural Organic Matter in Bottom Sediments from Western Lake Ontario. *Organic Geochemistry, 29(5–7)*, 1797–1812.

10. Kruge, M. A., Permanyer, A., Serra, J., & Yu, D. (2010). Geochemical Investigation of An Offshore Sewage Sludge Deposit, Barcelona, Catalonia, Spain. *Journal of Analytical and Applied Pyrolysis, 89(2)*, 204–217.

11. Maddock, C. J., & Ottley, T. W. (2007). Pyrolysis Mass Spectrometry: Instrumentation, Techniques, and Applications. in *Applied Pyrolysis Handbook*, 2 Ed., Wampler, T. P., Ed. CRC Press: Boca Raton, 47–64.

12. Gutteridge, C. S., & Norris, J. R. (1979). The Application of Pyrolysis Techniques to the Identification of Micro-Organisms. *Journal of Applied Bacteriology, 47(1)*, 5–43.

13. Meuzelaar, H. L., Windig, W., Harper, A. M., Huff, S. M., Mcclennen, W. H., & Richards, J. M. (1984). Pyrolysis Mass Spectrometry of Complex Organic Materials. *Science, 226*, 268–274.

14. Boon, J. J., Tom, A., Brandt, B., Eijkel, G. B., Kistemaker, P. G., Notten, F. J. W., & Mikx, F. H. M. (1984). Mass Spectrometric and Factor Discriminant Analysis of Complex Organic Matter from the Bacterial Culture Environment of Bacteroides Gingivalis. *Analytica Chimica Acta, 163(0)*, 193–205.

15. Ford, T., Sacco, E., Black, J., Kelley, T., Goodacre, R., Berkeley, R. C., & Mitchell, R. (1991). Characterization of Exopolymers of Aquatic Bacteria by Pyrolysis Mass Spectrometry, *Applied and Environmental Microbiology, 57(6)*, 1595–1601.

16. Goodacre, R., Shann, B., Gilbert, R. J., Timmins, É. M., Mcgovern, A. C., Alsberg, B. K., Kell, D. B., & Logan, N. A. (2000). Detection of the Dipicolinic Acid Biomarker in Bacillus Spores Using Curie-Point Pyrolysis Mass Spectrometry and Fourier Transform Infrared Spectroscopy, *Analytical Chemistry, 72(1)*, 119–127.

17. Qian, K., Killinger, W. E., Casey, M., & Nicol, G. R. (1996). Rapid Polymer Identification by In-Source Direct Pyrolysis Mass Spectrometry and Library Searching Techniques, *Analytical Chemistry, 68(6)*, 1019–1027.

18. Boon, J. J., Dupont, L., & De Leeuw, J. W. (1986). Characterization of a Peat Bog Profile by Curie Point Pyrolysis-Mass Spectrometry Combined with Multivariant Analysis and by Pyrolysis Gas Chromatography-Mass Spectrometry, In *Peat and Water*, Fuchsman, C. H., Ed. Elsevier: 215–239.

19. Remmler, M., Kopinke, F. D., & Stottmeister, U. (1995). Thermoanalytical Methods for Characterizing Hydrocarbon-Sludge-Soil Mixtures, *Thermochimica Acta, 263*, 101–112.

20. Pörschmann, J., Kopinke, F. D., Remmler, M., Mackenzie, K., Geyer, W., & Mothes, S. (1996). Hyphenated Techniques for Characterizing Coal Wastewaters and Associated Sediments, *Journal of Chromatography A, 750(1–2)*, 287–301.

21. Miketova, P., Abbas-Hawks, C., Voorhees, K. J., & Hadfield, T. L. (2003). Microorganism Gram-Type Differentiation of Whole Cells Based on Pyrolysis High-Resolution Mass Spectrometry Data, *Journal of Analytical and Applied Pyrolysis, 67(1)*, 109–122.

22. Schulten, H. R., & Gleixner, G. (1999). Analytical Pyrolysis of Humic Substances and Dissolved Organic Matter in Aquatic Systems: Structure and Origin. *Water Research, 33(11)*, 2489–2498.

23. Schulten, H. R. (1999). Analytical Pyrolysis and Computational Chemistry of Aquatic Humic Substance and Dissolved Organic Matter, *Journal of Analytical and Applied Pyrolysis, 49*, 385–415.

24. Leinweber, P., & Schulten, H. R. (1999). Advances in Analytical Pyrolysis of Soil Organic Matter, *Journal of Analytical and Applied Pyrolysis, 49(1–2)*, 359–383.

25. Letarte, S., Morency, D., Wilkes, J., & Bertrand, M. J. (2004). Py-MAB-Tof Detection and Identification of Microorganisms in Urine. *Journal of Analytical and Applied Pyrolysis, 71(1)*, 13–25.

26. Jeannotte, R., Hamel, C., Jabaji, S., & Whalen, J. K. (2011). Pyrolysis-Mass Spectrometry and Gas Chromatography-Flame Ionization Detection as Complementary Tools for Soil Lipid Characterization, *Journal of Analytical and Applied Pyrolysis, 90(2)*, 232–237.

27. Getty, S. A., Ten Kate, I. L., Feng, S. H., Brinckerhoff, W. B., Cardiff, E. H., Holmes, V. E., King, T. T., Li, M. J., Mumm, E., Mahaffy, P. R., & Glavin, D. P. (2010). Development of an Evolved Gas-Time-of-Flight Mass Spectrometer for the Volatile Analysis by Pyrolysis of Regolith (Vapor) Instrument, *International Journal of Mass Spectrometry, 295(3)*, 124–132.

28. Wampler, T. P. (2007). Analytical Pyrolysis: An Overview, In *Applied Pyrolysis Handbook, 2*, Ed., Wampler, T. P., Ed. CRC Press: Boca Raton, 1–26.

29. Wampler, T. P. (2007). Instrumentation and Analysis. In *Applied Pyrolysis Handbook*, 2 Ed., Wampler, T. P., Ed. CRC Press: Boca Raton, 27–46.

30. Jones, C. E. R. (2000). Gas Chromatography: Pyrolysis Gas Chromatography, In *Encyclopedia of Separation Science*, Wilson, I. D., Adlard, E. R., Cooke, M., & Poole, C. F., Eds. Academic Press: 282–287.

31. Davison, W. H. T., Slaney, S., & Wragg, A. L. (1954). A Novel Method of Identification of Polymers, *Chemistry and Industry*, 1356.

32. Giraud, A. (1970). Application of Pyrolysis and Gas Chromatography to Geochemical Characterization of Kerogen in Sedimentary Rock, *AAPG Bulletin, 54, (3)*, 439–455.

33. Irwin, W. J. (1979). Analytical Pyrolysis An Overview, *Journal of Analytical and Applied Pyrolysis, 1(1)*, 3–25.

34. Dembicki, H., Horsfield, B., & Ho, T. T. Y. (1983). Source Rock Evaluation by Pyrolysis-Gas Chromatography, *AAPG Bulletin, 67(7)*, 1094–1103.

35. Whelan, J. K., Hunt, J. M., & Huc, A. Y. (1980). Applications of Thermal Distillation-Pyrolysis to Petroleum Source Rock Studies and Marine Pollution, *Journal of Analytical and Applied Pyrolysis, 2(1)*, 79–96.

36. Hala, W. W. (1993). Organic Geochemistry of Contaminated Soil from an Abandoned Oil Field. MS Thesis. Southern Illinois University, Carbondale, .

37. Gallegos, E. J. (1975). Terpane-Sterane Release from Kerogen by Pyrolysis Gas Chromatography-Mass Spectrometry. *Analytical Chemistry, 47(9)*, 1524–1528.

38. De Leeuw, J. W., De Leer, E. W. B., Damste, J. S. S., & Schuyl, P. J. W. (1986). Screening of Anthropogenic Compounds in Polluted Sediments and Soils by Flash Evaporation/Pyrolysis Gas Chromatography Mass Spectrometry, *Analytical Chemistry, 58(8)*, 1852–1857.

39. White, D. M., Garland, D. S., Beyer, L., & Yoshikawa, K. (2004). Pyrolysis-GC/MS Fingerprinting of Environmental Samples, *Journal of Analytical and Applied Pyrolysis, 71(1)*, 107–118.

40. Saiz-Jimenez, C., Grimalt, J., Garcia-Rowe, J., & Ortega-Calvo, J. J. (1991). Analytical Pyrolysis of Lichen Thalli, *Symbiosis, 11*, 313–326.

41. Saiz-Jimenez, C. (1992). Applications of Pyrolysis-Gas Chromatography/Mass Spectrometry to the Study of Soils, Plant Materials and Humic Substances, A Critical Appraisal, In *Humus, Its Structure and Role in Agriculture and Environment*, Kubát, J., Ed. Elsevier: 27–38.

42. Saiz-Jimenez, C. (1994). Analytical Pyrolysis of Humic Substances: Pitfalls, Limitations, and Possible Solutions, *Environmental Science & Technology, 28(11)*, 1773–1780.

43. Van Heemst, J. D. H., Van Bergen, P. F., Stankiewicz, B. A., & De Leeuw, J. W. (1999). Multiple Sources of Alkylphenols Produced Upon Pyrolysis of DOM, POM and Recent Sediments, *Journal of Analytical and Applied Pyrolysis, 52(2)*, 239–256.
44. Richnow, H. H., Seifert, R., Kästner, M., Mahro, B., Horsfield, B., Tiedgen, U., Böhm, S., & Michaelis, W. (1995). Rapid Screening of PAH-Residues in Bioremediated Soils, *Chemosphere, 31(8)*, 3991–3999.
45. Schiavon, N., Chiavari, G., Schiavon, G., & Fabbri, D. (1995). Nature and Decay Effects of Urban Soiling on Granitic Building Stones, *Science of the Total Environment, 167(1–3)*, 87–101.
46. Schmidt, M. W. I., Knicker, H., Hatcher, P. G., & Kögel-Knabner, I. (1996). Impact of Brown Coal Dust on the Organic Matter in Particle-Size Fractions of a Mollisol, *Organic Geochemistry, 25(1–2)*, 29–39.
47. Abdel Bagi, S. T. (1996). Geochemical and Petrographical Characterization of Natural and Anthropogenic Sedimentary Organic Matter in Polluted Sediments from the West Branch of the Grand Calument River and Roxana Marsh, NW Indiana and NE Illinois, PhD Dissertation, Southern Illinois University, Carbondale.
48. Kim, M. G., Yagawa, K., Inoue, H., & Shirai, T. (1991). Determination of Sulfates in Urban Air by Pyrolysis Gas Chromatography with Flame Photometric Detection, *Journal of Analytical and Applied Pyrolysis, 20(0)*, 263–273.
49. Munson, T. O. (2007). Environmental Applications of Pyrolysis. In *Applied Pyrolysis Handbook, 2*, Wampler, T. P., Ed. CRC Press: Boca Raton, 133–173.
50. Fabbri, D., Trombini, C., & Vassura, I. (1998). Analysis of Polystyrene in Polluted Sediments by Pyrolysis Gas Chromatography Mass Spectrometry, *Journal of Chromatographic Science, 36*, 600–604.
51. Fabbri, D., Tartari, D., & Trombini, C. (2000). Analysis of Poly(Vinyl Chloride) and Other Polymers in Sediments and Suspended Matter of a Coastal Lagoon by Pyrolysis Gas Chromatography Mass Spectrometry, *Analytica Chimica Acta, 413(1–2)*, 3–11.
52. Fabbri, D. (2001). Use of Pyrolysis-Gas Chromatography/Mass Spectrometry to Study Environmental Pollution Caused by Synthetic Polymers: A Case Study: The Ravenna Lagoon, *Journal of Analytical and Applied Pyrolysis, 58–59(0)*, 361–370.
53. Moldoveanu, S. C. (2005). *Analytical Pyrolysis of Synthetic Organic Polymers*. Elsevier, 714.
54. Kusch, P. (2012). Pyrolysis-Gas Chromatography/Mass Spectrometry of Polymeric Materials, In *Advanced Gas Chromatography Progress in Agricultural, Biomedical and Industrial Applications*, Mohd, M. A., Ed. Intech, 343–362.
55. Kruge, M. A., & Permanyer, A. (2004). Application of Pyrolysis GC/MS for Rapid Assessment of Organic Contamination in Sediments from Barcelona Harbor, *Organic Geochemistry, 35(11–12)*, 1395–1408.
56. Faure, P., & Landais, P. (2001). Rapid Contamination Screening of River Sediments by Flash Pyrolysis-Gas Chromatography Mass Spectrometry (Pygc–MS) and Thermodesorption GC–MS (Tdgc–MS). *Journal of Analytical and Applied Pyrolysis, 57(2)*, 187–202.
57. Faure, P., Vilmin, F., Michels, R., Jarde, E., Mansuy, L., Elie, M., & Landais, P. (2002). Application of Thermodesorption and Pyrolysis-GC–AED to the Analysis of River Sediments and Sewage Sludges for Environmental Purpose, *Journal of Analytical and Applied Pyrolysis, 62(2)*, 297–318.
58. USEPA, Semivolatile Organic Compounds (PAHs and PCBs) in Soils/Sludges and Solid Wastes using Thermal Extraction/Gas Chromatography/Mass Spectrometry (TE/GC/MS). EPA Method 8275A. 1996, 23 P.
59. Stout, S. A., & Wang, Z. (2007). Chemical Fingerprinting of Spilled or Discharged Petroleum Methods and Factors Affecting Petroleum Fingerprints in the Environment, In *Oil Spill Environmental Forensics*, Wang, Z., Stout, S. A., Eds. Elsevier: Amsterdam, 1–53.

60. Douglas, G. S., Stout, S. A., Uhler, A. D., Mccarthy, K. J., & Emsbo-Mattingly, S. D. (2007). Advantages of Quantitative Chemical Fingerprinting in Oil Spill Source Identification, In *Oil Spill Environmental Forensics*, Wang, Z., Stout, S. A., Eds. Elsevier: Amsterdam, 257–292.

61. Buco, S., Moragues, M., Doumenq, P., Noor, A., & Mille, G. (2004). Analysis of Polycyclic Aromatic Hydrocarbons in Contaminated Soil by Curie Point Pyrolysis Coupled to Gas Chromatography Mass Spectrometry, an Alternative to Conventional Methods, *Journal of Chromatography A, 1026(1–2)*, 223–229.

62. Stout, S. A., Uhler, A. D., & Mccarthy, K. J. (2001). A Strategy and Methodology for Defensibly Correlating Spilled Oil to Source Candidates, *Environmental Forensics, 2*, 87–98.

63. Mudge, S. M. (2007). Multivariate Statistical Methods in Environmental Forensics, *Environmental Forensics, 8(1–2)*, 155–163.

64. Bujalski, N. M. (2010). Characterization of Contaminant and Biomass Derived Organic Matter in Sediments from the Lower Passaic River, New Jersey, USA. MS Thesis, Montclair State University, Montclair (NJ).

65. Meyers, P. A. (1997). Organic Geochemical Proxies of Paleoceanographic, Paleolimnologic, and Paleoclimatic Processes, *Organic Geochemistry, 27(5–6)*, 213–250.

66. Twichell, S. C., Meyers, P. A., & Diester-Haass, L. (2002). Significance of High C/N Ratios in Organic-Carbon-Rich Neogene Sediments under the Benguela Current Upwelling System, *Organic Geochemistry, 33(7)*, 715–722.

67. Micić, V., Kruge, M., Körner, P., Bujalski, N., & Hofmann, T. (2010). Organic Geochemistry of Danube River Sediments from Pančevo (Serbia) to the Iron Gate Dam (Serbia–Romania), *Organic Geochemistry, 41(9)*, 971–974.

68. Saiz-Jimenez, C., & De Leeuw, J. W. (1986). Lignin Pyrolysis Products: their Structures and their Significance as Biomarkers, *Organic Geochemistry, 10(4–6)*, 869–876.

69. Bennett, B., Lager, A., Russell, C. A., Love, G. D., & Larter, S. R. (2004). Hydropyrolysis of Algae, Bacteria, Archaea and Lake Sediments; Insights into the Origin of Nitrogen Compounds in Petroleum, *Organic Geochemistry, 35(11–12)*, 1427–1439.

70. Çoban-Yildiz, Y., Chiavari, G., Fabbri, D., Gaines, A. F., Galletti, G., & Tuğrul, S. (2000). The Chemical Composition of Black Sea Suspended Particulate Organic Matter: Pyrolysis-GC/MS as a Complementary Tool to Traditional Oceanographic Analyses, *Marine Chemistry, 69(1–2)*, 55–67.

71. Çoban-Yildiz, Y., Fabbri, D., Tartari, D., Tuğrul, S., & Gaines, A. F. (2000). Application of Pyrolysis–GC/MS for the Characterisation of Suspended Particulate Organic Matter in the Mediterranean Sea: A Comparison with the Black Sea, *Organic Geochemistry, 31(12)*, 1627–1639.

72. Çoban-Yildiz, Y., Fabbri, D., Baravelli, V., Vassura, I., Yilmaz, A., Tuğrul, S., & Eker-Develi, E. (2006). Analytical Pyrolysis of Suspended Particulate Organic Matter from the Black Sea Water Column, *Deep Sea Research Part II: Topical Studies in Oceanography, 53(17–19)*, 1856–1874.

73. Micić, V., Kruge, M. A., Köster, J., & Hofmann, T. (2011). Natural, Anthropogenic and Fossil Organic Matter in River Sediments and Suspended Particulate Matter: a Multi-Molecular Marker, Approach. *Science of the Total Environment, 409(5)*, 905–919.

74. Rampen, S. W., Abbas, B. A. S. S., & Sinninghe Damsté, J. S. (2010). A Comprehensive Study of Sterols in Marine Diatoms (Bacillariophyta): Implications for their use as Tracers for Diatom Productivity, *Limnology and Oceanography, 55*, 91–105.

75. Poerschmann, J., Parsi, Z., & Gorecki, T. (2008). Non-Discriminating Flash Pyrolysis and Thermochemolysis of Heavily Contaminated Sediments from the Hamilton Harbor (Canada), *Journal of Chromatography A, 1186(1–2)*, 211–221.

76. Poerschmann, J., Trommler, U., Fabbri, D., & Górecki, T. (2007). Combined Application of Non-Discriminated Conventional Pyrolysis and Tetramethylammonium Hydroxide-Induced

Thermochemolysis for the Characterization of the Molecular Structure of Humic Acid Isolated from Polluted Sediments from the Ravenna Lagoon, *Chemosphere, 70(2)*, 196–205.

77. Deshmukh, A. P., Chefetz, B., & Hatcher, P. G. (2001). Characterization of Organic Matter in Pristine and Contaminated Coastal Marine Sediments using Solid-State ^{13}C NMR, Pyrolytic and Thermochemolytic Methods: A Case Study in the San Diego Harbor Area, *Chemosphere, 45(6–7)*, 1007–1022.

78. Mansuy, L., Bourezgui, Y., Garnier-Zarli, E., Jardé, E., & Réveillé, V. (2001). Characterization of Humic Substances in Highly Polluted River Sediments by Pyrolysis Methylation Gas Chromatography Mass Spectrometry, *Organic Geochemistry, 32(2)*, 223–231.

79. Snyder, A. P., Dworzanski, J. P., Tripathi, A., Maswadeh, W. M., & Wick, C. H. (2004). Correlation of Mass Spectrometry Identified Bacterial Biomarkers from a Fielded Pyrolysis Gas Chromatography Ion Mobility Spectrometry Biodetector with the Microbiological Gram Stain Classification Scheme, *Analytical Chemistry, 76(21)*, 6492–6499.

INDEX

B

Back-extraction, 323
Background electrolyte, 114, 115, 307
Bailer pump, 235
Barcelona tap water analysis, 317
Basic measurement units, 10
Beer-Lambert law, 33, 35, 36, 366
Beer's law, 372, 377
Benzalkonium chloride, 506
Benzene, toluene, ethylbenzene and xylenes
　(BTEX), 376, 511
　benzene, 376
　ethylbenzene, 376
　p-xylene, 376
　toluene, 376
Bioaerosols, 383
Biochemical oxygen demand, 200
Biosphere, 191
Bismuth film electrode, 421
Bisphenol A, 311, 473
Black body radiation, 30
　continuum spectra, 30
Bladder pump, 235, 238
Bond angle, 48
　rocking, 48
　scissor, 48
　twisting, 48
　wagging, 48
Bragg diffraction, 74
Bremsstrahlung, 72
Brominated compounds, 481
Brominated flame retardants, 476
Bubble cell, 125
　capillaries, 125
　configuration, 125
Bubbles, 70
Bulk electrolyte, 161
Butler-Volmer equation, 159, 160

C

C-C bond energy, 42
Cadmium mercuric telluride, 49
Cadmium sulfide, 40
Calibration curve, 357
Calibration errors, 20
　graph, 20

Calomel electrodes, 143
Capillary electrochromatography, 126, 130
Capillary electrophoresis, 112
Capillary gel electrophoresis, 125, 126, 127
Capillary isoelectric focusing, 126, 128
Capillary isotachophoresis, 126, 129
Capillary zone electrophoresis, 114, 119,
　125, 306
Carbon cycle, 192
Carbon monoxide, 217
Carbon nanotubes, 510
Carbon paste electrodes, 145
Carrier gas, 89, 91
　helium, 89
　hydrogen, 89
　nitrogen, 89
Catastrophes, 202
Cathodic stripping voltammetry, 172, 173
CCD detector, 62
　CE Expert, 122Beckman Coulter Inc,
　122
CE instrument systems, 124
Cephradine, 380
Cethyltrimethylammonium bromide, 116,
　131, 515, 503
Cetyltrimethyl ammonium chloride, 282
Chain-of-Custody, 246
Changing solvent, 101
　gradient elution, 101
Charcoal carbon, 381
Charge capacitive discharged arrays, 67
Charge transfer, 158
Chelating reagent, 323
Chelation ion chromatography, 258, 263
Chemical analysis, 2, 5, 8, 9, 15, 18
　analytical data, 18
　basic tools, 15
　equipments, 15
　kinetic aspects, 8
　speciation, 9
　statistics, 18
Chemical ionization mode, 480
Chemical kinetics, 82
Chemical oxygen demand, 200
Chemical wastes, 225
Chemiluminescence, 353
Chlorfenvinphos, 420

Milton Keynes UK
Ingram Content Group UK Ltd.
UKHW030901141024
449569UK00025B/1279